THE ARRL
SPREAD SPECTRUM
SOURCEBOOK

Edited By

André Kesteloot, N4ICK
and
Charles L. Hutchinson, K8CH

Assistant Editor Joel P. Kleinman, N1BKE

Published by
The American Radio Relay League, Inc

Copyright © 1991 by

The American Radio Relay League

Copyright secured under the Pan-American Convention

This work is publication No. 121 of the Radio Amateur's Library, published by the League. All rights reserved. No part of this work may be reproduced in any form except by written permission of the publisher. All rights of translation are reserved.

Printed in USA

Quedan reservados todos los derechos

ISBN: 0-87259-317-7

$20.00

First Edition
Second Printing, 1993

Foreword

It is safe to say that amateur spread spectrum exists because of one unique entity—the Amateur Radio Research and Development Corporation. Based in the Washington, DC area, AMRAD played a key role in the development of amateur packet radio in the early 1980s, and is continuing to lead the way in the realm of amateur spread spectrum. Its members are dedicated to furthering the Amateur Radio art, and their accomplishments speak for themselves.

It is altogether fitting, therefore, that this book be edited by long-time AMRAD President André Kesteloot, N4ICK. He has done as much as anyone to spread the word about the intriguing intricacies—and very real benefits—that spread spectrum systems hold for the amateur service.

The League is proud to be associated with AMRAD's efforts to bring practical spread-spectrum systems to the amateur bands. This book brings together articles, papers and documents that provide an overview of how amateur spread spectrum evolved from a short notice in the *AMRAD Newsletter*. This newsletter provided the bulk of the early material on spread spectrum, but it wasn't long before *QST* and its then-new experimenter's newsletter, *QEX*, took up the cause. Nearly all the spread spectrum articles published in those three sources are reprinted in this book, as are the regulatory documents that made spread spectrum legal on the amateur bands. Toward the end of the book, two ground-breaking technical papers are published in their entirety. Overall, the book leaves little doubt that it's only a matter of time before the spread spectrum techniques become well established in the amateur service.

David Sumner, K1ZZ
December 1990

Please note: Some of the theory and experimentation reported in this book would be contrary to the current Part 97 regulations if it were conducted today. Future rules revisions may codify such experimentation at some point in the future. Also, some of the discussions are speculative in nature and do not necessarily deal with the practical implementation of spread-spectrum technology in the amateur bands.

At one time, the two reports that comprise Chapters 9 and 10 of this book were available from the Government Printing Office. Both are now out of print, however.

Introduction

Only a few years ago (in 1986), it was with delight and no small degree of surprise that I first heard the news that the FCC was planning to allow future use of spread-spectrum techniques in bands not dominated by military users. Not only that, but radio amateurs were to be allowed to send spread-spectrum signals.

Interestingly, I had been told a number of times through the years (since 1959 when I first became acquainted with spread-spectrum systems, and immediately began a love affair that has never abated) that "no licenses had ever been granted for spread-spectrum systems, no licenses were being granted for spread-spectrum systems, and none were ever going to be granted for spread-spectrum systems." Nevertheless, spread-spectrum systems remained alive and well over in the military world, though for many years they received less attention than they deserved.

Today, the story is very different. Every major military communications system being developed and/or deployed is a spread-spectrum system. (It might be interesting to note that these military systems are predominantly frequency-hopping systems, as opposed to direct-sequence systems, although some employ a combination of both techniques.) Many commercial and consumer-oriented applications of spread-spectrum techniques are also emerging. Some of these are oriented to position-location, local area networking, two-way radio, cellular telephony, meter reading, wireless thermostat control, and many other widely diverging uses.

Sadly, however, some of the currently available equipment designs are little more than thinly disguised attempts to take advantage of a one-watt unlicensed spectrum by doing everything possible to circumvent the spirit if not the letter of the FCC rules. These will undoubtedly disappear as real spread-spectrum systems become available, because real spread-spectrum systems will easily outperform them in every application.

Radio amateurs, I expect, will bring a new approach to applications of spread-spectrum methods, as they have done in many other areas. When André Kesteloot asked me to provide an Introduction to this ARRL book on spread spectrum, the decision was very easy. Though I am not a radio amateur myself, my sympathies have always been very strong in that direction. I always wish that I had taken the FCC examination years ago, when I was still in the Navy and my code speed was up. Already I have seen and heard of amateur approaches to spread-spectrum design that are fresh, innovative, and bred with an eye to economy and practicality that has often been lacking in conventional spread-spectrum approaches. (See for instance the May 1989 *QST* article reprinted in this book.)

What will the future bring? Although amateurs have so far been somewhat slow to capitalize on this new technology, I do believe that this situation will soon improve. I hope this book will serve to whet the appetite of those who read it, and that amateurs will apply their special talents to make spread-spectrum operation as common as other high-technology communications and data-transmission methods. Notwithstanding the fact that this book was prepared mainly by and for radio amateurs, I believe that design and maintenance engineers will find it equally interesting.

More power to you, radio amateurs, but let it be spread-spectrum-encoded for low interference, and may your signals be robust, so that others may never interfere with you.

R. C. Dixon
Palmer Lake, Colorado
December 1990

Table of Contents

Spread Spectrum Evolves

1 AMRAD's Perspective

Call for Experimenters	1-1
July 6 Special Interest Group Meeting	1-1
September 7 Special Interest Group Meeting	1-1
SIG Update	1-1
AMRAD STA	1-2
Literature	1-2
FCC SS Proposal	1-2
February AMRAD Meeting	1-3
AMRAD Comments in FCC Spread Spectrum NOI/NPRM	1-3
A Second Look at Spread Spectrum	1-4
AMRAD Files Reply Comments to FCC	1-5
New Experiment Proposed for the Next STA	1-5
It's a Jungle Out There	1-5
AMRAD to Talk at Milcom '82	1-5
Spread Spectrum at the ARRL National Convention	1-5
Mitre Corp Design Study	1-5
Spread Spectrum Legal?	1-5
Spread Spectrum NOPR	1-5
FCC Releases New Rulemaking Items	1-6
Amateur Spread-Spectrum Progress	1-6
Rule Making Update	1-7
New Spread-Spectrum Group	1-8
Spread Spectrum is Approved!!	1-8
Spread Spectrum Committee Meets	1-9
Bad Day at Black Rock	1-9

2 The ARRL Perspective

AMRAD Gets Special Waiver for Spread Spectrum Experiments	2-1
Special Temporary Authority	2-2
Experimental Radio Services	2-2
Spread Spectrum STA	2-2
FCC Grants Spread-Spectrum STA	2-3
FCC Proposes Part 15 and 90 Spread Spectrum	2-3
Spread Spectrum Day at the FCC	2-3
ARRL Spread-Spectrum Committee	2-3
Our New Spread Spectrum Rules	2-4
Spread Spectrum Systems in 902-928 MHz	2-5

3 The Regulatory Process

FCC Notice of Inquiry in General Docket No. 81-413, "Authorization of Spread Spectrum and Other Wideband Emissions"	3-1
FCC Notice of Inquiry and Proposed Rule Making in General Docket No. 81-414, "Amendment of Parts 2 and 97 to Authorize Spread Spectrum"	3-7
ARRL Comments in General Docket No. 81-414	3-12
ARRL Reply Comments in General Docket No. 81-414	3-15
AMRAD Reply Comments in General Docket No. 81-414	3-17

Early Experimentation

4 AMRAD's Contributions

Using Microprocessor-Controlled Rigs for Frequency Hopping	4-1
Experiment Update	4-2
Spread Spectrum Invades Packet Radio	4-3
10-Meter Frequency Hopping	4-6
SS Codes	4-7
Mobile Frequency Hopper	4-9
Self Enforcement	4-10
AO for the Future	4-10
Experiment #5	4-11
Civilian Uses of Spread Spectrum	4-11
Exploding a Myth	4-11
Shift-Register Update	4-11
Spread Spectrum News	4-12
A New Frequency-Hopping Experiment	4-13
Second STA Program Formulated	4-13
Drop-In Channel Assignments	4-13
Frequency Control of ICOM Portable Radios	4-13
Experiment Update	4-15
Frogs, or a Primer on Hoppers	4-16
Direct Sequence Rules	4-17
AMRAD Frequency Hopping Beacons	4-18
Thue-Moore Sequence Experiment	4-19
Repeaters and Spread Spectrum	4-19
Direct Synthesis Oscillator Works!	4-20
Good Books for the Experimenter	4-20
Hide Spread Spectrum Transmitter Hunt	4-20

5 The League's View of SS Experimentation

Spread Spectrum and the Radio Amateur	5-1
Spread Spectrum Techniques	5-4
Spread-Spectrum Applications in Amateur Radio	5-5
Notes on Spread Spectrum	5-10
Spread-Spectrum Communication: A New Idea?	5-11

Practical Applications/Hardware

6 The FCC Report and Order Legalizing Spread Spectrum

7 AMRAD Chronicles SS Progress

Skywave Propagation and Spread Spectrum	7-1
Some Notes on Authentication for the Packet Radio Command Channels	7-4
Will the *Real* LPI Please Stand Up . . .	7-6
First Steps in Direct-Sequence Spread Spectrum	7-7
A Spread Spectrum Repeater	7-10
Injection-Locked Synchronous Oscillator in a Single IC Package	7-12

8 Spread Spectrum Theory and Projects

Spread Spectrum Communications	8-1
A Digital Frequency Synthesizer with a Microprocessor Controller	8-11
Spread Spectrum: Frequency Hopping, Direct Sequence and You	8-25
Practical Spread Spectrum: A Simple Clock Synchronization Scheme	8-27
Experimenting with Direct-Sequence Spread Spectrum	8-31
Extracting Stable Clock Signals from AM Broadcast Carriers for Amateur Spread Spectrum Applications	8-36
Practical Spread Spectrum: Achieving Synchronization with the Slip-Pulse Generator	8-41
A Practical Direct-Sequence Spread Spectrum UHF Link	8-47
Practical Spread Spectrum: Clock Recovery with the Synchronous Oscillator	8-55
Practical Spread Spectrum: An Experimental Transmitted-Reference Data Modem	8-58
License-Free Spread Spectrum Packet Radio	8-64

Government Reports

9 Proposed Use of Spread Spectrum Techniques in Non-Government Applications
Walter C. Scales

10 Proposed Direct Sequence Spread Spectrum Voice Techniques for the Amateur Radio Service
J. Hershey

Bibliography 384

Chapter 1

AMRAD'S Perspective

It all started innocuously enough: a short note buried in the June 1980 issue of the AMRAD *Newsletter.*

There is some interest at the FCC: in Amateur Radio experimentation in wideband modulation schemes, or spread spectrum. As a possible type of usage it may be possible to overlay a new group of users employing spread-spectrum modulation over another group of users who already fully occupy a band of frequencies with narrow-band transmissions. Of course, the present rules do not permit this type of emission in the amateur bands, so a Special Temporary Authority or an Experimental License would be required. If you would like to experiment with wideband modulation techniques, and be part of an AMRAD project, please call Paul Rinaldo, W4RI.

By October of the same year, a special interest group (SIG) had been formed. It included Hal Feinstein, WB3KDU; Paul Rinaldo, W4RI; Glenn Baumgartner, KAØESA; Jim Osburn, WD9EYB; Terry Fox, WB4JFI and Olaf Rask, WA3ZXW. This group met several times to devise a series of experiments, and then applied for an FCC Special Temporary Authority (STA) to carry them out. The remainder of this chapter consists of reprints of AMRAD Newsletter *articles and news items that show how spread spectrum evolved. The FCC Notices that enabled this experimentation to take place on the Amateur Radio bands can be found in Chapter 3.*

August 1980

AMRAD IS GOING AHEAD WITH A SPREAD-SPECTRUM PROJECT, according to a meeting held on July 6. Present at the meeting were:
- Glenn Baumgartner, KAØESA
- Werner Fehlauer, WB2BRB
- Terry Fox, WB4JFI
- Paul Rinaldo, W4RI

It was thought at that meeting that it is premature to request an STA until we decide on the basic approach, i.e., whether to use direct sequence or frequency hopping. Also, everyone thought it would be a good idea to poll various sources for information on what had already been done by others. We are also interested in concepts for spread-spectrum systems that could be built on amateur budgets. Another thing we could use at the moment is a primer (or tutorial) on spread spectrum for the AMRAD Newsletter. We have also heard from the following individuals by mail:
- Mike Waters, Utica, NY
- Gerald Swartzlander, W8EPI, Fremont OH
- Richard Doering, WA6CFM, Argonne, IL

Anyone else interested in helping get this project off the ground is urged to call Paul Rinaldo, W4RI.

October 1980

THE SPREAD SPECTRUM SIG had its second meeting on September 7. Attending were: KAØESA, WB3KDU, K4KJ, W4RI and WA3ZXW. The purpose of the meeting was to try to determine who could do what in terms of experimental spread-spectrum (SS) systems so that we could nail down what to ask for in the STA request. If you're reading this and don't have the foggiest idea of what it's about, here is the story. The Federal Communications Commission let it be known that they might look favorably on a request for a Special Temporary Authorization for Radio Amateur experimentation in spread-spectrum techniques. Spread spectrum is a class of modulation systems which use bandwidths far in excess of what is needed to simply convey the intelligence... on the order of 10 to 100 times. The neat thing is that the total energy transmitted is spread throughout that band and is thus only a small fraction in any one hertz or kilohertz, maybe below the noise level. The SS receiver at the other end is synchronized with the transmitter and adds up all the pieces constructively. If you do it right, you can even get some gain and lots of privacy and immunity from interference while not bothering other users. Back to the meeting. WB3KDU agreed to act as secretary for the SS special interest group (SIG). WA3ZXW and WB3KDU are interested in modifying a pair of CB transceivers to 10 meters and developing a frequency-hopping scheme. KAØESA and W4RI are interested in working with WD9EYB on a direct sequence system for the 420-MHz band. If you are interested in joining this group of SS pioneers, please get in touch with Hal Feinstein, WB3KDU.

November 1980

THE SPREAD SPECTRUM SPECIAL INTEREST GROUP is making progress. The following information was supplied by the SSSIG secretary, Hal Feinstein, WB3KDU. The first draft of the request for Special Temporary Authorization (STA) was completed and is now being reviewed. Thanks to AMSAT for making the going easier by providing us with a copy of their recent STA request.

So far three experiments are being considered:
1) direct-sequence spread spectrum in the 420-MHz band;
2) frequency-hopping spread spectrum in the 10-meter band; and,
3) frequency hopping using a commercial rig in the HF bands.

Peter Driessen, VE7BBO, was in the Washington, DC, area recently and talked with Paul Rinaldo, W4RI, about packet radio, spread spectrum and other topics. Peter is a Director of CARF and is involved in packet radio experiments in Vancouver. Peter has suggested that it may be possible to frequency hop with the ICOM 701 which both Peter and Paul have. What we are looking at presently is a way of synchronizing and code generation. We are not sure at this point of the settling time on a new frequency, so there is much additional work to be done.

A few of the SSSIG people have been writing articles on spread-spectrum for various ham publications. Paul Rinaldo, W4RI, has written a nice article on "Spread Spectrum and the Radio Amateur," which has been accepted for publication in the November issue of *QST*. John Nagle, K4KJ, is working on an article for *Ham Radio*. We have a number of requests from other publishers; please drop me a line if interested.

I am planning to do an article on spread-spectrum coding techniques for the AMRAD Newsletter.

We received a letter from Jim Osburn, WB9EYB saying that he had to take time out to build a computer. Obviously, a computer will be of great help in developing spread-spectrum systems. We thought that might be the last we heard from Jim, judging by the way others have disappeared for a year or two after such a pronouncement. However, Jim said that he has finished building his computer. It is Z-80 based and has one 2716 for 2 k of ROM and two 2114's for 1 k of RAM. It's only five chips, but he has already programmed it to play music and to be an AFSK generator for RTTY.

Spread spectrum is a mode of radio communications originally developed for military applications. It features a unique ability to avoid detection by conventional receivers. This was the main motivation but as with some other well-known inventions, other interesting properties were observed after a time. These properties make spread spectrum an unconventional solution to today's crowded bands.

The spread-spectrum mode of transmission features a very wideband signal, one which is 10 to 100 times the information bandwidth. Normal modulation techniques try to pack as much information as possible into a narrow band. A conventional receiver limits its bandwidth to that of the incoming signal to improve the signal-to-noise ratio. Spread-spectrum spreads the signal over a large but controlled bandwidth.

Because the signal is spread out, narrow-band interference is still present but reduced in its effect. This is because the spread-spectrum receiver is gathering power from the entire bandwidth, not just the narrow window of a conventional receiver. This creates an antijamming (AJ) property for military applications and an anti-interference feature for anyone's applications. In a frequency-hopping (FH) system (one type of spread spectrum), the receiver's RF window is constantly hopping around in sync with the transmitter. If there is a narrow-band signal in the band, the window will fall on that signal only in a random fashion.

Interference between spread spectrum and conventional users will generally be very small. The spread-spectrum signal is designed to be invisible to normal receivers. Thus, there is a good chance that spread spectrum and conventional users can share the same spectrum space. Another feature of spread spectrum worth mentioning is that each spread spectrum unit employs a random-like code which is used to hop the frequency around. Two units that have different codes cannot hear each other. If two units have the same code, they have a kind of private channel, but the channel is not a fixed frequency—it is constantly moving. Thus, spread spectrum offers a new and interesting channelization scheme.

The challenge to Amateur Radio is to construct equipment that can use spread-spectrum modulation. It is not the kind of mode where you can order a rig from a mail-order house. It will require some real innovation to develop practical systems.

April 1981

THE FCC GRANTED AMRAD AN STA FOR SPREAD-SPECTRUM EXPERIMENTATION on March 6. Because the Commission considered this STA a significant one, we were invited to the office of the Chief of the Private Radio Bureau to receive the STA.

AMRAD's proposal requested permission to conduct 4 experiments:

#1. *HF Frequency Hopping* experiments on 80, 40 and 20-meter bands using commercial equipment made available via K2SZE.

#2. *10-Meter Frequency Hopping* using home-brew equipment, some built around converted CB transceivers.

#3. *UHF Direct Sequence* tests using home-brew equipment, including tests via the WR4AAG ATV repeater.

#4. *EME Spread-Spectrum* experiments using the 84-foot dish at Cheltenham, MD for moonbounce tests.

Currently covered by the STA are:
Glenn W. Baumgartner, KA0ESA; Robert A. Buaas, K6KGS; Thomas A. Coffee, W4PSC; Charles E. Conner, K0NG; Roger Cox, WB0DGF; Werner A. Fehlauer, WB2BRB; Hal L. Feinstein, WB3KDU; Terry Fox, WB4JFI; Michael J. DiJulio, WB2BWJ; Allan H. Kaplan, W1AEL; Richard J. Kessler, K2SZE; Joe H. Mehaffey, Jr., K4IHP; John J. Nagle, K4KJ; James P. Osburn, WD9EYB; Travis W. Pederson, N5TP; Raymond C. Petit, W7GHM; Charles O. Phillips, K5LMA; David H. Phillips, W3PJM; Olaf Norris Rask, WA3ZXW; Alexander H. Riccio, W2NBJ; Paul L. Rinaldo, W4RI; David B. Ritchie, N6DLU; John H. Sharpe, WD5JYF; Gerald W. Swartzlander, KG6Y; A. Brent White, W7LUJ; Glenn S. Williman, N2KHW; Stephen Ray Wimmer, WB0GGT; Caltech ARC, W6UE.

Theodore J. Brentlinger, WB9QFE, has completed a letter of participation which we held onto while the STA was being finalized. Any others wishing to make a serious contribution to this experimentation should get in touch with Hal Feinstein, Secy SSSIG. Any additional letters of participation should reach Hal by the end of April so we can submit them to the FCC as a group. Check with Hal for wording of the letter. Also let Hal know how you plan to contribute to the experiments.

August 1981
LITERATURE

There is quite a bit of spread-spectrum information in a new book *Principles of Military Communications Systems* by Dr. Don J. Torrieri, published by Artech House, 610 Washington St, Dedham, MA 02026. If anyone can find the $45 to buy it and the time to read it (or any other SS literature), please let us know your comments.

December 1981

The FCC is moving ahead with its desire to include spread spectrum as a new amateur modulation type.

The FCC is faced with several interesting issues in considering spread spectrum for ham radio. With the desire to deregulate and allow new advanced techniques comes the need to redefine and enhance traditional approaches to regulation and enforcement.

This point was underlined by Tom Clark of AMSAT several months ago at the ARRL's League Planning Meeting which was held in Falls Church, Virginia. Tom spoke mostly about AMSAT's Phase III program but foresaw the challenge of advanced modulation techniques facing the Commission.

Most important to the Commission is the possibility of interference from wideband "spread" or "swept" signals to both amateur and commercial users. This motivated the inclusion of several steps within the NOPR which should be pointed out.

Specifically, the Commission proposes to:

1) Authorize spread spectrum in three bands—50-54, 144-148 and 220-225 MHz.

2) Authorize spread-spectrum modulation only for Advanced and Extra Class licensees.

3) Include questions on spread spectrum in the question set for Advanced and Extra Class tickets.

The above items are configured to adhere to certain international agreements and sharing arrangements with the government. The Commission stated that they will wait to the end of the AMRAD STA before acting on this NOPR.

In looking this over, I would like to offer some comment about items 2 and 3. It goes without saying that the FCC is very concerned about interference possibilities. It appears from the NOPR that they surmise that interference will most likely come from "incorrectly or poorly designed equipment." The NOPR states that "use (of) spread-spectrum technique should be restricted to Amateur Extra and Advanced Class licenses only." The assumption I think the FCC is expressing is that the higher-class ticket holders will be less likely than General or Technician class holders to build "dirty" equipment. This forms a perplexing question: does the holding of a "higher-class ticket" influence the technical knowledge of the holder beyond a trivial noise factor or the care in which the licensee would conduct on-the-air experiments? Perhaps some light can be shed on the problem in light of the experience gained by some of the AMRAD members working under the spread spectrum STA. A lot of experimental activity has been conducted by hams who have higher-class tickets and by hams with General class tickets as well. The knowledge which allows them to plan and construct spread-spectrum gear is not correlated with the class of license.

The hams who were interested in spread spectrum enough to learn and digest the complex technical literature on the subject are experimenters. This is a special label for a ham to wear in this day. It requires considerable drive and initiative to be a self-starter with this complex technology.

We believe that experimenters are not readily identified by the class of ticket they hold. We do think that the Commission and the amateur community would be well served by allowing Technician and General class ticket holders of being experimenters as well. Limiting ham-radio experimentation to only higher-class tickets defeats this goal.

January 1982
FEBRUARY AMRAD MEETING

The February AMRAD meeting will be devoted to reviewing the activity of the spread spectrum SIG. As you know, March is the month that the STA will expire, and we are required to report to the FCC on the activities carried out under the spread spectrum STA.

Specific items that will be reviewed at the meeting are the completed experiment number one (hf frequency hopping) and experiment number five (vhf frequency hopping). Progress on experiment two (10-meter frequency hopping) is at the construction phase. We are looking forward to activity for experiment number three which is the first crack at direct-sequence spread spectrum

As you can see, most of our spread-spectrum activity has been with frequency hopping which seems to be technologically more intuitive than direct sequence and perhaps somewhat easier than direct sequence.

All members of the spread spectrum STA are urged to attend this meeting and get in their two cents' worth. We are also planning for a second STA which will be used to carry forward the work started in 10-meter frequency hopping and the 420-MHz direct-sequence work.

April 1982

Here for your information, are the comments that we filed with the FCC on their NOI/NPRM for permitting spread spectrum on VHF ham bands:

Before the Federal Communications Commission
In the Matter of

Amendment of Parts 2 and 97 of the Commission's Rules and Regulations to authorize spread spectrum techniques in the Amateur Radio Service.   GEN DOCKET NO. 81-414

COMMENTS

1) The Amateur Radio Research and Development Corporation (AMRAD) supports the FCC's proposal to authorize spread-spectrum techniques in the Amateur Radio Service, qualified as follows:

(a) AMRAD recommends extending spread-spectrum techniques to General and Technician licensees. In our experience, there is no reason to believe that the ability to competently handle spread-spectrum technology lies mainly in the higher-class licensees. Nor, is there reason to believe that the General and Technician class licensees will act less responsibly on the air with spread spectrum.

(b) AMRAD believes that the door should be left open for codes other than the specific shift-register configurations described in the Appendix to GEN DOCKET NO. 81-414. To restrict codes as proposed in the Docket would tend to limit experimentation. This objection would be removed if there were a provision for FCC registration or approval of additional codes written into the Rules.

2) *AMRAD* is an experimenters' radio and computer club of over 500 members. We applied for, and were granted on March 6, 1981 a Special Temporary Authority (STA) to conduct specified on-the-air experiments with spread-spectrum techniques. Briefly, these experiments were:

Experiment 1—HF Frequency Hopping (using commercial/military transceivers on the 80, 40 and 20 meter bands)

Experiment 2—10-Meter Frequency Hopping (using modified Citizen Band radios)

Experiment 3—UHF Direct Sequence (using modified amateur television (ATV) transceivers)

Experiment 4—Earth-Moon-Earth Spread Spectrum (using an 84-foot dish at Cheltenham, MD)

Experiment 5—2-meter Frequency Hopping (using modified commercial VHF-FM transceivers)

3) The above experiments are only partially completed. Only Experiment 1 has been concluded. Experiment 2 is in the breadboarding stage. Experiment 5 is ready for on-the-air testing except for a critical part due from a supplier. Little progress has been made on either Experiments 3 or 4. Lack of substantial progress on some experiments was due to the inherent reliance on voluntary experimentation by individual amateurs as well as competition of other activities. The results of AMRAD's spread-spectrum experiments will be detailed for the Commission in a separate report on or about March 6, 1982.

4) Because the above experiments include spread-spectrum techniques not permitted under the proposed Rules change, AMRAD intends to request a new STA after completion of the report and after consultation with the FCC staff.

5) There may be a problem specifying the use of the center frequency for station identification as in the proposed Rules change. Designation of certain "home" frequencies may be a more workable system. We have no objection to the center frequency initially with a future option of redefining this requirement based on experience. Further, we suggest that the proposed wording of Section 97.84(h) include a standard, concise format for identifying. E.g., "DE WD4IWG/SS 100 kHz," would mean that spread spectrum is being transmitted over a 100-kHz bandwidth centered on the identifying frequency. A brief standardized format would permit development of automatic methods for telegraphic identification.

6) AMRAD is somewhat concerned about spread-spectrum's potential for abuse if authorized for the Amateur Radio Service. On balance, however, we feel that the advantages to stimulating state-of-the-art experimentation by permitting spread spectrum outweigh the disadvantages. Nevertheless, it is incumbent on both the Commission and historically self-policing radio amateurs to closely monitor the progress of spread spectrum and take appropriate steps to minimize possible abuse. In order for the Amateur Radio Service to continue to be self-policing, amateurs need to develop the means of detecting and identifying spread-spectrum transmissions as well as locating sources of interference. Therefore, AMRAD plans to propose a program (in coordination with the FCC Field Operations Bureau and with the American Radio Relay League) to develop a capability for amateurs to self-police their spread-spectrum operations.

7) Below are our answers to the specific questions posed in paragraph 15 of GEN DOCKET NO. 81-414:

(a) The emission limitations specified in the proposed amendment to Section 97.73 are considered to be sufficient until such time that experience indicates otherwise.

(b) Although our experience is necessarily limited at this time, it appears that spread-spectrum interference to conventional amateur communications will not be a major problem. Of the various conventional amateur modes, radioteletype operation (including packet) appears to be the most vulnerable.

(c) It is not believed necessary for the Commission to have the capability to monitor the content of all amateur communications. However, it is definitely necessary for the Commission to be able to detect spread-spectrum signals and identify them. Further, it would be desirable for the Commission to be able to make pre-detection recordings and decode spread spectrum signals when required.

(d) The specific shift registers proposed in the amendment to Section 97.117 of the Rules are not considered to be an obstacle to self-monitoring by the amateur community

Respectfully submitted,

February 26, 1982

AMATEUR RADIO RESEARCH AND DEVELOPMENT CORPORATION

Paul L. Rinaldo, W4RI
President

May 1982

A SECOND LOOK AT SPREAD SPECTRUM

Spread spectrum (SS) is a communication technique which uses wideband modulation that may exceed the bandwidth of the information by more than ten times. In fact it is not uncommon to find as much as 10-MHz bandwidth to transmit 3 kHz of information.

Because of a property of invisibility under certain situations, it has been used by the military for signal hiding. SS cannot be properly received on a general ham radio receiver but needs special types of receivers to decode the SS transmissions. To decode an SS signal, the receiver needs a copy of the original coding sequence. This leads to many interesting properties which are currently being considered by both government and industry

We have finished the first year of a testing cycle on SS modulation techniques. The year saw the FCC granting AMRAD a Special Temporary Authority (STA) which allowed us to experiment with SS — a mode which was not legal under normal rules of the Amateur Radio Service. We have conducted several interesting experiments with SS using both gear developed for the military and home-brew rigs built by STA participants.

One of the big objectives of SS has been to domesticate it. It was developed for the military and has all the trappings, insights and jargon that one expects to find attached to military programs. There were several things that needed investigation. For example, a major concern was self policing by amateurs. Within AMRAD there have been some discussions which resulted in ideas on the self-policing and enforcement aspects of SS. We came up with a concept for a new type of receiving system, one which did not need to know the code sequence in order to receive a frequency-hopping signal.

We discovered that there are numerous codes which are complex yet do not have privacy properties. So, the number of codes that amateurs could use is larger than originally thought. In case a station illegally uses a complex code to hide the meaning, there are some workable concepts which could be used to detect this. TVI problems have not been encountered in our tests. However, some people seem to be concerned about it and had the idea that SS-generated TVI would be hard to identify. Study has shown that TVI complaints can be resolved in much the same way that normal TVI complaints are handled. So far, we haven't seen any practical results to indicate that SS is any worse a source of TVI than anything else.

All that AMRAD found out about SS in the first year was not totally favorable. The biggest unresolved problem is the near/far effect. In the near area, the SS signal is higher than the noise and is spread across the band. However, the signal strength falls below the noise some distance away. A rough bottom line is that SS can be worse in the near area and better in the far area than people usually imagine. We are not certain what can be done in the near area but are looking for approaches.

SS is an interesting technique and should be experimented with by amateurs. We feel that these experiments can be done in the amateur bands without undue interference with other users, including moonbounce and other weak-signal communicators. Hopping patterns and RF filtering can be used to protect the weak-signal listeners.

The FCC and commercial interests are looking to SS as a way of packing more signals into a given band. Hams look at it as a context for experimentation and innovation, not a routine operational mode.

AMRAD FILES REPLY COMMENTS TO FCC

On April 15, AMRAD filed its reply to the comments made by others to the FCC in response to the FCC's Gen. Docket 81-414. The docket proposed to make SS a legal emission under the FCC Rules. In it, the Commission proposed several restrictions on types of codes and classes of amateur licenses eligible to use SS.

In general, the comments on file appealed to the Commission to go slow. But, there was a general feeling that amateur experimentation with SS should be allowed provided that certain interests were protected. One of the more interesting suggestions was that the FCC consider a two-year test period after which the Rules would be reviewed.

Our comments were in favor of going ahead with the proposed Rules for SS and leave the protection of the special interests, such as weak-signal subbands, to be worked out within the amateur community.

NEW EXPERIMENT PROPOSED FOR THE NEXT STA

Chuck Phillips, N4EZV has told us that he wants to experiment in the 14- and 21-MHz bands with frequency hopping. The hopping speed is up to 80 hops per second. The HF equipment will be based on Chuck's 2-meter hoppers that he demonstrated at the February AMRAD meeting in Vienna, Virginia. Chuck has already lined up experimenters on the East and West coasts. He is writing up the STA test proposal for inclusion with the others. Meanwhile, if you have a specific SS test proposal, please mail it to me as soon as possible for inclusion in the new STA request. The new request will not be submitted until we finish writing the results of the first year's experiments, which should be done soon.

September 1982
IT'S A JUNGLE OUT THERE

It is distressing to read in various amateur (and professional) journals the restatement of various myths that have grown up around spread spectrum. Certainly reasonable amateurs when supplied with the real facts on spread spectrum will see this as another experimental mode—not as a legal jammer, invisible transmitter or the unbreakable coding system.

The FCC has plans to introduce spread spectrum as a normal amateur mode, above 50 MHz. We hope that any operating procedures mandated will not be at the expense of experimentation. It is our belief that experimentation will allow amateurs to develop proper operating habits to protect other users. This process will be furthered by getting validated information to the amateur community through various amateur publications.

AMRAD TO TALK AT MILCOM '82

Hal Feinstein, WB3KDU, and Paul Rinaldo, W4RI, were invited to speak at the 1982 IEEE Military Communications Conference in Boston on October 20. The paper, entitled "Spread-Spectrum Experiments for Civil Applications," will review AMRAD's on-the-air experiments with spread-spectrum systems which were conducted under a Special Temporary Authority from the FCC.

SPREAD SPECTRUM AT THE ARRL NATIONAL CONVENTION

WØIYH gave a talk on "Spread-Spectrum Applications for Amateur Radio" at the ARRL National Convention in Cedar Rapids in July to an audience of 100 or more. He sent us several copies of an excellent handout.

October 1982

Next month Paul Rinaldo, W4RI, and Hal Feinstein, WB3KDU, will be giving a paper at the IEEE/AFCEA Military Communications Conference (MILCOM '82) in Boston on October 17-20. A look at the conference agenda reveals that almost the entire conference is dedicated to military uses of spread spectrum.

Recently I was able to talk to Walter Scales of Mitre Corporation who has done some important studies in spread spectrum. I understand that Mitre is currently working on a design study for a distress system which will use direct-sequence spread spectrum and transmit the distress calls to the MARSAT maritime mobile satellite. Spread spectrum is being used because of its anti-jam properties. It seems that distress channels are in heavy use for things other than distress calls (such as a calling channel). The danger is of course that the ground stations would not hear a distress call employing standard modulation techniques when the channels are busy. Spread spectrum is expected to get through, because it is spread over 100 kHz, and its current design calls for keying in the user position. MARSAT is operated by a consortium of countries. The MARSAT system has attracted many ship users and has taken a lot of ship-to-shore traffic off the HF bands.

SPREAD SPECTRUM LEGAL?

On September 14, the FCC passed rule changes which will permit digital codes other than Baudot and ASCII for digital communications above 50 MHz. The new rules will allow speeds up to 56 kb/s or bandwidths up to 100 kHz, for example. Suppose that a slow data signal is being generated at the rate of 1 kb/s. Next let us make a pseudorandom code sequence by generating ASCII characters at the rate of 50 kb/s. Every six nonsense characters or so, one of the information characters is inserted into the stream. What do we have?

Because the stream is made up of a pseudorandom stream of characters, we should expect the spectral density to look much like our sin x/x curve. This is the spectral shape that we want. Inserting a "deterministic" character every once-in-a-while (by strict schedule known to all) should alter the spectral shape only a little. Food for thought.

August/September 1983
SPREAD-SPECTRUM NOPR

As you are probably aware, the Commission is considering incorporating spread-spectrum modulation techniques into the Amateur Radio Service as a fully legal technique. We think this is a good idea. The introduction of the home computer and the integrated circuit is causing a transition in Amateur Radio that is probably as significant as the transition from Morse code to amplitude-modulated voice techniques in the early part of this century.

However, as promising as this transition is, new ideas must be subject to regulation that permits all parties to enjoy the benefits of new technology without interfering with conventional modes of communication.

Herein lies a problem, the Commission is proposing that only holders of the Extra Class amateur license be allowed to use spread-spectrum techniques. The reason for this proposed restriction is that the Commission believes only Extra Class licensees have the technical expertise to operate without causing interference to other operators.

We understand the Commission's desire to limit the potential for interference, but we question the rationale of the proposed license restrictions. Our experience indicates that the class of license has little to do with actual technical ability. In fact, we found that amateurs who experiment with new modes of communication tend to have other than an Extra Class license.

One of the Commission's goals for the Amateur Radio Service is to promote experimentation and innovation at all levels of the hobby. We do not believe the Commission's interest can be served by implementation of this sort of restriction.

July-August 1984

This has been an active month for us spread-spectrum folks. First, the FCC has put out some new rulemaking items which amend the commercial spread spectrum proposed rules. The new stuff amends the items they already had issued to permit spread spectrum in commercial service. The new items seek to permit low-power spread spectrum in the 70 MHz and up band, at the same time it authorizes its use for police.

First, the low-power spread spectrum which has been authorized for use in the 70 MHz and up band is limited to 100 milliwatts at three meters from the antenna. This is essentially to permit a wireless form of local area network manufactured by Hewlett Packard to operate. The HP product performs the same type of local area networking as products such as SYTEK's base and broadband LAN, but uses spread spectrum encoding of the carrier for each node. The building where the system is installed is wired with radax, a special form of coax cable which is specifically designed to radiate and receive RF. This kind of cable is used in subway tunnels and large buildings to give fire departments access to their communication systems.

Each terminal, printer, facsimile machine or computer which desires to gain access to any of the other devices is simply outfitted with the spread-spectrum transceiver module. Each device is addressed by its own orthogonal code sequence. The system contains one or two more devices but this is the essential idea, no cables are needed between the terminal and the network.

The second part of the rules allows for police use of spread spectrum in such things as stakeouts, undercover operations and "tailing." Any of the three modes of spread spectrum are permitted: frequency hopping, time hopping and direct sequence; however, power is limited to between ten and fifteen watts. Frequencies available for police spread spectrum are the same as those currently authorized for them.

The police are interested in spread spectrum ostensibly to prevent the bad guys from overhearing them on scanners. In addition, when stake out cars are close-in to a target, the signal tends to be very strong. This again is a give-away to smart crooks. Partial relief was for police to use a good voice scrambler such as Motorola's Digital Voice Protection (DVP) or the Harris equivalent. These systems have larger bandwidths and transmit encoded, digitized voice using FSK. The signal sounds like a white noise on a scanner. A smart crook (successful smugglers fall into this category) will know exactly what's up when they hear this kind of signal on their scanners.

In theory, spread spectrum will solve the police's problems. It's supposed to be very hard to hear it without knowing the code sequence used, at least if they are at some distance from the receiver as well!

A while back someone spoke to me about a miniature transmitter that could be attached to a car which you could then trail all over the countryside using a direction finder. One problem, however, is that harmonics of the "beeper" often fall on popular FM radio channels which can compromise the beeper's trailing operation.

Spread spectrum, which is hard to receive if you don't know the code, would really not interfere with the local FM station.

Of course, there is the near far effect which means that a spread-spectrum signal right next to the antenna area will be rather loud. In fact, in the case of the above-mentioned beeper device the receiver and beeper are in the same car.

The lesson here is that, as with other kinds of spread spectrum, a variety of time-tested tricks must be used that worked well for non-spread spectrum devices of this kind. Combining them with spread spectrum gives a better product, since spread spectrum alone doesn't answer all the design requirements.

In fairness, police are suffering from a serious communication security problem (COMSEC) because of scanners. I have heard that some scanner manufacturers are about to bring out a scanner that will cover the cellular radio channels, up in the 900-MHz bands which will limit the privacy you can find there.

Some aspects of spread spectrum can aid the privacy problem, however, COMSEC is only part of the answer to law enforcement.

July-August 1984

AMATEUR SPREAD-SPECTRUM PROGRESS

The FCC is again interested in moving the amateur spread-spectrum issue into rulemaking and has been studying the issue of regulation, enforcement, and amateur self-policing. Here is some of the poop:

All amateurs will be eligible for Special Temporary Authorization to conduct any reasonable experiment which falls outside of the issued spread-65spectrum rules. The rules are consequently rather restrictive for general experimentation, but the Commission appears willing to consider any reasonable experiment as STA material.

HF spread spectrum will not be authorized. All experimentation will be done at 50 MHz and up. Some commercial lobbyists have been pushing to prevent 50-MHz experimentation by hams.

Three kinds of spread spectrum will be allowed: frequency hopping, time hopping and direct sequence. You must keep your signal inside the amateur bands. However, you don't have to use only the voice frequencies if you are using this form of keying.

Three code sequence generators are defined by the Commission which consist of simple linear feedback shift register sequences. The output of these three shift registers must be directly applied to the rig in an appropriate way for the type of spread spectrum being used. For direct sequence, the output must be mixed with the information source Modulo-two in the conventional way. For time hopping, the sequence must be used to gate the signal directly. Lastly, in frequency hopping some more sophistication is permitted.

For frequency hopping, the parallel output of the shift register is used to address an EPROM which is called the channel table. The channel table contains only the channels which are not occupied by fixed users such as repeaters. In effect, the channel table is used to bypass possible interference by only permitting hopping to vacant channels which are not used by fixed users.

The Commission is considering two different approaches to the channel table. The first proposal is that a nationwide channel table would be published by someone (perhaps ARRL) and would be used by all hams operating spread spectrum. Second, if a channel is occupied locally, you would simply lock out this channel and not switch the transmitter on for that hop, however you would have to dwell on that silent channel for the same length of time that you dwell on other channels. This is so that you don't mess up people who are trying to listen and don't know where your silent channels are. The third technique is to hop to a planned auxiliary channel for the occupied ones. However, you can only do this for about 25% of the channels. You can choose any channel you wish for the auxiliary channels, however, you must keep the number of sequentially visited auxiliary channels to under 5% in a single stretch. This will require the original channel table to be carefully thought out otherwise it will be difficult to implement.

The Commission is considering some upper bounds on dwell time and chip rate. I was not told of any numbers for these at this time.

Station identification is one of the really big issues. The Commission wants to make sure that hams can still self-police themselves in traditional ways. I don't know what will actually come out of this, but the scheme that was being floated at the time involved a lengthy ID sequence. The sequence would be transmitted in one of the standard hams' modes (AM/SSB/CW) and contains at least: (a) call sign, (b) spread spectrum mode, (c) seed value (initial load of the shift register), (d) sequence generator ID (there are only three authorized), (e) lowest and highest frequency in use, (f) dwell time or chip rate.

Some other things were considered but I don't know what their status is at present. At least a spread-spectrum user could look forward to ID with something like: W B 3 K D U W B 3 K D U/FH/11000/3/4500/7900/50/16000F3 K.

De-light-full, isn't it? Let me try to decode what it says: (1) the call of the station IDing, (2) the frequency-hopping mode is being used, (3) linear feedback generator three is being used, (4) the 19-stage sequence generator is initially loaded with hex 11000 (right justified), (5) the lower frequency is 144.500 MHz, (6) the upper frequency is 147.900 MHz, (7) the dwell time is 50 milliseconds per hop, (8) the modulation is normal narrow-band FM (16000F3).

I asked about automatic computer IDing of the kind used in packet radio. The commission didn't seem to be wild over this idea, however, it isn't ruled out as of this writing.

The ID frequency was also a sore point. Right now the theory is that a national ID frequency will be identified by some group (ARRL perhaps) for each band. All identification would happen on this frequency. Lastly, what kind of preamble would be authorized? Currently, only the falling edge of a 10 second or so tone will be used but there are a lot of other possible schemes that can be used. At this writing, only the falling edge synchronization (FES) technique is allowed.

Well, so much for the current FCC thinking on amateur spread spectrum. Some of the thinking is good, some needs further refinement, but it is at least a start in the right direction.

As a result of the new interest on the part of the FCC, we at AMRAD have started to put together an ad-hoc standards group which will include people from all around the country. I have a names list from past correspondence which I have received from various people and will try to select interested parties from it. In addition, I have spoken very informally with the ARRL about their possible participation in this group. They seemed positive about the idea. If you would be interested in participating in this ad hoc standards group, please drop me a line. I promise to answer all correspondence on this subject as soon as I can.

In summary, it seems that there is a new effort to push spread spectrum into regulatory form. Because of ways the regulators are trying to model regulation on conventional modulation, the attempts are awkward and cumbersome. It is not that the methods are wrong, the methods for regulating spread spectrum do not currently exist and must be created. In reality, this is empirical rulemaking.

January 1985
RULE MAKING UPDATE

In last month's column we spoke about the pending rule making for ham radio spread spectrum operations. The talk around Washington these days is that the Commission will consider a minimum of rules for amateur spread spectrum. The lengthy identification procedure and other cumbersome rules are out, just a minimum set of rules.

The spread-spectrum rules will probably require an identification, either in a conventional narrow-band modulation method, or by keying the spread-spectrum carrier off and on! Yes, you need only send your call signs by keying the carrier off and on. This applies to both types of spread spectrum: direct sequence and frequency hopping. Time hopping and the combination of two or more forms of spread spectrum (called hybrid systems) will not be permitted. In addition, if you identify yourself by this carrier keying method, you don't have to identify yourself by conventional modulation!

The carrier keying method was first used by a precision radio location system which is used on site in oil drilling rigs in the Gulf of Mexico. The system uses a form of chirp (linear FM) spread spectrum which sweeps its carrier many times a second over a specified band segment. The system operates in the 420-MHz amateur band and has not been a source of interference to local hams. This system should not be confused with one used by a different manufacturer on the West Coast of the US. The West Coast system operates on a different principle and has caused substantial interference.

The Gulf of Mexico positioning system is not typically heard by amateur receivers. In order to identify, the chirp rate is slowed so that tone is produced in a conventional receiver's bandpass. This is used to send a short Morse code identification sequence. The technique has found its way into the amateur rules for spread spectrum.

The properties of this form of identification are interesting. First it appears that the identification tone will be heard only near the transmitter, not at a distance. This is ideal for resolving interference problems but doesn't provide much help for a ham listening at a distance. A narrow-band identification will probably be needed to allow hams to work each other.

A second interesting property is how this form of identification will apply to other types of spread-spectrum, especially since chirp is not authorized. Would frequency hopping stations be required to produce this swept tone? How might this be done? This same question is open for direct sequence; however, direct sequence could simply key the entire carrier off and on, since it would probably be heard in the near field across the entire spreading width.

The rules also will probably restrict spread-spectrum communications to stations under FCC jurisdiction. This means no international spread spectrum will be permitted unless both governments agree beforehand. In addition, all spread-spectrum operations will be in VHF, consisting of 50 MHz, 144 MHz and 220-MHz bands. As reported in earlier columns, only three shift register sequences will be permitted. The longest of these is a nineteen stage register sequence which should be long enough for general communications.

Modulation of the spread-spectrum carrier can be done with any of the conventional techniques and can include such information as packet, digitized voice, digital television as well as AM, CW, SSB and FM. There is some speculation that many forms of phase shift keying will be permitted such as QPSK and MSK which is a new form of digital modulation for hams to experiment with.

There will be a down-side to all this however, stations running spread spectrum will be required to keep a logbook of such operations. If there is any question as to what you are doing with spread spectrum, the Commission may require you to record all your spread-spectrum contacts. Your tape should hopefully match the one which the Commission can make independently. In addition, the rules clearly call for use of plain language, publicly known codes and common protocols. The ground work for this type of "coded" message was laid when the Commission approved the use of ASCII and other codes by amateurs.

Lastly, a channel table will probably be called for frequency-hopping transmission. The channel table will consist of those channels which you choose to visit during a transmission. The table must be in ascending order and will be addressed by the parallel run of bits taken from one of the three authorized shift registers. Channel tables are useful to avoid hopping onto repeater and low-signal user areas and to have the channel available in a code format needed by the synthesizer, for example BCD.

While some of the rules do restrict experimentation, there are some good things here too. For example, the proposed rule restricting spread spectrum to Advanced and Extra Class hams has been dropped. There is the possibility that many new forms of modern modulation techniques will be made available to hams and of course spread-spectrum experimentation will be permitted.

For those interested in experimenting with different types of spreading codes, the Special Temporary Authorization (STA) is open to licensed hams. The whole rule-making package is slated to go before the Commission in February. Spread spectrum may be legal by the time you read the next newsletter!

May 1985

NEW SPREAD-SPECTRUM GROUP

I have a letter from Dick Bingham, W7WKR, Everett, Washington who tells me that his group will be applying for an STA to operate direct sequence. The group already has the system bench tested and is awaiting the STA to begin their experiment. They are planning a 2-meter repeater in which the uplink to the repeater will be spread (maybe on 2 meters) and the downlink from the repeater will be NBFM.

SPREAD SPECTRUM IS APPROVED!!

On Thursday, May 8 the FCC met and approved amateur spread-spectrum emissions for use in amateur bands 420 MHz and above. The approval gives amateurs the ability to experiment with both direct-sequence and frequency-hopping methods as well as various kinds of digital modulation techniques such as Bi-phase shift keying and QPSK (4 phases). While there is plenty of good news there is some bad news too.

There is a one-year delay before amateurs can actually use spread spectrum. During this time the FCC field engineering people will be gearing up to handle any ancillary enforcement problem that may crop up. Also, it looks like our effort to get on 220 has been deflected by the Commission. The reason stated in the rule hand-out prepared by the FCC states that there is not enough spectrum space to effectively spread a signal out. Let's examine this a bit: first, spreading the signal out is just one factor affecting spectral density; power is another and perhaps a better way to limit spectral density. Second, the rationale ignores the way frequency hopping can control its energy distribution and that FH can use small amounts of band and still preserve the desired low-profile transmission characteristics. Try to puzzle out for yourself why the Commission wants to limit more amateur activity on 220. I asked a lot of people at the Commission why, and didn't receive a satisfactory answer. To me this is an omen of what's to come. It's common knowledge around Washington that there are powerful lobbies at work trying to gain 220 for the land mobile people. Like a land-mobile industry friend of mine said recently: "Hal, you gotta realize something: MONEY is MONEY."

As much as I dislike the insider maneuverings over 220, I cannot support the ad hominen attacks coming from some amateur circles.

We also didn't get 2 meters or 6-meter operations either. In part, this was due to two factors: first, some broadcast lobbies were against 6-meter amateur spread spectrum—ostensibly to prevent amateur spread-spectrum signals from spilling out into TV Channel two. However, in reality many TV receivers are incapable of effectively rejecting legal signals anywhere near their assigned bandpass. Hence, most amateur 6-meter operations have been chased out of the higher end of 6 meters because of TVI. In fact the broadcasters have enjoyed a kind of 2-3 MHz wide guardband between 52 and 54 MHz.

Amateur frequency hopping and direct sequence would take advantage of the entire 4 MHz and this wrecks the gentlemen's agreement guardband.

Now, on two meters the Commission felt that it is already quite crowded and perhaps amateur spread spectrum might be the cause of interference. The interference angle dominated the decision on two meters. Now some good news, the Commission's original idea to limit spread-spectrum experiments to Advanced and Extra Class holders only was dropped.

Well, this is certainly a milestone for Amateur Radio. I know that there are a lot of folks who opposed these rules and a progressive group that was always for amateur spread spectrum. There is already interest in the commercial sector since in many cases amateur experimentation is a precursor to commercial uses of the technology.

Be sure to see the new rules. Remember that there is a one-year delay before they go into effect.

November/December 1985

SPREAD-SPECTRUM COMMITTEE MEETS

On December 14, 1985, the ARRL Ad Hoc Committee on spread spectrum met at the home of David Borden, K8MMO. AMRAD members presented a strong showing at the meeting, signing up to write papers on a variety of "Interoperability" topics. Our somewhat limited experimentation in the spread-spectrum area allows us to contribute some knowledge to the general amateur community which will be allowed spread-spectrum privileges on June 1, 1986. The papers will be brought together to form an interim standards document to be published by the June 1 deadline. While we realize it is a bit early to hammer standards in concrete (like AX.25), the general ham community needs some guidance on how to communicate using this new mode and can learn from our limited experience. The committee should provide a framework from which to produce standards of spread-spectrum communication.

BAD DAY AT BLACK ROCK

There is good news and bad news in the world of amateur spread spectrum. The good news is we have discovered the world of direct sequence devices thanks to Dick Bingham, W7KWR. Dick has sent an ingenious design for direct sequence which takes 2-meter FM in and puts FM direct sequence out. The whole thing has about 8 modules (prebuilt from minicircuits labs) and is not too complicated.

The mixed news is that the ARRL has formed an ad-hoc committee on spread spectrum to produce interoperability rules. Do we need interoperability rules? Like a bad cold! More on this later in the column.

Now the bad news. Totally unexpected the FCC has yanked our spread-spectrum STA privileges for any frequency below 420 MHz. What made them do this? It seems we interfere with the current Private Radio Bureau concept of what amateurs are all about. Why the hell should we experiment anyway? I can't think of a good reason to stay within the overregulated and constrained world of amateur regulations.

This sort of treatment is totally unjustified and convinces us that our experimentation will continue to be interrupted in the future. What is even more annoying is the clear lack of rationale in the letter. No explanation was given but some thinly reasoned and incomplete references to the new spread-spectrum rules. What does this have to do with our STA program? Nothing at all. Our experiments include investigation of the overlay concept in a crowded band—2 meters. Where else to test overlaying with crowded band conditions? It seems the reading coming from the FCC Private Radio Bureau is that 2 meters is a common carrier band and deserves protection against even occasional controlled experiments. Is this what ham radio is all about?

Of course the 220-MHz band will be carved up and given to the commercial interests as soon as politically convenient. In the meantime let's hope that amateur usage doesn't make it politically inconvenient. Do not authorize new amateur usage of the band because who knows where that will lead.

No 6 meters either, why?—heck if I know. HF is, of course restricted by treaty. No cryptography permitted. The problem is most people wouldn't know a real cryptographic system if it hit them in the face at 40 miles per hour. So no wonder everything that is "coded" scares the heck out of them. Look how long it took to get that most secret code ASCII into the amateur regulations.

No doubt, some FCC official will read this and create some rationale for this decision. After all, we don't know why the FCC took the action they did. We have found in official Washington that things are often done without the benefit of full information. After all, it's what your agency lawyers can make stick in court or before a trial board that counts; not understanding. To speculate that FCC actions are based on solid understanding of the issues would be to ignore the reality of how Washington is run.

We deeply regret the FCC's decision and wonder what light this casts on the basic premise of Amateur Radio. The message is that experimenters must look outside the framework of Amateur Radio in order to pursue our basic interest in experimentation. It is indeed sad that we have arrived at this juncture so soon in amateur spread spectrum's brief history.

Mixed news. The ARRL has formed an ad-hoc committee to invent interoperability standards. What are interoperability standards? They allow hams to "communicate" naturally. In doing so they also limit innovation. Consider that spread spectrum is new in the civilian domain. To date, only police and commercial development houses have been doing interesting things with it. This is a very new modulation technique for us and deserves time before it is calcified into a common carrier.

The complaint one used to hear about ham radio is that there was no innovation. In the 1940's hams were experimenters, in the 1960's they were "communicators," in the 1980's they have become consumers; something the FCC finally understands. Counter to what everyone was expecting from ham radio, digital techniques sneaked through and then came spread spectrum. The ARRL did a good thing in providing a standard for packet radio. The nature of packet is networking and networking requires agreement among many people.

To cast spread spectrum in this light is to misunderstand both the use of protocol standards and that spread spectrum is not inherently a networking art. In fact, spread spectrum is like other RF modulation techniques but with a wide diversity of possible characteristics. Rather than take the unintelligent way of calcified standards which impose minimal sets of common attributes; a sort of forced lowest common denominator; the ARRL committee should concentrate on providing a common experimenter vocabulary for spread spectrum, characterize measurements and show how they can be made on simple gear, describe the various signal waveforms and techniques, explore overlay and other spectrum management concepts and investigate different aspects of interference to conventional users—something the AMRAD spread-spectrum group has been doing for a few years now.

When a small group of FCC people got together to push through spread spectrum, they went against the common grain of opposition at the FCC. Since then this small group has essentially moved on and in their wake the *regime anciente* has reasserted itself. There is an unexpected and bizarre twist however. With deregulation, the FCC does not care to exercise detailed control; instead it relies on its alter ego—the ARRL, to provide this scrutiny.

Chapter 2

The ARRL Perspective

As spread-spectrum experimentation progressed, the League tracked its evolution in its membership journal, QST, *and in its "experimenter's exchange,"* QEX. *The first issue of* QEX *appeared in December 1981. Its editor, Paul Rinaldo, W4RI, continued his deep involvement with AMRAD's spread spectrum work.*

QST, May 1981

Happenings

Conducted By W. Dale Clift, WA3NLO

AMRAD Gets Special Waiver for Spread-Spectrum Experiments

The Federal Communications Commission has granted 28 Amateur Radio stations permission to conduct spread spectrum (SS) experiments on Amateur Radio frequencies. The radio amateurs involved are members of the Amateur Radio Research and Development Corporation (AMRAD), a nonprofit corporation and ARRL affiliated club comprised of over 350 members whose primary interest is developing skills and knowledge in radio and electronic technology. Spread spectrum is an unconventional form of modulation that uses 10 to 100 times the bandwidth needed to carry the same information via more conventional modes. However, its advantage lies in the fact that an SS signal's energy is spread throughout a band of frequencies so that the amount of energy at any particular frequency is much less than that produced by conventional narrow-band signals.

There are four basic ways an SS signal can be "spread": *direct sequence*, which is produced by modulating a carrier with a digitized code stream; *frequency hopping*, which is produced by jumping the signal to a number of different frequencies in an agreed sequence; *pulse-fm (chirp)*, where the carrier frequency sweeps over a wide band of frequencies at a known rate; and *time hopping*, which is a form of pulse modulation using a time-coded sequence to control the pulse. *Hybrid* spread spectrum systems use a combination of the four basic systems.

AMRAD's application to the Commission for the special temporary authority (STA) emphasized that spread-spectrum modulation techniques can conserve radio spectrum because different SS code sequences, or synchronizing SS signals at various points within one long code sequence, will permit many simultaneous communications free of interference from each other. Spread-spectrum receivers with the wrong code and conventional narrow-band receivers will be unable to demodulate the SS signal and may not even be able to detect the presence of the signal. In fact, depending on the transmitter power level

See "Spread Spectrum and the Radio Amateur," by Paul L. Rinaldo, in November 1980 *QST*, p. 15.

FCC Private Radio Bureau Chief Carlos V. Roberts (2nd from right), presents a Special Temporary Authority to experiment with spread-spectrum transmissions to Hal L. Feinstein, WB3KDU (far left) and Paul L. Rinaldo, W4RI. Dr. Michael J. Marcus (far right), chief, Technical Planning Staff of the Commission's Office of Science and Technology, witnesses the ceremonial occasion at the Commission's offices. Messrs. Feinstein and Rinaldo are two of the amateurs affiliated with the Amateur Radio Research and Development Corporation granted the STA. (*FCC photo*)

and the receiver's distance from the transmitter, the spread-spectrum signal may be below the noise level.

The FCC's STA gives AMRAD the authority to conduct four experiments. The first experiment will be conducted with commercially available hf frequency-hopping transceivers so that the group can quickly gain on-the-air experience with the technique. Test paths will be between Washington, DC, and Rochester, New York, on 3675-3775 kHz, 7050-7150 kHz and 14,100-14,200 kHz. Because of the crowding on these bands, particularly at certain times of the day, AMRAD plans to make prior announcements of the tests on these bands. W1AW will carry these announcements to the general Amateur Radio community.

The second experiment involves 10-meter frequency-hopping tests. To keep the costs low, the AMRAD group plans to modify several Citizens Band single-sideband transceivers for frequency hopping in the 28.1- to 29.3-MHz frequency range. The group will also investigate frequency hopping of the Inoue Communications Equipment Corporation (ICOM) IC-701 hf transceiver, one of a few amateur transceivers capable of external digital control. Because more spectrum is available in the 10-meter band than in those bands proposed for the first experiment, AMRAD will not have to make prior announcements of its tests via W1AW.

The third experiment involves the 420- to 450-MHz (70-cm) band for direct-sequence spread-spectrum tests. The group's initial plan is to use inexpensive television color-burst crystal modulators as the foundation for producing a phase-shift keyed (PSK) signal with shifts of 0° and 180°. One part of the experimentation, in cooperation with the Metrovision Amateur Television (ATV) Club in Alexandria, Virginia, is to operate SS concurrently with ATV transmissions to assess mutual interference potentials. Another part is to determine the feasibility of using an ATV repeater to pass SS signals, thereby obtaining the range advantage of a repeater. Also, the wide bandwidth of an ATV repeater assures small time delays for synchronizing SS signals sent through it.

The fourth and final experiment involves spread-spectrum techniques for enhancing earth-moon-earth (moonbounce) communications. Under the supervision of David Phillips, W3PJM, AMRAD will use an 84-ft parabolic dish antenna located at Cheltenham, Maryland. The AMRAD group believes that by spreading signal information over many different frequencies, it will eliminate the deep nulls that occur over narrow-band channels. The SS receiver, it is hoped, by correlating the SS transmitter's signal contributions across the band, will deal with the problems of libration fading and fading caused by wide-space reflections from the moon, rapidly adding to and then cancelling the reflected signal.

The Federal Communications Commission took the occasion of its granting the STA to AMRAD to encourage other Amateur Radio experimenters. In a news release, the Commission stated that it realized that in certain instances proposed experiments may conflict with existing amateur rules. It emphasized that it is willing to grant rule waivers for many different experiments, including: spread spectrum, packet-switching networks, radio-teleprinter codes (other than ASCII and Baudot, which are already permitted under the present rules), beacons for propagation studies, medium-scan television, frequency and/or amplitude compandoring, digitized voice techniques, digitized video techniques, trunked repeater systems and EME communications.

According to the Commission, radio amateurs wishing to conduct experiments within the amateur bands should first refer to the Commission's rules to determine if a rule waiver is required. If a proposed experiment will conflict with any of the Commission's rules, the licensee conducting the experiment must write to the Commission requesting a waiver of the specific rule(s). Waiver-request letters should be addressed to: Federal Communications Commission 334 York St., Gettysburg, PA 17325, Attention: Technical Section.

The waiver-request letter should cover complete details of the proposed experiment, including all technical parameters, specific frequencies to be used and a justification for the project. The Commission will approve or deny any request in writing, and no experimentation may begin until the written approval is received.

QEX, June 1982

SPECIAL TEMPORARY AUTHORITY

As an experimenter, the existing FCC Rules give you considerable latitude in which to conduct on-the-air experiments. However, there are limits to the Rules as written, and you may need special authorization to conduct certain experiments. So long as the tests you propose fit inside the frequency bands allocated to the Amateur Radio Service and generally follow the Rules for that service, you can request a Special Temporary Authority (STA) from the FCC.

Some recent examples of STAs are:
AMSAT's phase-shift keying tests on 10 meters.
AMRAD's spread-spectrum experiments.
AMTOR RTTY tests.

In order to apply for an STA, submit your request in letter form to the FCC, Gettysburg, PA 17325. Your request should be complete within itself so that the FCC approving official, (normally Richard H. Everett, Chief, Licensing Division) can get the entire story from the application. For more information or assistance, you can call the Licensing Division in Gettysburg on 717-337-1511. Here is the type of information usually needed:
Name, call and address of the applicant.
Purpose of the proposed tests.
Description of on-the-air tests.
Sections of the Rules to be waived.
Locations involved.
Operating frequency limits.
Types of emission.
Transmitter power (whether input/output).
Antennas, types, gain, height above ground.

Before submitting your STA request, it is a good idea to carefully check the FCC Rules to see if an STA is necessary. It is prudent to coordinate your tests with other amateurs who may also want to participate. A brief item in *QEX* is an effective way to get to many experimenters to see who would like to join in. Also, it will be useful for you to get in touch with amateur organizations that are already involved in such experimentation. It is worthwhile to contact Perry F. Williams, W1UED, the ARRL Washington Area Coordinator, telephone 203-666-1541, prior to finalizing your STA proposal, and certainly drop him a copy at ARRL Hq, 225 Main St, Newington, CT 06111.

Oh, there's a catch. At the end of the STA grant period, usually one year or less, you will need to submit a test report to the FCC.

In addition to notices of pending STA applications, please send copies of all STA proposals, grant letters and test reports to Editor, *QEX*. That will ensure that the Amateur Radio experimenters get a chance to give you some information and to share in your results.

EXPERIMENTAL RADIO SERVICES

There are occasions where proposed tests may involve transmissions which do not fall under the Amateur Radio Service, even with an STA. A recent example of this was in the case of beacon tests conducted in the new WARC bands, which are not yet allocated to the Amateur Radio Service. In such cases, the FCC can grant an experimental license.

The Experimental Radio Services are governed by Part 5 of the FCC Rules. You can purchase a copy from the Superintendent of Documents, Government Printing Office, Washington, DC 20402. FCC Form 442 (Application for New or Modified Radio Station Authorization Under Part 5 of FCC Rules Experimental Radio Services (other than Broadcast)) can be obtained from the Federal Communications Commission, 1919 M St NW, Washington, DC 20554.

If you would like to discuss the technical aspects of an experiment with someone at the FCC, you might try Dr. Michael J. Marcus, Office of Science and Technology, 2025 M St NW, Washington, DC 20554. Another helpful contact on experimental licensing is H. Franklin Wright, also in FCC/OST.

Again, coordination with the people in the Amateur Radio community mentioned under STAs is also recommended.

QEX, June 1984

SECOND SPREAD-SPECTRUM STA

The FCC just granted a second special temporary authority (STA) for experimentation in spread spectrum to the Amateur Radio Research and Development Corp. (AMRAD). This STA covers VHF frequency hopping. Look for more details in *QEX*.

QEX, July 1984

FCC GRANTS SPREAD-SPECTRUM STA

As mentioned in last month's *QEX*, the FCC has granted a second special Temporary Authority for spread-spectrum tests to the Amateur Radio Research and Development Corp. (AMRAD). Named in the STA are: Terry Fox, WB4JFI; David Borden, K8MMO; Robert Bruninga, WB4APR; Charles Phillips, N4EZV; Jim Elliott, K5KSY; Scott Schaefer, WR4S; Ted Seely, N4GFQ; Hal Feinstein, WB3KDU; William Hickey, WA5FXE; Joseph Crecente, KA3CHM; William Howard, K1LNJ; and Douglas Hardie, WA6VVV.

All tests proposed under the STA will use frequency hopping. Experiments 1 and 2 will use the bands 3675-3995, 7050-7295, 14100-14345, 21100-21345, and 28100-29300 kHz with service frequencies of 3725, 7100, 14150, 21150 and 28360 kHz, ±10 kHz for establishing communications and technical coordination. Prior announcements of tests on all bands except 10 meters will be made via W1AW bulletins.

Experiment 1 will use equipment designed by N4EZV. Experiment 2 may use the same equipment or the Yaesu FT-980 with external frequency control and will test a selective-addressing procedure where a unique frequency-hopping sequence is derived from a station's call sign. Experiment 3 will operate on the 144, 220 and 420-MHz bands and use an N4EZV transceiver as well as ICOM hand-held radios modified for external computer control.

FCC PROPOSES PART 15 AND 90 SPREAD SPECTRUM

The FCC has requested comments on its proposal to authorize spread-spectrum systems under Parts 15 and 90 of its rules. The FCC proposes to allow spread-spectrum systems to operate on any range of frequencies above 70 MHz without any restriction on their occupied bandwidth. The proposed rules are designed to minimize the likelihood of harmful interference so that it is comparable to that of existing Part 15 devices. The Commission also proposes that under Part 90 the Police Radio Service be able to use spread spectrum for physical surveillance, stakeouts, raids and other such activities on a secondary basis to operations of licensees regularly authorized on these frequencies.

The entire FCC Notice of Proposed Rulemaking is printed in this issue of *QEX*. This notice is recommended reading for all amateurs interested in (particularly weak-signal) operation at VHF and above. If you wish to comment to the FCC, your comments must be filed by September 14, 1984 and reply comments on or before October 12, 1984. As always, well-researched, logical and nonemotional comments will have much more impact than emotional hyperbole. Please send a copy of your comments to ARRL Hq.

QEX, May 1985

SPREAD-SPECTRUM DAY AT THE FCC

May 9 may be the day that the FCC authorizes limited use of spread spectrum under Parts 15 and 97 of the rules.

That afternoon, John P. Costas, K2EN, will give a talk on spread spectrum as part of the FCC Office of Science and Technology tutorial series.

QEX, June 1985

SPREAD-SPECTRUM DAY AT THE FCC

May 9 was indeed spread-spectrum day at the FCC. The Commission approved limited spread-spectrum operation in the Amateur Radio Service under General Docket No. 81-414. Permission to operate is delayed for one year "to give the amateur community time to develop initial voluntary interoperability standards as they have done recently in packet radio." The Washington, DC-based Amateur Radio Research and Development Corporation (AMRAD) has been the core group in spread-spectrum experimentation. They have already held an exploratory meeting about formation of a standards committee. League participation in, or sponsorship of, a spread-spectrum standards committee is a matter to be decided by the ARRL Board, which meets next in late July.

Spread spectrum will be limited to amateur frequencies above 420 MHz. Only frequency-hopping and direct-sequence modes are authorized. Hybrid techniques (involving both spreading techniques) or other spreading modes are prohibited. Power output is limited to 100 watts. Only certain code sequences which are authorized must be from the output of one binary linear-feedback shift register (which may be implemented in hardware or software). Only the certain code sequences specified in the Report and Order may be used.

AMRAD has been experimenting with spread spectrum since 1981 under Special Temporary Authority (STA) granted by the FCC. They have reported good results frequency hopping the Williams frequency synthesizer (*QST*, April, 1984).—*W4RI*

QEX, September 1985

ARRL SPREAD-SPECTRUM COMMITTEE

Are you interested in writing standards for Amateur Radio spread-spectrum communications? If so, send a letter addressed to Dr. Larry E. Price, W4RA, President, ARRL, 225 Main Street, Newington, CT 06111 outlining your qualifications. Be sure to mark the envelope "Spread-Spectrum Committee" and to get it in the mail as soon as possible.

The FCC has authorized spread spectrum in Amateur Radio bands above 420 MHz. In writing the new rules, the FCC specified only what they thought necessary to guarantee their ability to monitor such transmissions. They did not provide any technical standards needed for interoperability—that's a $10 government word for radios being able to talk to each other. They decided to delay the effective date of the rules until June 1, 1986 to give amateurs enough time to develop standards. So, at the July 1985 ARRL Board Meeting, an ad hoc committee on Amateur Radio spread-spectrum standards was established for a period of two years.

Here are the type of the standards to be considered by the committee: frequencies of operation, chip rate, the code, code rate, spreading function, transmission protocol(s) including the method of achieving synchronization, modulation type, type of information transmitted, and method/frequency(ies) for identification.

Happenings

Conducted By David Newkirk, AK7M

Our New Spread-Spectrum Rules

When we go on the air with conventional AM and FM emissions, the energy in each resultant signal is concentrated narrowly around a center frequency. Signal bandwidth usually increases with information rate. The hassle with such compact signals is that they're quite vulnerable to other similar signals at or near the same center frequency. (We bet you've already experienced somebody else calling CQ, or the bleatings of a "test pest," right atop the station you're working, for example.)

Spread-spectrum signals don't follow this rule of concentrating signal energy around a center frequency. Their bandwidth is not necessarily tied to data rate. The idea behind spread-spectrum work is the *intentional spreading* of signal energy over such a wide bandwidth that the signal's energy isn't very great at any one frequency for very long. What's the point? Great immunity to non-spread signals—like CW, SSB, RTTY—for one thing, and little likelihood of "collision" with other spread-spectrum signals spread according to differing binary sequences. And these techniques really work: Spectrum-spreading is a popular antijamming technique used by the military, for instance, because it's hard for anyone to interfere with a spread-spectrum signal who doesn't have its binary-sequence "key."

Radio amateurs are going to have their shot at spread-spectrum work as of June 1, 1986. That's when the rule amendments specified in FCC's *Report and Order* in GEN Docket 81-414, "Amendment of Parts 2 and 97 of the Commission's Rules and Regulations to authorize spread-spectrum techniques in the Amateur Radio Service," go into effect.

FCC had some ticklish questions to address in allowing us spread-spectrum operating privileges. For instance, since one of the main uses of spread-spectrum techniques has so far been the *hiding* of signals (implicit in the antijamming use of spread-spectrum techniques by the military), we couldn't just bring up our spread-spectrum rigs under present rules without transgressing prohibitions against use of codes and ciphers. FCC has specified the methods to be used in amateur spread-spectrum work closely enough so that the Commission is assured of being able to perform its monitoring and enforcement duties even when an amateur spread-spectrum station might be inaudible to those of us listening with "conventional" receivers. We will have to reaquaint ourselves with our old friend the logbook; FCC wants complete documentation of how and what we'll be doing.

There was also the concern that since authorized spread-spectrum signals might appear as broadband noise, spread-spectrum work should be limited to amateur bands offering plenty of "wide-open spaces" to keep intraservice interference possibilities to a minimum—especially while we're getting our feet wet with the new techniques. FCC concurred, and limited spread-spectrum work to bands 420 MHz and above.

We'll still have to *identify* our spread-spectrum transmissions with *narrowband* emissions, as FCC puts it, "so that CW, SSB and/or narrow-band FM receivers, which might be victims of interference, can receive the station identification." Frequencies used for such IDs will have been chosen to minimize interference to, while facilitating identification by, other operators.

Right off, spread-spectrum work will be limited to domestic communication (not international work), as other national administrations will have had to satisfy themselves of the achievement of proper safeguards against encryption and intraservice interference before *their* amateurs jump into the spread-spectrum swim.

This is really new ground for Amateur Radio—so new, in fact, that we have a number of decisions to make about exactly how to go about spread-spectrum work in ways guaranteeing station-to-station compatibility (see League Lines, December 1985 and January 1986 *QST*). FCC has not limited our choice of spread-spectrum options so narrowly that we can just press the button and go, although the final rules limit spreading methods to frequency hopping and direct sequence only.

The new Fifth Edition of *The FCC Rule Book* includes the following Part 97 changes. As of June 1, 1986, Section 97.3 will include a new paragraph (cc) as follows:

(cc) *Spread spectrum transmission.* An information bearing transmission in which information is conveyed by a modulated RF carrier and where the bandwidth is significantly widened, by means of a spreading function, over that needed to transmit the information alone.

Add a new Section 97.71 as follows:

Section 97.71 Spread spectrum communications

(a) Subject to special conditions in paragraphs (b) through (i) of this section, amateur stations may employ spread spectrum transmissions to convey information containing voice, teleprinter, facsimile, television, signals for remote control of objects, computer programs, data, and other communications including communication protocol elements. Spread spectrum transmissions must not be used for the purpose of obscuring the meaning of, but only to facilitate transmission.

(b) Spread spectrum transmissions are authorized on amateur frequencies above 420 MHz.

(c) Stations employing spread-spectrum transmissions shall not cause harmful interference to stations of good engineering design employing other authorized emissions specified in the table. Stations employing spread spectrum must also accept all interference caused by stations of good engineering design employing other authorized emissions specified in the table. (For the purposes of this subparagraph, unintended triggering of carrier operated repeaters is not considered to be harmful interference. Nevertheless, spread spectrum users should take reasonable steps to avoid this situation from occurring.)

(d) Spread spectrum transmissions are authorized for domestic radio communication only (communication between points within areas where radio services are regulated by the U.S. Federal Communications Commission), except where special arrangements have been made between the United States and the administration of any other country concerned.

(e) Only frequency hopping and direct sequence transmissions are authorized. Hybrid spread-spectrum transmissions (transmissions involving both spreading techniques) are prohibited.

(1) Frequency hopping. The carrier is modulated with unciphered information and changes at fixed intervals under the direction of a high speed code sequence.

(2) Direct sequence. The information is modulo-2 added to a high-speed code sequence. The combined information and code are then used to modulate a RF carrier. The high speed code sequence dominates the modulating function, and is the direct cause of the wide spreading of the transmitted signal.

(f) The only spreading sequences which are authorized must be from the output of one binary linear feedback shift register (which may be implemented in hardware or software).

(1) Only the following sets of connections may be used:

Number of stages in shift register	Taps used in feedback
7	[7,1]
13	[13,4,3,1]
19	[19,5,2,1]

(The numbers in brackets indicate which binary stages are combined with modulo-2 addition to form the input to the shift register in stage 1. The output is taken from the highest numbered stage.)

(2) The shift register must not be reset other than by its feedback during an individual transmission. The shift register must be used as follows.

(i) For frequency hopping transmissions using x frequencies, n consecutive bits from the shift register must be used to select the next frequency from a list of frequencies sorted in ascending order. Each consecutive frequency must be selected by a consecutive block of n bits. (Where n is the smallest integer greater than $\log_2 x$.)

(ii) For a direct sequence transmissions using m-ary modulation, consecutive blocks of $\log_2 m$ bits from the shift register must be used to select the transmitted signal during each interval.

(g) The station records shall document all spread spectrum transmissions and shall be retained for a period of one year following the last entry. The station records must include sufficient information to enable the Commission, using the information contained therein, to demodulate all transmissions. The station records must contain at least the following:

(1) A technical description of the transmitted signal.

(2) Pertinent parameters describing the transmitted signal including the frequency or frequencies of operation, and, where applicable, the chip rate, the code, the code rate, the spreading function, the transmission protocol(s) including the method of achieving synchronization, and the modulation type;

(3) A general description of the type of information being conveyed, for example, voice, text, memory dump, facsimile, television, etc.;

(4) The method and, if applicable, the frequency or frequencies used for station identification.

(5) The date of beginning and the date of ending use of each type of transmitted signal.

(h) When deemed necessary by an Engineer-in-Charge of a Commission field facility to assure compliance with the rules of this part, a station licensee shall:

(1) Cease spread spectrum transmissions authorized under this paragraph;

(2) Restrict spread spectrum transmissions authorized under this paragraph to the extent instructed;

(3) Maintain a record, convertible to the original information (voice, text, image, etc.) of all spread spectrum communications transmitted under the authority of this paragraph.

(i) The peak envelope power at the transmitter output shall not exceed 100 watts.

In **Section 97.84(g), Station identification,** add new subparagraph (5) as follows:

(5) When transmitting spread spectrum, by narrow band emission using the method described in (1) or (2) above; narrow band identification transmissions must be on only one frequency in each band being used. Alternatively, the station identification may be transmitted while in spread spectrum operation by changing one or more parameters of the emission in a fashion such that CW or SSB or narrow band FM receivers can be used to identify the sending station.

QST, May 1989

SPREAD-SPECTRUM SYSTEMS IN 902-928 MHz

A number of inquiries have been received from members concerning the local-area network (LAN) use of spread spectrum in the 902-928 MHz band as offered by Telesystems, Don Mills, Ontario, Canada; PA Consulting Group, Hightstown, New Jersey; and possibly other suppliers.

Spread-spectrum systems, for LANs or other purposes, are permitted at a 1-watt power level in the 902-928 MHz band as authorized under section 15.126 of the FCC's rules. Such operation is also permitted under this section in the 2400-2483.5 MHz and 5725-5850 MHz bands. Of concern to amateurs is the potential interference from such operations to amateurs operating in the 902-928, 2390-2450 and 5650-5925 MHz bands.

The rules place a number of limitations on spread-spectrum operation in these bands. The main point to bear in mind is 15.126(c), which requires that the spread-spectrum systems not cause harmful interference to any other operations which are authorized the use of these bands under other Parts of the Rules. Also, they must accept any interference from these sources. As far as the FCC is concerned, the Amateur Service, and the Amateur-Satellite Service where applicable, both have priority over the spread-spectrum systems despite our secondary status relative to a number of other services in these bands.

On the practical side, it is unlikely that these spread-spectrum devices will cause harmful interference to amateurs except when the amateur is physically close to the spread-spectrum transmitter. Interference from spread-spectrum devices, watt for watt, will be at least an order of magnitude less than from narrowband transmitters in the same frequency bands. Because the modulation rate of these transmitters is much higher than the audible range (66 kbit/s in the case of the PA Consulting system), any interference should not result in an audible output of an amateur radio (voice) receiver. Desensing could occur, but only in the near zone.

Spread-spectrum system designers are cautioned by FCC to keep the potential high EIRP emanations from government radars in mind when developing their designs for these bands, and are also cautioned that the 1-watt power limit on 902-928 MHz operation may be reduced in the future.

Chapter 3
The Regulatory Process

This chapter contains the complete text of the FCC documents that led to the authorization of spread spectrum emissions in the amateur bands. Reply comments filed separately by ARRL and AMRAD appear at the end of the chapter.

NOTICE OF INQUIRY IN GENERAL DOCKET NO. 81-413, "AUTHORIZATION OF SPREAD SPECTRUM AND OTHER WIDEBAND EMISSIONS"

Before the Federal Communications Commission FCC 81-289
Washington, D.C. 20554   29440

In the Matter of   )
) GEN
Authorization of spread spectrum and other   ) DOCKET
wideband emissions not presently provided   ) NO. 81-413
for in the FCC Rules and Regulations.   )

NOTICE OF INQUIRY

Adopted: June 30, 1981   Released: September 15, 1981
By the Commission: Commissioner Jones absent

Introduction

1. The Federal Communications Commission is initiating this Inquiry to gather information that will assist it in formulating policy regarding the use of wideband emissions. The Commission has before it a number of requests, both formal and informal, to authorize systems employing wideband modulation techniques for such diverse services as communications, radio location and telemetry operations. Because of the fundamental differences between wideband and narrowband types of modulation, the present Commission Rules and Regulations do not appear to provide the proper mechanisms for dealing with these requests. Indeed, our present rules implicitly forbid the use of some new technologies in this area.

2. Basically, a system employing wideband modulation takes an information signal of a small bandwidth and after modulation of a radio carrier, produces a radio frequency (RF) signal of a much larger bandwidth. Although conventional amplitude modulation (AM) and frequency modulation (FM) do result in spreading the information signal's bandwidth, the resultant RF bandwidth is still comparable to the bandwidth of the baseband signal. The modulation technique used in the FM broadcasting service increases the baseband signal's bandwidth by approximately a factor of sixteen. Although this is an example of wideband modulation, the FM broadcast signal's bandwidth still remains less than 240 kilohertz. In the context of this Inquiry, we shall concern ourselves with modulation techniques in which the baseband signal's bandwidth is greatly expanded; in some cases expansions of 1000 will be typical. This may result in RF bandwidths of many megahertz.

3. This inquiry is designed to serve two purposes. We hope to gather information to: (1) assist us in identifying specific radio services presently authorized by the Commission, as well as ideas for new services, where the authorization of wideband modulation techniques would serve the public interest; and (2) identify the technical parameters which characterize a wideband emission, including procedures used to measure these parameters, and identify technical standards necessary to insure operation on a minimum interference basis.

Spectrum Management Philosophy

4. Historically spectrum management policy has evolved around narrowband communication systems. In the early days of radio, there were few radio systems and much spectrum in which to operate. Users were assigned a carrier frequency and their transmissions were required to remain in a small band of frequencies (bandwidth) about the carrier. As time progressed more users wished access to the spectrum. Advancing technology provided relief in two ways: (1) it increased the upper limit of usable spectrum; and (2) it provided the means for reducing the bandwidth occupied by existing stations. So as the Commission's frequency allocations were pushed higher and higher, users were required to "consume" less and less of the spectrum. In the land mobile services, allotted bandwidths were actually halved a number of times.

5. The environment of advancing technology leading to smaller and smaller bandwidths has established a trend among radio regulatory agencies, including the Commission, of reducing bandwidths to achieve higher spectrum efficiency.[1] Obviously, reducing bandwidth is at least one way of increasing spectrum efficiency, but it may not always be the best way. In a 1959 paper, J. P. Costas showed the somewhat surprising result that decreasing bandwidth does not always increase spectrum efficiency, but that under certain communication conditions very large bandwidths are needed for efficient spectrum use.[2] Wideband modulation techniques in certain applications may actually increase spectrum efficiency over narrowband techniques, due to both natural interference and interference caused by other users.

Spread Spectrum Modulation

6. One class of bandwidth expansion techniques which is of particular interest is spread spectrum modulation. It was originally developed for military applications concerning covert communications and/or resistance to jamming. Generally spread spectrum systems transmit an information signal by combining it with a noiselike signal of a much larger bandwidth to generate a wideband signal. The spreading of the information signal over a wide bandwidth has obvious military advantages in that the resultant emission is more difficult to detect or jam than narrowband emissions.

7. Although basic spread spectrum theory was developed in the early 1950's and the military today is developing a number of operational systems, there are few civil applications of this technology. However, much of the earlier government

funded research has now been de-classified and is available for the general public.[3] Moreover, new advances in device technology, such as large scale integrated circuits, charged coupled devices, and surface acoustic wave devices, give promise to lower-cost spread spectrum systems. Current FCC Rules, however, implicitly forbid spread-spectrum's use in most services.[4] This alone may be inhibiting research and development in civil applications.

8. The CCIR defines a spread spectrum system as "one in which the average energy of the transmitted signal is spread over a bandwidth which is much wider than the information bandwidth."[5] Actually it appears that this definition could apply to any wideband modulation scheme. R. C. Dixon, in his book entitled *Spread Spectrum Systems*, narrows this definition by adding the requirement that "some signal or operation other than the information being sent is used for broadbanding (or spreading) the transmitted signal."[6]

9. There appear to be three basic types of spread spectrum techniques of interest to the Commission: a) direct sequence modulated, b) frequency hopping, and c) pulsed-FM. Hybrid systems may be formed from the combination of two or more of these basic types. A brief description of each type follows; more detailed information is available in an IEEE press publication entitled *Spread Spectrum Techniques* and an FCC funded report by the MITRE Corporation.[7]

10. Direct sequence systems, sometimes called pseudo-noise systems, employ a key generator to produce a high speed binary code sequence. An information signal is combined with this code sequence. This composite signal is then used to modulate an RF carrier. The code sequence determines the RF bandwidth, resulting in a spread spectrum signal. At the receiver another key generator produces a replica of the transmitter's code sequence, and the incoming RF signal is multiplied with this sequence. This collapses the RF bandwidth into a bandwidth which is commensurate with the information alone. Conventional narrowband demodulation techniques are then used to recover the information signal from the RF carrier. The critical problem in direct sequence systems is to synchronize the key generators in the transmitter and receiver.

11. Frequency hopping systems also use key generators at the transmitter and receiver. However, here the binary code sequence is used to control a frequency synthesizer. At the transmitter, the RF carrier is modulated in a conventional manner by the information signal. The carrier frequency is determined at any given moment by the code sequence; hence the carrier "hops" around in the frequency domain. Again the receiver and transmitter key generators are synchronized, and the receiver local oscillator tracks the changing frequency of the carrier. Conventional demodulation techniques are used at the receiver. Frequency hopping systems can be programmed to miss selected frequencies as they hop. They require only moderate accuracy in synchronization, as the speed of the code sequence generator is much less than in direct sequence systems.

12. Pulsed-FM or chirp systems are similar to frequency hopping systems in that the carrier frequency is varied. Frequency hopping systems shift the carrier among discrete frequencies, whereas in pulsed-FM systems the carrier varies smoothly across a band of frequencies. The transmitted signal is a swept-frequency pulse, similar to a signal produced by a laboratory sweep generator. This type of signal is relatively easy to generate today using a voltage controlled oscillator. Conventional narrowband modulation of the sweeping carrier is used to convey the information signal. At the receiver a dispersive filter accumulates and sums the transmitted energy received over a certain interval. By releasing this energy in one coherent burst, the filter compresses the signal into a narrow time slot and the signal behaves like a high power, narrow pulse.

Advantages and Applications of Spread Spectrum in the Civilian Environment

13. Spread spectrum modulation offers radio users a number of unique advantages. Only a receiver employing the same code sequence as the transmitter will be capable of decoding the transmitted signal and recovering the information signal. By assigning each receiver in a network a different code, the user may selectively address a particular receiver by employing the corresponding code at the transmitter.

14. By assigning different code sequences to different systems, i.e. the transmitters and receivers of each licensee, many such systems may be able to share a common frequency allocation. This could be done without the explicit coordination necessary for trunking, time division multiple access or frequency division multiple access. Transmitters will only be able to communicate with their intended receivers; in fact, each system should be unaware of the operation of other systems. This uncoordinated channel sharing is called code division multiple access (CDMA).

15. Selective addressing and code division multiple access could prove attractive to both radio users and the Commission. The message privacy and security inherent with coded transmissions would certainly be attractive to law enforcement agencies, mobile telephone users and perhaps some business users. If the Commission chooses to allocate new spectrum for a personal radio service, spread spectrum modulation with CDMA techniques could be used to restrict the allocation to its originally intended use; i.e. users could only communicate with units employing the same code structures. This would facilitate the use of the allocation for "personal business" and prohibit the transformation of the band into a "hobby service" as has happened in the 27 MHz Citizens Radio Service. Because each discrete address code is essentially a new channel, spectrum efficiency may be improved over conventional systems. This has been theorized in a number of papers, particularly in regards to cellular land mobile communications.[8]

16. In 1978 the FCC funded a study by the U.S. Department of Commerce, National Telecommunications and Information Administration (NTIA), on the spectrum efficiency of multiple independent spread spectrum land mobile radio systems. The study concluded that in this application spread spectrum was not as efficient as conventional FM modulation.[9] The report determined that in a band allocated exclusively to spread spectrum systems, base stations and mobile stations would have to operate on separate frequencies to prevent interference. When two mobile stations are transmitting at the same time, the station that is closest to the receiving station can saturate the front end of the receiver thereby preventing the far station from being received. As with conventional systems, this problem can be reduced by controlling the output power of all mobiles. But in the absence of power control, the "near-far" problem limits the spectrum efficiency of spread spectrum systems. However, in our opinion the report's conclusions apply only to direct sequence or fast frequency hopping systems, not all spread spectrum systems.

17. Another inherent property of spread spectrum modulation is the low power density of the transmitted signal. Because the transmitter output power is distributed across a wide band of frequencies, the power density (watts/hertz) is very small. In fact some spread spectrum systems can operate with the desired signal below the noise level at the receiver. This power density reduction can be used to advantage in applications involving covert communications, prevention of interference to other users, and privacy.

18. In Private Radio Docket 80-9 the Commission has considered the use of spread spectrum trailing devices.[10] These devices would allow law enforcement personnel to track moving vehicles without visual contact and their low power density would make them virtually undetectable to all others. Although the Commission has formally endorsed the concept of spread spectrum trailing devices, their use at this time is not permitted because the Commission has not addressed the issue of technical standards. Considering the low interference potential of these devices, it may be appropriate for the Commission to adopt only inband/out-of-band emission limitations. We invite comments indicating what standards will be necessary for the implementation of spread spectrum systems in covert trailing operations.

19. Spread spectrum systems also provide an interference rejection capability not possible with conventional narrowband systems. The strength of interfering signals at the receiver output is reduced by the system's "processing gain." This gain is approximated by the ratio of the transmitted RF bandwidth to the original information signal's bandwidth. Processing gain may suppress interfering signals by as much as 40 dB.

20. The low power density and interference suppression capability of spread spectrum systems suggests a unique application, that of band overlay. It may be possible in some circumstances to overlay spread spectrum systems on spectrum used by conventional services with little or no mutual interference. Obviously this would increase the spectrum efficiency of the affected band and could release additional spectrum for allocation to other services. Short range systems, such as cordless telephones, might prove ideal for such an application. Many of these telephones are carrier current devices operating on frequencies adjacent to the AM broadcast band. Current demand may already exceed the spectrum available for these telephones. If the United States decides to implement a WARC '79 decision to reallocate 1605 to 1705 kHz to the Broadcasting Service, cordless telephones will lose these frequencies. It may be impossible to find suitable additional spectrum to offset this loss. However, spread spectrum modulated telephones could possibly be overlaid on another frequency band. Not only would this relieve the problem of spectrum scarcity, but the orthogonal codes used in spread spectrum systems should prevent interference between neighbors' telephones without specific coordination. Surface acoustic wave devices, already employed in a number of television receivers, might be used to generate and receive spread spectrum signals at an affordable price.

21. Although theoretically band overlay is possible, more consideration must be given to interference from spread spectrum systems to conventional communications systems. The CCIR examined interference from direct sequence and frequency hopping systems to conventional AM voice, FM voice and FDM/FM voice signals.[11] The report gives signal to interference protection ratios for the cases considered and concludes that a potential for sharing exists, but further examination is required to determine detailed sharing criteria. The IIT Research Institute examined the performance of voice communications systems in the presence of spread spectrum interference in a report prepared for the Department of Defense.[12] The report concludes that a direct sequence interfering signal affects system performance similar to white Gaussian noise and a frequency hopping signal results in interference similar to that produced by a periodic pulsed signal.

22. NTIA has conducted several studies on the feasibility of overlaying spread spectrum systems on communication bands. The FCC funded a study to determine the effects of spread spectrum interference on TV. The study concluded that the amount of interference caused by a constant amplitude spread spectrum system should be about the same as that caused by a narrowband FM land mobile signal, as long as the spread spectrum RF bandwidth is less than 2 MHz.[13] If the spread spectrum RF bandwidth is greater than 6 MHz, the spread spectrum signal should have some advantage because of the out-of-band rejection capability of a TV receiver. This study suggests the possibility of overlaying very wide bandwidth spread spectrum signals on existing television bands.

23. Another study performed by NTIA examined the compatibility of spread spectrum and FM land mobile radio (LMR) systems.[14] It concluded that it would not be possible to overlay a spread spectrum LMR system onto a frequency band already occupied by conventional FM LMR systems without causing interference. The definition of overlay in this study was interpreted to mean the unrestricted operation of both spread spectrum and FM mobiles throughout the same service area. According to this report, the extreme range of propagation conditions encountered in a LMR environment can not be overcome by the reduction in interference obtained with a spread spectrum system. Interference can be reduced by increasing the RF bandwidth of the spread spectrum system but this reduction is not sufficient to compensate for the wide range of signal conditions in the LMR environment. However, the report did indicate that if a frequency hopping system was programmed to avoid frequency channels already in use, the signal suppression necessary for unrestricted operation might be achieved.

24. In 1978 the IIT Research Institute prepared another report on spread spectrum for the Department of Defense, this time developing procedures for analyzing interference caused by spread spectrum signals.[15] The report presents mathematical procedures for predicting interference conditions when conventional receiving systems are subjected to offending signals from spread spectrum transmitters. The procedures are generally applicable to all types of spread spectrum signals. This report seems to be a good foundation on which Commission procedures to analyze spread-spectrum interference could be based. We specifically request comments on the appropriateness of using this report as a basis for rule making.

25. An interesting characteristic of wideband systems is their ability to provide high resolution range measurements. Because the velocity of propagation of a radio signal is known, the distance between a transmitter and a receiver can be determined by measuring the time it takes a signal to propagate between them. A precise measurement of the signal's arrival time at the receiver is necessary. Because the uncertainty in measuring the arrival time is inversely proportional to the signal's bandwidth, wideband signals provide greater resolution than narrowband signals.

26. Del Norte Technology, Inc. a manufacturer of radiolocation equipment, has designed a radiolocation system which uses pulsed-FM technology. The system operates in the 420-450 MHz band and can determine distances up to 50 kilometers with an accuracy of plus or minus 2 meters. Del Norte has petitioned the Commission to amend its Rules to permit the marketing and use of this system.[16] A radiolocation system such as this could improve the accuracy of both aircraft and ship navigation.

27. A major concern with the use of spread spectrum is the Commission's ability to monitor and locate stations using this modulation technique. Because the emitted signal is both wideband and encoded, specialized receivers are necessary to demodulate it. The number of virtual "channels" realized with spread spectrum modulation may be very large, further limiting the Commission's ability to detect and monitor all transmissions. This problem is not necessarily restricted to spread spectrum systems; any system employing digital modulation techniques, whether for privacy or merely to facilitate communications, will pose a complex monitoring problem for the Commission. However, there are regulatory approaches that will mitigate this problem. Monitoring the technical parameters of a spread spectrum signal certainly poses much less of a problem than monitoring the same signal for message content. The Commission may choose to restrict the number of authorized user codes, assign them to licensees on a permanent basis or require that user codes be registered with the Commission. Spread spectrum systems could be authorized in services which had few enforcement problems. Only spread spectrum techniques which can be decoded with a conventional wideband receiver might be authorized. Finally, considering the low interference potential of spread spectrum emissions, it should be noted that spread spectrum signals strong enough to cause interference will probably be strong enough to locate.

Matters to be Addressed in this Inquiry

28. In December of 1979, the FCC issued a contract to the MITRE Corporation for the purpose of researching the potential use of spread spectrum techniques in non-government applications. MITRE has completed its study and submitted its report to the Commission.[17] The report presents a view of the potential benefits, costs and risks of spread spectrum communications and examines in detail many of the concepts briefly discussed in this Inquiry. Two examples of hypothetical implementations are considered: a slow frequency hopping system with FM voice and an intrusion detector using fast frequency hopping. We wish to direct attention to this report in the hope that it will stimulate further discussion and specifically request comments on it.

29. The Commission would also like to inquire about the measurement techniques used to evaluate spread spectrum systems. The Electromagnetic Compatibility Society (EMCS) of the Institute of Electrical and Electronics Engineers has written the Chief Scientist expressing their concern about broadband RF measurement and analytical techniques that may be required to evaluate the interference potential of broadband emissions to electronic equipment.[18] They are concerned about establishing a technical standard to promote good engineering practice in this area. The EMCS has suggested some basic steps to represent a model for evaluating the interference potential of broadband emissions. They believe that a model should be developed to, or in consonance with, the development of measurement and analytical techniques. This model would help define the technical characteristics that need to be considered. This Inquiry will afford an opportunity for all parties to assist in addressing our concerns regarding the measurement techniques that will be necessary to describe the interference potential and technical characteristics of spread spectrum and other wideband systems.

30. Comments are also invited on all of the issues discussed in this Inquiry. Suggestions of services which might benefit with the allowance of wideband modulation techniques are specifically requested. Comments on analytical procedures which the Commission could employ to evaluate the interference potential of wideband systems would be particularly helpful. In this regard, we again ask for comments on using the IIT Research Institute's report (reference 15) as a basis for Commission rule making. All parties interested in the development of wideband systems are urged to file comments in this proceeding.

31. The questions listed below are not exhaustive. They merely typify the Commission's areas of concern. Information not directly responsive to these questions but relevant to the general subject matter of the Inquiry is welcome and invited. To facilitate staff review each response should clearly state the precise topic or question being addressed.

32. Please provide answers and supporting data to the following questions:

   (a) What services can be accommodated by the use of spread spectrum systems? Can they be implemented by band overlay or will they require dedicated frequency allocations?

   (b) In the case of band overlay, should a spread spectrum system be required to operate on a non-interference basis with conventional systems? Should spread spectrum systems accept any interference they receive from other spread spectrum or conventional systems without protection from the Commission?

   (c) Should each wideband modulation technique be considered on its own merits as to its spectrum use and efficiency?

   (d) How should the power levels of spread spectrum systems be expressed? What power levels are necessary for operation? How can this power be measured?

   (e) What narrowband receiver characteristics should be considered in determining the interference potential of spread spectrum systems to conventional narrowband emissions? Do these characteristics affect the possibility of having spread spectrum systems and conventional systems co-exist in the same frequency band?

   (f) Will special test equipment be necessary to evaluate spread spectrum emissions? What type of detector, peak, quasipeak, or average, is best suited for measuring the interference potential of spread spectrum emissions?

   (g) Can systems which use different wideband modulation methods be evaluated by the same measurement techniques and the results?

   (h) If the Commission chooses to authorize spread spectrum systems, will detailed technical standards be needed? If so, what standards? Would establishing inband/out-of-band emission limitations be sufficient?

(i) Should the Commission consider authorizing spread spectrum modulated cordless telephones? How much more expensive than present cordless telephones would these units be?

(j) From a spectrum utilization standpoint, are there any capabilities or efficiencies which spread spectrum or other wideband systems possess that would allow the transmission of more information for a given frequency band than is now possible using conventional systems?

The following questions generally refer to ideas developed in the MITRE Corporation's report.

(k) Should the Commission authorize spread spectrum systems only if they utilize the spectrum more efficiently than conventional techniques or should spectrum efficiency be weighed along with the potential benefits of spread spectrum such as selective addressing, uncoordinated use, and secure communications? In such a case should spectrum efficiency be defined in terms of either instantaneous users/MHz, total users/MHz, or total users/MHz/square kilometer?

(l) Because the spectrum efficiency of spread spectrum appears to be high when transmitting low-rate data at microwave frequencies, what services could be implemented with this type of system?

(m) Which ISM bands might be suitable for spread spectrum overlay? How detrimental would this be to existing users? What sort of services could use ISM band overlay?

(n) Would the increased cost of spread spectrum equipment prohibit its acceptance by users? How much would equipment cost be expected to increase?

(o) Is it necessary for the Commission to monitor the message content of all types of transmissions? If spread spectrum is authorized, should the Commission require that equipment *not* have the capability of multiple user codes? Should user codes be assigned by the Commission? On a permanent basis? Should spread spectrum stations be required to identify themselves at some point in their transmissions with conventional modulation techniques, or some other means?

(p) Considering spread spectrum's low power density, would it be possible to design a police radar that could not be readily detected by motorists with monitors? How much more expensive than conventional police radars would such a unit be? What bandwidth would be required?

(q) Should the Commission consider a slow frequency hopping system with FM voice for any new personal radio service allocation? Could such a system be implemented today? At what increased cost over conventional systems? Would the potential advantages, such as privacy and uncoordinated channel access, outweigh the potential disadvantages, such as increased enforcement problems, if such a service was authorized by the Commission?

(r) In Appendix C of the MITRE Corporation's report, a model simulating the Citizens Band Service is developed. Is this model suitable for analyzing slow frequency hopping in the land mobile services? If not, what modifications would be necessary to make it suitable?

Procedural Matters

33. In view of the foregoing, we seek to obtain information from interested members of the public in order to assist the Commission in resolving the regulatory problems presented by wideband modulation techniques. We specifically request respondents to address the questions in paragraph 32.

34. Accordingly, pursuant to Sections 4(i), 4(j), 403 and 404 of the Communications Act of 1934, as amended, IT IS ORDERED that the aforementioned Inquiry IS HEREBY INSTITUTED.

35. In accordance with the provisions of Section 1.415 of the Commission's Rules, interested parties may file comments on or before March 15, 1982. Reply comments must be filed on or before June 30, 1982. Pursuant to the procedures set forth in Section 1.419 of the Commission's Rules, an original and five copies should be filed by formal participants with the Secretary of the Commission. Participants wishing each Commissioner to have a copy should include six additional copies. Members of the general public who wish to express their interest by participating informally may do so by submitting one copy. All comments are given the same consideration, regardless of the number of copies submitted. All comments should be clearly marked General Docket No. 81-413, and will be available for public inspection in the Public Reference Room at the Commission's headquarters. All written comments should be sent to: Secretary, Federal Communications Commission, Washington, D.C. 20554. For further information on this proceeding, please contact Michael Kennedy at (202) 632-7073. For general information on how to file comments, please contact the FCC Consumer Assistance and Information Division at (202) 632-7000.

FEDERAL COMMUNICATIONS COMMISSION

William J. Tricarico
Secretary

Notes

[1] Although there is no universally accepted measure of spectrum efficiency, it can be defined in general terms as the ratio of communications accomplished to spectrum used. These terms are usually difficult to quantify, but they may involve parameters such as: information delivered, users satisfied, radio frequency bandwidth occupied, geographical area covered and the time the spectrum is denied to other users. For a detailed discussion of metrics in this area see D. Hatfield, "Measures of Spectral Efficiency in Land Mobile Radio," *IEEE Transactions on Electromagnetic Compatibility*, Vol. EMC-19, No. 3, Aug. 1977, p. 266 and D. R. Ewing and L. A. Berry, "Metrics for Spectrum-Space Usage," Office of Telecommunications, OT Report 73-24, 1973.

[2] J. P. Costas, "Poisson, Shannon, and the Radio Amateur," *Proc. IRE*, Vol 47, pp. 2058-2068, December 1959.

[3] Current published searches of Federally funded research on spread-spectrum techniques include: "Spread Spectrum Communications," May 1980, PB 80-809726 and "Spread Spectrum Communications," May 1980, PB 80—809734. These and all other reports cited herein with "PB" or "AD" accession numbers are available from the U.S. Department of Commerce, National Technical Information Service, Springfield, VA 22161.

[4] On March 6, 1981 the Private Radio Bureau issued a Special Temporary Authorization (STA) to the Amateur Radio Research and

Development Corporation (AMRAD) for the purpose of conducting experiments on spread spectrum modulation. In order to permit spread spectrum, the STA waived two sections of the Commission's Rules. Thus, even the Amateur Radio Service, which is dedicated "to the advancement of the radio art," implicitly forbids spread spectrum modulation.

[5] International Telecommunication Union, International Radio Consultative Committee, *Recommendations and Reports of the CCIR, 1978, XIVth Plenary Assembly*, Kyoto, 1978, "Spread Spectrum Modulation Techniques," Report 651, Volume 1, pp. 4-14.

[6] R. C. Dixon, *Spread Spectrum Systems*, New York, Wiley-Interscience, 1976, p. 3.

[7] *Spread Spectrum Techniques*, ed. Robert C. Dixon, New York, IEEE Press, 1976. Walter C. Scales, "Potential Use of Spread Spectrum Techniques in Non-Government Applications," the MITRE Corporation, PB 81-165284, December 1980. The report by Mr. Scales will be inserted in the record of this proceeding.

[8] George R. Cooper, Ray W. Nettleton, and David P. Grybos, "Cellular Land Mobile Radio: Why Spread Spectrum?," *IEEE Communications Magazine*, Vol. 17, No. 2, pp. 17-24, March 1979; Robert P. Eckert and Peter M. Kelly, "Implementing Spread Spectrum Technology in the Land Mobile Radio Services," *IEEE Transactions on Communications*, Vol. COM-22, pp. 867-869, August 1977.

[9] L. A. Berry and E. J. Haakinson, "Spectrum Efficiency for Multiple Independent Spread-Spectrum Land Mobile Radio Systems," U.S. Department of Commerce, National Telecommunications and Information Administration, Report 78-11, PB-291539, November 1978.

[10] *Report and Order*, PR Docket 80-9, adopted January 8, 1981, FCC 81-1. For information concerning the ability of spread spectrum systems to share spectrum with conventional systems, see the following submissions:

Comments of Del Norte Technology, Inc. dated 3/31/80.
Reply Comments of Hewlett-Packard Company dated 5/15/80.
Reply Comments of American Telephone and Telegraph Company dated 5/16/80.

Copies of these documents will be inserted in the record of this proceeding.

[11] International Telecommunication Union, International Radio Consultative Committee, *Recommendations and Reports to the CCIR, 1978, XIVth Plenary Assembly*, Kyoto, 1978, "Considerations of Interference from Spread Spectrum Systems to Conventional Voice Communications Systems," Report 652, Volume 1, pp. 14-22.

[12] Leonard Farber and J. Cormack, "Performance of Voice Communications Systems in the Presence of Spread Spectrum Interference," IIT Research Institute, Report No. ESD-TR-77-005, AD A050844, December 1977.

[13] J. R. Juroshek, "A Preliminary Estimate of the Effects of Spread-Spectrum Interference on TV," U.S. Department of Commerce, National Telecommunications and Information Administration, Report 78-6, PB-286623, June 1978.

[14] J. R. Juroshek, "A Compatibility Analysis of Spread-Spectrum and FM Land Mobile Radio Systems," U.S. Department of Commerce, National Telecommunications and Information Administration, Report 79-23, PB-300651, August 1979.

[15] Paul Newhouse, "Procedures for Analyzing Interference Caused by Spread Spectrum Signals," IIT Research Institute, Report No. ESD-TR-77-003, AD A056911, February 1978. This report will be inserted in the record of this proceeding.

[16] Del Norte Technology, Inc. Petition for Rulemaking, General Docket 80-135, filed May 10, 1979.

[17] Walter C. Scales, *Supra*.

[18] Letter, dated September 12, 1979 to Dr. Stephen J. Lukasik from Jacqueline R. Sanoski. A copy of this letter will be inserted in the record of this proceeding.

FCC NOTICE OF INQUIRY AND PROPOSED RULE MAKING IN GENERAL DOCKET NO. 81-414, "AMENDMENT OF PARTS 2 AND 97 TO AUTHORIZE SPREAD SPECTRUM"

Before the Federal Communications Commission FCC 81-290
Washington, D.C. 29504

In the Matter of)
)
Amendment of Parts 2 and 97 of the) GEN
Commission's Rules and Regulations to) DOCKET
authorize spread spectrum techniques) NO. 81-414
in the Amateur Radio Service.)

NOTICE OF INQUIRY AND PROPOSED RULE MAKING

Adopted: June 30, 1981   Released: September 18, 1981
By the Commission: Commissioner Washburn concurring and issuing a statement; Commissioners Fogarty and Jones absent.

Introduction

1. The Federal Communications Commission is initiating this proceeding to propose changes in its Rules and Regulations to permit the use of spread spectrum modulation in the Amateur Radio Service. Although this modulation technique is not explicitly prohibited in the Amateur Service, its wideband spectral characteristics result in it being implicitly forbidden by the rules. The Commission feels that spread spectrum modulation is technically compatible with the present modes of operation in the Amateur Service. Allowing amateur experimentation with this relatively new modulation technique is therefore in harmony with the basis and purpose of the Amateur Radio Service.[1] Therefore, the Commission is proposing to amend Parts 2 and 97 of its Rules and Regulations to authorize the use of spread spectrum modulation techniques.

2. Although we feel comfortable proposing the authorization of spread spectrum modulation in the amateur service at this time, we recognize that its use may confront the Commission with a number of problems. In spread spectrum systems a coding process is generally used to spread the signal's energy across a wide band of frequencies. It is necessary to know the code sequence in order to demodulate the signal and recover its content. Wideband receivers may be required to make technical measurements on the transmitted signal. There is a potential for interference in overlaying wideband emissions on amateur bands already occupied with conventional narrow band signals, although this is limited due to the low power density of spread spectrum signals. For these reasons, this Notice also requests comments on problems associated with the enforcement and interference issues raised by authorizing spread spectrum modulation in the Amateur Radio Service.

Background

3. Spread spectrum techniques were originally developed for military applications concerning covert communications and/or resistance to jamming. The radio frequency (RF) signal transmitted in a spread spectrum system occupies a very large bandwidth, perhaps many megahertz, as compared to the information signal's bandwidth. This wide bandwidth provides for the military a signal that is very hard to detect or jam. Spread spectrum emissions possess qualities, however, that may also be desirable to civil users of the spectrum.[2]

4. All spread spectrum systems employ some function other than the information signal to spread the RF bandwidth.[3] Direct sequence modulated and frequency-hopping systems both use a high speed pseudorandom code sequence to achieve a large RF bandwidth. In direct sequence systems, this code sequence is combined with the information signal and this composite signal is used to modulate an RF carrier. The code sequence is used in frequency hopping systems to control a frequency synthesizer which generates the transmitter's many carrier frequencies. In pulsed-FM systems, a voltage function is used to sweep the transmitter's carrier over a wide frequency range. Although somewhat different in design, all three of these systems result in an RF signal with a very wide bandwidth. More detailed descriptions of the operation of these systems may be found in an IEEE press publication entitled *Spread Spectrum Techniques*.[4]

5. The Commission recognizes that spread spectrum may offer the communications user some unique advantages. Today the Commission adopted a Notice of Inquiry in the matter of wideband modulation techniques, including spread spectrum.[5] The Inquiry explores some possible civil uses of spread spectrum modulation and indicates some of the advantages associated with it. The coding techniques used in spread-spectrum systems, for example, allow message privacy, selective addressing and code division multiple access. These last two advantages provide the user the ability to selectively access only one or a fraction of the total number of receivers having the same spectrum. Spreading the signal at the transmitter and collapsing it at the receiver result in a low signal power density while simultaneously providing the ability of the system to reject interference. These features suggest the feasibility of overlaying spread spectrum systems on occupied spectrum. While the Commission is investigating the general question of spread spectrum desirability under that Inquiry, it may be that spread spectrum can be introduced into the Amateur Service with minimum potential for interference.

6. Spread spectrum techniques have not gone unnoticed in the amateur community. An article published in *QST*, the official publication of the American Radio Relay League, cited spread spectrum as a technology ripe for amateur experimentation.[6] The article indicated that the Amateur Radio Research and Development Corporation (AMRAD) would soon petition the Commission for Special Temporary Authorization (STA) to conduct tests using spread spectrum modulation. In March of this year the Private Radio Bureau, acting under delegated authority from the Commission, granted AMRAD an STA allowing the specific tests outlined in their petition.[7] Upon expiration of the STA, AMRAD must submit its findings to the Commission. The reports required by the STA will be inserted in the record of this proceeding and will be considered before final action is taken. If the results of the STA and the aforementioned NOI are favorable, the Commission will consider expanding the authorization of spread spectrum techniques. Now, however, the Commission proposes rule changes to permit broader experimentation with spread spectrum techniques in three bands.

Discussion and Proposal

7. Spread spectrum systems are inherently more complex than narrowband systems. Amateurs wishing to experiment with spread spectrum techniques will almost certainly have to construct their own equipment, due to the lack of com-

mercially available equipment. The principal technical problem will be constructing devices which maintain their parameters over wide bandwidths. Incorrectly or poorly designed equipment will increase the possibility of both intra-service and inter-service interference. It is for this reason the Commission feels that authorization to use spread spectrum techniques should be restricted to Amateur Extra and Advanced Class licensees only. Amateurs presently licensed in these two classes have been tested in the advanced phases of radio electronics; material covering spread spectrum techniques will be added in the future to the appropriate examination syllabuses.

8. The Commission proposes to authorize spread spectrum modulation in the following three frequency bands: 50-54 MHz, 144-148 MHz, and 220-225 MHz. The International Telecommunications Union (ITU) Regulations require that "transmissions between amateur stations of different countries...shall be made in plain language."[8] Since this might restrict the international transmission of spread spectrum signals, we propose frequency bands above 50 MHz so as to naturally limit propagation. Most of the frequency bands allocated to the Amateur Service above 225 MHz are shared with government users.[9] Amateurs operating in these bands must not interfere with government systems. Rather than coordinate spread spectrum systems with government users in these bands at this time, the Commission proposes restricting spread spectrum transmissions to frequencies below 225 MHz.[10] However, we welcome requests for STA's to perform limited spread spectrum experiments in the amateur bands above 225 MHz and will consider these on a case by case basis.

9. The Table of Frequency Allocations prohibits the use of pulsed emissions in a number of frequency bands including 50-54 MHz and 144-148 MHz.[11] It appears that this prohibition is no longer necessary, so we propose to discontinue it by deleting footnote US 1. Because WARC '79 adopted new emission designators, the use of which will be addressed in another proceeding, we will not at this time assign a designator to spread spectrum modulation. Until the Commission implements the decisions of WARC '79, we will use a footnote in the rules to indicate the authorization of spread spectrum modulation in a particular frequency band.

10. In order to allow amateurs maximum flexibility in the design of spread spectrum systems, the Commission intends to permit direct sequence modulated, frequency hopping, pulsed-FM, and hybrid systems.[12] We propose only that a system's authorized RF bandwidth be equal to or less than the width of the amateur band that the system is operating in and retained within that band. Therefore, amateurs may spread their transmitted energy across the complete band they are using. Although we do not anticipate interference problems, considering our limited operational experience with spread-spectrum transmissions we are proposing that the local Engineer in Charge be allowed to require stations transmitting spread spectrum signals to take whatever steps are necessary to resolve cases of interference, including terminating operation. We request comments on our belief that the interference potential in overlaying spread spectrum on the three bands proposed here is low.

11. Amateurs are presently forbidden to transmit coded messages in either domestic or international communications.[13] Because of the previously cited ITU regulation, international transmission of spread spectrum signals will not be allowed. However, we shall exempt spread spectrum transmissions between domestic stations from this prohibition.

12. In order to facilitate the demodulation of spread spectrum emissions, we propose to require that any pseudorandom sequence used in generating the spread spectrum signal must be the output of one of a number of specific linear feedback shift registers.[14] We propose specific shift registers in the appendix because they provide the minimum level of coding necessary to spread the signal. We realize that more sophisticated coding procedures could be employed, but their use would further obscure the meaning of the communications. This would appear to violate the intent of Section 97.117 of the present rules. Additionally, we will require that amateurs log the technical characteristics of their transmitted signal and that identification be given in telegraphy on the center frequency of their signal.

13. A major concern of the Commission in allowing amateur use of spread spectrum techniques is the Commission's, and the amateur's own, ability to monitor and locate stations transmitting wideband emissions. Presently the Field Operations Bureau (FOB) monitors the content of amateur transmissions to insure that amateurs are not transmitting communications which are prohibited by our rules.[15] This monitoring, coupled with the self policing afforded by the amateurs themselves, assures the bands remain usable for amateurs and are not usurped by business or clandestine activities. However, FOB currently has no capability to monitor spread spectrum emissions. We propose to amend Section 97.117 to include provisions to facilitate monitoring both by FOB and by other amateurs. We have precluded the use of esoteric encryption schemes and instead required that spreading codes be generated by linear shift registers. This will result in signals which may be received with only reasonable effort. This should facilitate self-monitoring, which has historically been very effective in the Amateur Radio Service. Additionally, we have proposed that amateurs log the technical characteristics of their signals and identify their transmissions in telegraphy. Considering that the characteristics of spread spectrum, e.g. low power density, are such that its transmission will not be disruptive to other users and the experimental nature of the Amateur Radio Service, we feel that we have imposed sufficient safeguards against its misuse. However, we request comments on these and other conditions the Commission could require to mitigate enforcement problems.

Questions

14. The questions listed below are not exhaustive. They merely typify the Commission's areas of concern. Information not directly responsive to these questions but relevant to the general subject matter of the Inquiry is welcome and invited. To facilitate staff review each response should clearly state the precise topic or question being addressed.

15. Please provide answers and supporting data to the following questions:
    (a) Are the emission limitations specified in the proposed amendment to Section 97.73 of the rules sufficient to prevent interference from spread spectrum systems to users in adjacent bands?
    (b) Will interference to conventional amateur communications be a major problem? If so, what steps can the Commission take to mitigate this problem? What types of other communications will be most vulnerable?

(c) Is it necessary for the Commission to have the capability to monitor the content of all amateur communications? If not, how can we enforce the limitations on the use of the amateur service and detect unlicensed transmission?

(d) Will the specific shift registers proposed in the amendment to Section 97.117 of the rules facilitate self-monitoring by the amateur community?

Procedural Matters

16. Accordingly, the Commission adopts this *Notice of Inquiry and Proposed Rule Making* (NOI/NPRM) under the authority contained in Sections 4 (i) and 303 of the Communications Act of 1934, as amended. The proposed amendments to Part 2 and Part 97 of the rules are set forth in the appendix.

17. Pursuant to Section 605 (b) of the Regulatory Flexibility Act, the Commission finds that this NOI/NPRM will have no effect on small businesses. The frequencies involved are assigned to the Amateur Radio Service, which by definition is used for non-business communications only. To the extent that organizations such as amateur radio clubs hold amateur licenses, the only effect of the proposed action will be to enable these clubs to experiment with spread spectrum techniques, presumably for wider dissemination to the amateur community.

18. For purposes of this non-restricted notice and comment rule making proceeding, members of the public are advised that *ex parte* contacts are permitted from the time the Commission adopts a notice of proposed rule making until the time a public notice is issued stating that a substantive disposition of the matter is to be considered at a forthcoming meeting or until a final order disposing of the matter is adopted by the Commission, whichever is earlier. In general, an *ex parte* presentation is any written or oral communication (other than formal written comments/pleadings and formal oral arguments) between a person outside the Commission and a Commissioner or a member of the Commission's staff which addresses the merits of the proceeding. Any person who submits a written *ex parte* presentation must serve a copy of that presentation on the Commission's Secretary for inclusion in the public file. Any person who makes an oral *ex parte* presentation addressing matters not fully covered in any previously-filed written comments for the proceeding must prepare a written summary of that presentation; on the day of oral presentation, that written summary must be served on the Commission's Secretary for inclusion in the public file, with a copy to the Commission official receiving the oral presentation. Each *ex parte* presentation described above must state on its face that the Secretary has been served, and must also state by docket number the proceeding to which it relates. See generally, Section 1.1231 of the Commission's Rules, 47 C.F.R. §1.1231. A summary of Commission procedures governing *ex parte* presentations in informal rule making is available from the Consumer Assistance and Information Division, FCC, Washington, D.C. 20554.

19. Pursuant to the procedures set forth in Section 1.415 of the Commission's Rules, interested persons may file comments on or before March 1, 1982 and reply comments on or before April 15, 1982. All relevant and timely comments will be considered by the Commission before final action is taken in this proceeding. In reaching its decision, the Commission may take into consideration information and ideas not contained in the comments, provided that such information or a writing indicating the nature and source of such information is placed in the public file, and provided that the fact of the Commission's reliance on such information is noted in the Report and Order.

20. In accordance with the provisions of Section 1.419 of the Commission's Rules, formal participants shall file an original and five copies of their comments and other materials. Participants wishing each Commissioner to have a copy of their comments should file an original and 11 copies. Members of the general public who wish to express their interest by participating informally may do so by submitting one copy. All comments are given the same consideration, regardless of the number of copies submitted. All comments should be clearly marked General Docket No. 81-414, and will be available for public inspection during regular business hours in the Commission's Public Reference Room at its headquarters in Washington, D.C. All written comments should be sent to: Secretary, Federal Communications Commission, Washington, D.C. 20554. For further information on this proceeding, contact Michael Kennedy at (202) 632-7073. For general information on how to file comments, please contact the FCC Consumer Assistance and Information Division at (202) 632-7000.

Federal Communications Commission

William J. Tricarico
Secretary

Notes

[1] Section 97.1(b) of the FCC Rules and Regulations illustrates in part the basis and purpose of the Amateur Radio Service as the "continuation and extension of the amateur's proven ability to contribute to the advancement of the radio art."

[2] See, for example, the following FCC funded report: Walter C. Scales, "Potential Use of Spread Spectrum Techniques in Non-Government Applications," the MITRE Corporation, Report No. MTR-80W335, December 1980. This report is available from the National Technical Information Service, Springfield, VA 22161, Accession No. PB81-165284.

[3] R. C. Dixon, *Spread Spectrum Systems*, New York, Wiley-Interscience, 1976, p. 3.

[4] *Spread Spectrum Techniques*, ed. Robert C. Dixon, New York, IEEE Press, 1976.

[5] *Notice of Inquiry*, General Docket 81-413, adopted June 30, 1981, FCC 81-289.

[6] Paul L. Rinaldo, "Spread Spectrum and the Radio Amateur," *QST*, November 1980, pp. 15-17.

[7] The STA was issued on March 6, 1981 for a period of one year. Although it permits spread spectrum modulation, only those techniques necessary to perform the indicated experiments were authorized. Also, the authorization is limited to those amateurs that were included in AMRAD's petition.

[8] Article 41, Section 2(1), *Radio Regulations Annexed to the International Telecommunication Convention* (Geneva, 1959). (See *FCC Rules and Regulations*, Part 97, Appendix 2).

[9] *FCC Rules and Regulations*, Part 97, Section 97.61 (b)(5).

[10] The 220 to 225 MHz band is also shared with government users. However, the Interdepartment Radio Advisory Committee (IRAC) has already indicated that they would not object to amateur use of spread-spectrum modulation in this band. Government radio-location systems will still be the primary users in this band.

[11] *FCC Rules and Regulations*, Part 2, Section 2.106, footnote US 1.

[12] Hybrid spread spectrum systems are created by combining two or more of the basic spread-spectrum techniques, e.g. a frequency hopping, direct sequence modulated system.

[13] *FCC Rules and Regulations*, Part 97, Section 97.117.

[14] For more information on linear code generator configurations, see R. C. Dixon's discussion in *Spread Spectrum Systems*, New York, Wiley-Interscience, 1976, pp. 60-64.

[15] See *FCC Rules and Regulations*, Part 97 Subpart E—Prohibited Practices and Administrative Sanctions.

APPENDIX

Part 2 of the Commission's Rules and Regulations is proposed to be amended as follows:

In section 2.106, remove footnote US 1 and all references to it in the Table of Frequency Allocations.

Part 97 of the Commission's Rules and Regulations is proposed to be amended as follows:

In §97.3, Definitions, add a new paragraph (aa) to read as follows:

* * * *

(aa) Spread spectrum techniques. Any of a number of modulation schemes in which, (1) the transmitted radio frequency bandwidth is much greater than the bandwidth or rate of the information being sent and, (2) some function other than the information being sent is employed to determine the resulting modulated radio frequency bandwidth.

In §97.7, Privileges of operator licenses, revise paragraphs (a) and (d) to read as follows:

(a) Amateur Extra and Advanced Class. All authorized amateur privileges including exclusive use of spread spectrum techniques and exclusive frequency operating authority in accordance with the following table:

* * * *

(d) Technician Class. All authorized amateur privileges, except spread spectrum techniques, on the frequencies 50.0 MHz and above. Technician Class licenses also convey the full privileges of Novice Class licenses.

* * * *

In §97.61, Authorized frequencies and emissions, add footnote number 1 to the 50-54 MHz, 144-148 MHz, and 220-225 MHz frequency bands as follows:

(a) The following frequency bands and associated emissions are available to amateur radio stations for amateur radio operation, other than repeater operation and auxiliary operation, subject to the limitations of §97.65 and paragraph (b) of this section:

Frequency band	Emissions	Limitations (See paragraph (b))
*	*	* *
50.0-54.0[1]	A1	
50.1-54.0	A2, A3, A4, A5, F1, F2, F3, F5	
51.0-54.0	A0	
144-148[1]	A1	
144.1-148.0	A0, A2, A3, A4, A5, F0, F1, F2, F3, F5	
220-225[1]	A0, A1, A2, A3, A4, A5, F0, F1, F2, F3, F4, F5	
*	*	* *

[1]Spread spectrum techniques for domestic communications only are authorized in this band.

In §97.73, Purity of emissions, redesignate existing paragraph (d) as paragraph (e), revise existing paragraph (c) and redesignate it as paragraph (d), and add a new paragraph (c) to read as follows:

* * * *

(c) The limitations specified in paragraph (b) of this section shall also apply to spread spectrum modulated signals except that for this purpose, "carrier frequency" is defined as the center frequency of the transmitted signal, and "mean power of the fundamental" is defined as the total emitted power.

(d) Paragraphs (a), (b), and (c) of this section notwithstanding, all spurious emissions or radiation from an amateur transmitter, transceiver, or external radio frequency power amplifier shall be reduced or eliminated in accordance with good engineering practice.

(e) If any spurious radiation, including chassis or power line radiation, causes harmful interference to the reception of another radio station, the licensee may be required to take steps to eliminate the interference in accordance with good engineering practice.

* * * *

In §97.84, Station identification, add a new paragraph (h) to read as follows:

* * * *

(h) When an amateur radio station is modulated using spread spectrum techniques, identification in telegraphy shall be given on the center frequency of the transmission. Additionally, this identification shall include a statement indicating that the station is transmitting a spread spectrum signal and the upper and lower frequency limits of that signal.

In §97.103, Station log requirements, revise existing paragraph (g) and redesignate it as paragraph (h), and add a new paragraph (g) to read as follows:

* * * *

(g) In addition to the other information required by this section, the log of a station modulated with spread spectrum techniques shall contain information sufficiently detailed for another party to demodulate the signal. This information shall include at least the following:

(1) A technical description of the transmitted signal. If the signal is modeled after a published article, a copy of the article will be adequate.

(2) The dates that the signal format is changed. Changing the center frequency of the signal does not constitute a change in signal format.

(3) The chip rate (rate of frequency change), if applicable.

(4) The code rate, if applicable.

(5) The method of achieving synchronization.

(6) The center frequency and the frequency band over which the signal is spread.

(h) Notwithstanding the provisions of §97.105, the log entries required by paragraphs (c), (d), (e), (f), and (g) of this section shall be retained in the station log as long as the information contained in those entries is accurate.

Revise §97.117, Codes and ciphers prohibited, to read as follows:

(a) The transmission by radio of messages in codes or ciphers in domestic and international communications to or between amateur stations is prohibited. All communications regardless of type of emission employed shall be in plain language except that generally recognized abbreviations established by regulation or custom and usage are permissible as are any other abbreviations or signals where the intent is not to obscure the meaning but only to facilitate communications.

(b) Spread spectrum transmissions between amateur stations of different countries are prohibited. However, for the purpose of the spread spectrum transmissions authorized between domestic stations in §97.7 and §97.61, pseudorandom sequences may be used to generate the transmitted signal provided the following conditions are met:

(1) The sequence must be the output of a binary linear feedback shift register.
(2) Only the following shift register connections may be used:

Number of Stages in Shift Register	Taps Used in Feedback
7	[7,1]
13	[13,4,3,1]
19	[19,5,2,1]

(The numbers in brackets indicate which binary stages are combined with modulo-2 addition to form the input to the shift register in stage 1. The output is taken from the highest numbered stage.)
(3) For direct sequence modulation the successive bits of the highest stage of the shift register must be used directly to modulate the signal. No alteration or other data may be used for the direct sequence modulation. For frequency hop modulation, successive regular segments of the shift register sequence must be used to specify the next frequency, and no alteration or other data may be used for frequency selection.
(4) The shift register(s) may not be reset other than by its feedback during an individual transmission.

In §97.131, Restricted operation, redesignate existing paragraph (b) as paragraph (c) and add a new paragraph (b) to read as follows:

*　　　*　　　*　　　*

(b) If the operation of an amateur station using spread spectrum techniques causes interference to other licensed stations, the Commission's local Engineer in Charge may impose conditions necessary to resolve the interference, including termination of operation, on the offending station.

(c) In general, such steps as may be necessary to minimize interference to stations operating in other services may be required after investigation by the Commission.

CONCURRING STATEMENT OF COMMISSIONER ABBOTT WASHBURN

re: Notice of Inquiry and Proposed Rule Making to Amend Parts 2 and 97 of the Commission's Rules and Regulations to Authorize Spread Spectrum Modulation in the Amateur Radio Service

The second paragraph of this item states: "we feel comfortable proposing the authorization of spread spectrum modulation." I do not share this feeling of comfort when the same document includes the following language: "A major concern of the Commission in allowing amateur use of spread spectrum techniques is the Commission's, and the amateur's own, ability to monitor and locate stations transmitting wideband emissions." So there is no real control. While new technology is to be encouraged, especially that which allows more use of the spectrum, this should not be at the price of interference. There must be assurance that expanded use of this technology will contribute to the effective and efficient use of the spectrum while at the same time not adding interference to the detriment of other technologies and spectrum users in general.

In a related Notice of Inquiry adopted today, the Commission initiated a proceeding seeking information regarding the future authorization of spread spectrum and other wideband emissions not presently provided for in our Rules. That inquiry requests information and data regarding the technical characteristics, efficient use of spectrum, possible standardization, potential applications of this technology, etc. It also raises questions regarding the measurement of interference potential, emission testing and the necessity for Commission monitoring. With issues such as these still in the initial and infant stage of information gathering, the Commission seems to be prejudging important issues by rushing to Rulemaking. A more prudent approach would be first to obtain the facts on monitoring and interference via the NOI, assess these facts, and *then* move to Rulemaking. However, we are assured by the staff that the risks involved will be minimal. I hope this proves to be the case.

ARRL COMMENTS IN GENERAL DOCKET NO. 81-414

Before the
FEDERAL COMMUNICATIONS COMMISSION
Washington, D.C. 20554

In the Matter of)
)
Amendment of Parts 2 and 97)
of the Commission's Rules) Gen.
and Regulations to authorize) Docket No.
spread spectrum techniques) 81-414
in the Amateur Radio Service.)

To: The Commission

COMMENTS OF THE AMERICAN RADIO RELAY LEAGUE, INCORPORATED, IN RESPONSE TO NOTICE OF INQUIRY AND PROPOSED RULE MAKING

The American Radio Relay League, Incorporated (the League), the principal spokesman for the over 400,000 licensed amateur radio operators nationally, submits the following comments in response to the *Notice of Inquiry and Proposed Rule Making* released September 18, 1981 (FCC 81-290, 46 Fed. Reg. 49617 *et seq.*) in the above-captioned proceeding. With respect to the Commission's proposal to permit spread spectrum modulation in the Amateur Radio Service, the League states as follows:

1. The League is gratified at the showing of support of the Amateur Service, reflected in the Commission's proposal to permit spread spectrum techniques to be used by amateurs by other than Special Temporary Authorization. Indeed, the Commission has exhibited an extremely cooperative and encouraging attitude towards the Amateur Radio Service with respect to the interests of amateurs in experimenting with spread spectrum modulation, especially in the context of issuance of Special Temporary Authority to conduct spread-spectrum experiments.[1] It is this type of cooperative encouragement which permits amateurs to remain at the forefront of technology pursuant to Section 97.1(b) of the Rules.

2. At this time, the interest of the amateur community in spread spectrum techniques is primarily experimental. While those techniques are of considerable interest to inquiring amateurs now that advances in technology have brought them within practical reach of individual experimenters, their major advantages do not particularly promote the *communications* objectives of the Amateur Service. Looking at the matter from the point of view of the amateur *qua* communicator, message privacy is not a desired feature of amateur communication. To the contrary, as the Commission has noted, techniques which provide privacy raise difficulties in monitoring and enforcement. Selective addressing and multiple access are desirable in certain situations, but may be achieved more easily by other means and with presently authorized modulation techniques. Thus, the primary motivation for amateur utilization of spread spectrum techniques is simply the desire to better understand and develop the concepts which make spread spectrum a useful communications medium. While it is unlikely that these techniques will be widely used in the Amateur Service in the near future, even a modest level of experimentation in this service will significantly expand the body of knowledge about spread spectrum techniques in non-government applications. For this reason, the League supports the introduction of spread spectrum techniques in the Amateur Service if accompanied by suitable safeguards. These safeguards should (1) minimize the possibility that amateur bands will be used for business or other non-amateur communications, and (2) minimize the possibility of interference to other amateur stations and other users of the radio spectrum.

3. Given that interest in spread spectrum techniques is primarily experimental in nature, the Commission's forward-looking proposal to include provisions for spread spectrum modulation in its Rules may, in fact, be somewhat premature. Owing to the limitations inherent in any program staffed by volunteers, the spread spectrum experiments already authorized by the Commission by Special Temporary Authority are incomplete. It is entirely possible that the number of amateurs *presently* interested in, and able to equip themselves for, spread spectrum experiments is sufficiently small that the Commission would assume a lesser administrative burden by simply issuing Special Temporary Authority on a case-by-case basis rather than by undertaking revisions to its Rules and Regulations. As the Commission itself has noted, adoption of the proposed Rules will not obviate the necessity for STA's, which will still be needed for experiments in other bands and for modulation techniques and coding sequences other than those proposed to be authorized.[2] Operation by Special Temporary Authority does have the advantage of placing the results of the work in the public record, in the form of the report which must be filed with the Commission at the conclusion of the operation. The STA also places other amateurs on notice that they may encounter interference from an unusual source, and gives them recourse if the interference proves to be harmful. It would be beneficial from the point of view of the Commission's Field Operations Bureau to have specific data on location, frequencies and modulation type of individual experimenters. The Commission is in a better position than the League to assess the relative merits of authorizing spread spectrum operation by STA or by changes to its Rules. The League encourages such experimentation either way. It merely wishes to point out, however, that the STA may have certain advantages at this time which tend to offset the fact that it is more burdensome to the individual experimenter. Should the Commission decide that the time is right to adopt specific rules governing spread spectrum techniques in the Amateur Service, certain safeguards will be required.

4. The Commission proposes to limit the use of spread spectrum techniques to Advanced and Amateur Extra Class licensees. While it is true that holders of these two classes of license have passed a more difficult technical examination than have the holders of Technician and General Class licenses, this does not in itself ensure that the licensees are more responsible or that they better understand spread spectrum techniques. As the Commission has observed,[3] up to now the amateur examinations have not included material on spread spectrum. Even if such material were added to the Advanced Class examination syllabus today, for the foreseeable future the vast majority of Advanced Class licensees (those who have already taken and passed the examination) would not have been exposed to spread spectrum in studying for the exam. Given the breadth of the material which must be covered in the Advanced Class syllabus, it is unlikely that more than one or two questions on the subject could be included on the 50-question exam. Furthermore, the Commission in Docket 80-252 noted that slow-scan television and facsimile privileges did not appear to be appropriate incentive devices, and expanded the privileges of its General Class licensees to in-

clude those modes on the high-frequency bands. There is even less of an argument for exclusivity at vhf, where the scarcity value of the spectrum is less than at hf. In the instant Notice, at Paragraph numbered 7, the Commission states:

> Amateurs wishing to experiment with spread spectrum techniques will almost certainly have to construct their own equipment, due to the lack of commercially available equipment.

The League agrees with this observation, and wishes to point out that if this indeed proves to be the case, the very process of construction will provide a much better education than could ever be provided through the examination process. Finally, it should be noted that the Technician Class license was originally created to encourage uhf amateur experimentation, and that many present Technician Class licensees continue to pursue the original objectives of the license. It would serve no useful purpose to deny them the opportunity to experiment with spread spectrum should they choose to do so. Thus, the League supports making spread spectrum techniques available equally to Technician, General, Advanced, and Amateur Extra Class licensees.

5. The Commission asks[4] whether it is necessary for it to have the capability to monitor the content of all amateur communications. The League's comments in Docket 81-699, concerning the use of additional digital codes in the Amateur Radio Service, are also appropriate here:

> The League is sympathetic to, and generally shares the Commission's reservations about any reduction in the Commission's ability to monitor coded transmissions for content, or the amateur's ability to monitor for purposes of self-enforcement. A combination of several factors should be sufficient to minimize the potential for abuse, however. First, the primarily local nature of communications on the vhf and uhf frequencies involved should limit the potential ability of unscrupulous business users to utilize the amateur bands to transmit data of a commercial nature, circumventing existing common carriers. Second, the vigor with which amateurs protect their allocated frequency bands against unauthorized interlopers, together with well-polished direction finding techniques, will in most cases result in investigation by amateurs of the source of repeated unusual or suspicious ...transmissions. The requirement of open identification procedures is a significant litmus for distinguishing legitimate amateur use from unlawful misuse of amateur frequencies...

6. The Commission proposes to require identification by telegraph on the center frequency of the spread spectrum emission. The problem with such a requirement is that it would be extremely difficult for someone monitoring with a narrowband receiver to determine the "center frequency" of a spread spectrum emission. Therefore, the League proposes that no more than two specific frequencies per band be designated for identification, and that either Morse code telegraphy *or* telephony be permitted.[5] While most experimenters may choose to use some means of automatic identification by telegraphy, as is the practice with radio-teleprinter emission, others may prefer to use voice and should be permitted to do so. The specific frequencies should not be in those portions of the 50 and 144 MHz bands which are allocated exclusively for type A1 emission. While the Commission may wish to designate these frequencies, the League prefers language in the Rules which would permit these frequencies to be designated by "commonly accepted band plans." This would permit greater flexibility, particularly if a selected frequency later proved to be undesirable for some reason not originally anticipated.

7. While the interference potential of spread spectrum to conventional modes is less than that of co-channel, narrowband emissions, there is some legitimate concern in the amateur community about its effects. There is also some room for concern with respect to services occupying bands adjacent to the amateur bands, because of the stringent filtering requirements which must be met if broadband emissions are to be confined in-band. Whereas two amateurs may operate in the same band in close physical proximity to one another if their narrowband equipment is of good design, if one of them were to use spread spectrum modulation, the other might well be unable to detect any but the strongest signals in the band. While the proposed Section 97.131(b) gives the Commission the authority to alleviate any interference which may occur, the League believes that a stronger statement of the responsibility of the spread spectrum operator is appropriate, particularly in view of the lack of available Commission field enforcement personnel. Thus, the League proposes the following wording for proposed footnote number 1 in Section 97.61(a):

> [1]Spread spectrum techniques for domestic communications only are authorized. Stations using spread spectrum techniques shall not cause harmful interference to stations of good engineering design employing other permitted emissions specified in the table.

The rationale for placing the burden of avoiding interference on the spread-spectrum operator is that the interference which may possibly be created cannot be avoided simply by selecting another operating frequency in the same part of the band.

8. The Commission proposes to permit operation in the 50, 144, and 220 MHz amateur bands. The League believes it is appropriate to exclude the narrow exclusive telegraphy segments, 50.0-50.1 and 144.0-0-144.1 MHz, from the range of permitted frequencies. This restriction would place no significant additional limitation on spread spectrum operation, as both segments are at the band edge and would have to be avoided in any case to avoid interfering with services in adjacent bands. Because of the practical limitations of filters which are available to amateurs, spread spectrum operation other than slow frequency-hopping systems will have to occur near the center of each band; this, coupled with the provisions proposed in the preceding paragraph, should provide sufficient protection for weak-signal amateur communication which normally takes place in the lowest several hundred kilohertz of each of the three bands. While of the three amateur bands proposed, the potential for interference is greatest in the 144 MHz band, because operation utilizing narrowband modes is greatest in this band, it is also true that equipment for experimentation with spread spectrum is most widely available for this band; thus, a flat prohibition on spread-spectrum operation in this popular band might have a chilling effect on amateur experimentation.

9. The Commission proposes to limit spread spectrum communication to domestic use. The League has no objection to this restriction at this time and understands the policy basis therefor. However, it is hoped that the Commission will permit international experimentation, should such prove to be desirable, and should both administrations concerned be agreeable to a limited waiver of the restriction. Such an agreement is permitted by virtue of paragraph 2734 of the 1982

edition of the ITU Radio Regulations.

In conclusion, the League supports and appreciates the Commission's efforts to introduce the radio amateur community to spread spectrum techniques. Should the Commission determine that it is time to provide for spread spectrum techniques by amendment of Parts 2 and 97 of its Rules, the League believes that the safeguards discussed herein are sufficient to protect the interests of other users of the radio spectrum, including amateurs using conventional techniques. The League believes that its comments and those of other interested amateur organizations and individuals will be a sufficient guide to the Commission in this proceeding despite the limited experience with spread spectrum techniques which the amateur community has amassed to date.

Respectfully submitted,

225 Main Street THE AMERICAN RADIO RELAY
Newington, CT 06111 LEAGUE, INCORPORATED

By _____
Christopher D. Imlay
Its Counsel

Booth & Freret
1302 18th Street, N.W.
Suite 401
Washington, D.C. 20036
(202) 296-9100

March 1, 1982

Notes

[1] See paragraph 6 of the *Notice*.
[2] See Paragraph 8 of the *Notice*.
[3] See Paragraph numbered 7 of the *Notice*.
[4] See Paragraph numbered 14(c) of the *Notice*.
[5] This would be no great difficulty, inasmuch as a separate, narrow band transmitter would have to be utilized for identification in any event in a multiplex arrangement.

ARRL REPLY COMMENTS IN GENERAL DOCKET NO. 81-414

Before the
FEDERAL COMMUNICATIONS COMMISSION
Washington, D.C. 20554

In the Matter of)
)
Amendment of Parts 2 and 97)
of the Commission's Rules)
and Regulations to authorize) Gen. Docket 81-414
spread spectrum techniques)
in the Amateur Radio Service.)

To: The Commission

REPLY COMMENTS OF THE AMERICAN RADIO RELAY LEAGUE, INCORPORATED

The American Radio Relay League, Incorporated (the League), the principal spokesman for the over 400,000 licensed amateur radio operators nationally, submits the following reply comments pursuant to the *Notice of Inquiry and Proposed Rule Making* released September 18, 1981 (FCC 81-290, 46 Fed. Reg. 49617 *et seq.*) in the above-captioned proceeding. The League's reply comments are limited to the concern expressed in the Comments[1] of the Association of Maximum Service Telecasters, Inc. (MST) that there is a significant risk of interference to television channel 2 (54-60 MHz) from amateur use of spread spectrum techniques in the 50-54 MHz band.

1. MST's concern, "supported" by the engineering study attached thereto, is speculative in the extreme, and fortunately unfounded for a number of reasons. First of all, as the concern of MST is limited to out-of-band amateur operation at 54-54 MHz, it ignores existing Commission rules which require signal purity. These rules are just as applicable to spread-spectrum emissions as they are to any narrow band technique. Section 92.73 requires that, as to most transmitters, the mean power of any spurious emission or radiation from an amateur transmitter, transceiver, or external radio frequency power amplifier being operated with a carrier frequency above 30 MHz but below 235 MHz shall be at least 60 decibels below the mean power of the fundamental. Paragraph (c) of that Section requires that, notwithstanding the above, all spurious emissions or radiation from an amateur transmitter, transceiver or external radio frequency power amplifier shall be reduced or eliminated in accordance with good engineering practice. Finally, paragraph (d) of that Section places the burden of eliminating interference caused by spurious radiation on the amateur licensee. Thus, ample rules exist to protect television channel 2 viewers from out-of-band amateur spread-spectrum interference. These rules are just as applicable to spread-spectrum emissions as they are to narrow-band amateur emission modes.

2. Further, despite MST's speculative arguments to the contrary, there should be no interference to television reception, on channel 2 or otherwise, from amateur spread-spectrum operation, as a matter of fact. The League, in testing the co-channel interference potential of non-government radiolocation use of spread-spectrum modulation techniques[2] against various forms of amateur radio and television, discovered that, in general, 420-450 MHz spread-spectrum modulation techniques did not disrupt co-channel wide-band amateur television operation. Those tests were repeated with essentially identical results at the Commission's Laurel, Maryland laboratory. Given the compatibility of spread-spectrum emissions and co-channel wide-band amateur television operation, MST's speculative concern about adjacent-channel interference commands little credence indeed.

3. MST appears to believe that the bandwidth of a spread-spectrum amateur transmission is inevitably broader than the 50-54 MHz amateur band. Such is hardly the case. The pulses of spread-spectrum emissions can be shaped by RF filtering, similar to the way keying is shaped in narrow-band transmitters to eliminate key clicks. It is not likely that amateur spread-spectrum transmissions would be at fault in any adjacent-channel interference situation. Indeed, perhaps what MST is trying to say, without actually admitting it, is that television receivers presently on the market have such poor selectivity and such a striking lack of adequate filtering that the receivers cannot discriminate between desired and undesired (i.e. adjacent channel) signals. If this is the case, then amateur spread-spectrum emissions are hardly the culprit. The League has tried for years to impress upon the Commission and the Congress the need for improvement in selectivity and filtering of home electronic devices, including television receivers, in the face of persistent resistance from home electronic equipment manufacturers. If amateur spread-spectrum or narrow-band transmissions are confined, in accordance with good engineering practice, to the 50-54 MHz band as per Section 97.73 of the Rules, there is no reason to expect adjacent-channel interference to television channel 2 in receivers of reasonable quality design.

4. Nor, despite MST's argument, need there be any difficulty in detection and identification of an interfering spread-spectrum amateur signal, should such a phenomenon ever exist. The League's comments suggested open identification procedures on two specific frequencies per band to facilitate rapid detection of the source of spread-spectrum emissions. Narrow-band telegraphy or telephony identification were suggested.

5. In summary, the League, following tests by its own technical department and at the Commission's Laurel Laboratory of the co-channel interference potential of spread-spectrum emissions to standard television reception, finds little merit to MST's speculative concern over adjacent-channel spread-spectrum interference.[3] Present emission purity rules for amateurs are sufficient to preclude interfering spurious radiation outside amateur frequencies. Identification of any amateur station using spread-spectrum emissions is possible, should any individual station not be operating in accordance with the rules.

Wherefore, the American Radio Relay League respectfully requests that, should spread-spectrum emissions be permitted for the Amateur Radio Service, such emissions not be precluded on the 50-54 MHz amateur band.

Respectfully submitted,

THE AMERICAN RADIO RELAY LEAGUE, INCORPORATED

225 Main Street
Newington, CT 06111

By _____
Christopher D. Imlay
Its Attorney

Booth & Freret
1302 18th Street, N.W.
Washington, D.C. 20036

April 15, 1982

Notes

[1] MST's Comments were filed in the captioned docket March 1, 1982.

[2] See Comments of Del Norte Technology in FCC Docket 80-135, filed on or about October 21, 1981.

[3] MST admits that very little data exists about interference from spread-spectrum signals and that questions as to spectrum-use efficiency of the techniques remain unanswered. This is one of the best reasons why the techniques should be permitted in the Amateur Radio Service, which has the best potential for experimentation and advancement of the radio art.

AMRAD REPLY COMMENTS IN GENERAL DOCKET NO. 81-414

Before the
Federal Communications Commission
Washington, DC 20554

In the Matter of)	
)	
Amendment of Parts 2 and 97 of the)	GEN
Commission's Rules and Regulations to)	DOCKET
authorize spread spectrum techniques)	NO. 81-144
in the Amateur Radio Service.)	

REPLY COMMENTS OF THE AMATEUR RADIO RESEARCH AND DEVELOPMENT CORPORATION

1. The Amateur Radio Research and Development Corporation (AMRAD) had read the comments filed under Gen. Docket No. 81-414 (and related Gen. Docket No. 81-413) and continues to support the FCC's proposal to authorize spread spectrum techniques in the Amateur Radio Service. A summary of our reply comments follows:

   a. Although not a panacea, spread spectrum has a definite, if limited, place in nonmilitary radio applications.

   b. The "near/far" effect of spread spectrum is key to understanding where it will work (and enhance spectrum management) and where it will not work (cause interference). Many comments failed to either understand this or to recognize its significance.

   c. Applications where there are no receiving stations using the same frequency band within the near area of the spread spectrum transmitting station are particularly suited to spread spectrum. Off-shore, isolated land, and aerospace transmitter applications fit this case.

   d. More experimental data is needed concerning civilian applications of spread spectrum. The Amateur Radio Service is an excellent experimental test bed. Spread spectrum experiments can be carried on in the amateur bands with minimal interference to other radio amateurs. The contention that amateur spread spectrum will somehow cause more interference to other services than existing amateur operations is without foundation.

   e. As for the administrative question of whether to (1) amend the Rules to permit spread spectrum or (2) liberally permit Special Temporary Authorities (STAs), we feel that both are required to gain the necessary experimental results.

   f. Neither the FCC nor the comments pointed out the differences in on-the-air behavior between the various types of spread spectrum modulation. It is likely that both direct sequence (DS) and frequency hopping (FH) modulation schemes would be good candidates; however, time hopping (TH) and chirp modulation schemes also exist and could be used. The behavior of each of these modulation techniques is so vastly different as to make the difference between interfering and not interfering with other users of the same band.

   g. Most of the comments were "worst-case," thus predictably negative and calling for tight control. The fact is that we need more experience. In this respect, we agree with the comments of John Maran, W2EM, that to make an omelet one must first crack eggs.

SPECTRUM MANAGEMENT COMMENT REPLIES

2. We find much to think about in the comments regarding compatibility of spread spectrum and conventional users in the 144-MHz amateur band. In this case, the band would be shared between spread spectrum and conventional types of modulation and has been called the overlay concept. In it, the spread spectrum users would maximize their bandwidth to the largest available. With the signal rarification that occurs spectral densities should fall below the natural noise floor except for those receivers in the near area. With such a low spectral density, conventional narrow band users should enjoy interference-free communications. Indeed, this concept can work but only if the number of spread spectrum signals is not excessive and that the proper spectral densities are maintained.

3. The near/far effect may be a considerable problem in areas where there may be a high usage of spread spectrum. If stations are located close to a spread spectrum emitter, then interference could be received from the spread spectrum emitter. Depending on the type of spread spectrum modulation used, the potential interference may fall off more rapidly with increased distance than for conventional narrow-band emitters.

4. If we consider three spread spectrum user densities, low, medium and high, we believe that low density spread spectrum will present little problem. Medium usage may add some interference to conventional users, but we find it hard to justify the comments of Mr. John Carrol referring to the prospect of communications degrading to the point of becoming a "nondeterministic" phenomena. Under high spread spectrum usage, we are aware of simulation studies which predict severe interference, and we must be candid in saying that we know of no actual operating experience which would either prove or disprove these studies. Again, any measure of user density should also include a near/far effect distinction to be meaningful.

5. We note some comments about the possibility of spread spectrum interfering with the following existing two-meter activities:

   a. Fm voice repeater operations.

   b. Long-distance communications in the 144.0-144.5 MHz portion of the band using ssb and which is characterized by weak-signal conditions.

   c. Moonbounce in the 144.0-144.5 MHz subband.

   d. Amateur Satellite Service operations in the 145.8-146.0 MHz portion.

6. These will be taken up below.

   a. *Repeater Interference*—AMRAD has run some short-distance tests, under dummy-load conditions, using a frequency hopper in the two-meter band. Several receivers, with antennas, were tuned to repeater frequencies. The frequency hopper used a group of discrete frequencies which included the repeater channels. The result was that the frequency hopper did not break squelch on the normal fm receivers, though the receivers would always activate when the repeater came on the air. This was due to design of normal fm receiver squelch circuits which require a certain strength of signal on for a certain length of time before the squelch is opened. Even when the receiver's squelches were opened (as a result of the repeater coming on the air or simply readjusting the squelch controls), little or no interference was noted. When it was, it was in the nature of quieting of the receiver for a time much shorter than a spoken syllable. We concluded that it is possible to operate a frequency hopper on the inputs of a number of area repeaters and never cause them to come on the air or to cause noticeable interference to repeater users except for an occasional syllable lost to quieting or a faint heterodyning. The actual on-the-air tests to finally put this matter to rest were not done due to delay in obtaining a critical part prior to expiration of our STA. However, we intend to request a new STA from the Commission in the near future. Nevertheless, the dummy-load tests were impressive—these were conducted in the near area.

b. *Weak Signal and Moonbounce*—It is the nature of weak-signal communications to want to hold the noise to the lowest possible. We agree that some type of interference protection is needed for these users but feel that the proper way to assure this is through time-proven self-policing of the Amateur Radio Service using "band plans," not through FCC Rules.

c. *Amateur Satellites*—Because the spacecraft are a considerable distance from the spread spectrum emitters, spectral density at the spacecraft should be small. Therefore, we do not expect any problems on the uplink side of satellite communications. However, the amateur ground station may experience interference if there are nearby spread spectrum emitters in the same band. Again, mutual interference can be controlled within the self-policing of amateurs, and FCC enforcement will not generally be needed.

7. We wish to point out that certain technical actions can be taken to resolve spread spectrum interference in the event of a conflict with satellite or weak-signal users.

a. Mutual interference can be reduced or eliminated by use of directional antennas and judicious selection of antenna polarizations.

b. It is possible to inhibit transmissions of a frequency hopper on certain specific frequencies by including a feature in the frequency-control logic to eliminate any use of those channels.

c. Another technique is spectral shaping, which is more applicable to direct sequence and hybrid spread spectrum systems. A sharp rf filter is placed in the transmitter output circuitry to reduce the amplitude of selected frequency components.

8. In view of the above, we must disagree with those comments that request the Commission to severely limit proposed operation of spread spectrum in the two-meter band or, in some cases, request that it not be allowed at all. We also do not agree with the call for specific spread spectrum suballocations such as the ones proposed by Mr. J. Carrol in his comments. However, the growth of spread spectrum could produce some unexpected side effects. Therefore, the Commission should consider the idea advanced by Mr. Carrol of authorizing spread spectrum subject to an additional NOI/NOPR in two years. A "sunset" clause might be considered, however five years may be needed to gather the depth of information needed for meaningful changes.

ENFORCEMENT AND SELF POLICING

9. Enforcement (by the Commission) and self policing (by radio amateurs) is important to both the Commission and radio amateurs. In light of this, we agree with the comments of John Maran to the effect that the Commission should concentrate on the interference aspect of spread spectrum.

10. Some comment has centered around the Commission's ability to monitor the content of all amateur radio communications. A capability to monitor the content of **all** messages would be an inefficient use of FCC resources if such resources were available, which they are not, even without spread spectrum. However, the ability to monitor selected transmissions for investigative purposes is not unreasonable.

11. There are several different factors which affect the ability to detect and recover spread spectrum transmissions. We note that the techniques used to detect and recover spread spectrum signals are known and can be applied when there is a need.

12. Factors which influence the ability to detect and recover spread spectrum signals depend upon the type of spread spectrum used. We would like to point out that the signal-hiding quality of spread spectrum rapidly decreases as one approaches the transmitter site. This is a direct result of the near/far effect. As spectral density rises, the spread spectrum signal can be detected with standard equipment.

13. It may be difficult to receive frequency hopping signals with conventional equipment. However, it is feasible to devise special wideband monitoring/direction finding equipment, with a directive antenna array, which can receive the frequency-hopping signal without knowledge of the hopping sequence.

14. Direct sequence provides a more difficult content monitoring problem. A wideband receiver and directional antenna array can be employed to provide accurate direction finding information. However, the information transmitted may be difficult to recover without knowledge of the key stream. If the cycle length of the keying stream is short, it may be possible to separate the key stream from the information stream with simple techniques. If the key stream is of moderate length and a number of cycles are available, statistical treatment may also be used to reveal the information content. Long codes may pose problems in recovery of content.

15. The code problem is central to spread spectrum enforcement and self policing. In Mr. Carrol's comments, he points out the usefulness of open experimentation with codes. He notes, for instance, that the NOPR does not propose the use of Gold codes which are very well known in the professional community.

16. We have noted some bounds to the detection and recovery of spread spectrum signals for amateur use. We wish to disagree with the comments offered by Mr. E. Miles Brown when he states that spread spectrum would be used for encrypting message contents in the Amateur Radio Service and that this practice would be significant. He also states that monitoring would be impossible. In practice, if unauthorized codes were used, that fact could be determined by the Commission whenever there is cause to monitor. If the amateur station is otherwise following the Rules, Morse code identification will be sent every 10 minutes. Also, the Commission possesses the direction-finding equipment necessary to locate the transmitter whenever it can be received.

17. Mr. Brown also states that "jammers" will use spread spectrum as a way to jam amateur signals and not be identifiable or will be provided with an excuse for these activities. We are uncertain as to the rationale Mr. Brown used. However, jammers can be located by techniques discussed above. Good engineering practices called for in the Rules can be policed by the Commission. Other amateurs, carrying out self policing, can see to it that offensive operating practices be curtailed. Failing these, the Commission can implement administrative or legal action against the jammer.

OUT-OF-BAND EMISSIONS

18. Some comments centered on the ability of amateur spread spectrum systems to maintain low out-of-band emissions. The comments of Maximum Service Telecasters seem to imply that significant energy will fall outside the amateur

bands thus causing interference to television broadcast channel 2. Mr. Luther Schimpf states that enforcement effort must increase at a rate which exceeds the rate of increase of the number of users of spread spectrum in the Amateur Radio Service.

19. In both of these comments, we are uncertain of the degree of consideration given by both parties as to good engineering practices which are already required by the Commission with respect to out-of-band emissions. The addition of spread spectrum does not in any way negate these requirements. We are confident that amateurs will continue to comply with these Rules.

20. There are a number of engineering techniques which can reduce or eliminate out-of-band emissions. One is rf filtering. Another is to shape the pulses much in the same way that amateurs now shape cw keying to remove key clicks.

21. Maximum Service Telecaster makes an interesting comparison between interference originating from a spread-spectrum emitter and weak signal reception of broadcast television. The premise under which Maximum Service Telecaster is operating is that some energy from an amateur radio spread spectrum station will spill out into the frequency authorized for TV channel 2. Their conclusion is that because spread spectrum is "noise like," it will introduce "snow" into the TV picture, and the viewer will decide that it is due to weak signal.

22. The above contention is incorrect on two counts. As for the amateur signal spilling out into TV channel 2, we have already pointed out that this is prohibited by the Commission's Rules. It is much more likely that the home TV receivers when tuned to channel 2 are so inadequately filtered that they will intercept amateur transmissions which are occurring totally within the 6-meter amateur band. On the second count, it is not possible to make a broad statement as to the extent or appearance of a spread spectrum signal on a TV screen without fully characterizing the spread spectrum signal as well as the TV signal field intensity and the adjacent-channel-interference susceptibility of the receiver. A similar assumption was made prior to laboratory tests which were conducted by the American Radio Relay League. The assumptions were not borne out in laboratory tests. To make the tests a worst case, the Del Norte Technology, Inc. spread spectrum transmitter was operating on the same frequency as a TV transmitter, not on an adjacent band which concerns Maximum Service Telecasters. Results of these tests were sent to the Commission under General Docket No. 80-135, Comments of the American Radio Relay League in Response to Further Notice of Proposed Rule Making, September 18, 1981.

23. Thus we cannot agree with the comments of Maximum Service Telecasters which recommend that six-meter amateur spread spectrum be disallowed.

MEASUREMENTS AND CALIBRATION

24. Mr. Luther Schimpf comments on the need for complex test equipment for proper maintenance and adjustment of the spread spectrum station equipment. Our experience with adjusting (and constructing) spread spectrum equipment indicates that much can be accomplished with traditional test equipment or simple instruments built for specific needs.

25. For example, the total output power can be measured with a filtered wattmeter. An oscilloscope provides visual indications of the information and keying sequences. A spectrum analyzer or "homebrew" panoramic adapter can display the level of in-band vs. out-of-band energies. As the individual problems surface, amateur ingenuity will be applied to effect the appropriate fixes. Amateurs will thus experiment and make contributions to the state of the art.

IDENTIFICATION

26. Because spread spectrum is difficult to receive on conventional receivers, the Commission has proposed telegraphic identification be made every 10 minutes as now done for radio-teletype operations. Mr. Brown raises an objection to the telegraphic identification proposed by the Commission. His suggestion is that the spread spectrum carrier be keyed on and off (in addition to the spreading keying or hopping) in a slow Morse code. We have not been able to discover the technical advantages to this form of identification procedure. On the contrary, it does not guarantee reception and/or accurate recognition on a conventional receiver, depending upon the type of spread spectrum used.

27. Mr. Carrol commented on the desirability of including some identification of the spreading algorithm used. This would be in addition to the identification proposed by the Commission. We feel that this suggestion has some merit and would endorse amateur spread spectrum operators adding this to their i-d transmission. However, we see this as an area within the competence of amateurs to police themselves, not one requiring more specific Rules issued by the Commission.

CONCLUSION

28. We feel that the Commission is to be congratulated for its forward-looking proposals to permit spread spectrum experimentation in the Radio Amateur Service. We believe that this proposal should go forward irrespective of the current disposition of General Docket No. 81-413 which deals with spread spectrum in the land mobile bands. In many respects, the experimentation in the amateur bands is needed to acquire the experimental base needed to fully utilize spread spectrum in other radio services in such a way as to enhance, rather than degrade, spectrum management.

Respectfully submitted,

AMATEUR RADIO RESEARCH AND
DEVELOPMENT CORPORATION

1524 Springvale Avenue
McLean, Virginia 22101
703-356-8918

Paul L. Rinaldo, W4RI
President

April 14, 1982

Chapter 4
AMRAD's Contributions

During the early 1980s, AMRAD-affiliated experimenters worked to bring amateur spread spectrum techniques from the blueprint to the feasibility stage. These experiments were documented in the AMRAD Newsletter.

June 1981

USING MICROPROCESSOR-CONTROLLED RIGS FOR FREQUENCY HOPPING

By Hal Feinstein, WB3KDU

A number of people have expressed interest in the idea of using microprocessor-controlled ham rigs as the basis of a spread spectrum frequency hopper. The idea is both interesting and appealing. In this article some of the requirements for such a setup will be examined.

First, let's take a closer look at what a frequency hopper does. The frequency hopper may employ standard modulation such as single sideband (SSB) in the primary information path. Other types of modulation are okay but may not provide the same level of interference to other users. Care must be taken to avoid creating another Russian woodpecker. The Russian woodpecker is an over-the-horizon radar which emits a train of short pulses. These pulses are very wideband and bother a large number of users.

Modulation techniques such as CW, AM or FM all have the capability to be a woodpecker if care is not taken. In this respect, SSB has advantages such as no carrier and an average amplitude of much less than maximum.

Some ham rigs get very sensitive about being mistuned. This may not be a great problem with some rigs that are either broadbanded or autotuned. Some antenna systems have very similar problems in that they must be peaked before you start operating. Of course, all this depends on the power you are running, sensitivity of your equipment and the limits of hopping frequencies.

In the AMRAD Special Temporary Authority (STA), the FCC allowed us to experiment with frequency hopping only within the U.S. amateur bands on a non-interference basis. Specifically, we are permitted to use 80, 40, 20 and 10 meters for frequency hopping tests. Ten meters seems prime for experiments because there is more room than in the lower-frequency bands.

Another thing that should be kept in mind is the so-called near/far effect. When you are operating a spread spectrum station, a receiver at a moderate distance should not hear anything unusual. The signal strength at this particular receiver will be low, and only a short piece of the signal can be heard every now and then. But suppose that the receiver was loaded into a vehicle and was driven in your direction. What would be the result? The signal strength of the spread spectrum signal would rise as the receiver approaches the QTH. When the receiver is very close it will receive a very strong pulse every so often and probably cause the automatic gain control (AGC) to reduce the receiver's gain. There could also be some desensing of the front end due to overload. (Note that the exact action of the near/far effect will vary according to the specific type of spread spectrum modulation used.) Your spread spectrum signal would hardly be invisible to that receiver. But the receiver's operator might not be able to tell what it is. If you cause television interference (TVI), the TV set owner will unlikely know the source of the interference. Some hams say that SSB is better for TVI because "they" can't tell who it is. With AM modulation, your voice might come clearly through the TV set. In any event, TVI is no blessing even if the victim can't tell who's talking.

What about all this invisibility stuff? It is true that normal receivers won't be bothered by your signal if you do it right. That means that your hopping pattern must appear to be very random and that the speed of hopping should be less than about 1/5th of a second. This is considered a minimum for invisibility and yet presents a problem for some standard synthesizers. There are two ways of generating a frequency for a frequency hopper. Both involve the synthesizer approach to life. The first uses a phase locked loop or PLL. In this approach, the local PLL oscillator is driven by an error voltage. The error voltage is produced by dividing down the VCO oscillator frequency and comparing that with a standard reference oscillator. By doing the VCO division right, the error voltage will trick the VCO into oscillating on a different frequency. The frequency in this case will be set by the divider circuit within the PLL. The divider is what you control from the frequency knob of the rig.

If your rig uses a synthesizer, you should investigate and verify that the time it takes the synthesizer to lock onto a given frequency is much less than 200 ms. For the hopping rate of 5 hops/s (or chips) your synthesizer must lock up much faster than this. Still, many say that 10 hops/s is the smallest that you should use, but this requires even faster synthesizer lock-up time.

How can you determine the lock-up speed? Many synthesizers have a transmit inhibit lead which is used to kill the transmitter when the rig is moved in frequency. This is required so that the VCO doesn't cause the transmitter to operate outside the band or put out a signal on an undesired frequency. The exact setup on your rig must be investigated. You can get a ball-park idea by just rotating up a bunch of frequencies and checking how much time is spent in an unlocked condition. Many synthesizers cannot be used because of the slow locking speed. CB sets, for example, usually don't do well in this respect.

Preamble and sync are the next things to think about. Let's take sync first. Synchronization is the process by which your rig and the other station's rig stay at the exactly right place in your hopping code pattern. If you or the other station fall out of sync no signal will be acquired. This means that some way is needed to keep the hopping code generators in the two rigs together. How can this be done? It can be difficult. The military approach is to use very expensive timing sources. An amateur approach, which has yet to be tried out, is to use the rise and fall of the signal (as the hopper switches frequencies) as a correction signal. You must have a timing source at your station and at the station that you are talking to. Because the timing source will be used to trigger a hop at your station, both stations must be synced. If you see that the other station's signal falls and you have not hopped then you can assume that your timing is a little slow. You can speed it up a little or slow it down depending on what is happening.

This is the first place that a microprocessor may be useful. The sync algorithm and clocking (using real-time interrupt facility) could be controlled in a very intricate manner with software. Some people expect that sync problems will stand in our way. Will it?

For two frequency hoppers to start a contact, they need a method that lets the other know that one has fired up. This may be done with a preamble. A preamble is simply a signal which the transmitter sends to alert the receiver to start hopping. The preamble can be a simple set hopping pattern which everyone uses. If this were used, every receiver would hear the preamble and make ready to hop. When the receiver is switched to its normal hopping code, only the receiver with the same hopping code of the calling station would acquire a signal. The other receivers would not acquire because they are set to a different hopping code. Perhaps the standard hopping code which follows the preamble should start with a few tenths of a second tone to indicate that the receiver is now locked. A receiver which did not acquire a tone after the preamble would resync to the preamble and wait for the next calling station.

There are some procedural problems that need to be dealt with, such as how two stations should act if they lose one another. Should they automatically retransmit the preamble? Should they go to some prearranged place in the hopping code and try again to achieve sync? These questions are open, and solutions are welcome.

The three areas mentioned above (sync processing, preamble processing and handling fadeouts) all require complex decision making in real time. This is a place where microprocessors can be used with good effect.

Coding of the frequency hopper is something that interests people. There are several possibilities here for different codes, but some form of standardization is needed. Without standardization, only a limited number of hams could communicate with each other. Remember that each code sequence is a new "channel."

AMRAD has decided to use the popular Gold codes for controlling the hopping pattern. Gold codes are easily generated and provide for a large number of possible channels to be generated.

There are two popular schemes for hopping, and both depend on the following idea. If you have several stations hopping around in (say) the 40-meter band, you would like to avoid hopping to the other station's frequency if you are trying to talk to someone else. In other words, you don't want to use a frequency in your hopping pattern if someone else has hopped there. How can you control this? The hopping codes (called PN codes) are designed to avoid collisions by using only those codes which have a small amount of collisions to start with. The Gold codes are one of these. M-length are another. Both codes are generated with linear feedback shift register circuits which can be loaded by the microprocessor. See *CMOS Cookbook* or *TTL Cookbook* for details.

Well, in this informal article we have explored some of the requirements of spread spectrum using microprocessor-controlled rigs. As you can see, many of the processes are complex and best implemented in software, with the exception of a shift register for code generation. Please let me know your comments.

July 1981

By Hal Feinstein, WB3KDU

Well, I can report some new things this time around for the spread spectrum effort.

EXPERIMENT #1

First, we had a series of very successful frequency hopping experiments carried out by Paul Rinaldo, W4RI, Dick Kessler, K2SZE (in Rochester, NY) and Olaf Rask, WA3ZXW (in Annapolis, MD). Experiment #1 of the AMRAD Special Temporary Authority (STA) called for tests with a commercial/military frequency hopper in the 80-, 40- and 20-meter bands. These rigs are capable of a frequency range of 2 to 15 MHz and hopping speeds adjustable from about 1 hop/s (1 chip for you units buffs) to about 20 hops/s. The hopping sequence was assumed to be nonlinear because this rig was meant for military purposes. Normal linear sequences of short duration are not useful for military applications.

Using these rigs is somewhat different from a standard single sideband (ssb) transceiver, but enough of the operations are the same that an amateur would feel right at home with this mode. I was able to sit in on a session that Paul had one evening and will describe what I heard.

First, both stations made contact using ssb (this was on 75 meters) on a *service frequency*. The next major item was making sure that the hopping sequence generators on both rigs were set the same way. The rig has a set of thumb wheels on the front panel which are used to control the hopping sequence. Both stations set the wheels the same way. If the wheels had been set differently, the hopping sequences would be different, and the two stations couldn't talk to each other.

Now that the hopping sequences were set the same way, one station would transmit a special fsk signal which the other station would receive. This fsk signal serves to alert the other station to start hopping as soon as the fsk signal stops. The fsk signal was generated by using the SEND SYNC switch on the rig. To set up the second station to use this fsk sync signal, there was a FAM (frequency-agile mode) switch position to enable the second station to lock on to the fsk signal.

When the fsk signal stopped, both rigs were in the hopping mode. The speed at which the hopping took place was 5 hops/s, which is slow as hopping goes but has many interesting features. The mode of transmission was ssb, and Dick's signal came in very well. I was surprised that the rig's synthesizer was right on each hop, which meant that the ssb signal was very clear. No "duck talk" was present, such as comes from being a little off with an ssb signal.

One of the main advantages of spread spectrum was the so-called *antijam* or interference avoidance feature which happens because the background QRM is being changed every hop. We were hopping at 5 hops/s with this rig, and I was able to observe this effect. Dick's voice was more readable than when in (non-hop) conventional ssb mode. We made a few experiments by moving the hopping sequence up a little to see what it would do against solid, congested ssb.

In the first part of the experiment, we were in the 80-75-meter band, which resulted in the hopping sequence varying mostly in the cw portion, but it would visit phone stations now and then. What did this sound like? Well, all the sounds that hams are used to were present, but every one fifth of a second the sound would change! What you would hear was a snatch of RTTY, a small burst of cw, a few sounds from some ssb station and some snap of QRN, each lasting only a fifth of a second.

It was easy to hear Dick's voice with this ever-changing background noise because Dick's signal was strong. There was some fading now and then, and this let us see what a weaker signal would sound like. When K2SZE's signal became weak, it was still readable even when it started to fade into the strange background sounds. My old cw training came back to me, and I mentally started to try and shut out the background, just as you do when receiving a weak cw station in heavy QRM. But this time the background sounds were changing!

I would say that there were about 30% ssb sounds in the mix with most being cw and RTTY. There was an advantage over normal ssb with this mix. Then we moved the hopping sequence into the 75-meter phone band. Both stations had to reset the service frequency higher and then go through the sync process again, which is not hard. This time we had a mix of 80% ssb and 20% cw and RTTY. With this mix Dick's signal was hard to hear. Because there was such a high percentage of ssb in the background, it was difficult to pick out Dick's signal from the rest. We tried to lock onto Dick's voice and ignore the rest. But this was difficult. Things were much better with more cw and RTTY in the mix and downright hard with a high percentage of ssb.

What conclusion can be drawn from this? With slow hopping (5 hops/s), ssb frequency hopping does well against cw-like signals and poorly against voice-type signals. The fact that we were hopping at 5 hops/s didn't allow us to take advantage of a property that fast hopping has. If we were hopping faster (say 25 times a second) very short snatches of the background noise would be received. As the frequency is changing so fast, the sound of each of these snatches would just meld together and form a kind of buzz. The ssb voice signal would still sound like an ssb voice signal, but the background would sound very constant. With the right filters to treat the background buzz, the voice signal should be readable. This would happen even with a strong mix of ssb signals in the background because the amount of signal from each frequency would be very small. This is what we would like to try a little later with Experiment #2 which allows us to build our own frequency hopper rig out of old CB sets.

As part of the experiment, Paul turned on his ICOM IC-701 so that we could hear what a conventional receiver would hear with the other rig transmitting in the frequency hopping mode. Paul picked out a place on 80 meters where it was quiet and let the receiver stay there. We didn't hear K2SZE's signal on this receiver, partially because it wasn't connected to an antenna. When we went to transmit mode, every once in a while you would hear a snatch of sound like "aup" or "thu," but it would be gone as quickly as it came. This was the result of the rig hopping in a random way.

EXPERIMENT #2

A few of the AMRAD Spread Spectrum Special Interest Group have been interested in doing something with old CB transceivers. Allan Kaplan, W1AEL in Richardson, TX has a bunch of good ideas. The last time I spoke with him, his group was in active search of a number of Hy-Gain surplus CB boards to use as the basis of a 10-meter frequency hopper.

I was able to get hold of two ssb CB transceivers which were modified for 200 channels and had seen extensive use in another "service." Both of these rigs have almost identical internals and use the standard uPD858C synthesizer chip. This chip has a number of components for a synthesizer built right onto the chip. External to the chip is an active bandpass filter, a main VCO chip (which produces the output frequency) and a number of mixers and oscillators used to mix the VCO output for use in the CB transmitter and receiver.

In order to frequency hop, there are three things that need to be done: (1) change the mixer crystals so that the rig operates in the 10-meter band (this is in progress now), (2) modify the feedback filter so that the synthesizer will lock up faster, and (3) hook the BCD frequency programming lines to a controller board.

The controller board will have a linear feedback shift register which has parallel output and will gate an IC which has 8 spst (single-pole single-throw) switches implemented in solid state. This IC will do the actual switching of the synthesizer programming lines for isolation's sake.

A second stage of the board will be used to sense/send a tone which will be used the same way that the special fsk signal is used by the commercial rig mentioned above. When the signal appears, the clocking associated with the shift register will go into a make-ready-to-hop state. Then, when the signal ends, hopping will start and continue until the stop-hopping switch is pressed. Lastly, a timing source is needed, and this will be supplied by a high stability crystal oscillator which will be on board as well.

K2SZE points out that timing is critical with fast frequency hopping. For now, we don't plan to hop very fast. Only when we have gained more experience with slow hopping will we try the faster stuff.

One problem being researched is that of how to get the loop filters of the synthesizer to react faster. The loop filter sits between the output of the phase comparator/charge pump and the VCO. Its job is to filter the correction voltage to the VCO to eliminate high-frequency components. So, the output of the phase comparator/charge pump goes through a low-pass filter before being applied to the VCO.

This loop filter in the two CB rigs consists of an active filter which is part of the uPD858C and a second filter which is an bandpass filter implemented as an active discrete transistor amplifier. The feedback network for the active filter within the uPD858C is via a resistor-capacitor combination attached to some of the 858's pins.

Both CB rigs are slated to be modified in the same way so that there will be two rigs which do the same thing right off. Differences in the synthesizer design or the controller board will probably produce a different hopping sequence. So, if you plan to do this al\ ays get at least two rigs of the same kind.

In Fig. 1 you will find a proposed logic sequence for the frequency hopper control board and a block diagram showing the basic functions.

SPREAD SPECTRUM INVADES PACKET RADIO

Recently, the packet gang at AMRAD have learned that some experimental packet radio systems use spread spectrum. The primary one is DARPA's (Defense Advanced Research

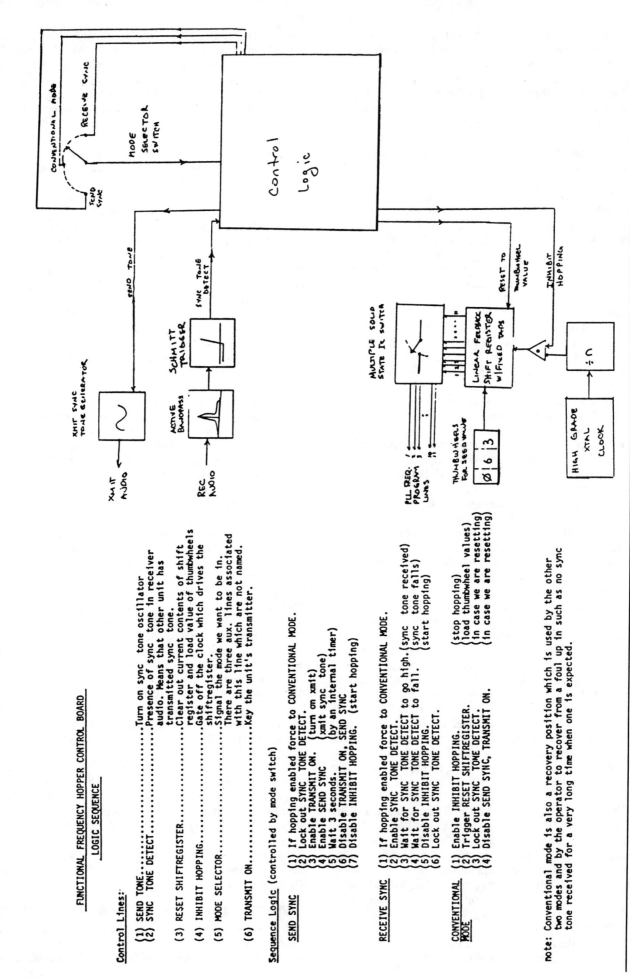

Fig. 1—Frequency-Hopper controller board basic functions.

Projects Agency) PRNET. Now, what can spread spectrum offer?

Well, all packet radio nodes transmit and receive on the same frequency. Each node tries to get its packets out as quickly as possible and then listen for packets addressed to it. As you might expect, sometimes two or more nodes try to send packets at the same time. The result is a collision. If a collision happens, both nodes will try again later. Some experiments revealed that when these nodes wildly try to transmit their packets, very few packets get through. So a number of plans have been cooked up to rule the channel and give order to an unordered universe of nodes. These plans go by different names like Aloha (yes, this was named by the University of Hawaii), Slotted Aloha, CSMA (carrier sense multiple access which is now in use by AMRAD), LWT, etc. All these plans try to get as little collision as possible while maximizing the number of packets transmitted.

Now spread spectrum can help this collision problem. Spread spectrum allows a station to lock out interference from all sources except the one which has the proper code sequence. Suppose that we assign a different code sequence to each packet node. In this case, a node would scan the list of sequences assigned to the other nodes and if it heard a transmission, it would quickly find the sequence of the node transmitting to it. Then it would lock onto that sequence and exclude any other node trying to interfere. There would be no collisions except when two stations started to send almost at the very same time.

Besides providing this anti-collision property, spread spectrum also provides privacy (in addition to a bitwise transposition cipher in PRNET for the data). As you might suspect, this technique has a four-letter acronym SSMA (spread spectrum multiple access).

I should point out that AMRAD has no plans to add spread spectrum to our packet networking planning. The main reasons are that spread spectrum would complicate things as well as increase the cost of getting packet radio going.

August 1981

By Hal Feinstein, WB3KDU

There has been some new activity at the FCC regarding spread spectrum. The Commission has taken a step forward with the idea of integrating spread spectrum techniques into civilian radio technology.

10-METER FREQUENCY HOPPING

Experiment #2 has been progressing at a good rate. The two CB transceivers that were acquired by Hal, WB3KDU have been modified for 10 meters, and some work has started on the hopping controller board.

A review of the divider capabilities of the 858 synthesizer chip leads us to believe that about 467 separate channel frequencies are possible. There is also a pin which is used for establishing a channel separation reference. Usually this is set for 10 kHz separation which is used in CB. Five kHz is also available from this chip, which may be better for hopping purposes.

While adjusting the synthesizer stage we ran into an interesting kind of feedback. When the synthesizer stages are not tuned properly, sometimes you would get motorboating sounds in the audio. As you increased the frequency, the pitch of the motorboating would increase. In transmit mode, with a frequency counter hooked to the transmitter, we found that these motorboating sounds were a result of the synthesizer being out of lock. The results read on the frequency meter showed that the transmitter output was jumping all over the hf spectrum.

There is a special circuit in most synthesized CB rigs which will kill the transmitter if the synthesizer selects a frequency outside the normal CB channels or if the synthesizer is out of lock as would happen when you are changing channels. This circuit is supposed to prevent out-of-band operation. One of the first steps in some conversion procedures is to cut out this circuit. Some people want to use these things out of band.

The direction that manufacturers have taken in the newer synthesizer chips for CB is to build a ROM (read-only memory) with the correct channel settings directly on the synthesizer chip. Then if there is an unlocked or out-of-band signal generated, an on-chip shutdown circuit is activated to cut the output of the chip.

This motorboating can be cured completely by simply adjusting the various tuned stages in the synthesizer to their proper values. Then the circuit operates without any motorboating over the entire range of ten meters.

One of the members of our group, Chuck Phillips, N4EZV (ex-K5LMA) has been playing around with a small range frequency hopper that has a dozen or so channels in it. It was designed to hop on the 2-meter band and uses standard tone decoder-encoder chips for much of the preamble and control work. The unit itself is small, and Chuck says that it doesn't draw much power. He also has a receiver that goes along with it. More later.

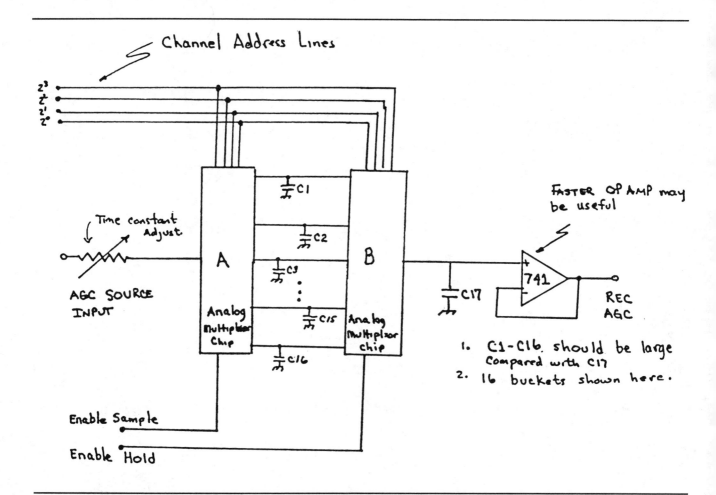

Fig. 2—Sample and hold for agc.

Joe, K4IHP sent in a circuit for dealing with the problem of "hard agc" which crops up in conventional superhet frequency hopping rigs. See Fig. 2. Dick Kessler, K2SZE says that it resulted from the fact that the AGC is sensitive to strong QRM on some channels. On one channel there is strong QRM and the agc responds by lowering the gain. When you hop to another channel without QRM the agc is too slow to recover in time. This results in the loss of an audible signal for a few hops, until the agc recovers. K4IHP has devised a solution to this problem for frequency hopping rigs where the number of channels is low.

Notes

[1] Buckets can be expanded cheaply in multiples of 16.
[2] To conserve the number of buckets, closely spaced frequencies can use the same bucket. This will happen automatically if you use just the high-order address lines of the synthesizer address lines.
[3] Cheap CMOS analog multiplexer chips can be used. (Try RCA.)
[4] Make sure that the maximum agc voltage does not exceed the plus power supply nor the minimum agc voltage exceed the negative power supply voltage.
[5] The principle of operation is that each of the capacitors (C1-C16) operates as an analog storage device. When multiplexer A is active it selects one of the links between the two multiplexer chips and connects it to input 1 (the agc source). The result is that the capacitor connected to that link is charged up to the agc source voltage. When chip A's enable goes low it resembles an open circuit with just a charged capacitor on it. Then, when the chip is enabled, the agc voltage level which is stored in the charged capacitor is made available to the receiver via the 741 op amp and capacitor C17.

September 1981

SS CODES

Codes are an essential part of spread spectrum (SS) communication systems. In fact, many of the more interesting qualities of SS can be traced to properties of the code used. With so many important features dependent on proper selection of a code a closer look may be useful.

Many of you know that a good SS code must be "random" and have good "cross-correlation properties," but what do these terms mean?

A code may be thought of as a string of numbers generated by some process. In the case of SS we will consider only strings of numbers in which the numbers assume two clues: one and zero. This then is a binary code with digits called bits.

Many processes can be used to generate a binary stream of bits. They can be divided into two kinds. First the bit stream may be random. That is there is no way to predict from a knowledge of the generating device or its past output what the next value it produces will be. This is a critical requirement for a truly random stream.

If you stop to think about it you'll find that any machine that you devise to produce truly random numbers (bit streams in our case) must have a source of *natural randomness* in it. Natural randomness means that the noise from a diode or the time that the next particles will be ejected from the nucleus of a radioactive atom are random. Why? Because no one has yet devised a way to calculate the values for these quantities.

There are a number of uses for pure randomness such as cryptography, simulation and measuring the characteristics of a communications system, but not in spread spectrum.

SS requires that both the transmitter and the receiver must have a copy of the random bit stream to command the synthesizer or modulate a direct spread carrier. This brings us to the next type of randomness which is not really random. In fact, this is a basic problem which mathematicians ran up against in defining randomness. Our first type of randomness uses a pure natural random source. We would expect it to pass any tests originally devised to check randomness, and it does. But, mathematicians also discovered a type of process which produces a stream of numbers which pass these randomness tests as well!

To this second type of randomness, the term "pseudo-random" was created. Another popular name in engineering circles which arose because pseudo-random numbers look so much like noise is "pseudo-noise" or PN sequences.

Anyway, PN sequences are not generated by a pure natural random source but by a physical device which has well understood rules. What is this magical device? It is the shift register (see Fig. 3). This little circuit implements what mathematics people call a recurrence system because some of the output is fed back to generate the next input. Thus the output continually is fed back and "recurs." You will also find this circuit called the linear feedback shift register (LFSR) because the two taps taken on the third and fifth cells are combined using exclusive-ORing (a linear operation) and then fed back.

Just as an aside from our main topic there are two more-

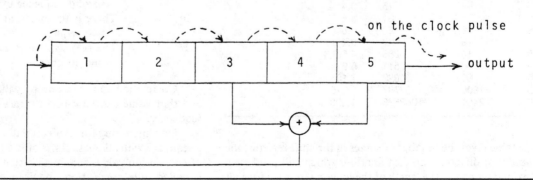

Fig. 3—A five-stage linear feedback shift register.

or-less familiar uses for recurrent systems that amateurs may have bumped into:

(1) *Fibinocci* numbers generated by the equation:

$F_n = F_{n-1} + F_{n-2}$

Next = Last + Next-to-last

This simply says take any two numbers for F_{n-1} and F_{n-2} starting values. Then add them together to find F_n. To continue, use the current value of F_{n-2} as your output then $F_{n-2} + F_{n-1}$; $F_{n-1} + F_n$. Repeat this loop by performing the next addition. Keep going; you are producing Fibinocci numbers! Fibinocci by the way, got these numbers named after him by cleverly solving a problem in which you put two rabbits in a cage and come back to count the number of offspring some time later!

As fate would have it, forms of Fibinocci number generators are the heart of most computer "random" number generators. These types of algorithms are one of the best known uses of Fibinocci numbers.

(2) *IC noise generators*—Some manufacturers have recently produced an IC chip which acts as an audio noise generator. The internals of the chip use a feedback shift register to produce noise-like voltages.

Now let us return to our discussion of SS codes. Because the shift register will be our generating device for these codes, what are some of its properties?

1. A shift register sequence repeats after a fixed number of outputs. The largest you can get this number to be is related to the feedback taps and the number of cells. The longest possible is found by length = 2^{N-1} where N = the number of cells. The taps must be connected to form a *maximal* sequence. Some maximal tap settings are given in Table 1.

2. A given maximal connection setup will produce a single long repeating sequence. A non-maximal setup will produce several sequences, depending on the initial load (called the *seed*). The summed length of all these shorter sequences will be equal to the length of a maximal sequence with the same number of shift register cells.

3. (For mathematicians only) How many maximal tap connections are there for a given size register? Answer is related to an important idea in mathematics called Euler's phi function (pronounced OILER's). Let us skip the calculation step and just list a few results:

Table 1

# of Stages	Possible Maximals	Length of Sequence	Sample Maximal Taps
2	1	3	1,2
3	2	7	2,3
4	2	15	3,4
5	6	31	3,5
6	6	63	5,6
7	18	127	6,7
8	16	255	4,5,6,8
9	48	511	5,9
10	60	1023	7,10
15	1800	32767	14,15
20	2400	1048575	17,20

The table gives the number of stages in the shift register, the number of different tap combinations which will produce a maximal sequence, the length of the longest sequence for that number of register cells and one maximal tap setting for each value.

4. N-Run Property. For a maximal sequence generated by a N stage LFSR, every run of N consecutive bits is different from every other run of N consecutive bits in the sequence.

5. Decimation property. If you take every jth bit out of the original maximal sequence which has maximum length L, then:

(a) if j (the sampling rate) *cannot* divide L without leaving a remainder, the new sequence formed by selecting every ith bit out of the initial maximal sequence *will also be maximal*.

(b) if j is an odd number and can divide L without leaving a remainder, the new sequence will be nonmaximal.

(c) if j is an even number and can divide L without leaving a remainder, then the resulting sequence will be a shifted version of the original.

6. Impulse response property. For a nonmaximal setup several different sequences will be produced. One of these sequences will be longer than the others. This sequence can always be produced by loading the initial seed to be all zeros except for the first or last cell which is set to 1. Now the length of any of the smaller sequences will always divide the length of this biggest impulse response sequence evenly.

7. Superposition property. By detailed mathematical analysis it can be shown that every maximal sequence is composed of the exclusive-OR of several small nonmaximal length sequences.

8. Cycle property. Because every sequence repeats itself after a given number of times, it is called a cycle. If any starting position of this cycle is chosen, the shift register will continue generating values which normally come next in that cycle. This is known as a phase difference because it is very much like choosing a sine wave at some advanced phase position on its curve. For example, if we select a value at random, say 26, and put this value in the generator, the generator will pick up by producing the next values which are 13, 6, 19 and so on. The same values are produced over and over again but from a different starting point.

Correlation techniques are essential tools in the study and characterization of random processes.

To understand some of the properties which SS derives as a result of using a coded carrier, we need to examine some correlation tools.

Basically correlation is the idea of determining how much likeness one set of data has with another. Correlation usually is measured between -1 and 1. These values have a well defined meaning.

Correlation Output	Properties
1	The first sequence matches the second sequence exactly.
0	There is no match at all between the two sequences.
-1	The two sequences are mirror images of one another.

Correlation can take on any value between 1 and -1, and that value can be used to measure likeness of the two sequences.

One interesting form of correlation is when you correlate a sequence with all phase shifts of itself. This results in a table of correlation values for the different shifts. This technique is called auto-correlation and is one of the tests used to measure how random a given sequence is. Pure random data should have a correlation equal to 1 for a perfect match with itself and a value close to zero for all auto-correlations with a phase shift other than zero. It turns out that most shift register sequences have this property and so pass this test for random data even if they are deterministic.

Correlation must always be done against two batches of data, to tell how much one is like the other. For autocorrelation, the same batch is measured against all phase shifts of itself. This results in a table of values, one value for each phase shift.

Crosscorrelation is a tool which measures the likeness of two sequences. One is considered a reference copy, while the other sequence will undergo various phase shifts. Like the autocorrelation, crosscorrelation results in a table of values, one for each phase shift of the nonreference sequence.

As we progress in our study of spread spectrum codes, we will see that crosscorrelation forms the heart of many SS properties. Codes are studied and chosen based on low crosscorrelation properties. Because the SS carrier is coded, the carrier takes on similar crosscorrelation characteristics. In fact, crosscorrelation is used by some SS receivers to choose or reject the properly coded carrier.

A very useful tool for measuring goodness of a sequence is the crosscorrelation function. This tool is easy to calculate and what it does is to report on the likeness of one sequence with another for *all* phase shifts.

This process can be done by this procedure:

1. Line up the two sequences, one under the other.
2. Find the number of places that the lower bit matches the upper bit, count these and save the count in a table.
3. Now shift the lower sequence by sliding it back one position and bring the last bit up to the front.
4. Perform steps 2 and 3 until the entire lower sequence has been shifted all the way around. The table you build by this process from step 2 is the output and result.

How can we use these tools to discover properties of SS? Let's examine some of the important properties of SS:

1. *Interference rejection*. The action of a SS receiver is like the crosscorrelation process described above. The receiver has a locally generated copy of the code which the transmitter also is using. Signals which don't look like the code are averaged out by the correlation process. This includes noise, cw and fm signals or other SS setups using a different code. If the code is secret, then jamming will also be reduced because the jammer cannot determine the code used.

2. *An abstract channel* (code multiplexing). A code can be considered as a channel because the correlation action is interested in receiving only the signal with the same code it has. Two users each using the same code can talk to each other. Different codes are rejected.

3. *Privacy*. Because receiving an SS signal requires that the receiver know the code, specifically chosen codes can provide privacy channels.

4. *Overlaying*. There is a theoretical ability to overlay other users with wide-band SS signals. To the conventional user, the SS signal would appear to be noise because of the random property of the PN streams. The less randomness a code shows, the more signal energy starts to collect in various places in the user spectrum. As randomness goes still lower, the wideband noise signal starts to take on the appearance of a simple high-speed digital signal which will interfere with conventional users.

Conclusion: Codes are essential to spread spectrum for many of its properties. We have listed several of the better known properties and note that many others exist which rely on more advanced math.

Now that the foundation has been laid, the next issue of this newsletter will explain some of the more important SS codes such as the Gold, JPL and M-length sequences.

A good write up on shift registers appears in the *TTL Cookbook* and *CMOS Cookbook*, both by Lancaster and published by Sams.

December 1981

MOBILE FREQUENCY HOPPER

A member of the AMRAD STA group, Chuck Phillips, N4EZV has been quietly working on design of a mobile frequency hopper. Now word comes that the unit is built and that some dummy-load testing has been completed.

As Chuck describes it, it is about as big as a Drake TR-22. It hops on setups which are multiples of 16 discrete frequencies. This means that the unit can be set to hop on 16, 32 or other multiples of 16 channels. Dwell time on any channel is 90 ms.

Power output is 2 watts. Synchronization between the two units has a simple form. There is a home channel where both units start. When the transmitter is activated on one unit, it sends an audio burst which the receiving unit decodes. This tells both units to start hopping. Hopping takes place over the channels to which both units are programmed.

There is also a look-ahead feature which basically looks a number of channels ahead and watches for speed differences in the hopping of the two units. If the receiving unit loses the transmitting one for a number of consecutive hops a resync procedure is activated and automatic resync is attempted.

Chuck describes some tests that he has tried with two units in his basement, operated on dummy loads. When one unit talks to the other, the sound is clear and not choppy. I have heard choppy sound with other spread spectrum rigs at this hop speed (90 ms). Next Chuck put on a standard vhf receiver which had one channel in the hopping sequence. When Chuck talked into one of the units, it didn't break the squelch on the vhf receiver. When taking the squelch off and listening to the empty channel, all that I was able to hear were a few unintelligible sounds once in a while.

It was interesting that there was a "solid armchair copy" on the FH transceiver but nothing on the vhf receiver. The next experiment was to move the channel separation of one of the FH units 5 kHz to see if the other FH unit would interfere. The answer was no. The two units could be in different worlds for all the interference that they caused each other.

What was amazing to me was how so much processing could be packed into so small a box. Anyway, Chuck is looking forward to some on-the-air tests in the 2-meter band, and Paul, W4RI is working on getting the STA amended for several stations to do this.

I would like to say that Chuck's work represents the first amateur spread spectrum rig built by an Amateur Radio experimenter. It marks an important first step for the spread spectrum experimenters' SIG. Good work, Chuck!

By the way, friends... Chuck has a *General Class* ticket as does this author.

January 1982

SELF ENFORCEMENT

By Hal Feinstein, WB3KDU

Self policing has been a traditional aspect of ham radio, but it faces some unique challenges with the introduction of new technologies. Both spread spectrum and packet are examples of nontraditional amateur communications modes which are not "decodable" on a conventional receiver system.

The STA group at AMRAD has received a number of comments about self policing of spread spectrum and feels that it is a valid point. The ability of a third party to monitor a spread spectrum transmission rests on knowledge of certain signal characteristics. If there is no knowledge of these signal characteristics available, then the job of trying to receive the spread spectrum signal is quite complex and requires both special training and special equipment.

To consider this point, a subgroup of interested parties within the SSSIG was set up to deal with this aspect of SS modulation. The purpose of the group is to devise a cost effective way that a "general purpose" spread spectrum receiver could be developed for the vhf amateur bands.

So far the group has been considering simple frequency hopping in the 220-MHz band as a model. The technique for receiving the frequency hopped signal without knowledge of the pseudorandom (PN) hopping sequence rules out anything like a "smart scanner" receiver as estimation of the PN sequence is quite difficult and usually requires a computer process for reconstructing the PN generator equation from some captured samples.

A second approach which seems more promising is to avoid the PN estimation problem by going to a wideband compressive receiver. There are two aspects to this technique. First is the fact that a frequency hopper will always be in a given band (for amateur applications). If no other signals are present in the band but just the frequency hopper, then by "compressing" the band down so that the compressed version is about the bandwidth of conventional fm will "focus" the frequency hopper signal right around a single frequency.

A compressive receiver is usually implemented as a very very fast scanning receiver. The scanning starts at the low end of the band and tunes upward toward the top of the band. This takes place over and over again at a very fast rate. The output from this receiver is at the i-f frequency and has a resultant bandwidth of the receiver i-f.

If only a frequency hopping signal is present then the scanning will always encounter it at some place within the bandpass. This will then be translated to the receiver i-f by standard receiver action. Because in this example the only signal present is the frequency hopper, it will be continually present at the receiver i-f.

Notice that this approach is useful for negating the effects of the PN hopping sequence. The PN code selects a new random discrete frequency for each hop which the system makes. By receiving every frequency within the range of possible hopping, the compressive receiver doesn't need to know the PN code.

The second step that is required is to throw away all signals which do not act like frequency hopping signals. This can be done by designing a window function which tosses out all signals which are present for longer than the dwell time of the frequency hopper and all signals which are significantly shorter than the dwell time. Both these signals will be things other than the frequency hopper signal.

This filter was dubbed a "time window," and at present the SIG subgroup is looking for an analog implementation. This time window filter is used to isolate the frequency hopping signal from interference before it is compressed by the compressive receiver.

The time window filter has a simple computer implementation and is currently part of a computer signal processing application. The trick is to get a low-cost purely analog approach to this.

AO FOR THE FUTURE

A major advance just over the horizon should make treating spread spectrum much easier. This is the field of acousto-optics (AO). This technology combines lasers, ultrasonics and microelectronics to produce some truly astounding devices.

AO techniques are based on the fact that when laser light passes through a lens system, the resulting diffractions can be described by transform mathematics. It turns out that transform mathematics is also used to describe what electronics does to a signal.

This happy fact allows us to modulate a laser beam and do very complicated processing by passing the beam through appropriate lenses. If a picture is placed in between the laser beam and the lens, the laser beam will take on the characteristics of the picture. This allows the picture to be processed by the lens system.

When a picture is processed by a computer (as NASA does) it is divided into pixels or picture elements. Many pictures have up to a million pixels in them. This represents a very large processing load for a computer, especially if there is a backlog of pictures that must be processed. Pictures are usually treated with complex processes as well which further increases the load.

The laser and lens system can do the same processing, but the lens does it with all the pixels at the same time! This means that results are obtainable almost at the speed of light.

The acoustic connection comes from an effect known in physics as the Bragg effect. Here if a lens is put under a varying pressure, the laser beam will be refracted such that the beam will break up into sub-beams with one for each spectral component of the pressure function.

An ultrasonic transducer (really a kind of speaker) can apply force by sound waves. If a transducer is attached to the lens and then a modulated voltage is applied, the lens will be subject to pressure at the modulating frequency. The result is that the laser light will break up and show the harmonic content of the modulating voltage.

What we have constructed here is a very versatile spectrum analyzer. Input to the spectrum analyzer is to the lens transducer. The output is taken from a light-sensitive array on which the laser beam shines. Each cell of the light-sensitive array represents a different frequency.

This device is called a Bragg Cell and has been implemented in one case to be a spectrum analyzer/receiver which received the entire band from 0 to 500 MHz at the same time! The resolution of the device is around 10 kHz. The output was put on a scope which displayed both the component and its amplitude, much as a conventional spectrum analyzer does.

Spread spectrum techniques will be greatly affected by this revolution in laser technology which seems geared for the present to meet the challenge of frequency-agile and low-probability-of-intercept (LPI) systems which are now in practice.

EXPERIMENT #5

Paul Rinaldo, W4RI, has handed the FCC an amendment to the STA. This experiment is for Chuck Phillips' mobile frequency hopping system in the two-meter band. The experiment calls for tests between different places in the Washington, DC and Annapolis, MD areas, including some ship-shore and ship-ship communications.

The frequency hoppers run up to 80 hops per second on a preset bank of channels which is variable in number but fixed for each test. Normal power for most tests will be 25 watts, but tests will include some long-range tests using up to 125 watts.

Those stations included in the STA amendment request are N4EZV, W4RI and WB3KDU.

From preliminary dummy load testing, very promising results are anticipated from the proposed on-the-air tests.

September 1982

CIVILIAN USES OF SPREAD SPECTRUM

By Hal Feinstein, WB3KDU

Spread spectrum is currently enjoying a wave of popularity mainly within military circles. It seems that almost every major system that is being acquired today has some spread-spectrum element in it. You can find spread-spectrum modulation as popular as ssb in some cases. The military, of course, was not the one who first brought spread spectrum to the civilian domain. To this end, we must credit NASA, who employed this mode for ranging of spacecraft.

Currently, civilian applications of spread spectrum are mainly involved with ranging, as in the Del Norte position-fixing system, and in aerospace. An interesting aerospace application is the use of process gain in a spread-spectrum system to make up for gain lost by using smaller earth-station antennas. Because one of the main costs in producing ground terminals for satellite communication is the antenna, any reduction in size also means a substantial reduction in cost of the earth station. But a smaller antenna size also means a drop off in gain. This gain must be made up in some other way.

A ground station can use process gain to offset the gain lost in reduction in antenna size. Such a system is currently operational on the West Coast, being employed to distribute teletex-style news copy. An added advantage to the operator is that the spread-spectrum transmission doesn't prevent the satellite transponder from being used by other narrowband systems as well. This is a plus for the low spectral density of spread spectrum.

There are some proposed uses of spread spectrum which have not hit the street quite yet. One interesting application, which is being proposed by Hewlett-Packard, is the use of a small direct-sequence transceiver of a few milliwatts to allow internal communications from a central computer to data terminals within the same facility. The terminals each have unique codes which act much as a privacy address. Such terminals would be connected only to a source of power and need no hard connection to a local-area network.

The Hewlett-Packard approach has its initial design aimed at the uhf ISM bands for the low-powered spread-spectrum signals. Another use of spread spectrum along similar lines but within a coaxial cable is to hook various robot sensors and "limbs" to a local-area network. Each sensor or limb would have a unique code. The approach is designed not so much to take account of spread spectrum's unique sharing capability but to provide a high degree of noise immunity. This invention is aimed at factory automation in which very high levels of shot and burst noise from the shop floor are the rule.

Spread spectrum has found its way into packet radio. Spread spectrum allows each node to have a unique code which acts as a hard address. Another node in the system can send data to that node by encoding that data with the spread spectrum address for the receiving node. Traffic for other nodes does not interfere because it would have a different code. Among the reasons cited for employing spread spectrum for packet switching are privacy, selective addressing, multi-path protection and band sharing. But it is interesting to note that a load is taken off the contention collision approach because now a single frequency is not in contention among the nodes wishing to transmit. The load is divided among different node addresses, and each that is interested in sending data to a target node competes for that node only.

Here is one final approach that you might enjoy. One cable operator has decided to try to use spread spectrum to transmit top-ten rock music albums to subscribers. To prevent pirating and to keep the cable band space free for narrow-band signals, he is using spread-spectrum direct sequence with a key to decoding the signal distributed every two weeks in the mail!

October 1982

EXPLODING A MYTH

Can frequency-hopped (FH) spread spectrum be received without knowing the code? The answer is almost always "yes." If you ask if FH can be received on a conventional ham receiver, the answer is almost always "no." The difference lies in the fact that FH is constantly changing frequency. It does so at a rate of many times a second. But something that doesn't change is the position of the spread-spectrum station.

The key to detecting FH is to use a wideband receiver which will encompass the hopping bandwidth and then point a very directional antenna at the FH station. Antennas have become very complicated in this day and age. For example, some are electronically steerable, and the reception pattern is electronically shaped to null out jamming and interference. For ham radio applications, a good yagi array will do nicely in most cases until such time that a band gets crowded with spread-spectrum signals.

The military has known all about the wideband-receiver/sharp-beam antenna trick and has gone to some trouble to see that their messages are protected. One system has several different FH transmitters set up at one location and rotates the data between each transmitter. When a transmitter doesn't have data, it is transmitting stuff that looks like valid data. Attempts to separate the individual signals will be difficult because the signals are coming from the same place.

SHIFT-REGISTER UPDATE

Every shift register has an equation which describes its actions. It turns out that if you are given the shift register equation you can always find the sequence that it generates by long division. To find the sequence, you merely need to find the reciprocal of the shift-register equation.

Suppose that you want to find the shift-register sequence for a shift register with the equation $1 + X + X^2$ (in the last term, ^ 2 is used to indicate the second power). Notice that if this is a maximal-length generator, we should expect the sequence to repeat after $2^3 - 1$ or 7 terms. Let's now take the reciprocal.

$$\begin{array}{r}
1 + x + x^3 + x^4 + x^6 + {}^7 \\
1 + x + x^2 \overline{\smash{\big)}\, 1} \\
\underline{1 + x + x^2} \\
x + x^2 \\
\underline{x + x^2 + x^3} \\
x^3 \\
\underline{x^3 + x^4 + x^5} \\
x^4 + x^5 \\
\underline{x^4 + x^5 + x^6} \\
x^6 \\
\underline{x^6 + x^7 + x^8} \\
x^7 + x^8 + x^9 \\
\underline{x^7 + x^8 + x^9} \\
0
\end{array}$$

The answer to this long division is:

$1 + x + (0*x^2) + x^3 + x^4 + (0*x^5) + {}^6 + x^7$

This means that the shift-register output is:

11011011

Notice that we are doing division a little different from the normal method. The main fact is that subtraction and division are the same. Why? Because we are doing all arithmetic modulo 2. In this system subtraction produces the same result as addition.

1 + 1 = 0
1 + 0 = 1   Modulo 2 addition is the same as
0 + 1 = 1   exclusive OR
0 + 0 = 0

August/September 1983

SPREAD SPECTRUM NEWS

By Hal Feinstein, WB3KDU

A lot has been happening in amateur spread spectrum during these last few hot summer months. Much of the activity comes from a few projects that are in the completion stages and a number of new projects just getting under way. I'll try to give a summary of some of these activities and some details on some of the experiments.

Chuck Phillips, N4EZV, has unveiled another one of his completely homebrew vhf frequency hoppers which I had a chance to examine in some detail. The units are about the size of a cigar box and run on 12 volts. These units hop in a small preset band in the 140 to 170 Mhz range and have a dwell time preset to 100 milliseconds. The hopping sequence is also preset in an onboard EPROM which is scanned out under the control of a pseudo-random number generator. The use of an EPROM gives the user the option of selecting which channels the units will visit and allows specific channels to be bypassed.

The front panel has few controls, which I found extremely easy to operate. Besides the volume, squelch and power, there were two additional controls which are used only in spread spectrum mode. These controls are used to set the frequency hopping control mode and the "key" value for the pseudo-random number generator.

The hopping sequence is controlled by three thumbwheel switches which allow the user to enter a seed value. This seed value is used to initialize the pseudo-random number generator and determines the order in which the channels will be visited (the hopping sequence).

The mode switch is used to select the technique by which the frequency hoppers will determine the starting channel for each transmission. There are three positions which can be selected. The first position of the mode switch cuts off the hopping circuits and allows the units to be used as conventional NBFM transceivers.

The second position of the mode switch selects frequency hopping with a single home channel. In this mode, each time a unit begins to transmit, it starts on a single, common home channel. The unit uses the home channel to send an encoded signal which alerts the other frequency hopping units that a frequency hopping transmission has started.

After the alert signal has been sent, the unit begins to hop through the frequency hopping sequence setup by the thumbwheel switches. The home channel is selected from the first hop of this sequence. Each time a frequency hopping unit transmits, the same procedure is used to synchronize the transmitter and receiver.

When a unit is not transmitting, it automatically resets itself to the home channel and begins to monitor it for the coded alert signal. This mode of synchronization always uses the same hop sequence for each transmission. In the mobile environment the average time for a mobile transmission is short and the same set of channels will be revisited for each transmission.

To remedy this situation, an additional mode is included in the frequency hoppers and is selected by the third mode switch position. This position is called "run" mode and differs from the second mode by using an always moving home channel.

The frequency hopping circuits run continuously when the unit is switched to run mode. In this mode, when the unit begins to transmit it chooses the next hop as the home channel and transmits the alert signal. The receiving unit's frequency hopping circuits are also continuously running; this has the effect of changing the receiving channel in step with the transmitter. Together the transmitter and receiver share a common channel, which allows the receiver to hear the transmitter's coded alert signal.

A built-in feature of the run mode is the use of a "look aside" channel which is automatically monitored by the frequency hopper. This channel is used when one of the two units falls out of synchronization due to some unusual conditions such as a power failure. The look aside channel is used to manually transmit a special encoded signal to the other frequency hopping unit which is still in run mode. When the unit detects this signal, they automatically fall back to the initial channel in the hopping sequence, which allows the units to resynchronize.

When a frequency hopper is done with a transmission it automatically signals this fact to the other frequency hopper unit by transmitting a special encoded signal on the last hop for which the microphone is depressed. The other unit detects this "end of transmission" signal and resets the hopping sequence to the home channel for the mode two transmissions. For the run mode of transmission, the end of transmission signal is not as important since both frequency hopping units are continuously hopping through the hopping sequence and do not reset to any home channel.

Networking with several frequency hopping transceivers can be done with Chuck's units. A net of frequency hoppers is established by three factors; first, the internal channel EPROMS must be set up to contain the same sequence of channels, which is done at "installation" time according to Chuck. Second, the control mode switches must be all set to the same control mode. Lastly, the thumbwheel switches must all be set to the same seed value.

When these three conditions have been correctly set, a number of units can hop in a net. Nets can use the same home channel and still remain in their separate nets if at least one of the three above mentioned factors are different.

Chuck's frequency hopping radios have attracted some commercial interest at this writing, mainly in the law enforcement sector. During the next STA period, we will be testing these radios to see how they fare in the amateur radio service. Anyone interested in more information on Chuck's radios, please contact him through AMRAD via AMRAD's address.

A NEW FREQUENCY HOPPING EXPERIMENT

What can you make with an ICOM-3A and a PET computer? Answer: A software controllable frequency hopper. This is exactly what AMRAD members Terry Fox, WB4JFI, Dave Borden, K8MMO, and Sandy, WB5MMB, did one afternoon. The bug bit them after the informal Saturday Pizza Hut get together more as a dare than anything. Well, schematics were examined, ICOM 3AT and 2AT were disassembled and reassembled and new cards were popped into the computer. By evening, an ICOM 2AT sat on Terry's workbench with all sorts of wires connecting it to a Z-80 computer in an old IMSAI 8080 frame. The first experiment was a success. The unit frequency hopped up to 50 times a second! Not bad for an afternoon of tinkering.

SECOND STA PROGRAM FORMULATED

A series of experiments has been devised for the second spread spectrum STA. The experiments will include three experiments: first, a proposal for an experimenter's frequency hopping beacon which will operate on 220 MHz, second, frequency hopping experiments in the 144 and 220 MHz bands and lastly a direct sequence experiment.

Each of these experiments is designed to explore some new aspect of spread spectrum in the amateur service. In the first STA, the experiments allowed us to gain the important initial experience with spread spectrum. Most of this activity was centered around frequency hopping experiments. In the second STA we are focusing on experimental circuits and procedures.

Dave Borden, K8MMO has pointed out the need for standardizing the acquisition and handshaking between spread spectrum units much as has been done with protocols for data communications. Spread spectrum requires accurate operations at high speeds between different self synchronized units. The issues are very similar to data communications; spread spectrum can lend itself to a protocol treatment. Hopefully we will be able to look forward to an experimenter's spread spectrum protocol.

Sandy, WB5MMB has been speculating on a very fast lock up, phase locked loop circuit for frequency hopping, which can make this part of the frequency hopping problem less of a task. Hal, WB3KDU has been experimenting with digitally controlled oscillators.

Computers are also making their appearance in our experimentation which will give us the flexibility to change parameters and protocol around in a much faster time than wiring a board.

In the second STA we would like to apply these ideas to actual on the air experimentation.

DROP-IN CHANNEL ASSIGNMENTS

A concept that is making its way around the frequency management circuit these days is the idea of drop-in channel assignments. Drop-in assignments are based on the notion that guard-band calculation between conventional channels are based on the capabilities of old equipment and techniques. Consider the guard-band situation between repeater channels. They are primarily based on the premise that NBFM receivers could not separate signals that are too close together.

What we are left with today are receivers that can separate those signals without getting even a little out of breath. This leaves spectral room between channels which is not wide enough for a conventional NBFM signal. However it may be wide enough for some of the new modulations techniques such as amplitude compandered sideband (ACSB) which is being tested currently at several commercial sites around the US and by the FCC.

Drop-in channel assignments are also being considered for frequency hopping spread spectrum. A frequency hopper can be designed to only visit frequencies which are in the guard-bands. The short time that the signal is actually present on any frequency is very small (1/30 second or less) which will appear as some form of low intensity shot noise. This should be ignored by the repeater and mobile units in an amateur NBFM operation. Even with a close in near-far effect, the actual experienced signal energy from a "drop-in" frequency hopper should be slight.

One advantage which a drop-in frequency hopper may have over a system such as ACSB is the fact that many commercial NBFM repeaters have a form of automatic frequency control (AFC) which will pull the receiver off the repeater center channel. This feature allows the repeater to hunt off frequency for a mobile which is accidentally off channel and requires alignment. A constant carrier signal on a drop-in frequency such as ACSB may trick the commercial repeater into thinking that it is hearing an off frequency mobile unit. A frequency hopper would appear to the repeater as a brief noise pulse, and the repeater would ignore the hopper.

August/September 1983

FREQUENCY CONTROL OF ICOM PORTABLE RADIOS

By Terry Fox, WB4JFI

Last year, AMRAD asked for and was granted an STA from the FCC to allow us to conduct experiments using spread spectrum in the amateur bands. Although I was listed on the STA as one of the participants, I was already heavily involved with packet radio and didn't have a lot of time to devote to this new endeavor. In addition, it seemed like spread spectrum might end up costing a lot of money to really do it right. Most of our emphasis was placed on HF, and I didn't want to buy an expensive HF rig and then immediately have to butcher it up for this new mode of communications. I ended up listening on my old NCX-5 to Paul doing some tests on 20 meters using a commercial spread spectrum radio.

This year may be different. Some of the AMRAD crowd gets together every Saturday for pizza at the local pizza place in Vienna. One Saturday we were discussing the various problems with spread spectrum as applied to amateur radio, and whether or not we should continue our experiments in this area. The biggest problem we had was that of finding a cheap radio that was easily available that could serve as a test bed for our experiments. By that time we had sort of decided to limit our efforts to a frequency hopping system, most likely on VHF and UHF. After talking about this during almost the whole pizza session through a sea of antennas (stubby duckies on our ICOM IC-2ATs, of which there were at least half a dozen), my subconscious finally realized that my conscious self wasn't going to figure out the obvious. It then started sending smoke signals about using these marvelous little radios in front of us as the test bed. The IC-2AT has become almost a standard in AMRAD, and since there are so many out there, it seems almost natural to make it the radio we use. After discussing this idea for a while, we left the pizza place and went to my house for an in-depth design conference.

The first order of business was for me to get out a screwdriver and demolish my IC-2AT, while Sandy, Hal and Dave looked at the schematics. After popping off the back of the radio, I quickly located the wires going from the thumbwheel switches to the programmable divider chip. Actually the connections are made with a flexible printed circuit board and not individual wires. This flexible board attaches through two sets of pins to the rear main board, with the pins straddling the programmable divider chip (see Fig. 4 for a diagram of the pinout of this connection). A quick look at the schematics revealed that the thumbwheel switches on the radio asserted a high on the appropriate line rather than grounding the lines. If the internal thumbwheel switches are set to the zero position, they are effectively removed from the circuit, and an external device could control the frequency by bringing the same lines high. This was confirmed by taking an external set of thumbwheels and wiring them into the radio, which worked fine.

The next step was to wire the IC-2AT to a computer. Having a spare parallel port on one of my S-100 boards, I decided to go that way. This turned out to take the most time of the whole project. Since I didn't have a manual for the board in question, I had to guess the port locations for the parallel port. After a lot of lost time finding the parallel port and its control port, we were ready for computer control of the ICOM. Using a BASIC interpreter, I sent various frequency commands to the IC-2, and checked the result with another IC-2 (since there were several in the room to use, as long as modifications weren't required). I was able to program the IC-2AT hooked to the computer to any frequency within its range in 10 kHz steps. Now the fun really began!

Being careful to hook the IC-2AT to a dummy load/power meter, I started writing programs to test the synthesizer by moving the frequency of the radio at certain rates. Due to the way the synthesizer is designed, there is no easy way of telling if it is out of lock or not, and once it is out of lock, the only way to get it to re-lock is to drop out of transmit. It does have an unlock pin, but this pin seems to be a direct output of the digital phase comparator, which means it has a lot of pulses whose width depends on the lock condition of the synthesizer. ICOM takes these pulses and integrates them over a long period of time (greater than a second). Obviously, this out-of-lock indicator is too slow for any real application, and even for our experimentation, so we came up with another way to test for lock, unscientific as it was.

We asked for a volunteer to donate the use of a second IC-2AT, which Sandy finally agreed to. We then wired his rig to the same port that my radio was hooked to. We now had two radios hopping to the same frequencies at the same time (I know a lot of people that wish the hopping synchronization problem could be cured this easily!). We then hooked a scope to the audio output of the second radio, allowing us to see what was coming out. As the radios were moved from one frequency to another, there was generally a short burst of noise followed by a sinusoidal (or ringing) pattern with decreasing noise riding on it (see Fig. 5). As the synthesizers locked, the ringing waveform decreased in level, until it disappeared completely, leaving a steady trace until the next time the frequency was changed. Another interesting phenomenon regarding this lock-up period was noticed. Almost as soon as the noise disappeared from the ringing, audio transmitted by one radio was recovered by the other, at least according to the scope. The obvious question at that point was whether or not the audio could be used once the noise was gone, but while the synthesizer was still locking, as indicated by the ringing waveform. My opinion is that it probably is usable, but only if the same type radios are being used at both ends. If the same type radios aren't being used, the synthesizer's time constants aren't going to be the same, so the frequencies the radios are operating on will not necessarily match during the locking period.

The time the synthesizer was stabilizing (as shown by the audio waveform discussed above) varied, depending on the distance between the old and new frequencies. Fig. 6 shows the rough times measured for the synthesizer to lock up, both for the noise period and the total lock-up (as indicated by the audio signal totally stabilizing). As can be seen, a 1 MHz hop can cause the radio to be out of lock for a relatively long period of time. The maximum dwell-time on any frequency

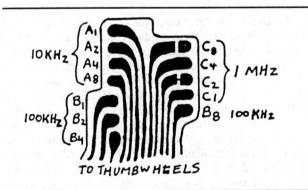

Fig. 4—Divider pinout

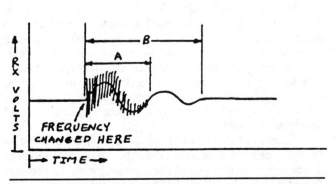

Fig. 5—Audio out during unlock.

Frequency Hop	"A" Period	"B" Period
1 MegaHertz	60 msec.	110 msec.
100 kiloHertz	40 msec.	80 msec.
10 kiloHertz	25 msec.	40 msec.

Fig. 6—Out-of-lock periods for IC2AT.

that most people consider effective is 100 milliseconds. Obviously, the ICOM radios cannot be expected to hop a megahertz within this dwell-time spec. and convey much intelligence. Therefore, the ICOMs cannot be used to hop over a megahertz without some help.

One of the first ideas we came up with when we saw what the audio output looked like during hopping was that if the audio was suppressed during the noise bursts (the most objectionable part of the hopping signal), the sudden noise bursts would be gone, and the human ear and mind would integrate out the blanked-out areas to a large extent. Since the human voice is very redundant, small losses in voice transmissions could be tolerated. Obviously, any mode of transmission that requires fully accurate reproduction (such as packet radio or slow scan television) could not use this system. AMRAD is working on several ideas to help reduce, or eliminate this lock-up period, both for the IC-2AT project, and for use in just about any other type of frequency hopping radio. We will report more on these ideas as they develop.

Since this first test seemed to work, we put both radios back together, and declared the work of the day a success. The next step was to find a couple ICOM HTs that we could leave hooked to a computer for an extended period of time. Dave graciously volunteered his pair of IC-3ATs. Sandy also volunteered the use of his spectrum analyzer. About a week after the initial test described above, the new equipment showed up at my house. Later that day, Hal and I purchased a couple of Commodore 64 computers, primarily to use in these frequency hopping tests. For the money, these computers seem to be a good buy, and plenty of documentation is available for them.

I then hooked one of the IC-3ATs to the parallel port of the 64, and let Hal write various BASIC programs to stroke the IC-3AT, while it was hooked to the spectrum analyzer. He came up with several iterations of random number generators to move the IC-3s frequency. The main thing to realize is that each digit of the frequency is encoded in BCD (0 to 9), while most random number generators can generate numbers over the whole binary range (0 to F hex). As a result, there are holes in the frequency control (when A thru F hex are sent to the radio) that can cause uncertain frequencies to be programmed in. The programmer should be sure that improper frequencies aren't accidentally programmed into the HT, as this could result in the synthesizer going out of lock, causing the transmitter to stop transmitting, which requires the transmitter to be turned off.

This is as far as we have gotten so far. We are looking for people to help us with the ICOM hoppers, and other spread spectrum experiments. There are many different areas of experimentation in the spread spectrum tests we are running. Just a few of these areas include the design of fast locking PLLs, special RF circuitry, computer programming, and audio processing. Anyone interested in helping out, please contact Hal, WB3KDU, or write AMRAD.

May/June 1984

EXPERIMENT UPDATE

By Hal Feinstein, WB3KDU

Recently, I was asked by a friendly official at the FCC about our experiments. He said, "You hams are just interested in making a secure (meaning secret) communications systems, aren't ya?" Well, I told him that of course us hams aren't interested in secret communications but in what new and interesting properties some of these techniques can provide. In particular, spread spectrum has some properties that allow us to shape the spectral profile of a radio signal, so that it can be made to fit any kind of space available on the spectrum. When I told that to the FCC man, he was perfectly satisfied with the answer and allowed us to go ahead with the experiments we are planning.

A few weeks after this conversation, the FCC issued us a special temporary authorization (STA) and a letter from their field engineering bureau (FOB) say that we could experiment with spread spectrum in the amateur bands, but we have to ring up their monitoring station located in Washington, D.C. before we begin.

Other than this desire to coordinate our experiments, no other major restrictions were levied by the Commission. So we are off to the lab to start experiments.

The first experiment being built as I write this is a channel selection logic board called the Fox Randomizing Appliance (FRA), after its designer. The board consists of a 2732 PROM which holds the channel visitation sequence and a bit of control logic to pull out sequential 8 bit entries from the PROM. The clocking is done with a 1 MHz xtal and a divide down circuit arranged so that we can select clocking speeds by DIP switches on the board. This is version one of the FRA, it will give us a non-repeating sequence of channels (in BCD) 4096 channels long. Programming the 2732 is the heart of the matter and must be done so that each 2732 location contains a randomly selected BCD number in the range of zero to one hundred. The BCD equivalent of the number is programmed into the PROM; otherwise, binary numbers will be supplied to the synthesizer channel lines which expect BCD. This results in invoking the don't care status of the synthesizer which in turn can generate unpredictable channels, perhaps outside the amateur bands.

The 2732 contains 4096 8-bit storage locations which can have a value between 0 and 255. In BCD, this equals about 10 bits of BCD or two and one fifth digits. The inventor of the FRA, Terry Fox, WB4JFI, suggests programming the binary to BCD converter on the output, giving a "pure" 2.2 digits of BCD to feed to the synthesizer. This seems like a good plan.

The Fox Randomizing Appliance will be used to drive the channels of a TEMPO-1 VHF handheld radio for a power profile test to be conducted at Dave's (K8MMO) house in a week or two. Simply, we are trying to verify the enormous balderdash that spread spectrum is both invisible and an unbearable noise source at the same time. The plan calls for putting the TEMPO-1 and the FRA in a car and driving around the area of Dave's house. Others will be watching on spectrum analyzers and listening on scanners and receivers at various locations in the surrounding area to note any interference or noise. We will coordinate this via the AMRAD repeater when it comes off.

Experiment number two is a two meter frequency hopping beacon which we are planning to put up at the Electronic Equipment Bank in Vienna, Va. EEB has donated the TEMPO-1 and some space to this project and Scott, WR4S, who works at EEB selling ham gear is going to be the beacon tender. The equipment for the beacon is currently being assembled but will include one C-64 Commodore computer with disk for control of the beacon, one Fox Randomizing Appliance Version 3 (programmable speed, fixed channel selection), the TEMPO-1 vhf handheld (2 meters) and a twenty five watt amplifier.

And now for the protocol section of this column. Since spread spectrum communications calls for highly coordinated and split second accuracy, some kind of protocol is needed to allow the two or more stations to communicate with each other. For the purpose of the beacon, we have come up with a protocol which we call P1. P1 says the following: transmissions start with an unmodulated carrier, then a CW ID of the beacon for the FCC, followed by a program indicator. The program indicator tells which version of the protocol is running and which subsequence of the protocol is currently active. The program indicator will be four digits sent in morse. Then a timing and synchronization sequence which consists of a train of transitions of an AFSK tone. The timing and synchronization train (TST) has two purposes, first it allows listeners to synchronize their clocks with the one used by the beacon, second, it indicates, by the falling edge of the AFSK tone, the actual moment that the hopping sequence starts. The hopping sequence will be published in future issues of this column for those people wanting to experiment and will be available on the AMRAD bulletin board for downline loading to your computer.

FROGS, OR A PRIMER ON HOPPERS

By Chuck Phillips, N4EZV

Frequency Hop Signals

The advantages of using frequency hop signals instead of conventional narrowband signals or other wideband signals include the following:
1. Processing gain.
2. Jamming resistance.
3. Traffic privacy.
4. Low probability of intercept.
5. Multiple access capability.
6. Short synchronization time.
7. Multipath rejection, and
8. Near-far performance.

These features are described in the following paragraphs.

Processing Gain

Frequency hop systems generally possess a large processing gain which allows the systems to operate with a low signal-to-noise ratio at the input of the receiver. The processing gain for frequency hop signals is

$$\text{Processing Gain} = \frac{\text{RF Bandwidth}}{\text{Information Bandwidth}}$$

For example, a frequency hop signal that has a 10 MHz RF bandwidth and an information bandwidth of 1 kHz has a processing gain of 40 dB.

$$\text{Processing Gain} = \frac{10 \text{ MHz}}{1,000} = 10,000$$

$$= 10 \text{ LOG } 10,000 = 40 \text{ dB}$$

Thus, the output signal to noise ratio of the frequency hopping demodulator will be 40 dB higher than the receiver input signal to noise ratio. This assumes no loss in the demodulator.

Jamming Resistance

Since a frequency hop signal generally has numerous frequency slots, the only time a narrow band jammer affects signal reception is when the signal hops to a frequency slot that is occupied by the jammer. If the frequency hopper has 500 frequency slots and a narrow band jammer interferes with the signal reception from one of the 500 slots, then only 1/500th of the signal might be jammed. Therefore, the low average power density combined with the pseudorandom frequency hopping make these signals difficult to intercept.

Multiple Access Capability

Frequency hop systems can be used for multiple access systems; time division, frequency division and code division multiple access systems can all employ frequency hop signals.

Frequency division multiple access systems assign a frequency band for each user. While most frequency division multiple access systems employ narrowband signals, wideband signals could also be used.

Short Synchronization Time

Frequency hop systems generally require a significantly shorter time to acquire synchronization than other types of systems having the same bandwidth. In frequency hop systems the receiver can usually synchronize with the transmitted signal within a small fraction of a second. Direct sequence systems, for instance, require about a second to achieve synchronization. For some applications like voice communications, the shorter acquisition time is highly desirable. If one or two seconds is required for synchronization, the transmitter has to be keyed at least two seconds before the voice will be received at the receiver. Therefore, a voice reply would be delayed by at least two seconds. By using frequency hop signals, the receiver can synchronize within a fraction of a second and no noticeable delay is encountered for voice.

Multi-Path Rejection

When the transmitted signal is propagated towards the receiver, several paths may exist which may cause interference due to phase cancellation at the receiver. This is called multipath propagation. If the signal is propagated via the ionosphere, the path delays can range from tens of microseconds to several milliseconds. Similar multipath delays can exist at VHF and UHF frequencies due to reflections from buildings, towers and other reflective materials.

If the hopping rate is adequately high, then the receiver

listens on a new frequency slot before the interfering paths have a chance to interfere with the direct path. For slow frequency hoppers, the path propagation times are too fast to allow the receiver to reject the interference. Thus, to be effective, the hopping rate must be kept higher than the inverse of any interfering path delay time.

Frequency Diversity

Frequency hopping systems provide frequency diversity since many frequencies are used in the system. If proper data coding is used, a severe fade at any one particular frequency will have little effect on the data transmission. At HF frequencies, signals fade independently of one another if the frequency separation is several kilohertz. For frequency hop systems to provide frequency diversity, the minimum separation between frequency slots should be greater than several kilohertz. Typically, the separation used is much larger than a few kilohertz, ranging from 20 kHz minimum separation for voice and data communications. Thus, frequency diversity is easily provided for.

Near-Far Performance

Frequency hop systems provide better near-far performance than direct sequence systems. Near-far performance describes the behavior of the spread spectrum system with other users both near and far away from the intended users.

Traffic Privacy

Frequency hop systems provide a great degree of traffic privacy. The low probability of intercept combined with pseudorandom frequency hopping make these signals difficult to demodulate for unintended receivers. If additional security is desired the intelligence may be further encrypted using additional techniques.[1]

Low Probability of Intercept

Frequency hop signals have a low average power density which can make these signals difficult to intercept. While the instantaneous power level of the frequency that is transmitted is high, the average power of that frequency is equal to the instantaneous power divided by the number of frequency slots. For example, a frequency hop system with 500 frequency slots transmits a specific frequency only a small fraction of the time (1/500th). If the instantaneous power from the transmitter is 100 watts, the average power in any one frequency slot is:

$$\text{Average power per frequency slot} = \frac{100 \text{ watts}}{500 \text{ slots}} = 0.2 \text{ watts}$$

A frequency hop system jumps to many different carrier frequencies in a time interval and filters the carrier frequency with the intermediate frequency filter. Users outside the filter's bandwidth are rejected and only the proper signal is demodulated. Since the I.F. filter passes only a narrow bandwidth, potential interferers are more easily rejected.

If the other users of the frequency band are near the frequency hop receiver and a frequency hop transmitter is far away, a frequency hop system can more easily reject the nearby interference than the direct sequence system can.

[1] Encryption is not allowed under Part 97 of the FCC Rules.

November/Deccember 1984

DIRECT SEQUENCE RULES

By Hal Feinstein, WB3KDU

In the last issue, amateur radio regulation for frequency hopping was discussed. In this issue, we continue the discussion with what the current thinking is for direct sequence.

At this time, the FCC is formulating rules for the amateur service to allow it to do spread spectrum communications. Making rules for spread spectrum is in itself an adventure, simply because so many properties of spread spectrum go against the well-worn traditional approaches of the rule makers. Specifically, there are two major areas that require completely new approaches. The first of these areas is a problem which occurs over and over again with spread spectrum—that of policing. As you no doubt remember from the last issue, the regulators are thinking of imposing a rather lengthy identification and transmission parameter sequence to be transmitted at standard intervals. This will allow the FCC to perform field monitoring and, perhaps more importantly, to allow hams to self police.

While the exact format of the identification and transmission parameter sequence will undoubtedly be modified before it turns up at a formal rule-making procedure, it forms a model for thinking about how stations using other forms of spread spectrum can identify themselves and provide transmission parameter information. Thus the current thinking is to include the above mentioned lengthy identification sequence each time you identify and, in addition, add parameters which are specific to the direct sequence type of spread spectrum.

The identification and transmission parameter sequence is long and it seems on the surface that it would be extremely inconvenient to listen to all these parameters dished out in "hand sent morse" which cranks along at twenty words per minute (per the FCC regulations) each time you ID. Therefore, the idea of automatic transmitter identification (ATI) was advanced as a possible substitute for hand sent morse. ATI will allow the amateur station to automatically transmit all this information in computer readable format (machine readable) and allow other hams and the FCC to receive all these parameters and the station ID.

At this time, however, nothing is settled on the issue of ATI but it certainly seems one workable way to give the FCC all the information they desire while not overloading a transmission.

A second issue which is pending is the issue of sequence selection and phase. There are currently three sequences which are slated to be authorized for spread spectrum amateur work. The sequence in use would be indicated by an indicator or token digit during the transmission identification sequence. This is merely a single digit; however, the phase of the sequence, which refers to the starting point within the sequence, must be indicated with the codeword with which the sequence generator register is originally loaded. The code word is in reality a binary string which is the length of the register itself. Transmission of this codeword could be a rather lengthy affair since in the case of the nineteen stage register the codeword is nineteen bits long.

Some sort of compressed sequence will be required for this codeword. Imagine trying to send a string of zeros and ones in morse! Again, the automatic transmitter identification technique can be considered here.

More importantly, the problem of determining the working sequence of an amateur station is a nonobvious process. Part of the problem comes from the process of modulating a direct sequence signal itself. The direct sequence signal is a fast keyed phase shift keyed (PSK) signal which is composed of a noiselike spreading sequence (called the working sequence) mixed with a digital stream containing the information to be sent. You must determine what the working sequence is and where in the working sequence the current intercepted signal is. To find the working sequence and displacement into the working sequence, a number of procedures are available. The first and simplest procedure is to turn off the information stream some small amount of time to allow for "pure" working sequence to be transmitted. The time at which this would occur would be standardized to take place at well defined times. For example, one time could be just after the start of transmission, a second time might be every minute. The length of the "working sequence window" need only be as long as it takes to determine which sequence is being used and what displacement is currently appearing. The number of bits needed for this is found from a well known result in the mathematical theory of shift register sequences to be twice the number of bits in a codeword of the sequence.

A second, and perhaps less desirable, way to determine the working sequence and displacement is to perform varying amounts of mathematical analysis on the received signal and derive the sequence and displacement as products from this analysis. While the process holds theoretical interest to the mathematically inclined amateur, it requires an amount of sophistication which is not found among the typical ham radio operator. In addition, indications are that the FCC may favor the simpler approach of inhibiting the information stream, yielding the "pure" working sequence and displacement in a straightforward manner.

What is significant about the sequence identification procedure is that it had been a sticking point in previous attempts to create a regulatory framework for amateur spread spectrum. The fact that three FCC defined sequences are now "accepted" as part of this framework and the proposed use of working sequence windows provides us with a possible solution to the enforcement problem.

The question of how to monitor a transmission which is using a nonauthorized working sequence (rogue sequence) is not difficult. It is treated as a transmission which requires standard FCC enforcement investigation which would involve direction finding and a site inspection if appropriate. The main point is that with the defined sequences and working sequence windows, a station employing a rogue sequence can be quickly identified from those employing authorized sequences. This allows enforcement and self policing by amateurs.

So far we have discussed two areas where good thinking allowed us to approach and solve two regulatory problems. The third area is more difficult and one which is still pending, both among the amateur radio community and at the Commission. This is the problem of interference from a direct sequence signal. It is a complicated topic and I will only touch on it in this issue of the newsletter. In the next issue, I will present in detail some of the problems and possible solutions which are currently being considered.

One of the most intriguing things about spread spectrum is the ability to operate the spread spectrum signal below the noise level. This amazing property comes from the fact that the signal is spread out over a wide bandwidth. The amount of energy present in a receiver's bandwidth is less than the amount of atmospheric noise being received and the amount of noise contributed by the receiver internally. For our analysis and thinking we have been using a field strength of 0.1 microvolt. If the spread spectrum signal presents less than 0.1 microvolt to a receiver (at the antenna terminal) we think of it as being less than the strength at which the receiver can separate it from the noise discussed above. Therefore, whenever we talk about a signal below the noise we mean a signal which is being observed at a 0.1 microvolt or less to a conventional, narrowband receiver which is amplitude sensitive.

The problem comes when we try to consider how much spreading bandwidth is available in each of the vhf-uhf amateur bands. For example, in the 1.25 meter band (220 MHz) we have 5 megahertz to spread over. Forgetting for a moment about the distribution of energy resulting from PN-PSK assume for the sake of argument it is flat across the band. Basically, the energy acquired by the receiver is the transmitted power divided by the spreading bandwidth (yielding power per hertz) divided by the receiver bandpass (3 kHz). Of course this must be converted to voltage per meter. Next the effect of distance must be figured into the equation. Since the power drops off as the inverse of the square of the distance, the energy at the receiver will, at some distance, go below our 0.1 microvolt level and this will make it fall below the receiver's noise floor.

Receivers with similar bandpass characteristics will not "hear" the spread spectrum signal and will not be interfered with. However, receivers closer in the 0.1 microvolt contour will hear the direct sequence signal and consequently will experience interference.

To add to the problems, suppose that the transmitting station used a beam antenna for the spread spectrum signal. The beam antenna has the effect of concentrating the spread spectrum transmitter's RF in a beamwidth, along which the signal strength will be considerably higher. Therefore, the 0.1 microvolt contour will extend far out in the direction of the beam.

A station using a beam for receiving will experience a similar problem since its narrow beam width will provide gain and will stretch the 0.1 microvolt contour out, perhaps now encompassing the receiving station as well. Note also that the height above terrain of both the spread spectrum station and the receiver (intended or otherwise) will affect the signal strength and hence, the level of interference from the DS signal.

In summation, it is unclear how to predict the interference level resulting from direct sequence spread spectrum on other than a case-by-case basis. The question of how to fit this into an amateur radio framework will be discussed in the next issue of the newsletter.

January 1985

AMRAD FREQUENCY HOPPING BEACONS

Two different FH beacons are operating in the Washington D.C., Northern Virginia area for experimental use. The first beacon is operated by Chuck Phillips, N4EZV from Falls Church, VA at a power of twenty five watts into an omni-directional antenna. The beacon operates with 100 ms dwell and hops randomly on channels 25 kHz spaced from 144.9 to 147.8 MHz. A preamble tone burst is transmitted at the start of each transmission sequence on 144.9 for synchronization purposes. The last hop also contains a tone burst to signal end of transmission. The hopping sequence is modulated with "VVV DE N4EZV" in morse using NBFM.

The beacon itself is based on a portable FH transceiver that Chuck designed for the law enforcement community. Chuck indicated that the beacon will be up for periods when his supply of the FH transceivers allows him to devote one to beacon use. He is willing to leave the beacon up longer if you are interested in trying to receive it.

The second beacon is operated by Scott, K4RS at the Electronic Equipment Bank Inc. in Vienna, Va. and was built by Dave Borden, K8MMO, Terry Fox, WB4JFI, Elton Sanders, WB5MMB, Bernie Stuecker, K4XY, and Hal Feinstein, WB3KDU. The beacon operates with an output power of 80 watts into an omnidirectional antenna.

The beacon operates as follows: on the five minute mark, the beacon comes up on the AMRAD repeater with an ID consisting of "K4RS FH BEACON V1.1" in 20 WPM morse. Following the morse is a five second preamble tone. The beacon begins hopping at that point, which is indicated by the fall of the tone in the repeater's input. The tone remains on, in fact, but since the beacon has started hopping, it no longer is on the 147.81 input channel for the repeater. The dwell time of the hopping depends on the sequence mode being used. Mode one is called the linear mode and consists of visiting every channel from 144.0 to 147.790 using a dwell of 1 millisecond. The two meter band is swept at the rate of once every four seconds and the beacon performs 50 sweeps in a single transmission. Mode two employs a pseudo-random sequence of channels which are fixed for each transmission. Each channel in the two meter band is visited just once but in random fashion. The dwell time for mode two is 100 ms. At this rate, interference to repeaters and other fixed users is possible so the channels which they use are removed from the channel table. In addition, for repeaters which are close to the Electronic Equipment Bank one channel either side of their input frequency has also been removed.

At the end of FH for either mode one or mode two, the beacon switches to 145.630 and transmits an information message in 20 WPM morse: "FOR AMRAD UNDER FCC STA V1.2 MODE 1 K." The purpose of this transmission is to allow you to receive the beacon in narrowband mode to measure signal strength, fix beams, indicate which version of the beacon is running and what the mode is. Remember, receiving the beacon is a simplex operation, without the aid of repeaters.

The beacon has some anti-interference features: it will wait for the repeater to drop before sending its ID on the AMRAD repeater. If 145.630 is busy, the beacon will skip over the morse message and wait for the next five minute mark to begin transmission.

The beacon is built from an ICOM-2AT and a Commodore 64 computer. The beacon program is written in BASIC with the ICOM driver and the morse code program written in machine language.

A number of people have started developing receivers for the beacon, most notably K8MMO using a XEROX 820 in "C" and an ICOM 2N, N4ICK is looking at the Commodore 64 and an ICOM as a receiver, Sandy, WB5MMB, is midway on building a 10ms dwell pseudo-random receiver from two GLB synthesizers. Sandy's receiver is based on the Motorola MC145151 CMOS LSI PLL synthesizer chip. A complete synthesizer based on this chip is offered by GLB Electronics for about $40. We will feature Sandy's project in a future issue of the AMRAD newsletter.

Another project which is currently being wired is something we call the EAR or extremely agile receiver. It is a combination of techniques which should permit us to receive FH signals well below the 1 ms dwell barrier we are now living with.

THUE-MOORE SEQUENCE EXPERIMENT

The Thue-Moore sequence (TMS) is a simple sequence that has many synchronization properties, is easy to generate and more importantly, it is easy to decode. Dr. John Hershey is a mathematician who specializes in the study of sequences to solve communication system problems. He has been studying the properties of this sequence for many years for use as a synchronizing sequence for spread spectrum.

Dr. Hershey was in touch with AMRAD recently concerning a proposed experiment to transmit and decode the TM sequence using personal computers. The hitch is that Dr. J. is located in Boulder, Colorado, and most (but not all) AMRAD facilities are in the Washington D.C. area. In addition, Dr. J. doesn't hold an amateur radio license. The solution being proposed is that AMRAD transmit on HF the Thue-Moore sequence using slow keying (probably several bits a second) and that Dr. J. will receive the transmission and decode it on his personal computer. The reception would be done using some simple gear that AMRAD can easily build. The second proposal is that AMRAD obtain an Experimental License from the FCC which would allow operation outside the ham bands. Being carried under this license, Dr. J. could communicate with us back in Washington DC for the experiment.

The Thue-Moore sequence is a very simple sequence to generate. It consists simply of the odd parity of the parallel output of an n-stage stage binary counter. Consider the following example:

counter	TM sequence
0001	1
0010	1
0011	0
0100	1
0101	0
0110	0
0111	1
1000	1
1001	0
1010	0
1011	1
1100	0
1101	1
1110	1
1111	0

We're still in the proposal stage of this experiment. Anyone interested in transmitting or decoding this kind of sequence from the radio is invited to write to me for the details. If things start to move on this experiment, we will probably form a sub-group to the current special interest group.

March 1985

REPEATERS AND SPREAD SPECTRUM

By Hal Feinstein, WB3KDU

As many of you know, the AMRAD spread spectrum group set up a frequency hopping (FH) beacon in the Washington, D.C. area to give people a signal-of-interest to try to receive. Moreover, we were interested in initial testing of a spectrum management concept called "overlay" in which a spread spectrum transmitter can share the same band with conventional users.

Effective spectrum management tries to fully allocate the band while holding interference to a minimum. In conventional modulation techniques, this can commonly be done by first channelizing the available frequencies. Channel spacing is usually dependent on low cost technical means to reject adjacent channel interference and the modulation bandwidth.

Wideband FM with its large guardbands gave way to narrowband FM. Now, narrowband FM is under pressure to accommodate the even narrower amplitude compandered single sideband (ACSB). ACSB is a form of SSB which has the audio compandered for better intelligibility and a pilot tone to aid in automatic SSB tuning.

The concept of spread spectrum overlay is another approach to utilizing the unused guardbands and moreover, the intermittently used channels commonly found with repeater operations. Spread spectrum's wideband signal can be viewed as a form of frequency diversity in which the signal energy is distributed across the band. A spread spectrum receiver can utilize the energy in the clear places within the band as an offset against the frequencies which have a conventional signal present. This is the concept of process gain used in anti-jamming applications to offset interference from conventional narrowband users. Under general conditions, the spread spectrum signal should not experience much interference after it is "despread."

The AMRAD beacon was used to test several aspects of the overlay concept. First, could general amateur radio equipment be easily modified for spread spectrum use? The beacon itself was built from a Commodore-C64 and an ICOM 2AT. Second, what was the slowest hopping speed that would not interfere, and third under what conditions could we avoid activating repeaters?

The answer to the first question showed many PLL type synthesizers can be modified for slow frequency hopping, and that the settling time of some limits the pseudo-random hopping to no less than 100 ms per hop. Surprisingly, we found that many of the common PLL synthesizers would stay in lock if a "sweep" mode was used. In the sweep mode, each frequency starting from 144.0 is visited to the top of the band at 147.990. We found that 2000 hops per second were possible with the sweep mode! Looking at the output on a spectrum analyzer revealed that, while the unlocked condition was not reached, the synthesizer was not stationary but was ringing around the channel attempting to settle.

At 1 ms hops this effect was still present but to a lesser degree; 10 milliseconds was found to be useful for the beacon tests but still had some percentage of ringing.

We have found that the amount of interference to be expected by conventional users is dependent on several factors including the FH power output, path loss, use of directional antennas, polarizations, the hop rate and the number of channels to hop. The larger the number of channels, the less often the hopper visits and hence, the less power per channel. Moreover, the faster the hopping, the less power per channel. Both the speed and the number of channels are most significant for reducing the power per channel and hence the spread spectrum signal heard by a conventional channelized receiver.

Hop speed also affects the amount of energy observed by a conventional user. Hop times of less than 1 ms seem to still be visible to conventional receivers. We have not found the point that the signal becomes invisible yet, but we speculate that once the rate of hopping passes the IF bandwidth of a conventional receiver, the energy should be essentially rejected.

Lastly, the problem of a repeater triggering on the FH signal was investigated. We tried 10 milliseconds and found that almost all repeaters were not triggered due mainly to the anti-kerchunking delays. Several repeaters lack this delay and would come up every time the signal landed on their input. There was a fix for this problem which we applied. The repeaters which were subject to this quick trigger effect were removed from the beacon table so that they would not be triggered.

We ran a one hour test to determine if the fix worked and found that indeed it had. Next to test is the hopping speed at which even these repeaters do not report our presence. In order to do this we must move to our next synthesizer design: the Fred Williams direct synthesis oscillator.

DIRECT SYNTHESIS OSCILLATOR WORKS!

The gang here has obtained one of the Fred Williams direct synthesis oscillators from A and A Engineering. It was assembled and is currently undergoing testing at Sandy, WB5MMB's QTH. The output signal of the oscillator contains some harmonics due to the staircase reconstruction of the sine wave within the D to A converter. The signal is filtered by a two stage LC filter to perform some additional smoothing to the signal. Without loading the increment register which determines frequency, the oscillator appears to "free run" at about 500 kHz.

The next step is to modify the shift register serial to parallel converter within the oscillator to accept plain parallel output from the Xerox 820 or the C-64. The expected hop speed of the direct synthesis oscillator is around 100,000 per second maximum.

GOOD BOOKS FOR THE EXPERIMENTER

Good books are hard to find on spread spectrum subjects if you are not a PHD in math. Well, Mr. Dixon has just published the second edition of his excellent book *Spread Spectrum Systems*.

The book is published by John Wiley and Sons in the Wiley-Interscience series and sells for about $37.50. It is clearly written and has a minimum of advanced math.

Here is the table of contents:
1. The Whats and Whys of Spread Spectrum Systems
2. Spread Spectrum Techniques
3. Coding for Communications and Ranging
4. Modulation and Modulators: Generating the Wideband Signal
5. Correlation and Demodulation
6. Synchronization
7. The RF Link
8. Navigation with Spread Spectrum
9. Applications of Spread Spectrum Methods

HIDE SPREAD SPECTRUM TRANSMITTER HUNT

As you know, we are planning a spread spectrum transmitter hunt for the fall or early summer. No date has been set yet, but a number of people have expressed interest at this point. Often, they say: how can you DF a spread spectrum signal when you can't even hear it! Then they look at you like you are in the slower learner class. Well, you are dealing with one of two types of people here. The DOGMATIST has read this kind of nonsense in some military publication meant for the rear-guard, or perhaps one of the numerous popular magazines. I've found that there is no way to discuss anything complicated with such folk because they rely on simple slogans such as: "bigger IS better," or "time wounds all heels."

Anyway, having dispensed of fifty percent of the voting public, we turn to the people who are in a position to know. Yes, however, we have a new problem. People who KNOW: —can't SAY.

Well, here is some real poop:

(1) Chuck, N4EZV, has a modified doppler direction finder which can lock up on a 10 millisecond or more frequency hopping signal.

(2) Wideband equipment is available at hamfests for reasonable prices. We got a 55-260 MHz wideband receiver (tubes mind you) at a recent hamfest for $65. However, Sandy, WB5MMB, says all he can get on it is country music.

(3) A set of sharp beam antennas can bring an FH signal out of the "noise" and eliminate some of the other interference. So now it can be processed on the above mentioned receiver.

(4) Linear shift register sequences are easy to decode, see Dixon appendix 5.

(5) DFing a frequency hopping signal is easier than trying to listen to what it says. Listening to what it says may not be that hard either.

(6) Well then, what about Low Probability of Intercept (LPI). Answer—depends who you talk to. For example: "You can't hear 'em if they're below the noise (a sometimes true statement)." "You got to be far enough away from a receiver so that they can't hear ya otherwise you don't have LPI." I ask, "How far away is far enough?" They say, "depends what THEY are receiving with. If it's a cat's whisker or lead crystal or even an ATWATER KENT receiver, you got LPI." So I ask, "What if they use an HP spectrum analyzer or even a WWII BUSHIPS ECM receiver with the four MHz IF feature?" They say, "BOY, you're askin' the WRONG questions," after a moment of thought THEY added, "if you can see the signal on a spectrum analyzer THEN you can get the information back and if you can get it out of noise by using lots of antennas, amplifiers and what have you, well, then you can get 'em and they don't have LPI."

One author has gone as far as to imply that LPI is wrongly named. Instead, he suggests calling it Low Probability of Recognition (LPR). A receiver will certainly intercept the signal even if it's below the noise, but the operator may not recognize its presence. Like the famous line from a Gershwin musical "It ain't necessarily so..."

Chapter 5

The League's View of SS Experimentation

QST, **November 1980**

Spread Spectrum and the Radio Amateur

Spread-spectrum signals are unlike any emissions presently used by radio amateurs. But we stand at the threshold of what may be a new mode for amateur communications.

By Paul L. Rinaldo, W4RI

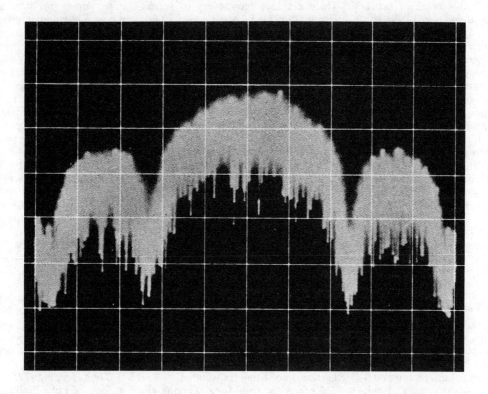

A modulation technique that has been in development since the late 1940s, spread spectrum (SS) has, until recently, been virtually unthinkable for use by radio amateurs for a number of reasons. First, SS occupies bandwidths far in excess of the necessary bandwidth; that would be illegal! By using a pseudo-random digital sequence to scatter energy over a wide band, there is only a small amount of energy in any one hertz; that would make it an unauthorized code. SS systems have been complex and expensive; that would be beyond the resources of radio amateurs. Much of the development has been conducted under government contract; most hams knew little or nothing about the subject. There were more than enough reasons to deter hams from even dreaming about an SS rig in their shacks.

The situation has changed greatly in recent years! SS technology has progressed to the point where affordable systems can be built for amateur and other nongovernmental uses. The replacement of the Federal Communications Commis-

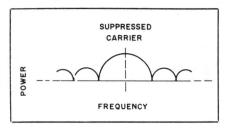

Fig. 1 — Power vs. frequency for a direct-sequence-modulated spread-spectrum signal. The envelope assumes the shape of a $\left(\frac{\sin x}{x}\right)^2$ curve. With proper modulating techniques, the carrier is suppressed.

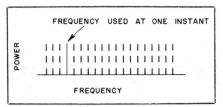

Fig. 2 — Power vs. frequency for frequency-hopping spread-spectrum signals. Emissions jump around in pseudo-random fashion to discrete frequencies.

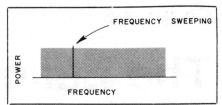

Fig. 3 — Power vs. frequency for chirp spread-spectrum signals. The carrier is repeatedly swept, continuously, from one end to the other in a given band.

sion's Office of the Chief Engineer with the Office of Science and Technology (OST) carried with it the mandate to encourage the use of new technology. The FCC's OST sees the Amateur Radio Service as a test bed for new techniques. Some at the FCC feel that the long-term retention of amateur frequencies, in competition with other radio services, depends largely on continued technological advancements by amateurs. We may be entering an "experiment or expire" era.

Why the sudden interest in SS? The reasons are many. First, there is the simple technical imperative, meaning that the technology is there as a result of many years of government-sponsored development, so why not use it for civilian applications? Another reason is that a number of SS users, say in the Land Mobile Service, could be overlaid on top of an existing band already "full" of mobile users employing conventional frequency modulation. Similar overlays could be tried by amateur experimenters in the ham bands. If this is done with care, the preexisting users wouldn't even detect the presence of the SS overlay. Yet another possibility is the creation of new bands, maybe a 900-MHz band, which would use SS exclusively to accommodate thousands of users. Moreover, SS could afford these users both privacy and immunity from interference through proper code settings. In general, SS offers possibilities for more extensive sharing of frequencies while minimizing interference.

Spread-Spectrum Fundamentals

SS systems employ radio-frequency bandwidths that greatly exceed the bandwidth necessary to convey the intelligence. Bandwidths for SS systems generally run from 10 to 100 times the information rate. By spreading the power over a wide band, the amount of energy in any particular hertz or kilohertz is very much smaller than for conventional narrow-band modulation techniques. Depending upon the transmitter power level and the distance from the transmitter to the receiver, the SS signal may be below the noise level.

SS systems also use coding sequences to modulate and demodulate the transmission. Receivers with the wrong code will not demodulate the encoded SS signal and will be highly immune to interference from it. On the other hand, receivers with the right code are able to add all the spread energy in a constructive way to reproduce the intended modulation. In fact, the use of coherent correlation can yield some process gain. Changing the code to another sequence effectively creates a new "channel" on which a private conversation can take place. Many good code combinations could be made available on a single chip and selected by means of thumbwheel switches on the SS transceiver.

Types of Spread Spectrum

There are four basic types of spread spectrum: direct sequence, frequency hopping, pulse-fm and time hopping. In addition, there are hybrids consisting of combinations of two or more of the above basic types.

Direct Sequence (DS): Direct sequence SS is produced by modulation of a carrier with a digitized code stream. This type of modulation is also known by the terms pseudo-noise (PN), phase hopping (PH), direct spread, or direct code. Phase-shift keying (psk) is usually used to produce the marks and spaces, but frequency-shift keying (fsk) could also be used. The wide rf bandwidth arises from the use of a high-speed code. Of course, if the transmitter were allowed to rest on the mark frequency, there would be a steady carrier in one place whenever there is no modulation. This would produce interference to a narrow-band user on that frequency. It would also pose problems for other SS users of the same band, particularly if they did the same thing. So it is conventional for SS systems to include techniques to continue a pseudo-random code sequence even during intervals when intelligence is not being transmitted.

The power spectrum for a DS signal (as might be seen on a spectrum analyzer) is not uniform across the band, but has a main lobe and sets of sidelobes as illustrated in the title photo and in Fig. 1. The bandwidth of the main lobe as measured from null to null is two times the clock rate of the code sequence. The bandwidth of the side lobes is equal to the clock rate. To receive a DS signal, the receiver must collapse or "despread" it to the original bandwidth of the information. This is done by using a replica of the code sequence used by the transmitter.

Frequency Hopping (FH): As the name implies, frequency hopping is simply jumping to a number of different frequencies in an agreed sequence. The code sequence is usually at a slower rate than for direct sequence and is normally slower than the information rate. The hopping rate may also be determined by practical considerations, such as how long it takes for a particular frequency synthesizer to settle down on a new frequency.

Actual modulation of the frequencies uses normal narrow-band techniques such as frequency modulation. At any instant, an FH transmitter is emitting all of its power on a specific frequency slot and potentially could interfere with someone else using a narrow-band system on that frequency. However, the FH dwell time on that particular frequency is so short that most narrow-band users would not be bothered. Mutual interference between two or more FH users sharing the same band could be extremely low, depending upon the design of the code sequences. Fig. 2 illustrates the power spectrum for an FH signal.

Pulse-FM (Chirp): A chirp spread-spectrum system sweeps its carrier frequency over a wide band at a known rate. Again, conventional narrow-band modulation of the sweeping carrier is used to convey the intelligence. The receiver uses a matched, dispersive filter to compress the signal to a narrow band. Chirp systems typically do not use a code sequence to control the sweep generator. Sweep time can be largely independent of the information rate. Normally, a linear-sweep pulse is used, similar to that produced by a sweep generator. The power spectrum for a chirp system is illustrated in Fig. 3.

Time Hopping (TH): Time hopping is a form of pulse modulation using a code sequence to control the pulse. As in other pulse techniques, the transmitter is not on full time and can have a duty cycle of 50% or less. Several systems can share the same channel and function as a time-division multiple-access (TDMA) system. TH is more vulnerable to interference on its center frequency than other SS systems. Seldom seen in its pure form, TH is typically used in hybrid systems using frequency hopping as well.

Hybrids: In addition to the TH/FH hybrid system just mentioned, there are also DS/FH and DS/TH combinations. Hybrid systems are typically designed to accommodate a large number of users and to provide a higher immunity to interference. They also produce better results at practical code sequence rates governed, for example, by how fast a frequency synthesizer can be switched. Also, hybrids can produce greater spreads than those which are practical for pure SS systems.

Some Considerations

Synchronization: In the design of a spread-spectrum system, usually the toughest problem is synchronization of the code sequence at the receiver with that of the incoming signal. If sync is not attained, even just one bit off, nothing but noise can be heard. The problem becomes worse when more than two stations are trying to communicate in a net. This is because of the different propagation delays between stations; i.e., it takes a different time for a signal to travel over paths A-B, A-C, or B-C if the stations are not equidistant. These differences may be only slight but just enough to degrade the signal-to-noise ratio of the received signal. In addition to the time uncertainty related to propagation, there is also a frequency uncertainty in trying to keep oscillators at two or more stations from drifting.

Because the stations cannot be expected to synchronize on their own with no reference, it is normal for at least one station to transmit an initial reference for sync purposes. Upon reception, the receiving stations can generate the code sequence at a rate different from the code sequence used at the transmitter. Eventually, the two code streams will slide into phase with one another and may then be locked up. After initial synchronization, maintaining sync presents another problem which can be solved in different ways. One is to use a code sequence preamble at the beginning of each transmission. Another is to use ultra-stable clocks at all stations to ensure that the code-sequence clock frequency does not change. Numerous other schemes have been devised and implemented with varying degrees of difficulty. The exception is that chirp systems do not have this problem because the matched filter used in demodulation inherently achieves sync on each pulse transmitted.

Transmitter and Receiver Design: One difference between SS and conventional rf equipment is that SS requires transmitters and receivers that have 10 to 100 times the bandwidth of narrow-band systems. That may pose some problems at lower frequencies, but in the 420-MHz band the amateur television (ATV) experimenters already have equipment that can handle wideband signals. The transmitter design, which should be well within amateur capability, amounts to taking care in broadbanding the rf stages after modulation to maintain amplitude linearity, and in keeping the antenna system VSWR very low. Receivers must not only have wideband front ends but must have good dynamic range and linearity to handle both the desired signal and any interference. Where an i-f is used, the frequency chosen must be higher than for conventional transceivers. In practice, 70 MHz is a common SS i-f. Components (such as filters) are available for this frequency to build SS i-f modems (modulator/demodulators).

Amateur SS Experimentation

The Amateur Radio Research and Development Corporation (AMRAD) has formed a group to experiment with several different types of SS systems. Before on-the-air tests are conducted, it will be necessary to obtain a Special Temporary Authorization (STA) from the FCC. Readers wishing to participate should contact AMRAD via the author.

The continued existence of the Amateur Radio Service depends, in part, on amateurs' contributions to the state of the art through experimentation. Spread spectrum is fertile ground for amateur investigation. While SS has been developed extensively for military and other governmental applications, civil uses are virtually unexplored. Hams have the capacity to build SS systems which are practical and inexpensive. There is no guarantee that SS will prove itself worthy of regular use in civilian radio services, but the technology is ripe for Amateur Radio experimentation.

[The title photo, a spectrum analyzer display of a direct-sequence spread-spectrum signal, is reprinted through the courtesy of Robert Dixon and John Wiley & Sons, Inc. The photo appears on the cover of *Spread Spectrum Systems*. — Ed.]

Selected Bibliography

Reading material on spread spectrum may be difficult to obtain for the average amateur. Below are references that can be mail ordered. Spread spectrum papers have also been published in IEEE Transactions on Communications, on Aerospace and Electronic Systems and on Vehicular Technology.

Dixon, *Spread Spectrum Systems,* 1976, Wiley-Interscience, 605 Third Ave., New York, NY 10016, $29.50.

Dixon, *Spread Spectrum Techniques,* IEEE Service Center, 445 Hoes La., Piscataway, NJ 08854, IEEE member prices $19.45 clothbound, $12.95 paperbound; nonmembers $29.95 clothbound.

Brumbaugh, et al., *Spread Spectrum Technology,* a series of papers presented at the 1980 Armed Forces Communications Electronics Association show printed in the August 1980 issue of *Signal,* available from AFCEA, Skyline Center, 5205 Leesburg Pike, Falls Church, VA 22041.

Current published searches on spread spectrum are available from the U.S. Department of Commerce, National Technical Information Service, Springfield, VA 22161 for $30 each:

Spread Spectrum Communications (99), May 79 NTIS/PS-79/0494/9.

Spread Spectrum Communications (188), May 79 (EI) NTIS/PS-79/0495/6.

Glossary of Spread Spectrum Terms

Chirp — Same as pulse-fm.
Code sequence — A series of 1 or 0 bits arranged in a known pattern.
Direct code — Same as direct sequence.
Direct sequence — A type of spread-spectrum modulation using a code sequence to modulate a carrier, normally using phase-shift keying.
Direct spread — Same as direct sequence.
Frequency hopping — A type of spread spectrum which employs rapid switching between a large number of discrete frequencies.
Hybrid — A spread spectrum system that combines two or more basic types of spread spectrum.
Phase hopping — Same as direct sequence.
Pseudo-noise — Same as direct sequence.
Pulse-fm — A type of spread spectrum that uses a swept carrier.
Spread spectrum — A class of modulation types that produce bandwidths far in excess of the bandwidth necessary to convey the intelligence.
Time hopping — A type of spread spectrum using a form of pulse modulation in which the pulses are controlled by a code sequence.

Technical Correspondence

Conducted By Gerald L. Hall, K1TD

The publishers of QST assume no responsibility for statements made herein by correspondents.

SPREAD SPECTRUM TECHNIQUES

☐ The article on spread spectrum Amateur Radio in November 1980 *QST*[1] was very interesting. Author Paul Rinaldo, W4RI, is to be commended for discussion an extensive subject with such a concise treatment. The essential features and characteristics of spread spectrum were mentioned. However, there are some potential misconceptions.

In the article Rinaldo mentions that spread spectrum signals can be overlaid on top of existing operating frequencies and states, "If this is done with care, the preexisting users wouldn't even detect the presence of the SS overlay." As Rinaldo implies, under the proper circumstances this is completely true. However, if the spread-spectrum transmitter is close to a conventional receiving station, that receiving station will certainly be jammed on all channels in that band! On the converse, when the spread-spectrum station is receiving and the nearby conventional station is transmitting, the reverse will occur: The spread-spectrum station will experience severe bleed-through of the undesired signal. This arises from using limited bandwidth spreading ratios. The equation for jamming rejection is:

Jamming rejection = 10 log (bandwidth ratio) (Eq. 1)

where bandwidth ratio is the rf bandwidth divided by the information (audio) bandwidth. Jamming rejection is the amount of suppression of undesired signals in the spread-spectrum receiver. Eq. 1 also holds true for the degree of covertness (noninterference with conventional receivers) when the spread-spectrum station is transmitting. As can be seen, the amount of bandwidth spreading required for modest rejection is significant: For 30 dB of rejection a 1000:1 spread would be required, and for 40 dB, 10,000:1. A 1000:1 spread of filtered audio is 2.5 to 3 MHz of rf bandwidth, and that only gives 30 dB of dynamic range. But it is common for signal strengths to exceed a dynamic range of 80 dB when considering in-band transmitters within a couple of miles of a receiving station. The author gives typical values of 10 to 100 for spreading ratios. These are clearly inadequate.

The author also states that changing the code allows another private channel. Again, this is completely true under the proper circumstances. Here, there is a better relationship for off-channel rejection:

Channel rejection = 20 log (bandwidth ratio) (Eq. 2)

Since this is a 20-dB-per-decade relationship, the channel rejection builds up at a faster rate than jamming rejection. For the same 2.5 MHz, 1000:1 spread spectrum voice signal, 60 dB of off-channel rejection is possible. But that is with a very limited number of codes; with a wider choice, the channel rejection falls short by several decibels.

In addition, the suggestion of spread spectrum for the Amateur Radio Service is not new. J. P. Costas, W2CRR, proposed it back in the 1950s, just when ssb was in its infancy.[2] Costas contributed a circuit, the Costas loop demodulator, which is very valuable to many spread-spectrum systems today. It was designed to demodulate double-sideband suppressed carrier (usually abbreviated dsb) transmissions. Dsb is the simplest spread-spectrum signal, with a spreading ratio of 2. Dsb also has some significant advantages over ssb.[3] The most notable is that the transmitter peak-to-average power requirements for non-sinusoidal modulating signals is much tamer. For example, according to the reference,[4] if a square wave is transmitted with ssb, the peak-to-average power requirement of the transmitter approaches infinity, but for dsb the peak-to-average power requirement is 1:1. For a sine wave, the peak-to-average requirement for both ssb and dsb is 2:1. In the 1950s it was quite difficult to make good ssb transmitters, even compared to the added complexity of the Costas loop demodulator needed for a dsb receiver. Even today this is a significant problem with speech processors being used to help alleviate the wild peak-to-average requirements of nonsinusoidal (voice) modulation. Even with these significant disadvantages, ssb became the standard modulation format. There is a simple reason for ssb winning out over dsb: adjacent-channel rejection. It is easier to increase adjacent-channel rejection to the levels needed (80 dB and greater) with fancy receiver filters than it is to spread the bandwidth of the transmission.

I don't want to sound like a conservative who doesn't want change. Actually I would like to see more experimentation with advanced modulation techniques. It is with the kind of exposure that Rinaldo is giving that this can take place. However, I would not like to see significant misconceptions spoil it by creating distrust between various groups of amateurs. That isn't likely if the fine details are understood in advance. Then the problem becomes (as it is in the military), "Where can enough available spectrum be found for such a service?" Perhaps the answer is at 900 MHz, as suggested by Rinaldo.

Incidentally, the article by J. P. Costas makes very interesting reading. Even though it was published in the IRE literature, it is still readable by persons without a PhD in mathematics, with only a couple of equations per page. Moreover, it reads as if it were written in 1979, rather than 20 years earlier. His comments about congestion and solutions are very perceptive, if not prophetic. Of particular interest are his evaluations of ssb versus spread-spectrum performance in congested conditions. "As the congestion becomes worse it will be impossible to avoid reducing the data rate per circuit. The important point here is that the broad-band philosophy ACCEPTS INTERFERENCE AS A FACT OF LIFE and an attempt is made to do the best that is possible under the circumstances. The narrow-band philosophy essentially denies the existence of interference, since there is an implied assumption that the narrow-band signals can be placed in non-overlapping frequency bands and thereby prevent interference." It is my perception that this philosophy remains prevalent today. — *Ken Wetzel, WA6CAY*

[1] P. L. Rinaldo, "Spread Spectrum and the Radio Amateur," *QST*, Nov. 1980, p. 15.
[2] J. P. Costas, "Poisson, Shannon, and the Radio Amateur," *Proceedings of the IRE,* Dec. 1959, pp. 2058-2068.
[3] J. P. Costas, "D.S.B. vs. S.S.B.," Technical Correspondence, *QST,* May 1957, p. 42.
[4] *Reference Data For Radio Engineers*, Fifth Ed., p. 21-5.

Spread-Spectrum Applications in Amateur Radio

Through the properties of their coded modulation, spread-spectrum systems can provide multiple-access, low-interference communications to radio amateurs.

By William E. Sabin, WØIYH

Traditionally, the emphasis in Amateur Radio has been to make a transmitted signal as narrow in bandwidth as possible. Also, receivers are made as narrow and as interference-immune as possible. In this way, many signals can occupy a ham band successfully. This approach has been successful, to a point. But if a group of stations is on one frequency (the pileup!), the system does tend to break down, with disastrous results.

A new approach is being advanced that amateurs should take a look at. Military and commercial organizations are developing spread-sprectrum systems. Such systems deliberately occupy a wide band of frequencies, as part of a strategy to make communications more reliable and more secure, or private. (The word "privacy," as used here, has a special meaning in Amateur Radio, which will be considered later.)

To be more exact, a transmitter sends its message in such a way that a wide spectrum is used, according to a very carefully designed plan. The receiver has the ability to use this same plan in reverse, to convert the signal back to narrow-band form. By performing these actions in the right way, privacy and interference immunity are improved. Fig. 1A shows a conventional transmitter output of, say, 1 kW. With spread-spectrum operation, this same power is spread out as shown in Fig. 1B. There is no strong carrier at any one frequency. Within a 3-kHz band, the amount of signal is greatly reduced. In fact, it may be less than the noise level. But after "despreading," the signal once again looks like the signal at A.

This scheme is different from wide-band fm in that the message itself does not produce the spread spectrum. Instead, another agent is employed to spread the signal. Also, there is no carrier, as in an fm system.

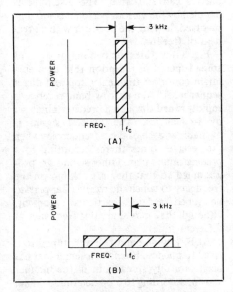

Fig. 1 — The power distribution of a conventional communications signal (A) versus a spread-spectrum signal (B). The same total power is contained in both signals.

Why Amateur Spread Spectrum?

Consider the network shown in Fig. 2. Using spread spectrum, stations A and B can communicate privately, while C and D do the same. Or, A can address all stations. Also, a member of this net can address an entirely different net or a single member of the other net. Or, B could address all nets simultaneously. In a particularly large region, these nets all use the same frequency band, and no equipment retuning is needed for any of the above operations.

In any of these operations, a degree of "privacy" is achieved, in the sense that communications are programmed according to the requirements of the moment. This "selective calling" is achieved by using the microcomputer-based "protocol," or message-routing procedures. The use of spread spectrum is an enhancement of the "packet radio" techniques that advanced amateurs are now experimenting with. By adding to this packet system a carefully managed spread-spectrum protocol, it should be possible to greatly reduce "collisions," avoid interference and add significantly to the repertoire of the packet system.

This extra element of spread-spectrum management, in addition to time management, makes it possible to reduce the guesswork with respect to frequency selection, which is a major problem in Amateur Radio. The difficulty is that a clear frequency at my station may not be clear at

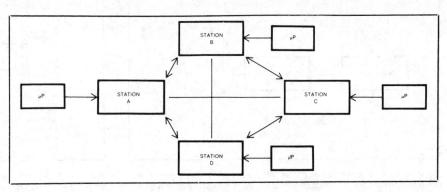

Fig. 2 — A network of stations using μP or microprocessor-based protocols. Network and internetwork communication may be obtained without equipment retuning.

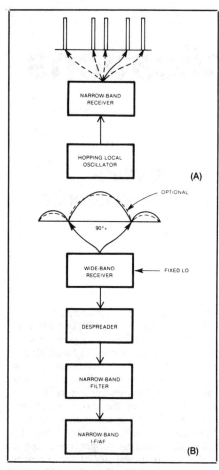

Fig. 3 — The two major types of spread-spectrum operation. At A, the frequency-hopping mode is illustrated; at B is the direct-sequence mode. The spreading is caused by applying a coded modulating signal at the transmitter that is independent of the intelligence modulation. The same predetermined code is applied in reverse at the receiver to despread the signal, and the intelligence is then demodulated in conventional fashion. If different spreading codes are appropriately chosen, interference-free operation may be obtained with different stations sharing the same frequency spectrum.

your station. We need a way to improve this situation. The communication is not restricted to data. Digitized voice messages can be sent, stored at the receiving station and converted back to speech at the completion of the message.

The information below summarizes the possible advantages of combining spread spectrum with packet-network protocols. There is no particular argument in favor of using spread spectrum by itself.

- Voice or data
- Simultaneous net combinations
- Network privacy
- Reduced collisions between
  A) spread and nonspread systems
  B) other spread systems
- Combines well with packet protocols
- Enhances packet repertoire

The important extra element is the voluntary "discipline" to which the various players can subscribe. The discipline is handled by the various personal computers involved, leaving the operator with a great deal of freedom.

Fig. 3 illustrates the two major types of spread-spectrum reception that amateurs might consider: frequency hop and direct sequence. At A, a narrow-band receiver is rapidly tuned through a predetermined set of frequencies. The desired signal is available at each of these frequencies when the receiver tunes there, according to a predetermined plan. Other signals are programmed so that they are seldom on the frequency to which the receiver happens to be tuned at one particular moment, although they may share this frequency at different times.

At B, a wide-band receiver, tuned to a fixed frequency, listens to a signal that has been carefully spread out in the manner shown, according to a predetermined plan. The receiver possesses the key by which to despread this signal and put it through a narrow filter. All other signals using the same frequency band are essentially ignored. A different spread-spectrum signal fails to despread. A conventional signal is likewise unable to penetrate the narrow filter, because it is converted to a wide-band signal by the receiver.

On the hf bands, frequency hopping would be better because of its narrow-band nature. At uhf, direct sequence offers advantages. By carefully selecting the kind of modulation in direct sequence, the spectrum is improved as shown in the dashed line in Fig. 3B. This could be msk (minimum-shift keying) modulation.

Direct-Sequence Spreading

Fig. 4A shows a conventional phase-shift-keyed data signal and the transmitted spectrum it produces. The width of the main lobe is twice the data rate, and in an amateur RTTY system would be less than 1.5 kHz wide. In Fig. 4B, each data bit is modulated by a PN code, where PN means pseudo-noise. There may be thousands of PN code bits for each data bit. The data bit has the effect of inverting, or not inverting, a group of PN bits, depending on a data 1 or 0. After combining these code streams, a mixer with a fixed-frequency local oscillator produces the spread-spectrum rf signal. The width of the main lobe is twice the PN code rate.

The design of the PN code is critical to the performance of the spread-spectrum system, and will be covered in the next section. The spectrum in Fig. 4B shows that, when listened to by a conventional receiver, a weak, hissing, noise-like signal is heard. It is not truly random noise, because of the nonflat nature of the spectrum, and because the PN code does repeat itself after some long time interval. The expression "pseudo" noise is therefore appropriate.

Fig. 5A shows a PN code generator. A shift register with N stages, initially loaded with all 1s, generates an output that is

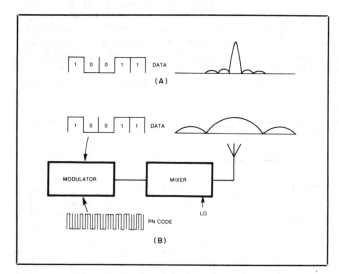

Fig. 4 — At A, a conventional phase-shift-keyed (psk) data signal and the transmitter spectrum that it produces. At B, the same data signal and a pseudo-noise (PN) code are combined to produce a direct-sequence psk signal with its much broader spectrum.

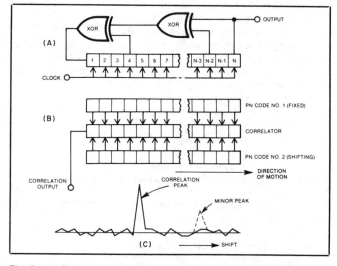

Fig. 5 — A PN code generator, A, and a correlator, B. The strings of blocks represent shift registers. At C is shown the correlation output with shifts of data in the lower shift register. This output peaks when synchronization is obtained.

influenced by the XOR feedback gates. After as many as $2^N - 1$ pulses, the code repeats itself. Code length can be from, say, 10 bits, to 10 million or more. If my transmitter and your receiver use the same PN generator circuit, we can communicate.

Fig. 5B shows a correlator. One input is a fixed (or static) PN code that has been stored. The other input is the received signal, which ripples through the bottom section. The output of the correlator, Fig. 5C, is very small (± 1 or 0) except when the two codes exactly coincide, at which time the output equals N. When this peak occurs, the two codes are "synchronized," or correlated.

A well-designed code has only one of these sharp peaks. Other codes will have other minor peaks that can produce "false" synchronization. Also, a short segment of a good code may not be so good in this respect. Therefore, the register length should be as long as the code, or at least as long as possible. A good possibility for amateur use is to combine two short Gold codes.[1] The resultant codes have good correlation properties and are easy to generate.

Fig. 6 shows two types of direct-sequence receivers. In Fig. 6A, the despreading is done at i-f. A narrow-band crystal filter lets only the despread signal get into the data or voice detector. In B, a double conversion takes place, and the correlator, operating at baseband, performs the despreading. A low-pass filter then passes the desired signal only, which can be derived from the correlator output. Some receivers use both methods, that at A for data detection, and that at B for synchronization.

Direct-Sequence Synchronization

Fig. 7 shows, in a very general way, how synchronization is achieved and maintained. One way is to slow down or speed up the PN code generator in such a way that it can search backward and forward in time to "acquire" the incoming code. Once the code has been acquired, a tracking operation takes place, so that the PN code stays closely aligned with the desired signal. Acquisition is greatly aided by the use of very stable clocks and by prealignment of the codes, so that only a small amount of searching is needed. For example, using WWV, all code generators could be initialized each hour.

Another method uses a special, short preamble that the receiver quickly recognizes. This recognition starts the tracking operation. In an amateur application, this would be easier and cheaper to implement. In this case, the prealignment

[Editor's Note: Gold codes are a family of codes named after the developer, R. Gold. The significance of the family is that a large number of codes may be obtained with relatively short shift registers in the code generator.]

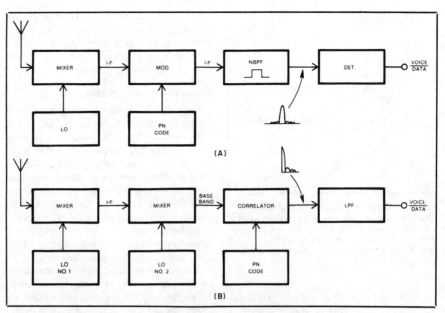

Fig. 6 — Block diagrams of two types of direct-sequence receivers. See text.

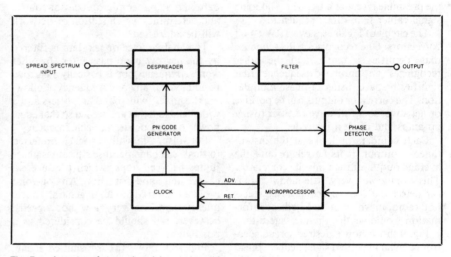

Fig. 7 — A system for synchronizing and tracking with direct sequence.

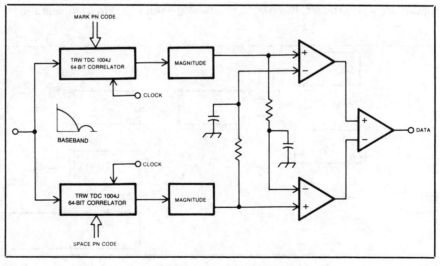

Fig. 8 — A correlator that operates at baseband with direct sequence.

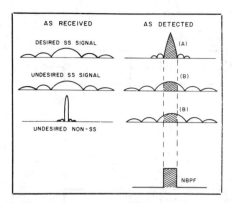

Fig. 9 — Illustrating interference rejection in direct sequence. Processing gain is obtained with despreading, and in decibels is equal to 20 log (A)/(B), where (A) and (B) are the amplitudes depicted in the shaded areas at the right.

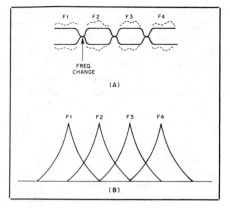

Fig. 11 — At A, an amplitude-versus-time representation of a frequency-hopping signal. The broken lines indicate possible variations arising from amplitude modulation of the signal. At B, an amplitude-versus-frequency representation of the same signal; any predetermined arrangement of frequency hopping may be used.

of codes and time-of-day clocks would not be needed. When the preamble is sent, it is immediately followed by the start position of the long PN code. The receiver performs the identical operation. Sometimes, the preamble is sent as a frequency-hopping signal rather than direct sequence.

The circuit of Fig. 8 uses two TRW 64-bit correlators. One recognizes a PN sequence that identifies a mark, and the other recognizes an unrelated sequence that signifies a space, using bi-phase modulation. The correlator outputs can be positive or negative, so a full-wave rectifier (using op amps and diodes) is needed.

Each comparator looks at the instantaneous output of its correlator and the average output of the opposite correlator. This circuit, with very little modification, can detect marks and spaces in a frequency-hop radio, in which up to 64 bits of information would signify the message bit.

Fig. 9 shows how a desired spread spectrum and an undesired signal, either spread or narrow-band, are interpreted by a spread-spectrum receiver. The narrow-band signal is "smeared" by the receiver so that little energy is passed by the narrow band-pass filters. If the undesired local signals are strong enough, however, they can still override the desired distant signal, even after it has been despread, as shown. This "near-far" problem in direct sequence is an important limitation. Frequency hop seems to be more immune.

Frequency-Hop Spreading

Fig. 10 is a block diagram of a frequency-hop transmitter. A fast-frequency-change local oscillator is needed. The time that it takes to settle on the next frequency should be less than 10% of the dwell time on that frequency. Hop rates of 10 to several thousand per second are feasible with today's technology, and amateurs should experiment with speeds over the entire range. For example, a real opportunity exists for innovation in a low-cost, fast-hopping synthesizer design.

Shortly before changing frequency, the signal is smoothly attenuated, as shown. After the hop, the signal is brought up again, smoothly. This is necessary to reduce the transmitted spectrum (key clicks) as much as possible and to allow the LO time to make its frequency change. The hopping pattern is all under microprocessor control and determines which station or network will be addressed.

The analog voice or fsk data is filtered by the narrow band-pass filter. In other words, frequency hop is basically a narrow-band mode at any one frequency. Following the mixer, wide-band amplifiers and a wide-band antenna are needed so that equal power output occurs at each frequency.

For voice, a-m (with carrier) is preferred to single sideband because ssb causes phase jumps between hops, which produce excessive noise and distortion. Analog voice is not a preferred mode, in general, in frequency hop. At very low hop speeds, however, ssb should be considered as a possibility.

Fig. 11A shows the transmitted signal, smoothly attenuated between hops. The dashed lines show possible amplitude variations with a-m operation. Fig. 11B shows the spectrum. The tapered Christmas-tree shape at each frequency is caused by the turning on and off (Fig. 11A), and also by the mark/space information or a-m sidebands. To minimize interference, the drop-off in the spectrum should be as fast as possible, consistent with good communication. We can also visualize that the receiver bandwidth should be only wide enough to receive, say, 90% of the total signal energy at each frequency.

A frequency-hop receiver is shown in Fig. 12. An antenna input switch controls the turn-on at each frequency in a way that reduces intermodulation with strong, undesired signals on nearby frequencies. After mixing, a narrow filter leads to the signal detectors. The outputs of these detectors provide signal information to the microprocessor to control the synchronization algorithm and to determine that sync has occurred. When synchronization is achieved, a tracking operation is started in which the hop clock rate is adjusted momentarily. Later paragraphs cover this topic more thoroughly.

Instead of a single filter plus discriminator, consider two narrow filters, one for mark and one for space. The outputs are rectified and compared to determine the

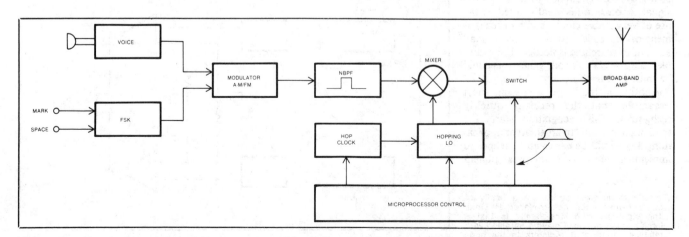

Fig. 10 — A frequency-hop transmitter. The hopping LO must be capable of making fast frequency changes with short settling times.

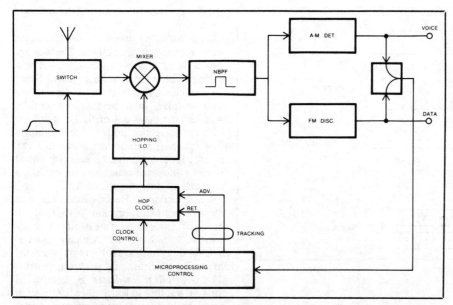

Fig. 12 — A frequency-hop receiver. The antenna switching reduces intermodulation from strong adjacent-frequency signals.

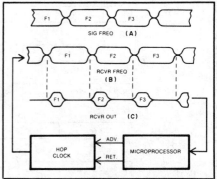

Fig. 13 — Illustrating frequency-hop sync searching or tracking.

mark or space condition. For data reception, each filter has a 3-dB bandwidth that in hertz is about 1.25 times the number of hops per second. Linear-phase filters, called "matched" filters, are needed.

Frequency-Hop Synchronization

One way to synchronize is to use very stable hop-clock oscillators. The hopping code patterns are then all initialized at some time, say each hour, using WWV as a time reference. Then, only a slight amount of searching back and forth in time is needed to align the receiver with the signal.

At the start of each reception time, a small sync adjustment is made. A block of data would be preceded by a special segment that sets up the receiver to copy data. A latecomer in a net would also need opportunities to get fully synchronized.

Sync searching is illustrated in Fig. 13. We see that the incoming signal, at A, does not completely coincide in time with the receiver tuning, at B. The signal switches to frequency f2 before the receiver is ready to switch to f2. The result is that the receiver output has a signal only during the interval shown at Fig. 13C. The receiver uses this information to advance or retard the hop clock slightly until the overlap has improved sufficiently. In the example shown, the receiver hop clock would be speeded up.

In another synchronization method that is somewhat more complicated, the receiver slow hops until the computer recognizes that sync has occurred. Then, the receiver fast hops in sync with the signal. The advantages of this method are that very stable hop-clock oscillators are not needed, and no prealignment of hopping codes is needed.

An additional method is to reserve certain frequencies as "start" or "stop" frequencies. When the start frequency is detected, hopping automatically begins according to the hop pattern plan. A stop frequency advises that the net control is available.

Interference Rejection

In Fig. 14, two kinds of interference are shown. In one case, a steady signal appears on one of the frequencies. In the second case, another hopping signal occasionally occupies the same frequency at the same time. Also shown in Fig. 14 is a common occurrence, signal fading from time to time on various channels.

Ways to combat these conditions are (1) make the receiver bandwidth narrow, to reduce interference; (2) design the hop codes to minimize "collisions" between nets and between net members; (3) use a fast-responding agc; (4) use a lot of message redundancy, i.e., repeat everything on several frequencies; (5) use error-correction codes; and (6) delete occupied frequencies and insert clear frequencies. Possibly certain frequencies would be reserved for backup use only.

Network Protocol

How could frequency hop be used to enhance an amateur packet network? Consider the format in Fig. 15.

Block 1: The net control has a hop code that addresses the net. The stations in the net are using this hop code to monitor the net control station and synchronize to it.

Blocks 2 and 3: The net control also sends data that contains the information in blocks 2 and 3.

Block 4: Having acquired frequency-hop sync, the net members acknowledge.

Block 5: The net control listens for my message using my frequency-hop code. Other stations in the net communicate with each other at the same time, using predetermined hop patterns that do not interfere with other members.

Block 6: The stations use two kinds of code simultaneously: frequency-hop pattern and mark/space code. This makes it possible, for example, to address a particular station and to identify the caller at the same time, or to inform the recipient how to respond to the call. Once the net members become synchronized, it should be possible to maintain sync for a long period of time, with tracking adjustments as required. If a member requires sync, he can send a sync-request message to which the recipient responds.

These ideas are offered for illustrative purposes and do not represent any known system in use. The important thing is that

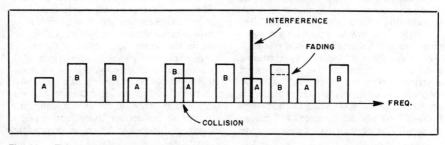

Fig. 14 — This drawing depicts interfering signals and fading in a frequency-hop system. The letters A and B identify emissions from two different transmitters sharing the spectrum, but with different hopping codes. The two frequency sets are almost orthogonal. The coding system is designed to reject interference, fading and collision. The same hop frequencies may be used by both stations, A and B, but at different times.

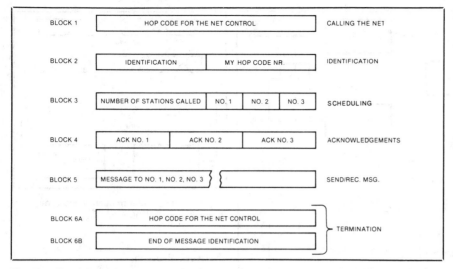

Fig. 15 — Possibilities for network protocol.

action is needed to improve the message reliability.

In a fast-hop system, the dwell time on each frequency might be 1 ms. The message rate is the same as with slow hop, but now each data bit is repeated on several different frequencies. After each bit is received, a vote is taken and the majority decides whether the bit is a 1 or 0. The addition of error correction adds to the reliability. The fast-hop method should be a better scheme for amateur use because, with good design of hopping code patterns, collisions and loss of data should be less.

The Amateur Radio Research and Development Corporation (AMRAD) is a group of amateurs who are dedicated to advanced technology in Amateur Radio. Within this group are subgroups interested in spread spectrum and packet networks. The *AMRAD Newletter* is helpful to anyone wanting to learn more, or to get in touch with others having similar interests. (Another interesting newsletter, *QEX*, an experimenter's exchange, is published by the ARRL.) AMRAD is also in touch with the FCC and has obtained special permits to do various kinds of experimental work. The address is AMRAD, P.O. Drawer 6148, McLean, VA 22106. Terry Fox, WB4JFI, is president of AMRAD. Hal Feinstein, WB3KDU, heads the spread-spectrum subgroup, and Dave Borden, K8MMO, heads the packet-protocol group.

amateurs are free to devise schemes that are right for them.

The key to success in frequency hop is to repeat the message often and to provide error correction. This means the message rate must be reduced to improve reliability. In frequency hop, the possibilities of interference are strong. This means that any part of the message must be repeated on several different frequencies. In a slow-hop system, a block of data is sent on each frequency and then the frequency is changed. A typical dwell time on each frequency might be 0.1 second.

It is important to have some way of detecting "bad" frequencies and moving away from them. The network protocol should include an avoidance strategy. The key is to find out if excessive errors occur on certain frequencies. This frequency management is under computer control. My computer tells your computer what

QST, November 1983

NOTES ON SPREAD SPECTRUM

☐ The article by William Sabin, WØIYH, on spread-spectrum techniques (July 1983 *QST*) raised two questions about implementing spread-spectrum communications. First, except for the intrinsic charm of using a pseudo-random hop sequence, it seems that only privacy is gained by using this technique. If interfering signals occur randomly on the hop frequencies, there seems to be no additional signal-to-noise ratio gain from use of a pseudo-random sequence rather than a sequential-hop sequence. Because of limited spectrum allocations in the hf bands, a long random sequence must map into a small number of available frequencies. As an alternative hopping strategy, I suggest that the frequencies chosen by the operators be visited sequentially, rather than randomly. This technique yields advantages of simpler equipment and minimal synchronization problems.

Second, following Mr. Sabin's suggestion of using a-m rather than ssb, there should be consideration given to heterodyne problems. An a-m carrier will mix with any interfering carrier and produce audio beat notes, which can be different for each hop frequency occupied by an interfering signal. This is certain to produce annoyances.

One possible solution is to use ssb, with the carrier reinserted at a level that allows phase locking at the receiver. Any phase-locked loop (PLL) frequency synthesizer suitable for transmission could then be used on reception. The acquisition range of the PLL can be restricted to prevent accidental lock to an interfering carrier. A loss-of-lock indicator used to signal a move to the next hop frequency could simplify the sync problem. For initial sync, the out-of-lock condition could allow fast jumps through the hop frequencies until lock is achieved. Once in lock, the hop rate would be controlled by the active transmitter. — *Paul Selwa, N9CZK, Brownsburg, Indiana*

☐ In response to Mr. Selwa's letter, consider 50 frequencies, sufficiently separated so that there will be no adjacent-channel interference. Fifty groups of stations, each group with two or more stations, could share these frequencies with no interference between groups. All of the groups would be part of a packet-network system. Protocol would include time and frequency management, and various control terminals would have rapid signal deployment capabilities for both domains. As an extension of this idea, consider a group of 250 frequencies. Using frequency-hop, each group could search for a clear operating frequency. When such a frequency is found, the group could elect to stay there until conditions deteriorate. Some members would want to hop on a set of frequencies in order to simultaneously address several groups.

The frequency hopping discussed above is done under protocol control. In addition, frequency hopping of the type described in my article could be used to provide increased reliability of communication during fading or interference. That is, some group(s) could hop on a set of the available frequencies mentioned above. (This hopping could be done sequentially, as Mr. Selwa suggests, rather than randomly.) If 50 frequencies were reserved for fast hopping, 50 groups could hop on these with no mutual interference when using protocol frequency management. The main advantage of non-sequential hopping is that errors from certain patterns of fading and interference (called burst errors) are more easily avoided. These hop sequences are easy to generate and supervise.

My article *does* mention ssb in a slow-hop system. Nbfm should also be considered. "Ancient modulation" is not a very attractive mode, I agree. Primarily, I would like to see an amateur data-packet network using spread spectrum. — *William E. Sabin, WØIYH, Cedar Rapids, Iowa*

SPREAD-SPECTRUM COMMUNICATION: A NEW IDEA?

☐ Spread-spectrum communication is the intentional "broadbanding" of signals by scattering RF energy over a range of frequencies, with only a small amount of RF on any one frequency at any given instant. Reception is achieved by synchronous "descattering" of the signals. The technique was first proposed for military use in the 1940s, but it has recently received attention from Amateur Radio operators. Is this a new idea?

The following is an excerpt from *Radio for Everybody,* by Austin Lescarboura, published in 1922. Lescarboura wrote about radio in the 1920s and '30s. He worked for Hugo Gernsback publications and also wrote speeches for David Sarnoff of RCA:

> The question of secrecy is an important one, for nobody cares to be talking to a relative or friend while one hundred thousand other persons are listening in. Just so long as the messages stay on the wires, they are private, but at the present stage of radio telephony, the moment these same messages are passed through the radio transmitter they become public property. However, this question of secrecy can be solved — and will be solved, in the very near future. There are several ways in which this end can be attained. Perhaps the ultimate solution will be obtained by a system of double or triple waves employed simultaneously for the transmission of speech so that unless a person has a receiving set which intercepts a certain combination of waves, only a small and almost unintelligible part of the conversation will be detected. It is also quite possible to use a device at the transmitting end which continually alters the wave length of the transmitted radiotelephone waves, while the receiving set is also provided with a means of altering its wave length in step with the transmitter.

That paragraph, written 62 years ago, proposes a spread-spectrum communications concept. The reasons stated may be different from those of today, and his timing prediction is certainly inaccurate, but the essence of the technique is there. — *Ken Johnston, W7LIX, Helena, Montana*

Chapter 6

The FCC Report and Order Legalizing Amateur Spread Spectrum

Before the
FEDERAL COMMUNICATIONS COMMISSION
Washington, DC 20554

In the matter of)
)
Amendment of Parts 2 and 97 of the) GEN
Commission's Rules and Regulations to) Docket
authorize spread spectrum techniques) No. 81-414
in the Amateur Radio Service.

REPORT AND ORDER
(Proceeding Terminated)

Adopted: May 9, 1985 Released: May 24, 1985

By the Commission:

INTRODUCTION

1. The Commission adopted a *Notice of Inquiry and Proposed Rule Making* ("Notice") on 30 June 1981 on its own motion[1] proposing changes in its Rules and Regulations to permit the use of spread spectrum emissions[2] in the Amateur Radio Service. Comments were received until 1 March 1982 and reply comments until 15 April 1982. In this *Report and Order* we are adopting our proposals as modified below and terminating the proceeding.

2. It is the Commission's intent to provide licensees in the Amateur Radio Service with the opportunity to experiment with and take advantage of this technology which has been, until now, almost entirely limited to costly military systems. Spread spectrum transmissions have been implicitly prohibited by Commission's Rules without regard to the potential benefits which might accrue to the public interest from their use. These benefits may include:

- Reduced power density thus reducing interference to narrow band communication systems;
- Significant improvements in communication under conditions with poor signal to interference ratio;
- Improved communication performance in selective fading and multipath environments;
- Multiple, nearly independent communication channels functioning simultaneously in the same spectrum.

The final rules adopted today authorize amateurs to develop, test, and operate low cost spread spectrum systems. This action is consistent with the basis and purpose of the Amateur Radio Service expressed in Section 97.1 of the Rules. By removing regulatory barriers to innovation, we believe that technical advances in radio technology can be stimulated, forwarding our goals as stated in Sections 7(a) and 303(g) of the Communications Act of 1934, as amended.

3. While this proceeding deals only with the use of spread spectrum transmissions in the Amateur Radio Service, we expect that the experience gained, especially on the subject of compatibility between spread spectrum transmissions and conventional narrow band systems, can be used in our more general rule making dealing with spread spectrum[3] and will be a stimulus to the general radio technology community.

DISCUSSION

4. Commenting parties focused primarily on the issues identified in the questions posed in the Notice. These dealt with interference and monitoring. In addition, comments were received dealing with frequencies available and eligibility of licenses to use spread spectrum transmissions.

5. *Intra-service interference*. Concerns were raised by commentors about the potential for extensive, broadband interference from spread spectrum transmissions. Mathematical modeling and various experiments[4,5,6,7,8,9] indicate that, in some cases, there is a risk of causing intra-service interference. However, they also indicate that there are conditions under which spread spectrum transmissions can be compatible with existing modes of communication. We believe that amateurs should be permitted the freedom to experiment and to add to the body of knowledge on this subject. The Amateur Radio Service has a long history both of experimentation and frequency sharing among licensees. We believe that this tradition is adequate to prevent intra-service interference in most cases. However, to emphasize both the experimental nature of spread spectrum as well as some of the potential benefits associated with it (see paragraph 2), we are authorizing such transmissions on the condition that they not cause harmful interference to and accept all interference from stations operating with emissions previously authorized. Additionally, we are persuaded by the comments that the broad allocations to the Amateur Radio Service above 420 MHz lend themselves better to power dispersing techniques than lower frequency bands, thus reducing the interference potential of spread spectrum transmissions. (See paragraph 12.) Therefore, frequency bands above 420 MHz will be authorized.

6. *Inter-service interference*. The National Association of Broadcasters (NAB) raised the subject of interference to television channel 2 (54 - 60 MHz) related to spread spectrum operation in the adjacent band (50 - 54 MHz) allocated to the Amateur Radio Service. NAB's principal concern was that uncontrolled amateur transmissions might fall outside the allocated band into channel 2. The Commission believes that NAB's concerns are not well founded. First, rather simple transmitter output filters can be used by amateur licensees to prevent positively out of band emissions. Second, licensees in the Amateur Radio Service have had no significant

history of operating outside the allocated bands. However, NAB's concerns are moot, as the Commission is not now authorizing spread spectrum transmissions in the 50-54 MHz band. In other frequency bands where the Amateur Radio Service has successfully shared allocations with different services, we expect no worsening of interference since the power density associated with spread spectrum transmission is much lower than the power density from currently existing narrow band transmissions having the same total effective radiated power.

7. *Monitoring.* Several parties expressed concern that the Commission and amateur licensees might not be able to monitor readily the station identification and content of spread spectrum transmissions. In PR Docket No. 81-699, the Commission dealt with the related matter of non-standard digital codes in the Amateur Radio Service.[10] In that *Report and Order* we stated,

> "In balancing our objectives of encouraging new technologies against ensuring our enforcement capability, it must be recognized that there is an incompatibility between authorizing experimentation with 'exotic' technologies and the employment of channel monitoring as an enforcement tool. Our ability to verify that the content of messages complies with our rule requirements will be hindered by the broad relaxation of regulatory constraints that we are ordering in this proceeding. However, the Commission [agrees]...that special provisions we are including in the final rules, as well as existing provisions that identification be made in plain English or the international Morse code, should, when combined with the zealous effort of the amateur community to protect their allocated frequency bands, provide adequate protection against unauthorized operation in the service."

In this matter, we are authorizing a new "exotic" technology, spread spectrum, with certain constraints intended to reduce the conflict between experimentation and monitoring. These are requiring station identification that can be received with common narrow band receivers (see paragraph 9) and limiting the spreading sequences and methods (see paragraph 10).

8. We recognized that more detailed standards than those given in the adopted rules are needed for amateur interoperability and self-monitoring on a convenient basis. We are reluctant to adopt more detailed monitoring standards as they might inhibit experimentation. However, we are delaying the effective date of the rule change by one year in order to give the amateur community time to develop initial voluntary interoperability standards as they have done recently in packet radio. In this interim period we are continuing our policy of granting STA's to those who wish to experiment in this area.

9. *Station identification.* As in the case of unspecified digital codes discussed in Docket No. 81-699, common narrow band methods of station identification will be required. Based upon our experience with Del Norte[11], we have added an additional option for spread spectrum transmissions. Stations will be permitted to identify by varying the emission while in the spread spectrum mode of operation so that CW, SSB, and/or narrow band FM receivers, which might be victims of interference, can receive the station identification. We expect amateurs transmitting spread spectrum to select the frequency used for narrow-band station identification taking into consideration the need to minimize interference to other amateur operations and to facilitate reception of station identification by other amateurs.

10. *Spreading sequences and methods.* Section 97.117 of our Rules prohibits "the transmission by radio of messages in codes or ciphers in domestic and international communications to or between amateur stations." The intent behind this rule is to prohibit encrypted communications in this service. A spread spectrum transmission that uses an unknown spreading sequence is intrinsically an encrypted communication. We proposed to allow as spreading sequences only the output of one of three specified linear feedback shift registers.[12] Thus a party interested in monitoring a transmission could do so by a short process of elimination with appropriate equipment. In order to eliminate ambiguity in the implementation of spread spectrum that might complicate monitoring, we have added a few details beyond what was given in the notice.[13] We are satisfied that our limitations of spreading sequences are sufficient, for domestic purposes, to see that spreading functions remain within the realm of codes and do not overstep the bounds of becoming ciphers which are prohibited by Section 97.117. Use of additional, more complex spreading sequences including Gold codes suggested in the Comments by Leon Scaldeferri, overstep the bound of becoming a cipher and, therefore, are not authorized. We proposed no limitations on spreading methods. However, after further consideration, we feel that a spread spectrum transmission that uses a complex method of dispersing the transmitted energy is also intrinsically an encrypted communication and, thus, oversteps the bounds of becoming a cipher. Therefore, the final rules limit spreading methods to frequency hopping and direct sequence only.

11. *International communication.* In addition to the clear message requirements of Section 97.117 of our Rules, the ITU Radio Regulations require that all international communication between radio amateurs be in plain language.[14] The Commission has, in the past, applied a more stringent test of what is and what is not plain language for purposes of international communications than for domestic communications. In Docket No. 81-699, the Commission recognized the desirability of international communication, but limited the codes authorized to internationally recognized codes recommended by the CCITT. Later, AMTOR, a commonly recognized amateur version of a CCIR recommended communication protocol, was authorized.[15] Until such time as spreading functions and spread spectrum transmissions become well recognized internationally, the Commission cannot consider authorizing its use for general international communication. Accordingly, we are authorizing spread spectrum transmissions for domestic communication. However, in the future the US Government may conclude arrangements with other administrations as permitted by the ITU Radio Regulations.[16]

12. *Frequencies.* In the *Notice*, we proposed that spread spectrum transmissions be limited to the 50, 144, and 220 MHz bands. E. Miles Brown and Luther G. Schimpf pointed out in their comments that these bands would allow only about 30 dB effective reduction in power density due to their limited bandwidth, probably not enough to prevent interference from strong spread spectrum transmissions in most cases. John A. Carroll observed in his comments that the large amateur frequency allocations above 420 MHz are more attractive places for spread spectrum operations because they provide significant additional bandwidth over which to dis-

perse the transmitted power. For these reasons, we are modifying our original proposal and authorizing spread-spectrum transmissions in all amateur allocations above 420 MHz except as previously noted.

13. *Eligibility*. In the *Notice*, we proposed to limit eligibility for spread spectrum transmissions to Advanced and Amateur Extra licensees. Several parties commented that this was overly restrictive and unjustified. We agree with these observations and are authorizing in the Final Rules spread spectrum transmissions for all licensees eligible for the frequencies where spread spectrum is permitted.

CONCLUSION

14. Accordingly, IT IS ORDERED that Part 97 of the Commission's Rules and Regulations, 47 CFR Part 97, is AMENDED as set forth in the Appendix effective 1 June 1986. It is further ORDERED that this proceeding is TERMINATED. This action is taken pursuant to authority contained in Sections 4(i) and 303 of the Communications Act, as amended. Further information on this matter may be obtained from Dr. Michael Marcus, Chief, Technical Analysis Division, Office of Science and Technology, Washington, DC 20554, telephone (202) 632-7040.

Federal Communications Commission

William J. Tricarico, Secretary

APPENDIX

Part 97 of the Commission's Rules and Regulations, 47 CFR Part 97, is amended as follows:

In Section 97.3, Definitions, add a new paragraph (cc) as follows:

* * * *

(cc) *Spread spectrum transmission*. An information bearing transmission in which information is conveyed by a modulated RF carrier and where the bandwidth is significantly widened, by means of a spreading function, over that needed to transmit the information alone.

Add a new Section 97.71, Spread spectrum communications, as follows:

Section 97.71 Spread spectrum communications.

(a) Subject to special conditions in paragraphs (b) through (i) of this section, amateur stations may employ spread spectrum transmissions to convey information containing voice, teleprinter, facsimile, television, signals for remote control of objects, computer programs, data, and other communications including communication protocol elements. Spread spectrum transmissions must not be used for the purpose of obscuring the meaning of, but only to facilitate communication.

(b) Spread spectrum transmissions are authorized on amateur frequencies above 420 MHz.

(c) Stations employing spread spectrum transmissions shall not cause harmful interference to stations of good engineering design employing other authorized emissions specified in the table. Stations employing spread spectrum must also accept all interference caused by stations of good engineering design employing other authorized emissions specified in the table. (For the purposes of this subparagraph, unintended triggering of carrier operated repeaters is not considered to be harmful interference. Nevertheless, spread spectrum users should take reasonable steps to avoid this situation from occurring.)

(d) Spread spectrum transmissions are authorized for domestic radio communication only (communication between points within areas where radio services are regulated by the US Federal Communications Commission), except where special arrangements have been made between the United States and the administration of any other country concerned.

(e) Only frequency hopping and direct sequence transmissions are authorized. Hybrid spread spectrum transmissions (transmissions involving both spreading techniques) are prohibited.

(1) Frequency hopping. The carrier is modulated with unciphered information and changes at fixed intervals under the direction of a high speed code sequence.

(2) Direct sequence. The information is modulo-2 added to a high speed code sequence. The combined information and code are then used to modulate an RF carrier. The high speed code sequence dominates the modulation function, and is the direct cause of the wide spreading of the transmitted signal.

(f) The only spreading sequences which are authorized must be from the output of one binary linear feedback shift register (which may be implemented in hardware or software).

(1) Only the following sets of connections may be used:

Number of stages in shift register	Taps used in feedback
7	[7,1]
13	[13,4,3,1]
19	[19,5,2,1]

(The numbers in brackets indicate which binary states are combined with modulo-2 addition to form the input to the shift register in stage 1. The output is taken from the highest numbered stage.)

(2) The shift register must not be reset other than by its feedback during an individual transmission. The shift register output sequence must be used without alteration.

(3) The output of the last stage of the binary linear feedback shift register must be used as follows.

(i) For frequency-hopping transmissions using x frequencies, n consecutive bits from the shift register must be used to select the next frequency from a list of frequencies sorted in ascending order. Each consecutive frequency must be selected by a consecutive block of n bits. (Where n is the smallest integer greater than $log_2 x$.)

(ii) For a direct sequence transmissions using m-ary modulation, consecutive blocks of $log_2 m$ bits from the shift register must be used to select the transmitted signal during each interval.

(g) The station records shall document all spread spectrum transmissions and shall be retained for a period of one year following the last entry. The station records must include sufficient information to enable the Commission, using the information contained therein, to demodulate all transmissions. The station records must contain at least the following:

(1) A technical description of the transmitted signal.

(2) Pertinent parameters describing the transmitted signal including the frequency or frequencies of operation and, where applicable, the chip rate, the code, the code rate, the spreading function, the transmission protocol(s) including the method of achieving synchronization, and the modulation type;

(3) A general description of the type of information

being conveyed, for example, voice, text, memory dump, facsimile, television, etc.;

(4) The method and, if applicable, the frequency or frequencies used for station identification.

(5) The date of beginning and the date of ending use of each type of transmitted signal.

(h) When deemed necessary by an Engineer-in-Charge of a Commission field facility to assure compliance with the rules of this part, a station licensee shall:

(1) Cease spread spectrum transmissions authorized under this paragraph;

(2) Restrict spread spectrum transmissions authorized under this paragraph to the extent instructed;

(3) Maintain a record, convertible to the original information (voice, text, image, etc.) of all spread spectrum communications transmitted under the authority of this paragraph.

(i) The peak envelope power at the transmitter output shall not exceed 100 watts.

In Section 97.84(g), Station identification, add new subparagraph (5) as follows:

* * * *

(5) When transmitting spread spectrum, by narrow band emission using the method described in (1) or (2) above; narrow-band identification transmissions must be on only one frequency in each band being used. Alternatively, the station identification may be transmitted while in spread spectrum operation by changing one or more parameters of the emission in a fashion such that CW or SSB or narrow band FM receivers can be used to identify the sending station.

Notes

[1]See *Notice of Proposed Rule Making*, 87 FCC 2nd 972 (1981), 46 FR 49617 (7 October 1981).

[2]Spread spectrum systems were originally developed for military applications where covertness and jam resistance were sought. In a spread spectrum system, an information signal is combined with a much wider bandwidth noise-like signal to yield a transmitted signal which is both broad band and noise-like. At the receiver, a copy of the original noise-like signal is used to derive the information signal. Because the energy of the transmitted signal is dispersed in the spreading process, it is less likely to cause interference in narrow-band receivers than a conventional signal of the same power. For further information, *see* "The ARRL Handbook for the Radio Amateur" (1985 edition), pp. 21-6 through 21-9.*See also* Spread Spectrum Techniques, Report 651-1, CCIR Volume 1, Kyoto, 1982.

[3]*See Notice of Inquiry* [in GEN Docket No. 81-413], 87 FCC2d 876, 46 FR 51259 (19 October 1981), and *Further Notice of Inquiry and Notice of Proposed Rule Making*, 49 FR 21951 (24 April 1984). The *Further Notice* proposes to authorize spread spectrum transmissions in three contexts: certain law enforcement applications, low power unlicensed (Part 15) devices above 70 MHz, and moderate power devices in three ISM bands.

[4]Feinstein, Hal and Paul Rinaldo, AMRAD Report to the FCC on Spread Spectrum Experiments, 3 October 1983.

[5]Report of Experience of Del Norte Technology, Inc., attached to Reply Comments of Del Norte Technology, Inc. in GEN Docket No. 80-135 [In the matter of revisions of Parts 2 and 90 of the Commission's Rules and Regulations to permit inland assignment of frequencies in the 420-450 MHz band for non-Government radiolocation, 47 FR 34415 (9 August 1982)].

[6]Comments of the American Radio Relay League in Response to Further Notice of Proposed Rule making, Appendix A, in GEN Docket No. 80-135.

[7]Report of Brian Elliott, Ph.D., on Hewlett-Packard's experience with wireless data terminals, 26 October 1984, attached to the reply comments of Hewlett-Packard Company in GEN Docket No. 81-413.

[8]Memorandum to Chief, Research and Analysis Division from Chief, Research Branch on the subject of Del Norte Spread Spectrum Demonstration on January 29, 1981, dated 19 February 1981.

[9]Considerations of Interference from Spread-spectrum Systems to Conventional Voice Communications Systems, Report 652, CCIR Volume 1, Kyoto, 1982.

[10]*See* PR Docket No. 81-699, In the matter of the use of additional digital codes in the Amateur Radio Service, 47 FR 42751 (29 September 1982).

[11]See footnote [5].

[12]Information about linear feedback shift registers may be found in: Dixon, R. C., *Spread Spectrum Systems*, Chapter 3, John Wiley & Sons, Inc., New York, 1976; Pickholz, Raymond L., Donald L. Schilling, and Laurence B. Milstein, "Theory of Spread-Spectrum Communications - Tutorial," *IEEE Transactions on Communications*, Vol. *COM-30*, pp. 855-884, May 1982.

[13]These details describe how the spreading sequences are used to increase the information rate of the signal to be transmitted. In the case of direct sequence transmissions, the licensee may select a value of m to be used in m-ary modulation. In the case of frequency hopping, the licensee may choose the number of frequencies, x. (See the new Section 97.71(d) in the Appendix.) While both m and x are the choices of the licensees, their values can be determined by observing transmitted signals.

[14]No. 2732 of the ITU Radio Regulations, Geneva, 1979.

[15]*See* RM-4122, *Order*, In the matter of Authorization of the digital code "AMTOR" for use by stations in the Amateur Radio Service, 48 FR 7457 (1983).

[16]No. 2734 of the ITU Radio Regulations, Geneva, 1979.

Chapter 7

AMRAD Chronicles SS Progress

Having published a considerable number of articles on spread spectrum theory, the AMRAD Newsletter *in 1985 set about to turn theory into practical applications. The articles that follow confirm that, although amateur spread spectrum is still evolving, it's far more than an abstract theory.*

July 1985
SKYWAVE PROPAGATION AND SPREAD SPECTRUM

Skywave propagation is one of the more enjoyable modes of amateur communications chiefly because it supports long distance communications. Direct line-of-sight communications is characteristic of the VHF bands and is the predominant mode of propagation for our discussions of spread spectrum. In this edition of the column we depart from this norm and consider HF spread spectrum.

A short while ago, some of us began to wonder about spread spectrum in the HF bands and in particular, the possibility of DX using spread spectrum modes. What we found is that the ionosphere is intrinsically tied to skywave spread spectrum in ways other transmission modes are not. The ionosphere is a much more complicated and troublesome "reflecting" media than most radio amateurs suspect.

The difficulties with the ionosphere for spread spectrum are not the DXer's problem of determining when the band will be open to Danger Island or Lundi but in getting even a simple one hop skip out of it. The ionosphere is not "well behaved" as mathematicians would call it. It is like an ever changing lens which bends and distorts the RF passing through it, or like looking at the moon reflected from a waveswept lake. The ever changing nature of the ionosphere becomes evident when you attempt to transmit high speed signals which are synchronized with a receiving clock on the ground. At low speeds, you can only detect gross behavior changes in the ionosphere. With high speed systems you are depending on the microstructure of the ionosphere to help by being stable; it does not and that is today's topic.

Most amateur transmission techniques do not involve a precise, synchronized clock running at the receiver. We begin to see such things in high speed data transmission where the system works because the signal energy arrives with a fixed predictable transmission delay. In order to establish a constant timing, it is assumed that the path length between the transmitter and receiver will not alter. In line-of-sight transmissions this is commonly the case for fixed stations. If the path length does not alter, then the time it takes for energy to reach the receiver will be more or less a constant.

In skywave propagation, common modulation techniques such as SSB and RTTY (even packet) appear to be "reflected" back to earth in the way familiar to HF operators. In this simple model knowing the takeoff angle of the transmitted energy and the height of the "reflecting" layer allows the skip

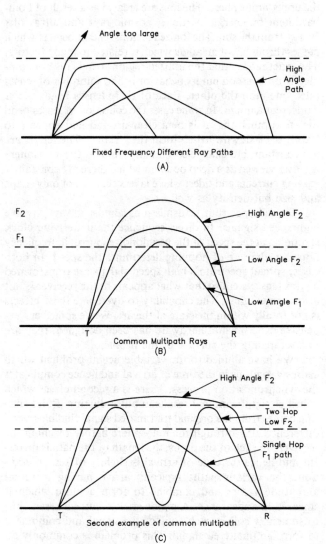

Fig 1—At A, different ray paths are illustrated. B shows common multi-path rays. C is a second example of common multi-path.

distance to be computed from a simple triangle model of the reflective process. Yet, if a precise high speed clock were transmitted along with the signal we would find (ignoring multipath for a moment) that the clock would wander significantly while the overall signal appeared nearly constant.

The clock wandering identifies the fact that the travel distance of the signal is not constant. The clock allows us to have a kind of radar. It shows height changes in the ionosphere and as in other radars, the smaller degrees of motion can be detected by increasing the clock speed. The variations tell us that there is a lack of significant stability in the height of the "reflecting" layers of the ionosphere. There is some gross regularity in the reflecting layers' height as observed by the fact that we can communicate to fixed locations but when we put the high speed clock on our transmitted signal we have devised a kind of ionospheric probe which is reporting back unsuspected detail about what is really going on up there.

The ionosphere is not a simple collection of static "reflecting" mirrors but a complex and continually changing region rich with ions dislodged from the various gases of the upper atmosphere. The ions are created as a result of bombardment by energic particles, cosmic rays and ultraviolet energy from the sun. The ionized regions tend to bend rf which passes through it at an angle which is related to many factors. The exact reason that the ionized regions bend rf is not completely understood but its behavior is like a number of lenses piled one atop the other. Each of these lenses bends the rf a different amount. In some cases the collection of lenses bend the rf so much that it is bent clear around and appears to "arc" back downward to form the "skip." The lenses are not constant, but change continually. In fact, the ionosphere is better viewed as a deep ocean in which there are constantly varying currents and tides; where layers change not only shape and size but density as well.

Spread spectrum transmissions almost always involve high speed signals. In direct sequence the underlying clock is running at the speed of the pseudonoise source; in frequency hopping, the rate of hopping determines the speed. In both cases, spread spectrum's high speed signals are transformed by this lens system so that what appears at the receiver is not always intelligible. The capability to overcome these effects is not totally within the state of the art. While some very expensive experimental hardware has been designed, they are just scratching the surface of the problem.

We have alluded to the variable height problem which causes variations in the time of arrival and hence complicates the synchronization process. There is a second effect which is just as significant and results in pulse slurring. Pulse slurring comes from the original transmitted signal finding several different paths through the ionosphere before reaching the receiver. In each of the paths, the length of the path is different and hence the time of arrival is slightly offset. In addition, the various paths typically arrive having different amplitude values and combine to form a pulse which is smeared in appearance. A major goal of current research is to somehow collect each copy of the signal and combine it to form an unaltered signal. This problem is commonly attacked with digital signal processing techniques and is very expensive.

To add to the true dimension of the issue, recall that the "reflecting" structures in the ionosphere are not fixed but vary in height over time. Thus each copy of the signal has its path length changing and often in an independent fashion from each other! So if you built a digital filter which could recombine the copies of the signal it would have to be very clever and be capable of following each copy of the signal as it wandered around. Remember also that the amplitude would also tend to change as the signal wandered.

Now, having considered the wandering heights and multipath effects let us describe just one more significant effect. The height at which a "reflection" takes place is frequency sensitive. A frequency hopper with a wide hopping range would find that hops spaced far enough apart would travel different amounts of time. While one hop would arrive in short order by virtue of it having been reflected from a lower zone, other hops would take much longer and travel farther up before being reflected back down. Indeed, at some fast hop speed it is possible for two hops to trade places; the first hop traveling far up into the ionosphere before being sent back down, the second one reflecting off a relatively lower layer and appearing first.

Consider how you might design a receiver to overcome this problem. One obvious way is to restrict the hopping bandwidth so that all hops occur within a small range of frequencies and reflect from the same layer. Perhaps you would just make restrictions on the hop speed so that these effects are not significant. This would put a restriction on hop speed to about 10 hops a second and certainly not more than 30. Then perhaps you might design lagging receivers which "hung around" while the hops were returning to earth, or a variable hopping system which automatically adjusted its speed depending on the height a pulse goes. As you might expect, there are no simple answers to this problem.

One experimental spread spectrum system combines very expensive customized hardware, a built-in ionospheric sounder and a bunch of 68000 microprocessors to actually adapt to and follow all these effects.

Because all these effects are accurately modeled by "linear" systems it is possible to filter most of the effects out if you can construct an "inverse filter." Since the effects are changing in time, you will have to continuously recompute the inverse filter. This is a computationally difficult job and the computer can stay ahead of the problem only if just a few adjustments are needed in a minute's time or so. Experiments in HF have shown that gross features of the ionosphere stay similar for about five minutes time in a quiet ionosphere. This means that only the smaller variations would have to be taken care of, which vary much more frequently. When the ionosphere is not quiet, such as the dawn and dusk transition or other solar event, the gross features do not stay put for long. The inverse filter must be computed much more often and the workload typically outpaces what the computer can do. Remember, even after you have an inverse filter computed, you must now apply it to the incoming signal. This requires still another computer and this computer must keep up with everything the other computers are doing and also apply the inverse filter to the incoming signal.

This part of the column will give some idea of the type of effects to be found in the ionosphere.

The zones within the ionosphere are called by letters of the alphabet. Those involved in rf bending are typically the D, E and F. It is common to distinguish two F layers called F1 and F2. The E layer comes into play occasionally as sporadic-E propagation. D layer is prominent for night time broadcast band skywave and the F1 and F2 are responsible for common HF skywave propagation.

While propagation is commonly attributable to the D, E, F1 and F2 regions, there are other modes. One such mode is called ionoscatter which is an incoherent scattering of energy which occurs above the maximum usable frequency (MUF). This mode is not well known and is almost never used in ham communications.

Signals which are bent back to earth have been continuously bent so that they "arc" back down. Fig 1 shows this effect. The amount of bending is frequency dependent and

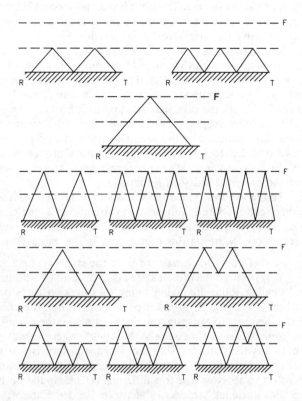

Fig 2—Typical multi-hop paths.

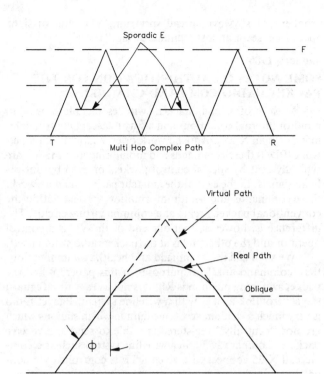

Fig 3—Multi-hop complex path.

as the operating frequency approaches the MUF, less and less bending occurs. Above the MUF, bending still occurs but is insufficient to "arc" the signal back down to earth. Instead, the signal may shoot off into space at some angle or it may get bent at an angle which projects it into another zone of the ionosphere. When this occurs we have multimode propagation and the signal may reemerge almost anywhere. This is the mechanism behind ionoscatter propagation.

So much depends on the angle at which the signal hits the ionosphere (Fig 2). Many antennas including some beam antennas simply push a signal out in a given direction. Energy hits the bending layer of the ionosphere at many different angles. Some angles will cause the signal to travel far up into the layer before being reflected. This will result in a very long skip zone and is a strange characteristic of an almost 90 degree launch angle. Low angles offer the next best distance and higher angles are equally good but with shorter distance. At some angle which is almost 90 degrees bending is insufficient to arc the signal back to earth and the signal may shoot off into space or another layer to begin a multimode path.

It is important to remember the two factors which determine bending: arrival angle and frequency. These two can interact in interesting ways. Fig 3 shows such an unusual but not uncommon path. A signal hits a layer too sharply to be bent back to earth. Instead, it is bent at some angle and goes up to a higher layer where it is bent back down. So the signal travels down, but in the process, it hits the upper side of the first reflecting layer and is sent back up! Now it bounces back and forth between the upper layers until it comes to a place were the sun is not visible. Here the MUF has dropped and so the lower layer cannot bend the signal around and send it back up. Instead it just applies a little bending before the signal passes down to the ground somewhere totally unexpected.

When sporadic-E is present it contributes to these effects as another reflecting layer does. It is important to remember that all the effects described here are not always present all the time.

It is interesting to speculate on the ionosphere. It seems that signals with low information bandwidths only are affected by slow changing gross variations. Occasionally two or more paths will arrive out of phase resulting in "fading." In fact if you watch the fading on your HF receiver from some shortwave station you can see the ionosphere at work. The fading seems to "drift" back and forth. In fact, many times this is the sum of two signal paths shortening and lengthening so that the two signals are varying just around 180 degrees out of phase. This is sometimes called selective fading.

Another interesting effect is that often the ionosphere sends back two copies of a single path. The first is called the ordinary path and the second is called the extraordinary path. Typically, they are offset in distance but the extraordinary path is out-of-phase with the ordinary path. Occasionally, the two overlap and this causes them to cancel each other out.

Since the ionosphere seems to affect low bandwidth signals less than high bandwidth signals we could say it is a kind of low pass filter. Whenever spread spectrum signals utilize wideband components, they will be affected by this "filter" and will emerge not unchanged.

It is ironic that hams have been locked out of HF experimentation since this appears to be the place where the most unsolved problems with spread spectrum are. The rules governing amateur spread spectrum forbid HF operations except by STA from the FCC. Yet, in light of the intricate problems of skywave spread spectrum, VHF line-of-sight operations seem almost "uninteresting."

January 1986

SOME NOTES ON AUTHENTICATION FOR THE PACKET RADIO COMMAND CHANNELS

Most professional packet switches include a way to monitor internal conditions and a way to accept control commands. In an X.25 private network or even a public data network (PDN) these commands and monitoring information are typically sent by special control packets or even by out-of-band signals. In the case of the amateur packet radio network, these commands and health information are sent within the conventional packet stream as a common virtual circuit. The difference is, however, that one end of the VC is a control operator and the other end is at the packet radio switch itself.

We shall call this command and health communications the "command link" for purposes of this paper. When the packet switching medium is RF, it is not as easy to safeguard the control link as it is with a wireline system. Packet radio is very much a random access medium in which stations which are not "centrally" registered are able to send and receive packets. Comparatively, in a wire line system such as a commercial X.25 connect to a given DTE is controlled via connection security control (not part of X.25 but sometimes present in special networks), a name server facility or a closed user group. With these facilities it is easy to monitor and control special "sensitive" DTE addresses located in the network.

Even above the use of special connection security techniques many networks go further and use authentication techniques to prove that the connection is originated and owned by an authorized caller. Since the current formulation of the amateur packet radio system lacks many of the special controls mentioned above, we proposed a simple authentication scheme for control operator-to-packet radio switch. This method is reviewed is more detail in the *ARRL Amateur Radio Computer Networking Conferences 1-4*[1] and I will give a simplified discussion of it here.

1. What is authentication anyway?

Authentication is a method by which end parties (sometimes called the "principals" in banking circles) can be sure that messages which they receive from each other in fact have originated from the other party and have not been tampered with.

There are different types of authentication depending upon the aspect of the problem to be solved; however, in each case we employ a cipher and secret key as a form of electronic signature.

2. What are the different aspects of the authentication problem?

First, guaranteeing that the message comes from the other party is the primary objective. If we didn't guard against this, we could get fooled by a "false origination" attack and couldn't tell if a message was from our other party or not. Consider the case of the stock broker who receives a cable instructing him to sell a certain stock at a low, even though the plan was to sell it at some specific high. Some time later the true principal gets a check for the proceeds of the sale which he did not initiate. The stock broker is innocent because there was no previously agreed upon authentication signal used to prove the originator's authenticity.

The second type of attack is to take an authenticated message and tamper with the contents. Suppose the authenticator was a secret password affixed to the letter which changed each day depending upon a codebook which both parties keep secret (i.e., both the principal and his stock broker). An attacker could intercept the letter, retype it using the password-of-the-day contained on the genuine letter and send this one forward to the stock broker. Again, the stock broker would have no way to know that the contents were false, but only that the principal did send a message.

To answer this type of attack requires message content authentication. Simply, this involves computing a checksum over the message and then encrypting the checksum value with a secret key. More on this aspect later in our discussion.

3. Are there any other ways to trick the stock broker?

Yes, suppose the principal wants to do a lot of wheeling and dealing today involving many messages to the stock broker. One thing the attacker could do is switch the order of the messages to cause havoc. For example, in the morning the principal sends a message to the stock broker instructing him to buy ten thousand shares of Borden Enterprises at 15 and a half per share. The attacker intercepts this message and holds on to it. Now due to subtle market manipulation by the "institutional" traders, a share of Borden Enterprises begins to rise and by later afternoon it has risen an amazing five points. Now the attacker releases the message allowing it to be delivered to the stock broker.

The stock broker, acting on instructions contained in this message which is both originator authenticated and content authenticated buys ten thousand shares of Borden Enterprises at 20 dollars-per-share. Now, the stock broker's institution goes to draw the required twenty thousand dollars from the principal's ready cash account to find only fifteen thousand available. Very irregular to say the least. What with a man of the principal's reputation—indeed!

This illustrates the need for time-of-origination authentication. It is easily provided by time stamping the message and placing the exact time as part of the message text. Hopefully, the message transmission system will not take too long to deliver the message, but if this is the case, then some execution time window is needed.

4. OK, so far we've blocked false origination, message content tampering and delivery. That's all there is. Right?

No, in fact there is one more attack that needs to be guarded against. Returning to the last example, suppose that the principal decides to sell Borden Enterprises at twenty-one dollars per share. The attacker could simply decide to lose the sell message. Moreover, the fact that the stock was never sold might not be discovered at all for some time.

This is called a message deletion attack but it is easy to guard against. Besides a time stamp we also introduce a message number as part of the message text. The stock broker then checks to see if the message number is the next one he was expecting, and if so, executes the order. If it is some higher number, then some messages were "lost" and a recovery procedure, commonly using the telephone is employed.

5. Is there some way the principal can get an acknowledgment that his orders were executed?

Yes, here we would have to enter the world of end-to-end acknowledgment. Computer protocols are resplendent

---

[1] P. Newland, AD7I, "A Few Thoughts on User Verification Within a Party-Line Network," *ARRL Amateur Radio Computer Networking Conferences 1-4*, ARRL, Newington, 1985, p 4.89.

with approaches to this problem, yet most are less than satisfying. It is almost trivial to send back an acknowledgment but suppose it is lost, accidentally or deliberately. Do you send another order to the stock broker assuming he didn't get the first one? Moreover, acknowledgment can be faked and so, you have an authentication problem within an authentication problem.

Taking this to its illogical conclusion, you would discover that you have created something like a picture of a rabbit pulling itself out of a hat! Or as the country folk say around here, "you've gotten yourself wrapped around the axle."

6. Now I think I understand: we have to guard against false origination, message tampering, timely delivery and message sequence. Do banks really do this?

Well, more are doing it. In fact, the biggest banks are designing authentication standards. Specifically, the American Banking Association (ABA) has developed an authentication standard called FIMAC (Financial Institute Message Authentication Code) which is now ANSI X9.9. This authentication standard solves a lot of the problems discussed above and is being offered as a commercially developed box that would go with an electronic funds transfer (EFT) terminal. In fact the US Treasury Department is very active in promoting X9.9 to safeguard EFT transactions.

7. Now I know about authentication, but how does this apply to packet radio?

Every packet network contains some kind of command and status monitoring channels. These channels can take different forms, for example some systems use internal virtual circuits to connect the monitoring processors with individual network equipment. Commands arriving over these channels command the network equipment to perform different types of functions necessary for network operations.

A common usage of the command channel is to transmit specialized routing tables which can be used to reconfigure the network in case of outages. These tables are used by individual packet switching nodes to route traffic.

A second common usage is to trigger network components to perform self-diagnostics and in some cases equipment reconfiguration. If these channels could be accessed by anyone, it would be quite possible to crash the network either by accident or maliciously. Authentication is a way to keep unauthorized users out of the sensitive equipment.

8. What types of authentication are needed on the network?

It depends on the type of operation. Most simple short commands fit the transaction model of communications. Transactions are commonly composed of a single message and should contain its own authentication. Session oriented communications consist of a bi-directional stream of messages requiring protection of the message stream.

For most amateur radio applications, authentication can be limited to user authentication unless intelligent spoofers or jammers are present. Alteration of message content is possible but requires a lot of work to get right. Insertion of a false message is possible, so some form of message number would be required. Delaying or deleting (perhaps using jamming) should be detected by the semantics of the channel. That is, the structure of the command sequence used on the channel should detect missing messages. Positive acknowledgment by the packet switch would alert the operator to missing messages.

9. What type of sequence can be used for user authentication?

Typically, a secret password of some kind. For example the following sequence will do the trick:

Packet switch:
  seed = RANDOM-NUMBER;
  packet token = ENCRYPT(seed,password);
  send(seed);

Operator:
  receive(seed);
  ENCRYPT(seed,password);
  send(seed);

Packet switch:
  receive(seed);
  does seed = packet-token?
    yes: Operator has good password—give him control.
    no: User has invalid password and should not get control.

This sequence authenticates the operator without revealing the secret password.

10. What sequence can be used to authenticate individual messages?

There are several different ways to do this. One method simply uses some sequence of things which only the operator and packet switch know. One such sequence is:

  s1 = ENCRYPT(seed,password)
  s2 = ENCRYPT(s1,password)
  s3 = ENCRYPT(s2,password)

Using the seed from the individual authentication step described in 9 you can get an unpredictable number (s1, not the seed). Since these depend on the password, hopefully only the packet switch and the operator should be able to generate the sequence of s-values. Placing these at say, the back of a message will prevent someone from inserting a phony message.

11. What about protecting the text?

Once more, there are a few ways to do this. In most cases protecting the text means generating a checksum. A little thought is required for checksums used to detect text modification. Consider that simply adding up the hex value of each character and taking the sum modulo 256 (8 bits of sum) means there are many different combinations of text that will add to the same thing. The banking people (ANSI X9.9) use a scheme in which they combine each block of text (8-character blocks) with the last encrypted 8-character block. The result is then encrypted in a sort of feedback arrangement. The contents of the message are not replaced by the encrypted data; instead, the very last encrypted 8-character output is used as the checksum and inserted at the back of the message.

The receiver simply repeats the above described feedback chaining operation and compares his checksum with the one transmitted. A match indicates the message is good; however, a mismatch indicates something has gotten altered.

A simpler checksum which does not have all the elegance or power of the banking approach is simply to add up the hex values using a 16-bit sum and add it to the seed to generate the unpredictable value. You might want to experiment with different checksum procedures. Sums of the kind I have described are susceptible to attack by substituting text values with equivalent checksum values. Error detecting codes are certainly one way to generate a good checksum.

The above procedure is known as the modification detection indicator (MDI) and is necessary if you think that an opponent is smart enough to pull off a spoof where he replaces the text with another.

12. What are the hidden problems?

Probably the most important is error recover and resynchronization. You may want to do a little thinking about this, since it can spell the difference between a workable system and one that just helps jam the channel.

13. Are there other approaches to authentication?

Yes, the one discussed here is based on a cipher in which both the switch and the operator share the same secret key. Two things that are desirable in networks are to allow many authorized users to each have their own key and second to somehow protect the key in the switch from what is called the "midnight" attack. The midnight attack means someone sneaks into the switch and steals the key.

July 1986
WILL THE *REAL* LPI PLEASE STAND UP...

One of the most intriguing aspects of spread spectrum is the so-called low probability of intercept (LPI), which gives spread spectrum the ability to be difficult or impossible to receive by conventional radio receivers. More than just an intriguing feature, it is this property that will allow the amateur spread spectrum user to share a ham band with narrowband users.

Ever since our infamous two meter beacon test a year or so ago, I have been thinking about the LPI property and its place within amateur spread spectrum. As was reported in this column, for the beacon test the FCC field operations people brought out a few enforcement trucks to monitor and participate in a hidden transmitter hunt. We hid the frequency hopper in a Chevy van (we were mobile, you see) and drove up to a local motel, right beside this huge radio tower with lots of microwave and land mobile stuff on it. Our reasoning was that all those signals would make DFing our signal difficult.

Well, it took the FCC people about 10-15 minutes to drive up to the motel parking lot, and another minute or two to figure out which car we were in. We had deliberately taken a different route to the motel so that we would not drive past the operations command post at the Bob's Big Boy in Vienna, VA. This way we were reasonably sure the FCC folks had not seen our vehicle.

After the hunt, we had a debriefing session. The FCC people said they would have been there sooner except the traffic in this Washington suburb is heavy on weekdays.

Well! so much for undetectable transmitters. We had a chance to see what the signal looked like from the FCC's spectrum analysis truck. It was clear that the signal wasn't hidden at all but, in fact, stood out rather vividly against the "grass" on the spectrum analyzer.

Other results were also available from the beacon test. First, we had a homemade repeater located directly across the street from our 50 watt beacon. This repeater consisted of two hand-held radios tied back-to-back and is used as a local repeater. I like to think of this as the worst possible case. Indeed repeater users reported hearing a "double-click" every time the beacon swept through its channel table (about 10 seconds for 400 channels).

What was this double-click? Upon analysis we found that one click was the hopper landing on the repeater input and the second click was the hopper landing on the repeater output! Remember that this repeater was just across the street. While this was the worst case we discovered, we found that the key to getting LPI for conventional narrowband receivers is to eliminate the click somehow.

A ham located a few miles away didn't notice the click was present until it was pointed out to him. Still, on his regular equipment, it was just barely noticeable. In order to get a better idea of what was going on, he used his OSCAR two meter equipment. This consisted of a side-by-side set of beams and a more sensitive than average receiver. The result was a very noticeable click every time the hopper passed by.

Now I want to mention just two more reports before giving you my thoughts on the subject. Located about 45 miles out from our transmitter site in suburban Washington is the Blue Ridge Mountain chain. Not very high by Western standards they typically get about 3000 or so feet. The Blue Ridge is home to a number of hams, one of whom sent us a report. Up on his mountaintop, he was amazed to discover one day that there was some kind of clicking going on in his two meter receiver. Upon investigation he found it was, sure enough, all over the band!

This ham's equipment was standard, and it came as no surprise that he was able to get a strong signal from our hopping beacon. You see, he gets strong signals from all over the area, having a line-of-sight to most of them.

The last observation is from our own monitoring effort in the two meter band conducted during the beacon's operation. We spent about six hours, late one Saturday night, locating repeaters that were triggered by our hopping signal. We then removed their frequencies from the beacon's table.

Now, not all repeaters responded to the hopper's signal. In fact, most repeaters have an impulse noise feature which allows them to ignore one shot impulses. Yet a group of repeaters were not so equipped and got triggered by our hopping beacon. One was a good distance away, and upon careful review it was discovered that it had line-of-sight to our beacon location by virtue of the fact it was on the tallest building in the county.

There is a lesson here about LPI which is no real surprise. If you transmit energy, its strength at a receiver will be related to how much signal is lost getting to the receiver, and in addition, to how long the signal is available to the receiver. For a frequency hopper, the signal is available to the receiver for just an instant. So, the signal received will be a function of transmitter power and path loss.

In the case of a ham located a few miles away, as in our second example, the signal from the hopper was weak, forming a mild click. When he used his OSCAR equipment the signal became much more noticeable. Using an antenna with gain (the side-by-side beams) and a more sensitive receiver "amplified" our hopper signal, just as though we had increased power.

The mountaintop ham or the repeater on a high building, both with line-of-sight to our transmitter site, experienced little signal lost in the usual terrain loss of the above example. These hams heard our signal loudly when it visited their frequency.

Thus, in order to be unnoticeable, you must keep the total transmitted power down below some low threshold. Yet, whether a receiver will hear a spread signal depends more on the unintended receivers than on things you do.

Receivers close-in to the transmitter (a mile or so) have not experienced enough terrain loss to reduce the signal energy sufficiently, so they hear the signal as a loud click. Outside this "radius" the signal has been sufficiently attenuated to be less noticeable.

By adding gain through beam antennas or a more sensitive receiver, it is possible to raise the equivalent power to the receiver. Thus it would receive as much signal from the hopper as if it were closer and the signal had experienced less attenuation due to terrain loss. Thus: adding gain is just like moving the receiver closer to the transmitter.

Lastly, much of the terrain loss is not present with our mountaintop ham so he sees a much stronger signal naturally, without the aid of antenna gain. From our terrain loss model the mountaintop ham is very close indeed to our beacon.

In somewhat more abstract terms, you can view a radio system as composed of gain elements and loss elements. Antennas, amplifiers, and even what modulation scheme you are using can provide gain. The transmission line, path loss, receiver noise, interference, and QRM are loss elements. All radio systems have different amounts of these loss and gain elements which commonly indicate the performance of the system. In turn, performance is measured by the quantity of information transmitted error free for a given amount of system gain.

A second lower limit can be calculated from Shannon's channel capacity theorem which states that for a given channel noise and rate of information transmission (bits/sec) you must use a defined level of system gain. If you use less gain, your error rate increases. The only way to lessen your error rate is to slow down the transmission rate. Notice that using error correcting codes slows down your transmission rate by cleverly adding redundancy (as in error correcting codes) or retransmission of wrong data (as is done in ARQ).

This implies that we cannot continue to reduce the system gain without at some point making the signals too weak to use!

The trick in spread spectrum is to design the system with an acceptable maximum error rate for the intended receiver (the spread spectrum receiver) while keeping the error rate to the unintended receiver acceptable. This might be termed a minimum interference criteria which could be used in industrial applications but, perhaps, not in ham radio.

The reason is that hams are used to clean radio spectrum. In the VHF bands hams are used to clean signals and interference free signals. There are many stations inside our minimum terrain loss radius and numerous ones up at high points and using beams and sensitive receivers. It seems that no matter what signal energy your spread spectrum station transmits, it will be received as a click by one of these stations.

Well then, what is LPI all about? I am coming to the conclusion that there is no future in slow hopping in the ham bands. Because of the wide geographic and gain distribution of ham stations, we can have no real control over this aspect of the system equation. Increasing the hopping speed and thus reducing the interval upon which power can be collected by a conventional receiver is still open to exploration, but it may require a few hundred hops per second or perhaps a thousand!

Fast hopping produces large sidebands because of the pulse-like behavior of the signal. Some of the total energy of the signal is "spread" in these sidebands and so the maximum energy in the carrier's "center" frequency is reduced. Thus, in addition to a reduction effect in a conventional receiver due to the shorter dwell time, the "spreading" of energy to the sidebands lowers the peak energy at the center frequency and further reduces gain. At what point this will produce a usable system is open to experimentation.

I would like to close with an observation that answers the opening question of the column. Will the real LPI please stand up? LPI of spread spectrum is not a way around the laws of physics. Instead, spread spectrum is a signal which is more easily camouflaged, by QRM and existing interference, from attempts to discover it. It changes the appearance of a narrowband signal to something which is more easily hidden.

The way to think of LPI is more along the lines of camouflage, attempting to make the camouflaged signal less recognizable among other signals and noise which surround it. But, successful camouflage hides by presenting the viewer with exactly what he expects to see. On the VHF bands, hams are used to seeing clear frequencies. Hence, there is little for the camouflage artist to work with.

January 1987
FIRST STEPS IN DIRECT-SEQUENCE SPREAD-SPECTRUM

By André Kesteloot, N4ICK

At the September 1986 meeting of AMRAD, I demonstrated a simple direct-sequence spread spectrum modem operating at a carrier frequency of 2 MHz.[1] The purpose was to experiment with some simple circuits, making use only of readily available components and test equipment, so that the average radio amateur (i.e. one who does not necessarily have access to an elaborate laboratory) might be encouraged to develop his own concepts.

A carrier frequency of 2 MHz had been chosen at the time so that the various waveforms could be displayed on any low-frequency (5 MHz bandwidth) oscilloscope. Of course, in practice, only UHF transmissions are presently authorized by the FCC. Before venturing into UHF frequencies however, it seemed advisable to investigate some of these basic concepts at 144 MHz. The purpose of this article is (a) to describe some recent investigations conducted in October and November 1986, and (b) to discuss some possible new concepts.

Fig 4 shows a basic direct-sequence set-up: an RF oscillator produces a carrier at a frequency Fo, which is mixed in a doubly-balanced mixer (DBM) with a pseudo-random noise generator (PN), generating a square pulse at random intervals. The voltage vs. frequency representation (by Fourier transform) of a square pulse is a (sin x/x) function. Since power is proportional to the square of the voltage, the output of the DBM, in the frequency domain, is a (sin x/x) squared spectrum. Single "0" and "1" pulses at the clock frequency

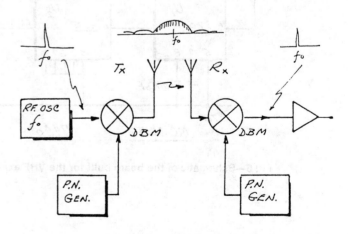

Fig 4—A direct-sequence block diagram.

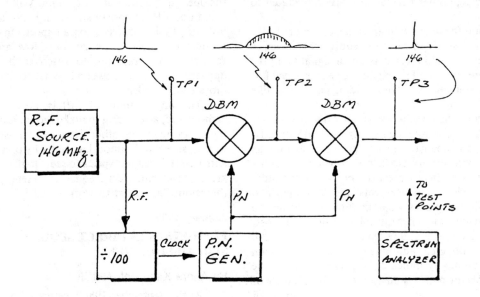

Fig 5—Block diagram of the first VHF experiment.

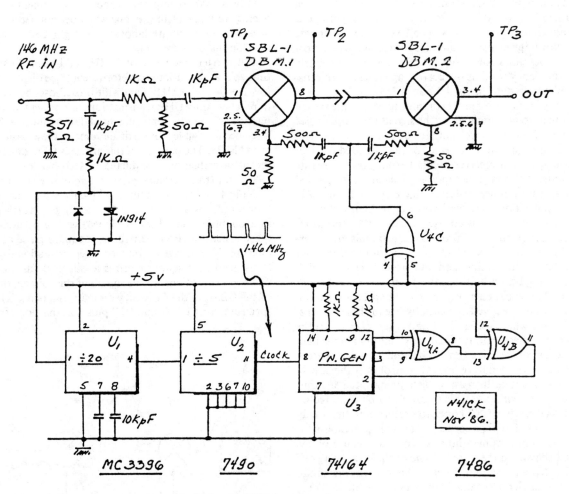

Fig 6—Schematic of the board built for the VHF experiment.

occur more often than any other sequence[2] and this creates a first null at [Fo + Fc] and [Fo – Fc], where Fc is the frequency of the clock pulse driving the PN generator. The transmitter output power (which was concentrated at a single frequency Fo) is now spread over a wide bandwidth.

By mixing again an identical PN sequence with the output of the transmitter, it is theoretically possible to "despread" the spectrum, and thus recover the original frequency Fo. (Note that if Fo were frequency modulated, this information would also be available upon despreading. This type of modulation scheme would not, however, meet present FCC regulations, and particularly Part 97.71.e.2.)

For our first VHF experiment, we decided to use the arrangement shown in Fig 5. The output of a 146 MHz RF source is fed to one input of a DBM, and is also applied to a divide-by-100 stage, the output of which clocks a PN generator. The PN output is applied to the second port of the DBM, and the result is visible at Test Point 2 (TP2). This represents the output of our "transmitter" and is connected to the input of another DBM (our "receiver"). The original PN sequence is also fed to this second DBM, and the original 146 MHz carrier is recovered at the output.

Fig 6 shows the schematic of the board built for the experiment. The output of a 146 MHz source (in this case a 2-meter hand-held) is fed to the board and terminated into 51 ohms. Two 1N914s are connected as clipping diodes to limit the input signal of U1 (an MC3396P) to 0.7 V P-P. This stage divides the RF carrier by 20, and feeds U2, a 7490 connected as a divide-by-5 stage, which produces narrow (0.1 μSec), positive-going 1.46 MHz clock pulses. These clock pulses drive a 74164/7486 combination, (U3 and U4) connected as a 7-stage shift register of the type previously used in my frequency-hopping synchronization scheme (see Ref 3 for details).

The PN sequence at the output of the shift-register is applied to DBM1 (an SBL-1 manufactured by MCL). Per MCL's suggestion (ref 4), the PN sequence is applied to what is usually considered the IF port, and the output signal is taken from what is normally the LO port.

The signal at TP2 is indeed a spread spectrum signal, as can be seen in Fig 5. Note that there is still a fair amount of energy at Fo (146 MHz), most probably due to the "open construction" approach used to build the test board as well as to poor IF port termination. Note the first two nulls at (146.00 + 1.46) = 147.46 MHz and (146.00 – 1.46) = 144.54 MHz. The figure also clearly shows the small "subcarriers" created by the various recurring frequencies of the PN sequences (and affectionately referred to by Hal WB3KDU as "stalactites"). Using a 7-stage PN generator is simple from a hardware point of view, but has the disadvantage that the sequences, and hence the subcarriers, recur very often thus possibly putting out too much energy at finite places in the spectrum (which could lead in certain cases to interference with other users of the band.) This potential problem could be alleviated by using longer PN sequences (the FCC authorizes amateurs to use 7, 13 and 19-stage shift registers.)[5]

The distance between these vertical lines is equal to the PN code repetition rate: in our model the clock rate is 1.46 MHz and the PN code length is 2 to the seventh power, or 128. The code repetition rate is thus (1,460,000 / 128) = 11,406 per second, and the separation between these lines is 11,406 Hz. Had I chosen a 19-stage shift register, the PN code length would have been 524,288 and the code would have repeated only (1,460,000 / 524,288) = 2.78 times per second.

The output of DBM1 is then fed to the input of DBM2. The same PN sequence is applied to the LO input of the DBM, and at its output (TP3), the original signal can be recovered (Fig 6). Again, although most of the spread spectrum signal

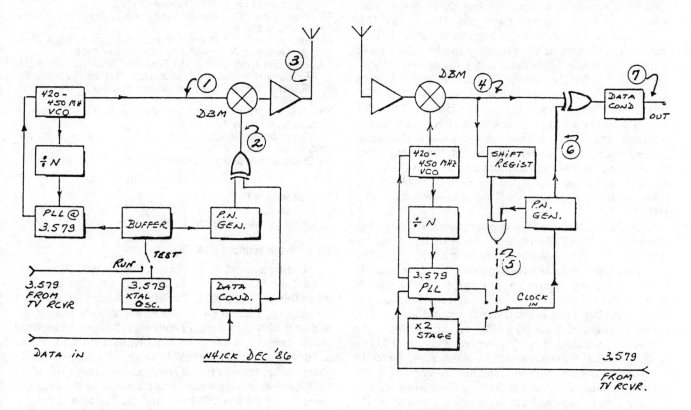

Fig 7—A proposed scheme whereby two amateur stations would derive their DS clocks from a local TV station's color burst information.

has been "despread," a residual portion of the input signal can be seen, probably feedthrough due to poor shielding.

Two important aspects of spread-spectrum have not yet been addressed in this article. The first is how to modulate this spread-spectrum carrier. A practical scheme was shown in Ref 1, where data was modulo-2 added to the PN sequence prior to multiplication with the RF carrier. The second is the problem of synchronizing the two PN sequences (at the transmitting and receiving ends). In this experiment, I have used the same sequence to drive both DBMs. In "real life," these two could be miles apart and the two PN generators would need to be synchronized. As indicated in Ref 3, I believe that we radio-amateurs should accept, at least at the beginning, to synchronize ourselves to an external reference, a solution suggested by William Sabin W0IYH[6]. To quote Robert Dixon: "More time, effort and money has been spent developing and improving synchronizing techniques than in any other area of spread spectrum systems. There is no reason to suspect that this will not continue to be true in the future."[7] and again "To give an idea of the magnitude of the synchronization problem (...) one needs only to note that most systems descriptions assume synchronization and avoid any discussion of how it is to be achieved or maintained."[8]

In light of the above remarks, a scheme whereby two amateur stations would derive their direct sequence clocks from the color burst information of a local TV station, for instance, would seem to me to be perfectly satisfactory as a first approximation.

Such a scheme, planned to be implemented soon, is described in Fig 7. At each end of the radio link, a color TV receiver will be needed, and the 3.579545 MHz color burst will be extracted from the color burst amplifier. If the two TV sets are tuned to the same TV station, the two bursts will be in phase. The amateur transmitter consists of a 420-450 MHz VCO (voltage controlled oscillator), the output of which is divided by N. (In our next experiment, N will equal 121, as this is easily feasible by cascading two divide-by-11 integrated circuits, such as the 11C90 or the MC12013.) The output of the divider will be fed to a PLL (phase-locked loop), along with 3.579 MHz from the TV set. Once the PLL locks and the system reaches equilibrium, the output of the VCO at point [1] will be 3.579 × 121 = 433.125 MHz. The color burst will also be used as the clock for the PN (pseudo noise) generator, and the data will be modulo-2 added to the output of the PN generator, as described in Ref 1. The output of the VCO and the PN generator and data at point [2] will be mixed in a DBM, and then fed to a broadband amplifier. (At point [3], the first two nulls will occur at 429.5 and 436.7 MHz.)

At the receiver site, the opposite process will be implemented: a 433.125 MHz signal will be derived from the color burst just as was done at the transmitter site, and will be mixed in a DBM with the incoming signal. (This is essentially a direct conversion receiver.) At point [4] will appear the PN sequence clocked at 3.579 MHz on which is superimposed the data. At the beginning of the transmission, before any data is sent, the signal at point [4] will be the same as that at the output of the transmitter's PN generator. This signal will be loaded in a shift register, the output of which will be fed to a correlator (similar to that described in Ref 1). A PN generator similar to that used at the transmitter (same number of stages, and same feedback taps) will be clocked by the output of a multiply-by-two stage, the input of which is the color burst. Thus the correlator will receive two signals, identical except for the fact that one will be clocked at twice the rate of the other. The fast sequence will catch up with the slow one, and there will be a moment when the two sequences will be synchronous. At that moment, the correlator output at point [5] will switch the PN clock input from the output of the ×2 stage to the PLL output. The two identical PN sequences (at the transmitter and at the receiver) will now be in phase. The output of the receiver's PN generator (which is now the same as the transmitter's) can now be XORed with the signal at point [4], and the original data will appear at point [7].

Finally, for those who might not want to be tied to color TV receivers, a variation of the above idea could be implemented using as external reference the carrier of a local AM radio station in the 1,500 kHz area of the AM band. A ferrite rod and a phase-lock loop could be used to extract the carrier information. The divider chain would now have to be able to divide by any integer between 281 and 299. (In practice, it might be prudent to operate in the middle of the amateur band. A 11C90 divide-by-11 stage, followed by another divider by 27 would yield an Fo of 445.50 MHz, while the use of pulse swallowing techniques would allow for N to be any number, including primes.)

Acknowledgments

The board of Fig 6 was tested on David K8MMO's spectrum analyzer on 29 November 1986, with Hal WB3KDU and Sandy WB5MMB of the AMRAD Gang in attendance to offer advice, solace and moral support. My thanks also go to Chuck Phillips N4EZV, for discussions of some of the concepts and circuitry of Fig 7.

References

[1]A. Kesteloot, N4ICK, "Experimenting with Direct Sequence Spread Spectrum." *QEX*, December 1986.
[2]R. Dixon, *Spread Spectrum Systems*, 2nd Ed., John Wiley & Sons, NY 1984, p 61.
[3]A. Kesteloot, N4ICK, "Practical Spread Spectrum: A Simple Clock Synchronization Scheme." *QEX*, September 1986.
[4]Mini Circuits Laboratories, PO Box 166, Brooklyn, NY 11235. "RF/IF Signal Processing Handbook 1985/1986," Vol 1, p 25.
[5]D. Newkirk, AK7M, "Our New Spread Spectrum Rules," *QST*, April 1986, p 45.
[6]W. Sabin, W0IYH, "Spread Spectrum Applications in Amateur Radio," *QST*, July 1983, p 16.
[7]R. Dixon, *Spread Spectrum Systems*, 2nd Ed., John Wiley and Sons, NY 1984, p 214.
[8]R. Dixon, *Spread Spectrum Techniques*, IEEE Press, 1976, p 321.

November 1989

A SPREAD-SPECTRUM REPEATER

By André Kesteloot, N4ICK

During a recent visit to Chuck Phillips N4EZV's shack, I could not help but notice a new rack containing, among other things, a 900 MHz duplexer. It is part of his latest design, a repeater for direct-sequence spread spectrum transmissions. How does it work? Referring to Fig 8, the duplexer output feeds (1) a down-converter with an output centered at 70 MHz. Then (2) the first IF stage splits the IF into two channels separated by the 1st IF output frequency. This is fed to (3) the 2nd IF stage which takes the output of the separated channels, combines (XOR) the two, filters and limits the output. The latter drives a PSK demodulator (in our case a Costas-loop) which converts the 2nd IF signal to a digital out-

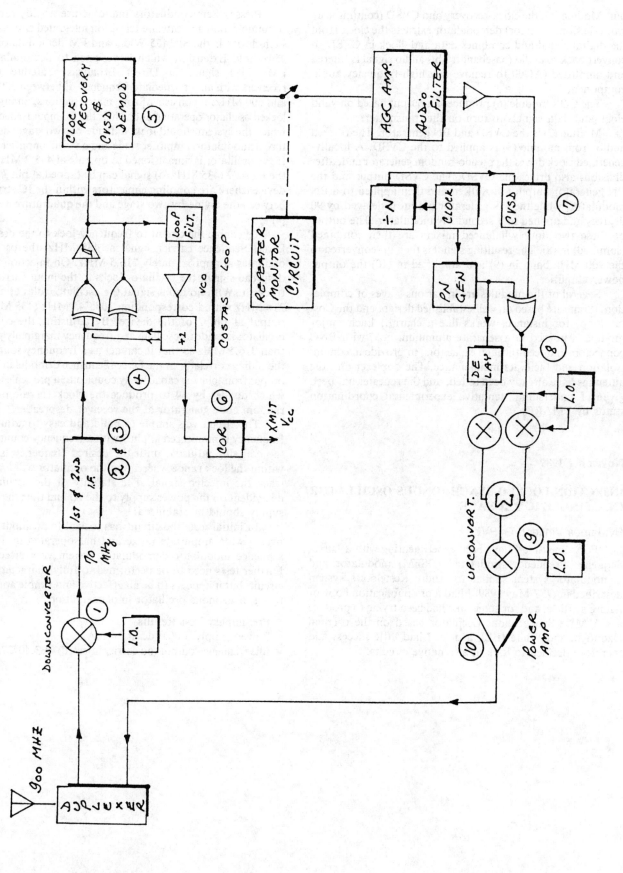

Fig 8—Block diagram of the N4EZV direct sequence repeater.

put. Module (5), the clock-recovery and CVSD (continuously variable slope detector) demodulator extracts the clock from the digital signal and combines data and clock in CVSD to convert back to audio (baseband). This audio signal is filtered and amplified (AGC) to remove any high-frequency audio component.

The COR module (6) produces an output based on valid clock-and-data signals to turn on the transmitter.

Module (7) is the CVSD and PN generator. The filtered audio from module (5) is applied to the CVSD. A locally-generated clock drives the pseudo-random generator and, after division, also drives the CVSD. The CVSD output and the PN generator output are XORed and then applied to a PN modulator, while the PN generator output is delayed by 90 degrees and applied to a second PN modulator. The output of these two doubly-balanced mixers are then combined (summed) in (8). The resulting signal is then up-converted to the 900 MHz band in (9) and amplified in (10) the output power amplifier.

Several of the modules are in various stages of completion. (Some are based on well-established design, and the Costas loop, for instance, works like a charm.) Each major module is housed in a separate aluminum box, with BNC connectors and feed-through capacitors, to provide maximum isolation and facilitate maintenance. The duplexer and the antennas have already been tested, and the repeater has been given, I understand, tentative experimental coordination status by TMARC.

November 1989

INJECTION-LOCKED SYNCHRONOUS OSCILLATOR IN A SINGLE IC PACKAGE

By James Vincent, G1PVZ

For some time I have been experimenting with a Direct Sequence Frequency Modulated (DSFM) modulator and demodulator system similar to André Kesteloot's system described in *QST* May 1989. I had a pre-publication copy of André's article, and for some time had been trying to produce a 435 MHz Synchronous Oscillator based on the original Uzunoglu and White IEEE article. I had little success and therefore decided to look at alternative circuits.

Plessey Semiconductors manufacture a wide range of communication and satellite-television integrated circuits. One such device is the SL1455 Wideband FM demodulator with Threshold Extension which is designed to demodulate the FM video signal in Direct Broadcast Satellite (DBS) receivers with an intermediate frequency of between 300 MHz and 600 MHz. It consists of four major sections, an injection locked oscillator operating at one half the RF input frequency to which the system should lock, a divide-by-two stage, quadrature demodulator/amplifier stage and a video amplifier stage. If the oscillator is dimensioned to operate at 435.2 MHz then the F osc/2 (435 MHz/4) signal can be tapped at pin 7 of the device (there are coupling capacitors within the IC structure between the divide-by-two stage and the quadrature network pins).

To enable the circuit to frequency-lock to the despread Direct Sequence carrier signal at 435 MHz, the oscillator oscillates at approximately 216.5 MHz. On injection of the despread signal, the oscillator locks to the input carrier and tracks it. With no input signal the oscillator idles at around 216 MHz with a corresponding 108.745 MHz (435 MHz/4) output at pin 7 of the device. By adjusting the coil, the oscillator is made to oscillate at a frequency marginally greater than 216.5 MHz so that it can act as a frequency source for the sliding correlator in the Direct Sequence demodulator. The output from pin 7 is capacitively coupled to a pre-scaler device which divides by 64 to produce the clock for the pseudo-random code generator at the receiver despreader.

The circuit was simple to build and easy to adjust, the frequency is monitored at pin 7 with a frequency counter and the coil slug adjusted until the desired frequency is input within the lock range whereupon the oscillator will lock and track the injected signal. The stability of the oscillator is dependent on the power supply to the IC and thus the power supply should be stabilized.

In initial tests the circuit has locked to an input signal over a wide amplitude range and has operated in a direct sequence modulator/demodulator system very effectively. Further tests need to be performed to fully characterize the circuit, but it appears to be an effective and simple approach to a synchronous oscillator implementation.

Preliminary Test Results
Power Supply 5.00 Vdc at 34 mA
Idle frequency output (no carrier injected) = 3.407279 MHz

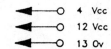

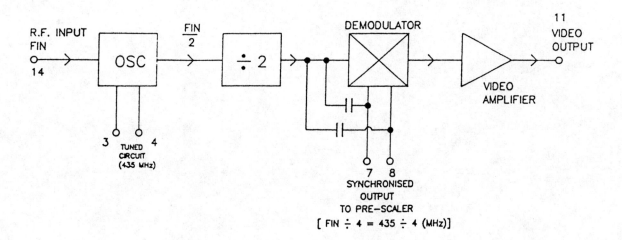

Fig 9—Block diagram of SL 1455-based synchronous oscillator.

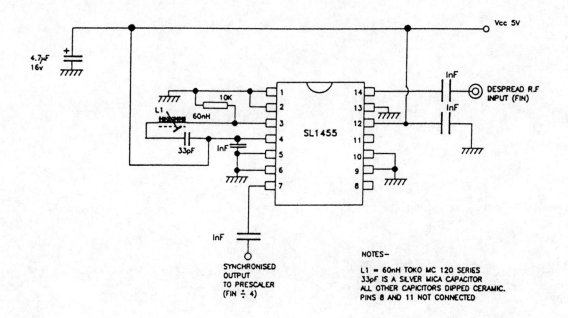

Fig 10—Circuit diagram of 435-MHz synchronous oscillator.

Chapter 8

Spread Spectrum Theory and Projects

The 1991 ARRL Handbook for Radio Amateurs

Spread-Spectrum Communications

The common rule of thumb for judging the efficiency of a modulation scheme is to examine how tightly it concentrates the energy of the signal for a given rate of information. While the compactness of the signal appeals to the conventional wisdom, spread-spectrum modulation techniques take the exact opposite approach — that of spreading the signal out over a very wide bandwidth.

Communications signals can be greatly increased in bandwidth by factors of 10 to 10,000 by combining them with binary sequences using several techniques that will be described later. The result of this spreading has two beneficial effects. The first effect is dilution of the signal energy so that while occupying a very large bandwidth, the amount of power density present at any point within the spread signal is very slight. The amount of signal dilution depends on several factors such as transmitting power, distance from the transmitter and the width of the spread signal. The dilution may result in the signal being below the noise floor of a conventional receiver, and thus invisible to it, while it can be received with a spread-spectrum receiver!

The second beneficial effect of the signal spreading process is that the receiver can reject strong undesired signals—even those much stronger than the desired spread-spectrum signal power density. This is because the desired receiver has a copy of the spreading sequence and uses it to "despread" the signal. Nonspread signals are then suppressed in the processing. The effectiveness of spread spectrum's interference-rejection property has made it a popular military antijamming technique.

Conventional signals such as narrow-band FM, SSB and CW are rejected, as are other spread-spectrum signals not bearing the desired pseudonoise (PN) coding sequence. The result is a type of private channel, one in which only the spread-spectrum signal using the same pseudonoise sequence will be accepted by the spread-spectrum receiver. A two-party conversation can take place, or if the code sequence is known to a number of people, net-type operations are possible.

The use of different binary sequences allows several spread-spectrum systems to operate independently of each other within the same band. This is a form of sharing called code-division multiple access. If the system parameters are chosen judiciously and if the right conditions exist, conventional users in the same band space will experience very little interference from spread-spectrum users. This allows more signals to be packed into a band; however, each additional signal, conventional or spread spectrum, will add some interference to all users.

Benefits for Spectrum Managers

Effective spectrum management attempts to use a band as fully as possible while keeping interference to a minimum. A limit exists as to how many signals can be put in a band. When the allocation is used up, additional stations that use conventional modes may cause interference that degrades or blocks communications by other users. Additional spread-spectrum signals, however, may not cause severe interference; they just raise the background noise level. The limit to the number of spread-spectrum signals that can occupy a band is sometimes called a "soft" limit because the effects of over allocation are not as severe as the rapid degradation caused by over allocation of conventional users.

Overlay is a spectrum-management concept that takes advantage of the spread-spectrum signal rarification and interference-rejection effects to share a band with conventional modulation users. In bands that are channelized and are fully allocated by assigned users such as repeaters, there are few ways to accommodate new users. Yet, viewing such a band on a spectrum analyzer reveals that much of the spectrum is only lightly used because of the intermittent use of many repeaters and the presence of numerous guard bands between fixed channels.

In the overlay concept, the spread-spectrum signal is continuously spread over a shared band. It can exist in both the unused guard bands and on intermittently used repeater channels. It is, in effect a form of frequency diversity that takes advantage of whatever unused spectrum space is available within its spreading width.

Spread spectrum possesses a number of additional advantages. Spread spectrum pro-

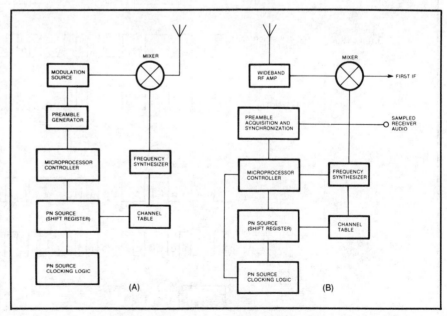

Fig. 15 — The block diagram of a frequency-hopping (FH) transmitter is shown at A. The channel table contains frequencies that the FH transmitter will visit as it hops through the band. These channels are selected to avoid interference to fixed band users (such as repeaters). The preamble employs the falling edge of an audio tone to trigger hopping. Conventional modulation, such as SSB, is employed. The block diagram at B is for an FH receiver. Preamble acquisition and synchronization trigger the beginning of the hopping mode and keep the receiver channel changes in step with the transmitter.

vides a degree of protection against fading (a large variation in received signal strength caused by reflections, as in TV "ghosts"). Frequency-selective fading typically affects a relatively narrow band of frequencies. The spread carrier offers a form of frequency diversity that can offset the faded frequencies by those within the spread-spectrum signals that do not experience the same fading. The effects of reflective or multipath interference, which result from several parts of the same signal arriving at the spread-spectrum receiver at slightly different times, can be largely reduced. Signals that arrive late at the spread-spectrum receiver will not match the spreading code currently being used to decode the signal. Hence they are rejected as interference.

Finally, spread spectrum can be used to construct very precise ranging and radar systems. The spread carrier, modulated with the PN sequence, permits the receiver to measure very precisely the time the signal was sent; thus, spread spectrum can be used to time the distance to a transmitter for ranging, or to an object as in the case of a radar reflection. Both applications have been commonly used in the aerospace field for many years.

Types of Spread Spectrum

There are numerous ways to cause a carrier to spread; however, all spread-spectrum systems can be viewed as two modulation processes. First, the information to be transmitted is applied. A conventional form of modulation, either analog or digital, is commonly used for this step. Second, the carrier is modulated by the spreading code, causing it to spread out over a large bandwidth. Four spreading techniques are commonly used in military and space communications, but amateurs are currently authorized to use only the frequency hopping and direct sequence techniques.

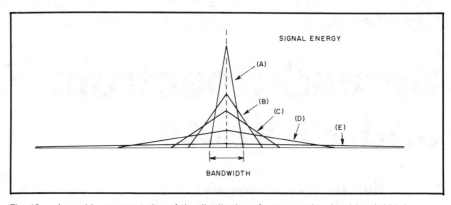

Fig. 16 — A graphic representation of the distribution of power as the signal bandwidth increases. The unspread signal (A) contains most of its energy around a center frequency. As the bandwidth increases (B), the power about the center frequency falls. At C and D, more energy is being distributed in the spread signal's wider bandwidth. At E, the energy is diluted as the spreading achieves a very wide bandwidth. Bandwidth is roughly twice the bit speed of the PN code generator.

Frequency Hopping

Frequency hopping (FH) is a form of spreading in which the center frequency of a conventional carrier is altered many times a second in accordance with a pseudo-random list of channels. See Fig. 15. The amount of time the signal is present on any single channel is called the dwell time. To avoid interference both to a conventional user and from conventional users, the dwell time must be very short, commonly less than 10 milliseconds.

Direct Sequence

Direct sequence (DS) is a second form of spreading in which a very fast binary bit stream is used to shift the phase of an RF carrier. This binary sequence is designed to appear to be random (that is, a mix of approximately equal numbers of zeroes and ones), but is generated by a digital circuit. This binary sequence can be duplicated and synchronized at the transmitter and receiver.

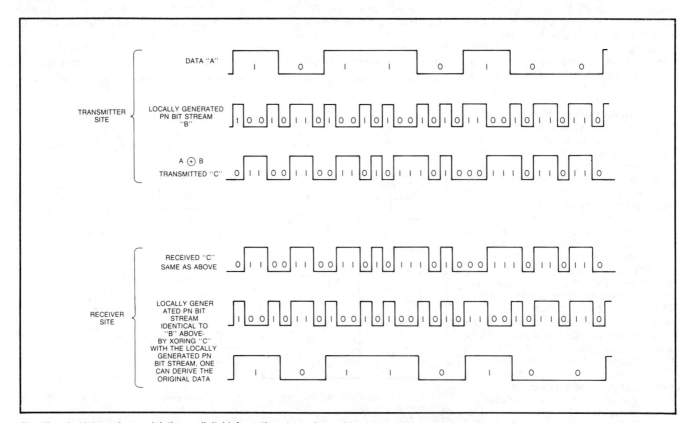

Fig. 17 — In bit-inversion modulation, a digital information stream is combined with a PN bit stream, which is clocked at four times the information rate. The combination is the exclusive-OR sum of the two. Notice that an information bit of one inverts the PN bits in the combination, while an information bit of zero causes the PN bits to be transmitted without inversion. The combination bit stream has the speed characteristics of the original PN sequence, so it has a wider bandwidth than the information stream.

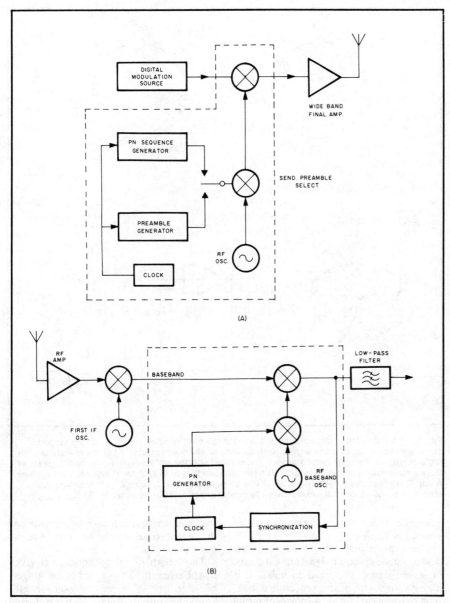

Fig. 18 — The block diagram of a direct sequence transmitter is shown at A. The digital modulation source is mixed with a combination of the PN sequence mixed with the carrier oscillator. The PN sequence is clocked at a much faster rate than the digital modulation; a very fast composite signal emerges as a result of the mixing. The preamble is selected at the start of transmission. Part B shows a direct sequence receiver. The wideband signal is translated down to a baseband (common) frequency. The processes form a correlator, mixing a baseband oscillator with the PN source and then mixing the result against the incoming baseband RF. The synchronization process keeps the PN sequence in step by varying the clock for optimal lock. After mixing, the information is contained as a digital output signal and all interference is spread to noise. The low-pass filter removes some of this noise. Notice that the transmitter and receiver employ very similar designs, one to perform spreading, the other to despread.

Such sequences are called pseudonoise or PN. See Fig. 16. Each PN code bit is called a *chip*. The phase shifting is commonly done in a balanced mixer that typically shifts the RF carrier between 0 and 180 degrees; this is called binary phase-shift keying (BPSK). Other types of phase-shift keying are also used. For example, quadrature phase-shift keying (QPSK) shifts between four different phases.

DS spread spectrum is typically used to transmit digital information. See Figs. 17 and 18. A common practice in DS systems is to mix the digital information stream with the PN code. The result of this mixing causes the PN code to be either inverted for a number of PN chips for an information bit of one or left unchanged for an information bit of zero. This modulation process is called bit-inversion modulation for obvious reasons. The resulting PN code is mixed with the RF carrier to produce the DS signal.

Chirp

The third spreading method employs a carrier that is swept over a range of frequencies. This method is called *chirp* spread spectrum and finds its primary application in ranging and radar systems.

Time Hopping

The last spreading method is called *time hopping*. In a time-hopped signal, the carrier is on-off keyed by the PN sequence, resulting in a very low duty cycle. The speed of keying determines the amount of signal spreading.

A hybrid system is formed by combining two or more forms of spread spectrum into a single system. Typically, a hybrid system combines the best points of two or more spread-spectrum systems. The performance of a hybrid system is usually better than can be obtained with a single spread-spectrum technique for the same cost. The most common hybrids combine both frequency-hopping and direct-sequence techniques.

Spreading Sequences

One of the most important aspects of spread spectrum is the PN sequence used to spread the RF carrier. The spreading sequence determines how well the various properties of spread spectrum will perform.

The spreading sequence takes a form geared to the type of spread spectrum being used. For frequency hopping, the spreading code is a stream of numbers that represent the channels to which the frequency hopper will travel. In a DS system, the spreading code is a very fast bit stream generated by a digital circuit. In both cases, the PN code is generated to resemble random activity and passes many of the tests devised to identify random sequences. Codes that have random-like properties are called pseudorandom or pseudonoise sequences.

Correlation

Correlation is a fundamental process in a spread-spectrum system and forms a common method of receiving signals. Correlation measures how alike two signals are; that is, how similar in appearance they are to each other. The degree of likeness is often expressed as a number between zero and one. A perfect match is typically indicated by a one, while no match may be indicated by a zero. Partial matches yield values between one and zero, depending upon likeness.

In a spread-spectrum receiver, correlation is often used to identify a signal that has been coded with a desired PN sequence. Correlation is usually done with a circuit known as a correlator. A correlator is typically composed of a mixer followed by a low-pass filter that performs averaging. The mixer is where the two signals to be compared are multiplied together. A match yields a high value of output; but if the two mixed signals differ, the output will be lower depending on how different the signals are. The averaging circuit reports the average output of the mixer. This value is therefore the average likeness of the two signals.

In a DS system, the correlator is used to identify and detect signals with the desired spreading code. Signals spread with other PN codes, or signals not spread at all, will differ statistically from the desired signal and give a lower output from the correlator. The desired signal will have a strong match with locally generated PN code and yield a larger output from the correlator.

Notice that the averaging circuit of the correlator gives the average mixer value over time. If noise or interference is present, some of the received signal will be corrupted. After mixing, the interfering signals are spread and resemble noise, while the desired signal is despread and narrowband. The averaging circuit of the correlator then performs a low-pass filter function, thereby reducing the noise while passing the desired narrowband information. This is the heart of the DS interference-rejection process.

Correlation action in an FH system is implemented differently, but the concept is the same. In a frequency-hopping system, the transmitter carrier frequency is being moved about many times a second according to the spreading sequence. The receiver uses the same spreading sequence to follow the transmitter, moving from channel to channel in exact step with the signal. If the receiver is out of step with the signal, it cannot recover the information being transmitted.

FH signals that are under the control of a different PN sequence will be received only randomly, by the chance that both the desired and undesired PN codes have a channel in common at that moment. Narrowband signals will be visited occasionally by the hopping signal and should not be a cause of interference.

Despreading and Detection

In DS systems, collapsing the spread-spectrum signal by removing the effects of the spreading sequence is called despreading. If the DS signal is viewed as a signal with two types of modulation impressed on it (one for spreading and the other containing information) then despreading is a demodulation step aimed at the spreading sequence. What remains after the removal of the spreading sequence is the digital information stream.

The despreading process is in fact performed by a correlator containing a mixer and an averaging circuit. In the correlator, like signals produce a high value while unlike signals produce lower values. Thus signals that are the same produce high outputs because the signals reinforce each other. Signals that are unlike cannot reinforce each other and therefore form lower-valued products.

Bit-inversion modulation is detected by this correlator action since an information bit of "1" caused the PN code to be inverted, while a "0" leaves the PN code unchanged. In the correlator, a "0" information bit generates an uninverted PN code stream that closely correlates with the local PN code. Comparatively, a "1" results in a complete decorrelation since the PN code is inverted for this information bit. Through this action the correlator recovers the transmitted information.

Undesired signals in the DS receiver passband are not correlated with the local PN code. Within the correlator, the undesired signals randomly fall in and out of match on a bit-by-bit basis. Here the mixer output resembles noise that is filtered by the correlator low-pass filter. In comparison, the mixer will produce a stream of ones for an uninverted PN sequence or a stream of zeros for the inverted PN sequence.

Thus, the mixing process despreads the desired spread-spectrum signal and causes the undesired signals to spread to noise. It is mainly through the mixing process that interference is rejected within a spread-spectrum receiver.

Spreading Sequence Generators

Generation of the PN sequence is commonly done with linear-feedback shift registers (LFSR). See Fig. 19. The shift register consists of a number of one-bit memory registers that shift their contents to the right with each clock pulse. New values are introduced at the left-most stage while output is commonly taken from the right-most stage.

The shift register can be used to generate a wide variety of sequences. Generation of a linear sequence is typically done by tapping the values of certain cells and combining the values using an exclusive-OR (one-bit addition without carry) gate. The combined output is then fed back to the input of the left-most cell.

The LFSR sequence is a repeating sequence whose maximun length is related to the number of stages (N) in the shift register by the equation: length = $2^N - 1$. This equation determines the maximum length; not all tap choices produce the longest sequence, however. There are two cases to consider: the maximal-length sequence or m-sequence and the nonmaximal-length or composite sequences.

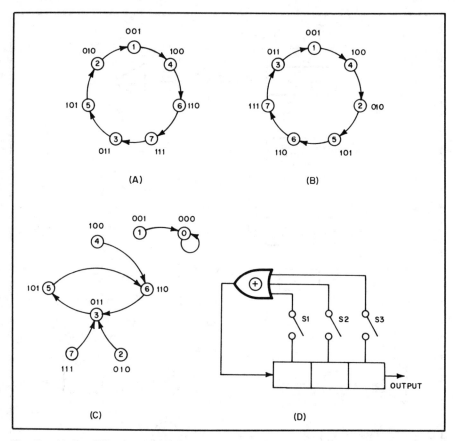

Fig. 19 — Various PN sequences. At A, a three-stage shift register with (1,3) stages tapped. There are seven unique code words, each as wide as the shift register. At B, a three-stage shift register with (2,3) stages tapped produces a somewhat different arrangement of code words. At C, two nonmaximal-length cycles result from the tap choices of (1,2) for the three-stage register. A and B are maximal-length cycles. At D is a three-stage linear feedback shift register. The stages to be combined with exclusive-OR to produce feedback are selected by S1-S3.

The length of an m-sequence is given by the equation $2^N - 1$ and is the longest possible for a given register length. Since the output repeats after producing $2^N - 1$ bits, the entire sequence can be viewed as a cycle (see Fig. 19).

Initializing the register with any sequence of consecutive bits from the cycle will generate the entire cycle from that point. A cycle is commonly started by an initial value of all zeros followed by a one such as 0000001 for a seven-stage register. Other starting values still generate the same sequence but are said to be phase shifted from the zeros followed by a one initial starting point. A run of bits selected from the sequence that is the same length as the generating shift register is called a code word of the sequence. All code words from an m-sequence are unique with no repeats, and no code word contains all zeros.

Comparatively, the length of the nonmaximal sequence is always shorter than $2^N - 1$, and its outputs form a number of distinct short cycles. Code words are not unique for this type of sequence. The nonmaximal sequences are called a composite sequence because a composite sequence can be expressed as a combination of shorter maximal-length sequences. The nonmaximal-sequence length is in fact the product of the length of the smaller m-sequences of which it is composed.

Determining whether a given tap sequence will produce a maximal or composite sequence is quite complicated. Most efforts have been carried out by computer search methods, and a number of excellent tables are available for the designer.

A maximal-length sequence seven-stage shift register can easily be built with two ICs, as shown in Fig. 20. A more complete description of this shift register appeared in October 1986 *QEX*.

Orthogonality

An important consideration in choosing a sequence for a spread-spectrum system is the amount of statistical similarity a sequence has with conventional signals and with sequences employed by other spread-spectrum systems. The greater the degree of similarity, the less able the spread-spectrum receiver will be able to reject interference.

The ideal sequence will show a very low correlation when compared with undesired sequences. Moreover, the PN sequence should also have a low correlation with shifted versions of itself. A measure of a sequence's self-correlation is measured by its autocorrelation function (ACF). The ACF measures how much interference will be received from other spread-spectrum units using the same PN sequence but with different starting points within the PN code cycle.

Two code families that have found common usage are the m-sequences and Gold codes. Gold codes are a family of spreading codes generated by exclusive-OR combining the output of two "preferred" m-sequence generators. Many other PN codes can be used with spread spectrum — Dixon lists a number of such codes.

Synchronization

Synchronization is the most difficult issue for spread-spectrum systems. For a spread-spectrum receiver to demodulate the desired signal, it must be able to synchronize the locally generated PN code reference with the one used by the transmitter. This operation commonly takes place at very high speeds. It is usually viewed as two processes: rough synchronization which searches and acquires the receiver's PN code within one bit or channel of the transmitter's, and fine synchronization which maintains bit or channel dwell timing.

In rough synchronization, the receiver attempts to line up the local PN code as close as possible to the transmitters. There are two methods. The first is called epoch synchronization, in which the transmitter periodically sends a special synchronization sequence. Commonly, this synchronization sequence is a unique short bit sequence chosen because it is easily detected by a simple correlator known as a digital matched filter. This filter consists of a shift register that clocks the received sequence by one bit at a time and compares it against the unique sequence. The output of the matched filter is highest when the unique word is found in the input bit stream.

A second method of synchronization is known as phase synchronization. Here, the receiver attempts to determine which of the $2^N - 1$ phases the m-sequence could be in. To determine the phase, and thereby the synchronization, at least N bits would have to be received without error. Noisy signals may require more than N bits depending on the signal-to-noise ratio.

PN codes with short cycle times will exhibit several repetitions in a short amount of time. A digital matched filter can be constructed to signal the presence of a unique code word from this short spreading code sequence. This is essentially a form of epoch synchronization.

Longer sequences require more-complex synchronization methods. One method uses a sequence based on the time of day. The receiver sets up a digital matched filter with a value several seconds ahead of the current time and waits for this unique code word to appear. The procedure may be useful for net operations in which stations can enter or leave a net at will.

There are a number of specially developed sequences available for rapid synchronization; however, they fall beyond the scope of this brief discussion.

Preamble

A preamble signal is commonly sent by the spread-spectrum transmitter immediately before the transmitter enters the spread mode. The preamble signal alerts the receiver to set up its synchronizing procedure to acquire the spread signal. The format of the preamble is different from system to system.

In slow frequency hopping, which has hop rates of less than 100 times a second, a tone appearing on a prearranged frequency or home channel can be used. The falling edge of the tone signals the beginning of hopping. At faster speeds, the precise instant the tone falls may be difficult to measure accurately; other preamble methods are typically employed. One such method calls for the frequency hopping receiver to examine a specific set of channels continually. A frequency-hopping transmitter would hop on these channels a prearranged number of times to allow the receiver to synchronize.

In direct-sequence transmissions the preamble performs three functions. First, the RF carrier must be acquired by the receiver. This may be done by transmitting a run of all zero bits, which reveals the carrier's center frequency to the receiver. Second, local clock synchronization is established by transmitting a sequence of alternating ones and zeros. The receiver will detect these transitions and derive bit timing from them. Last, the spreading code itself must be synchronized. This can be done by an epoch synchronization procedure.

Once rough synchronization is established, the spread-spectrum receiver must track and maintain bit timing or channel dwell timing. There are several techniques to accomplish fine synchronization; however, the underlying mechanism is a feedback loop that attempts to minimize the timing error between the transmitter and receiver PN code at the fraction of a bit time.

The feedback loop itself commonly consists of a differencing circuit that calculates the actual bit difference and a feedback path that adjusts the receiver spreading code clock. In one technique called dithering, the spreading-code clock speed is continually rocked back and forth around the synchronization point, causing the receiver to periodically move in and out of synchronization a small amount. The point at which best synchronization is achieved is used to synchronize the receiver clock. Dithering can track a spreading code whose timing may be changing, perhaps because the transmitter is mobile, or in the case of HF spread spectrum, with changes in the height of the reflecting layers of the ionosphere.

Spectrum of Spread Spectrum

The spectrum of each type of spread-spectrum signal depends on several factors, such as the speed at which the spreading code is clocked, the type of spreading code used, whether frequency hopping or direct sequence is being used, the modulation bandwidth and the method of modulation.

Fig. 21 is the spectrum of a BPSK DS sequence. The signal is symmetric around the center frequency and contains several peaks that are called lobes. The main lobe is maximum at the center frequency but falls rapidly. The point at which the main lobe falls to its low point is called the first zero; subsequent lobes are called spectral sidelobes. The main lobe of a DS signal contains the majority of power, about 90 percent, while the remaining 10 percent is distributed over the side lobes. In many systems, the side lobes

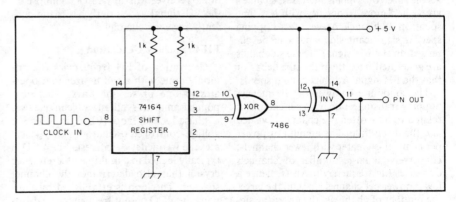

Fig. 20 — A maximal-length sequence seven-stage shift register. A complete description of this circuit appeared in October 1986 *QEX*.

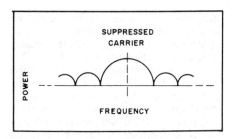

Fig. 21 — Power vs. frequency for a direct-sequence-modulated spread-spectrum signal. The envelope assumes the shape of a sin x^2 divide by x curve. With proper modulating techniques, the carrier is suppressed.

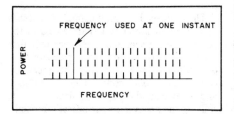

Fig. 22 — Power vs. frequency for frequency-hopping spread-spectrum signals. Emissions jump around in pseudorandom fashion to discrete frequencies.

are clipped since they tend to extend out over a large span of spectrum but carry little of the DS signal's power.

The chip rate determines the overall spread of the DS signal. The overall size of the DS carrier expands as the chip rate increases, while lower chip rates collapse the signal into a nonspread conventional PSK signal. In addition, the chip rate determines the size of the DS main lobe and the location of the zeros. The main lobe is 2/chip-rate wide, centered about the DS carrier's center frequency. Zeros occur at multiples of 1/chip-rate, symmetrically on either side of the center frequency.

The spectrum of an FH signal is shown in Fig. 22. The spectrum structure is not as complex as in the DS signal, but depends on the spreading bandwidth and type of modulation used. The spectrum of FH signals consists of a carrier that moves pseudorandomly among many channels. As the speed of hopping increases, channels are visited more frequently with less time spent on a channel. At very fast hopping speeds, significant sidebands can be observed on the FH signal. These sidebands are generated from the pulse-like behavior that the FH signal exhibits at high speeds.

The amount of power the frequency-hopping transmitter delivers per channel is related to how often the channel is visited and the dwell time. The amount of power per channel is greater with fewer channels; conversely, a large number of channels decreases the frequency of visits, hence a lower power per channel. In sum, the larger the number of channels, the lower the signal power per channel.

Last, the method of modulation has a marked effect on the spectrum of an FH signal. For example NBFM produces a constant-amplitude signal and generates a relatively flat spectrum. SSB, however, emits energy only when there is information to transmit, and hence less energy is emitted overall.

Near vs. Far-Field Strength

Spread-spectrum signals exhibit some unusual but logical signal effects. Close to the transmitter, a spread-spectrum signal may be observed readily on a conventional receiver; however, at a distance (50-100 miles) the signal may be noticed only with careful measurement. This results in a property called *low probability of intercept* (LPI). Some authors dispute the notion of LPI, saying that spread-spectrum signal energy is in fact intercepted by the receiving station's antennas and receivers; however, there is a low probability that the listener will recognize the presence of spread spectrum. The term *low probability of recognition* (LPR) has been suggested in place of LPI.

This radical difference in signal strength between close-in and distant observers is quite logical. The power of a narrow-band signal is typically concentrated about a center frequency; hence, a conventional narrow-band receiver will be in a position to collect much of the original power subject to path loss. In the DS spread-spectrum case, the same power is distributed over a band of frequencies, giving less power per hertz. A narrow-band receiver can collect only as much of this distributed power as the width of its IF passband; consequently, it registers a much weaker signal. A spread-spectrum receiver's passband is quite wide, allowing it to receive the entire bandwidth containing the spread-spectrum signal. This power is concentrated by the despreading process, making the output signal-to-noise ratio at least equal to the narrow-band signal's.

In the case of FH spread spectrum, a full-power narrow-band carrier is hopped among many channels. The concentrated power of the carrier can be observed at a distance equal to that of a nonhopping narrow-band signal; however, a conventional receiver will only catch a glimpse of the FH signal as it briefly visits the currently received channel.

FH Frequency Generation

Generation of FH frequencies is commonly done with a synthesizer that operates in one of several ways. The most popular amateur synthesizer technique uses a phase-locked loop (PLL) in which a voltage-controlled oscillator (VCO) generates the frequency of interest. The VCO is typically locked to a multiple of a reference crystal that also determines the channel spacing. The loop is established by sampling the VCO output frequency and dividing it down so its divided frequency matches the reference-oscillator frequency. The error voltage generated by the phase comparator holds the VCO on the desired multiple of the reference oscillator.

This approach is limited because of the time required for the loop to settle. Various modifications are possible such as replacing the loop's low-pass filter with an active low-pass filter; however, the settling-time problem is difficult to overcome. One successful method has been to use two or more synthesizers, alternately allowing one to settle while using the other to control the current channel. The increase in hopping speed can be found by calculating one minus the number of synthesizers multiplied by the ratio of dwell to settling time.

Another enhancement to the basic PLL synthesizer comes by inserting a D/A converter into the PLL. The D/A takes a digital word from a computer which represents a voltage feed to the VCO. The voltage has been selected so the VCO will produce a desired frequency. The PLL and reference oscillator are used to lock the VCO output to a precise channel; however, the large settling time, attributed mainly to the loop tracking large jumps in frequency, is absent.

A different approach to FH generation can be found in the direct-synthesis approach. Here, a large number of samples are taken of a sine wave that is stored in a ROM. A signal is generated by feeding these samples into a wide-band D/A converter, producing an RF sine wave. The frequency of the RF sine wave is determined by the rate at which the samples are fed to the D/A. A synthesizer of this type is capable of hopping many thousands of times per second and commonly has a resolution of 1-hertz steps.

Self Policing and Enforcement

Spread spectrum belongs to the general class of wideband signals that are new to many Amateur Radio operators. Traditionally, amateur experience has been with narrow-band signals that can be easily received with conventional receiving equipment. Wideband signals require larger IF bandwidths and some new ways of thinking.

Spread spectrum employs a carrier with a very wide bandwidth. In addition, the carrier is coded with a PN sequence. These two factors make conventional amateur receivers unsuitable for spread-spectrum reception. There are, however, many techniques available that recognize, locate and recover the transmitted information for amateur spread-spectrum users.

Amateur operators are interested in experiments that enhance communications; hence, system parameters are chosen to facilitate communications. Amateur systems are designed so they can be received without too much difficulty. Conversely, military users are interested in signal hiding — to prevent reception by unauthorized users — so military system parameters are set with this objective in mind.

When a spread-spectrum signal is above the noise level, it can be received with a wideband receiver, and direction-finding techniques can be used to locate the signal source. Interference to other amateur operations or TVI complaints are resolved easily because a signal strong enough to cause interference is far enough above the noise that it can be located with conventional DF techniques. In particular, a directional antenna can be connected to the receiver that is being interfered with. Rotation of the antenna can be used to help locate the source. Moreover, the periodic identification required of all amateur signals can aid the interference-determination process.

Currently, the FCC has authorized two types of spread-spectrum communications for amateurs. The user must identify with narrow-band transmissions on one frequency in the band being used. Alternatively, the station ID may be transmitted while in spread-spectrum operation if the transmission is changed so that CW, SSB or narrow-band FM receivers can be used to identify the station. One proposed identification procedure is for the stations involved to send call signs by on-off keying of the spread-spectrum carrier. This will provide a method to readily identify an amateur spread-spectrum user.

Sometimes it is necessary to monitor the contents of an amateur spread-spectrum transmission to determine if the amateur station is being used properly. This is possible when the signal is above the noise; a wideband receiver and appropriate detector to demodulate the signal are necessary. Other signals in the passband can cause interference, so a sharp beam antenna is required to reduce the effects of the undesired signal.

In DS spread-spectrum transmissions, both the spreading PN code and the transmitted information are digital in nature and are often mixed before being used to modulate the RF carrier. The signal must be despread before the digital information can be recovered; however, the PN code must be known in advance. In the spread-spectrum rules, the FCC defines three m-sequence PN codes for amateur use. These codes are generated by three shift-register configurations: a seven-stage register tapped at stages (7,1); a 13-stage register tapped at stages (13,4,3,1); and a 19-stage register tapped at stages (19,5,2,1). These three sequences are the only ones that amateurs are allowed to use. It is not difficult to test all three sequences against a received amateur DS signal to determine which is in use. Once the sequence is identified and synchronized, the transmitted digital information can be recovered.

Amateur SS Experimentation

Spread-spectrum development for jam-proof military communications began in the late 1940s. John P. Costas, W2CRR, was the first person to recognize non-military applications for SS. Costas presented a paper entitled "Poisson, Shannon, and the Radio Amateur," in the *Proceedings of the IRE* for December 1959. (Poisson and Shannon developed mathematical models for communications systems — all analytical studies in communication theory and information theory are based on their results.)

In this paper, Costas explained that congested band operation presents an interesting problem in analysis that can be solved by statistical methods. He showed that in spite of the bandwidth economy of SSB, there are definite advantages to using very broadband techniques. Costas concluded that broadband techniques would result in more efficient use of the spectrum and an increase in the number of available channels. At the time, Costas' statements were revolutionary, for they challenged the conventional theory that congestion in the radio spectrum can be relieved only by the use of smaller transmission bandwidths.

In 1980, Dr. Michael Marcus of the FCC Office of Science and Technology (OST) suggested that radio amateurs experiment with spread-spectrum modulation techniques. The rationale was that (a) the civil radio services could take advantage of the spread-spectrum pioneering of the military, (b) design of spread-spectrum systems by the private sector was slow because of the high cost of development vs. return on investment, (c) more experimentation was needed in areas such as designing for low-cost and on-the-air testing in congested frequency bands, and (d) radio amateurs could perform useful experiments without the need for either governmental or industrial research and development money.

The Amateur Radio Research and Development Corporation (AMRAD) requested, and the FCC granted, a Special Temporary Authority (STA) to permit spread-spectrum tests in the amateur bands by a small number of amateurs, for one year beginning in March 1981. Under the STA, the first Amateur Radio SS tests were conducted by W4RI in McLean, Virginia and K2SZE in Rochester, New York. Later, WA3ZXW in Annapolis, Maryland ran additional on-the-air tests with K2SZE. The equipment used was capable of hopping over a frequency range up to 100 kHz at rates of 1, 2, 5 and 10 hops per second. RF power output levels of 100 and 500 watts were used into dipole antennas.

These particular radios functioned best at 5 hops per second. This was subjectively judged on the basis of least-bothersome interference from the various signals at the different hopping frequencies. It was observed that frequency hopping was more successful in the presence of heavy CW interference than it was in the presence of heavy SSB interference. In comparison, conventional SSB usually provided better communications than frequency-hopped SSB whenever a single clear channel could be found for the conventional SSB. However, the conventional SSB could be disrupted by strong interference on that channel. While hampered by cyclic interference when busy frequencies were revisited, the frequency-hopped link could be maintained despite band congestion.

Although the tests were announced beforehand in Amateur Radio publications and on the air from W1AW, no correspondence was received indicating that the frequency-hopping tests either interfered with, or were heard by, other Amateur Radio stations. The only exception was that several amateurs in the Northern Virginia area could recognize the presence of the frequency-hopped transmissions on conventional SSB receivers after learning what the signal sounded like. All were within five miles of W4RI and were able to hear both ends.

AMRAD member Chuck Phillips, N4EZV, built a VHF frequency-hopping radio in 1980. The design consisted of a modified VHF-Engineering scanner board capable of switching between four different crystal oscillators. By tripling the clocking speed, the transmitter was made to hop at 12 hops per second. The oscillators were arranged to operate continuously, and were switched in and out by using high-isolation solid-state switches. Sometime in 1981, the Motorola synthesizer chip showed up in a GLB add-on synthesizer board. N4EZV modified a couple of those boards and managed to reach 15 hops per second. Later that year, N4EZV tried a new synthesizer, working at lower frequencies, by modifying some Heathkit HW2031s. The synthesizer signal was then heterodyned to the desired output frequency. Both the transmitter and receiver were equipped with identical frequency "look-up tables" stored in EEPROMs. When the push-to-talk microphone switch was depressed, a short audio tone burst was transmitted and then decoded at the receiver. At the end of the burst, both transmitter and receiver started hopping to the frequencies pre-loaded in the look-up tables. The hopping rate was derived by dividing down from the reference crystal used to pilot the whole setup. When the microphone push-to-talk button was released at the end of a transmission, another tone was sent, which stopped the hoppers. If the receiver detected only noise, (caused by interference), it would automatically revert to its "home frequency" where it would wait for another transmission. This system seems to operate well up to about 100 hops per second over a 4-megahertz frequency range.

Experiments run by AMRAD in 1985 used a Commodore computer to control the frequency of an ICOM IC-2AT. Very little interference to conventional 2-meter users was created by the spread spectrum transmissions. The signals were inaudible to amateurs involved in regular QSOs, but the transmissions did cause some interference by keying some repeaters that did not have a carrier-sense activation delay.

This interference affected only a few repeaters out of many in the Northern Virginia area, and the problem was cured by removing the appropriate repeater input frequencies from the table of frequencies used in the frequency-hopping scheme.

In 1987, AMRAD member Elton Sanders, WB5MMB used the direct-synthesis oscillator described by Fred Williams in the February 1985 issue of *QST* to build a frequency hopper. The frequency of the direct-synthesis oscillator was determined by a program written by David Borden, K8MMO, to run on an IBM® PC-XT clone, and the oscillator output was limited to frequencies within the 6.15 to 6.25 MHz band. This output was then multiplied by 72, and fed to a 70-centimeter output amplifier. (Hopping rates ranging between 250 and 1000 hops per second were obtained by this method.)

Initial discussions with the FCC concerning the second STA revealed that the FCC Field Operations Bureau was interested in conducting monitoring and direction-finding exercises against amateur spread-spectrum transmissions. In February 1985, FCC personnel conducted a "fox hunt" to find an amateur spread-spectrum transmitter operating at an undisclosed site. Using the standard FCC spectrum analysis van and enforcement car, FCC personnel located the transmitter within 25 minutes, proving that they can DF amateur spread-spectrum transmissions if necessary.

On May 9, 1985, the FCC amended the rules to permit SS operation on the amateur bands above 420 MHz at power output levels no greater than 100-W PEP for domestic communications. These rules became effective in June 1986. The amateur community is working toward establishing a set of standards for amateur SS operation. The current rules authorize three PN sequences, and both FH and DS are allowed. Hybrid spread-spectrum techniques are not permitted. Bandwidths may be as large as the amateur band of operation. Also, SS users are specifically prohibited from causing interference to other users of the band.

Under these new rules, amateurs engaging in SS operation must maintain a log with the following information.

1) A technical description of the transmitted signal.

2) Signal parameters including the frequency or frequencies of operation, the chip rate, the code, the code rate, the spreading function, the transmission protocol(s) including the method of achieving synchronization, and the modulation type.

3) A general description of the type of information being conveyed; for example, voice, text, memory dump, facsimile, television and so on.

4) The method and, if applicable, the frequencies used for station identification.

5) The beginning date and ending date of use of each type of transmitted signal.

Throughout 1986 and 1987, AMRAD continued to devise and run experiments designed to bring this spread-spectrum technology within the reach of the average radio amateur, by utilizing only parts and technologies readily available to everyone. In June 1986, AMRAD member Andre Kesteloot, N4ICK, demonstrated a way to derive clock signals (suitable for frequency-hopping) by using broadcast TV signals. The complete experiment is described in October 1986 *QEX*, pp 4-7. With the equipment described, the trailing edge of an audio tone burst was used to derive original synchronization, and the two sets hopped at 15 hops per second for more than one hour without losing sync.

In September 1986, N4ICK demonstrated an experimental board allowing for the transmission of data with direct-sequence spread spectrum. Synchronization was obtained by using a sliding correlator, fully described in the December 1986 issue of *QEX*. In 1987, N4ICK developed a method for deriving precise clock signals from AM radio broadcast transmitters (see October 1987 *QEX*) which was used in an interesting experiment conducted in January 1988. This experiment (see May 1988 *QEX*) can be summarized as follows (see Fig. 23): N4ICK used two hand-held transceivers operating on 445 MHz. At both sites, PN generators were driven by clock pulses derived from the same AM radio transmitter. These two PN sequences were thus driven by synchronous clocks (it was still necessary to correlate them). At the receiver end, additional pulses were fed into the clock pulse stream, making that clock operate slightly faster than the transmitting clock. The receiver output was connected to a DTMF decoder. In the absence of a valid tone, the receiver clock would continue to operate faster than the transmitter clock. If a DTMF tone was sent at the transmitter for a certain duration of time, there would necessarily be a moment when the transmitting and the receiving PN sequences would be in phase. At that moment, the output of the receiver's doubly balanced mixer would be correlated, and a signal would reach the DTMF decoder. A valid output from the DTMF decoder would be used to slow the receiver clock-pulse generator down to its normal speed. Although the overall synchronization process was relatively slow

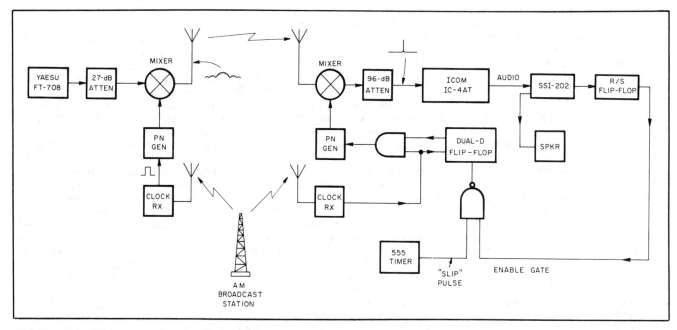

Fig. 23 — Block diagram of one of N4ICK's spread-spectrum experiments.

(it could take up to 45 seconds to reach sync) once synchronization was reached both stations would remain in lock for hours.

These AMRAD experiments all relied on a clock derived from an external reference. A major problem, one common to all spread-spectrum systems, remained. How can the transmitter clock signal be recovered directly from the received signal? Because the received signal appears to be noise (Fig 24B) this can be a formidable challenge. In the May 1989 issue of *QST*, Andre Kesteloot described his design of a complete UHF direct-sequence transmitter and receiver, offering a remarkably simple solution to the problem of synchronization. (Because of its simplicity, the solution does not offer all the anti-jamming properties of more sophisticated systems, but this should not be of concern to Amateur Radio operators.)

The block diagram is shown in Fig 25. At the transmitter site, the output of a 446 MHz generator is fed to one of the input ports of a doubly balanced mixer (DBM). The other input port of the mixer is connected to a PN sequence generator, which is clocked at a submultiple of the transmitter carrier frequency. Since the carrier and the clock are harmonically related, recovering either one will be sufficient to derive the other.

At the receiver end of the link, a sliding correlator is used as follows: The output of a free-running "synchronized oscillator" is divided to create a clock signal at a frequency close to that of the transmitter clock. This locally generated clock is used to drive a PN generator identical to that used at the transmitter site. The resulting PN sequence is mixed with the incoming RF in a doubly balanced mixer. The free-running frequency of the synchronized oscillator can be adjusted to run the receiver PN sequence slightly faster or slower than the transmitter oscillator (which is fixed). Although they are not synchronized, the transmitter and receiver PN sequences are identical. They do differ in speed, however, so the receiver sequence "slides by" the transmitter sequence, somewhat like two wheels rotating at different speeds on the same shaft. One of the wheels can thus be thought of as "catching up" with the other. There will necessarily be one instant when the two PN sequences are momentarily in phase. At that moment, correlation, or "despreading" occurs, and the output of the receiver DBM duplicates the original carrier, as shown in Fig 24C. This carrier is fed to the input of the synchronized oscillator, which locks on the incoming signal. Since the recovered carrier is now synchronized to the transmitter oscillator, the receiver PN generator (which is a submultiple of the carrier) remains in phase with that at the transmitter, and the loop is closed. The original clock has been regenerated and the system stays locked. With a seven-stage shift register as described in the article, search-and-lock at the receiver is almost instantaneous (on the order of a few milliseconds).

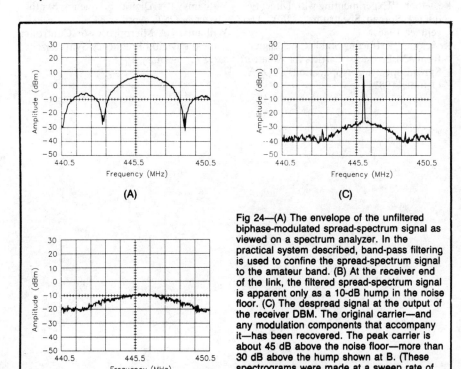

Fig 24—(A) The envelope of the unfiltered biphase-modulated spread-spectrum signal as viewed on a spectrum analyzer. In the practical system described, band-pass filtering is used to confine the spread-spectrum signal to the amateur band. (B) At the receiver end of the link, the filtered spread-spectrum signal is apparent only as a 10-dB hump in the noise floor. (C) The despread signal at the output of the receiver DBM. The original carrier—and any modulation components that accompany it—has been recovered. The peak carrier is about 45 dB above the noise floor—more than 30 dB above the hump shown at B. (These spectrograms were made at a sweep rate of 0.1 s/div and an analyzer bandwidth of 30 kHz; the horizontal scale is 1 MHz/div.)

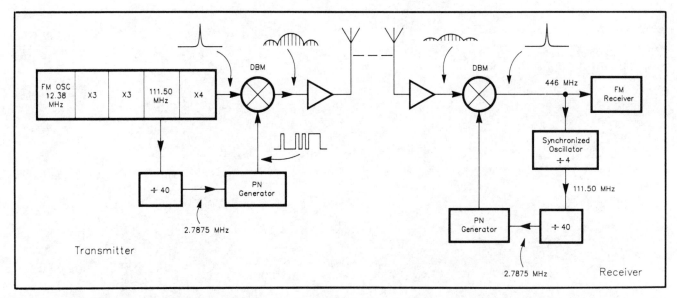

Fig 25—A block diagram of the practical spread-spectrum link. The success of this arrangement lies in the use of a synchronized oscillator (right) to recover the transmitter clock signal at the receiving site.

Selected SS Bibliography

Reading material on spread spectrum may be difficult to obtain for the average amateur. Below are references that can be mail ordered. Spread-spectrum papers have also been published in *IEEE Transactions on Communications, on Aerospace and Electronic Systems* and *on Vehicular Technology*.

Dixon, *Spread Spectrum Systems*, second edition, 1984, Wiley Interscience, 605 Third Ave., New York, NY 10016.

Dixon, *Spread Spectrum Techniques*, IEEE Service Center, 445 Hoes La., Piscataway, NJ 08854.

Golomb, *Shift Register Sequences*, 1982, Aegean Park Press, Laguna Hills, CA.

Hershey, *Proposed Direct Sequence Spread Spectrum Voice Techniques for Amateur Radio Service*, 1982, U.S. Department of Commerce, NTIA Report 82-111.

Holmes, *Coherent Spread Spectrum Systems*, 1982, Wiley Interscience, New York.

The *AMRAD Newsletter* carries a monthly column on spread spectrum and reviews ongoing AMRAD experiments. In addition, the following articles on SS have appeared in Amateur Radio publications:

Feinstein, "Spread Spectrum — A report from AMRAD," *73*, November 1981.

Feinstein, "Amateur Spread Spectrum Experiments," *CQ*, July 1982.

Kesteloot, "Practical Spread Spectrum: A Simple Clock Synchronization Scheme," *QEX*, October 1986.

Kesteloot, "Experimenting with Direct Sequence Spread Spectrum," *QEX*, December 1986.

Kesteloot, "Extracting Stable Clock Signals from AM Broadcast Carriers for Amateur Spread Spectrum Applications," *QEX*, October 1987.

Kesteloot, "Practical Spread Spectrum: Achieving Synchronization with the Slip-Pulse Generator," *QEX*, May 1988.

Kesteloot, "A Practical Direct-Sequence Spread-Spectrum UHF Link," *QST*, May 1989.

Kesteloot, "Practical Spread Spectrum: Clock Recovery With the Synchronous Oscillator," *QEX*, June 1989.

Rhode, "Digital HF Radio: A Sampling of Techniques," *Ham Radio*, April 1985.

Rinaldo, "Spread Spectrum and the Radio Amateur," *QST*, November 1980.

Sabin, "Spread Spectrum Applications in Amateur Radio," *QST*, July 1983.

Williams, "A Digital Frequency Synthesizer," *QST*, April 1984.

Williams, "A Microprocessor Controller for the Digital Frequency Synthesizer," *QST*, February 1985.

The 1987 ARRL Handbook for the Radio Amateur

A Digital Frequency Synthesizer with a Microprocessor Controller

Frequency synthesis can be accomplished three ways: direct synthesis, indirect synthesis and digital synthesis. This article describes the digital-synthesis technique. It includes a complete design for a high performance digital synthesizer, shown in Figs. 22 through 33.

The project, designed by Fred Williams, was first described in *QST*, April 1984 and February 1985. Stas Andrzejewski, W6UCM, became involved when his firm, A & A Engineering, decided to design circuit boards for the project.[1] He overcame several parts-procurement problems and made a few circuit changes to improve the reliability and operating ease of the synthesizer and controller. The parts and diagrams shown here include the changes made by A & A Engineering.

Digital Synthesis

Advances in integrated circuits that perform arithmetic functions and convert signals from digital to analog form have made this kind of synthesizer possible. The digital synthesizer does not have the drawbacks of the other two types, such as a large number of separate oscillators or very sharp filters for a direct synthesizer, or the phase noise common with phase-locked-loop indirect synthesizers. These advantages over the other methods should make the digital synthesizer very popular.

Many readers are probably unfamiliar with the branch of engineering called digital-signal processing, so a few of the fundamentals are discussed first. That will help your understanding of the operation of this digital synthesizer.

Let's think about a sine wave. Even if it's only one complete cycle, there are an infinite number of points at which it can be evaluated. Unfortunately, even modern high-speed ICs can only process a limited number of points in a second, so something has to give. Intuitively, it seems reasonable to represent a sine wave by using the amplitudes at some number of points, as shown in Fig. 23. To get the original signal back, we can use a low-pass filter to interpolate, or smooth out, between these points.

The questions then arises, "How many points do we need to represent a sine wave and still get a good reproduction?" The answer is really quite surprising: any number greater than or equal to two. On the average, this will give a perfect sine wave at the output of a low-pass filter. For example, 2.1 is quite sufficient, if the filter is good enough. The number of sampled points per cycle doesn't have to add up to a whole number. Fractional (or even irrational) values are allowed (for example, 2.9, 4.7, 4.8). This requirement for two or more samples per cycle is called the Nyquist limit, named after its discoverer, an engineer at Bell Labs.

Suppose we take a perfect 2-MHz sine wave, which might come from a crystal oscillator. Let's look at it every 50 ns, as illustrated in Fig. 23. The time between samples will be determined by the maximum frequency we want to represent and the speed of the ICs that are available. The synthesizer described later uses a 59.6-ns time between samples. Since we have sampled more than two points per cycle, there will not be any problem with getting the original signal back when we need it, as shown in Fig. 24.

But if we are going to process the signal digitally there is another problem: We don't have infinite precision. Specifically, we only have eight "bits" (a "bit" is a binary digit — a number that can be only a 0 or 1) or about 0.4% accuracy, in the arithmetic section of a typical synthesizer.

What is this effect going to be? Without delving deeply into the mathematics, the only effect caused by this lack of precision is to increase the harmonic content somewhat, and give a low-level broadband noise. What we have done so far is to reduce our 2-MHz sine wave to a sequence of numbers. If we want the sine wave back, we feed the sequence of numbers into a digital-to-analog converter (DAC) and follow it with a low-pass filter. It stands to reason, therefore, that to synthesize a sine wave, all we have to do is come up with the same sequence of numbers that we would get by measuring a sine wave, and feed that sequence into a DAC.

Let's look at another graph of a sine wave (Fig. 25). This one is not much different than the previous one. All we've done is change the X axis from time to phase angle. The value of the sine wave varies from +1 to −1, depending on the

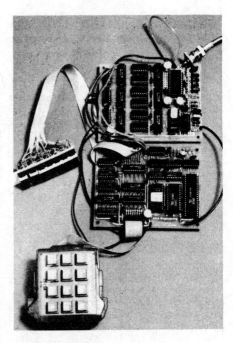

Fig. 22 — The synthesizer and controller boards, along with keypad and display, form a small package.

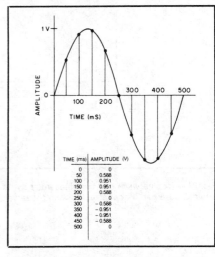

TIME (ms)	AMPLITUDE (V)
0	0
50	0.588
100	0.951
150	0.951
200	0.588
250	0
300	−0.588
350	−0.951
400	−0.951
450	−0.588
500	0

Fig. 23 — A 2-MHz sine wave is shown, with its amplitude sampled every 50 nanoseconds.

measurement angle. So, one way of getting that series of numbers is to find the angle of a sine wave at a particular point in time, and convert it to the amplitude value. This is done by storing the values for the sine wave in a read-only memory (ROM). But

---

[1]Circuit boards and complete parts kits for both the synthesizer and the controller portions of the project are available from A & A Engineering. Contact A & A Engineering for current pricing and availability of the parts you need.

Spread Spectrum Theory and Projects 8-11

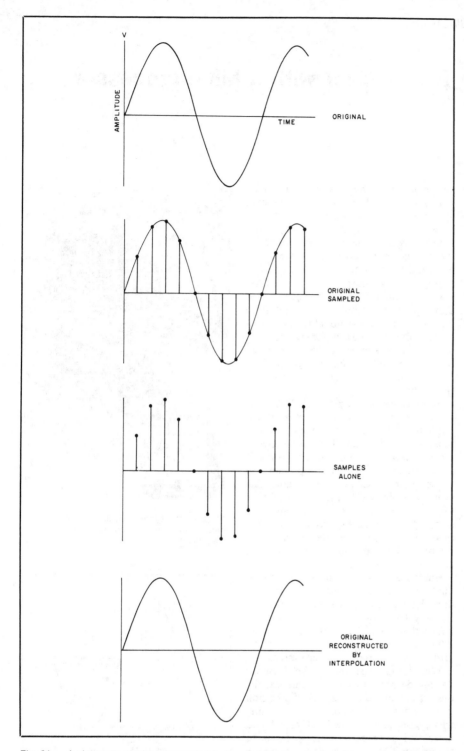

Fig. 24 — A sine wave can be represented by its value at a few points, and a replica of the original wave can be constructed from those values.

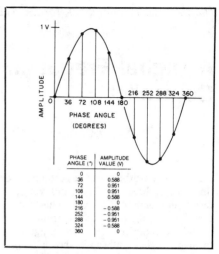

Fig. 25 — Any sine wave can be sampled at various phase angles (in this case, every 36 degrees) and the values stored for later reconstruction of the wave.

Table 8
Examples of "Overflow" Condition in Decimal or Binary Addition

Decimal			Binary		
00	47	94	0000	0110	1100
+47	+47	+47	+0110	+0110	+0110
47	94	141†	0110	1100	10010††

†"Overflow" condition; the result is 41.
††"Overflow" condition; the result is 0010.

if we look at the phase angles, we notice something quite remarkable: They are all multiples of the same number.

The phase-angle values may look as though they can get huge, but every time we pass 360 degrees we can subtract 360, so that the phase angle is always less than that. One way of getting this specific series of numbers is to take an adder and a storage register and just add the same number to the value in the register every 50 ns. If we program the read-only memory to contain one cycle of a sine wave, the number in the register will be a fraction of 360 degrees.

A storage register will store a binary number of a given size; the register used in the synthesizer described here can store 24 bits. This is about equivalent to eight decimal digits. What happens when the sum that we get is larger than 24 bits? Table 8 shows both decimal and binary examples of this condition, called "overflow." If we just ignore the carry digit, the effect is the same as subtracting a number one larger than the biggest number that can be stored.

Since the number in the register is a fraction of 360°, every time the register overflows it has the same effect as subtracting 360°. This combination of an adder and a register is called a "phase accumulator," and the number that is added repeatedly is called the "phase increment."

So these are the blocks of a digital synthesizer, shown in Fig. 26: a phase accumulator, a ROM, a DAC and a low-pass filter. To make the synthesizer easier to use, a controller may be added to allow a dial or keypad to provide the required phase increment. As designed, the digital synthesizer requires a number with 24 binary digits (bits) to specify the frequency.

There are several possible ways of controlling the synthesizer. A personal computer with three serial I/O ports or one parallel port can provide the proper sequence of data. Individual switches can be used to enter the frequency directly into the

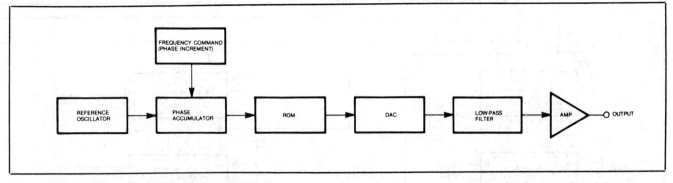

Fig. 26 — Block diagram of a digital frequency synthesizer.

accumulator. This would not require the shift registers. It may be okay for experimentation, but not on a regular basis. Similarly, a multiple-position switch could be used to select a set of diodes, which would enter the frequency directly into the accumulator, again not using the shift registers. This project uses a dedicated controller with a microprocessor IC.

The frequency is entered as a 24-bit binary number, shifted in most-significant bit (MSB) first. If you are using the shift register, three signals are needed; data, clock and load. The data must be valid during the rising clock edge, and the load control must go high after the last data bit has been shifted in. The clock may be continuous or gated, as long as each data bit is present at the appropriate rising edge and the load control goes high before the next rising clock edge.

Circuit Description

The schematic diagram for the digital synthesizer is shown in Fig. 27. It really is just a straightforward implementation of the block diagram. To get a 24-bit adder, six 4-bit adder chips are cascaded. The speed of the circuit is determined by Fairchild 74F283 adders. Slower units will not work at a 16-MHz clock rate. The 24-bit register is made out of three 8-bit registers. These don't need to be cascaded, just clocked at the same time. The nine most-significant bits of the output from the register go to read-only memory, which contains the stored values of the sine wave. This data tells the ROM which phase value to send. The ROM can be any ROM with a short enough access time.

From the ROM, the signal goes to a video-speed digital-to-analog converter. A TRW TDC1016B7C8 IC is used in this part of the circuit because of its speed. This chip has a 75-ohm output over a wide frequency range, and makes the design of the low-pass filter, which interpolates the points, much easier. It can also provide ½-V peak-to-peak signals into a 75-ohm load without an external amplifier. (Many video-speed DACs provide a current output, which places some of the burden of getting good response on an external amplifier. These amplifiers can be more expensive than the DACs that drive them!)

Fig. 28 is a parts-placement diagram for the synthesizer board. Double-sided circuit boards with plated-through holes are available from A & A Engineering. (See note 1.)

Some Design Criteria

Since people think and work in decimal, some way is necessary to convert from the decimal form that people use to the binary numbers that digital circuits use. That job is performed by the controller section. Since the synthesizer and the controller (which provides for keypad entry of the desired frequency) are built separately to make debugging easy, the design connects these two units with three wires — a data line to carry the information one bit at a time, a clock line to tell the synthesizer when to look at the data line, and one line to tell it that the new number is ready to use. This allows the use of different kinds of controllers. That's the reason for the three 8-bit shift registers. If we didn't mind a broadband burst of garbage every time the frequency changed, the intermediate latches could be eliminated.

Now let's focus on how some of the numbers were chosen for this circuit. Some of them may seem arbitrary; however, there are good reasons for every choice. A synthesizer using the block diagram of Fig. 26 can be designed for different frequency steps. The maximum output frequency is determined through the formula

$$f_{out} < \frac{f_{clock}}{2}$$

Actually, it is best to leave some room for the low-pass filter to cut off. A good rule of thumb is:

$$f_{max} = \frac{f_{clock}}{2.2}$$

The master clock frequency also determines the size of the phase accumulator (in bits) and the size of the frequency steps that you can get. There's a very simple formula for this:

$$\Delta f = \frac{f_{clock}}{2M}$$

where

Δf = the frequency step size
M = the number of bits in the phase accumulator

This design uses 24 bits and a master clock frequency of 16.777216 MHz — which happens to be 2^{24}. This gives a frequency step of 1 Hz.

These formulas work for any digital synthesizer that has been designed from this block diagram. If you want an audio synthesizer, a smaller clock frequency and phase accumulator size can be used to give the same performance over a more limited range. For example, a clock frequency of 65.536 kHz (2^{16}) and an M value of 16 bits will give 1 Hz steps from 1 to about 30 kHz. Standard 74LS-series parts could be used for everything in that case. If a step size other than 1 Hz is used, the phase increment is the desired frequency divided by the step size. Of course, this number must also be in binary form.

The ultimate stability of a digital synthesizer is a function of the stability of the reference oscillator. The one used in the synthesizer for this article is not perfect, and a better oscillator is not difficult to build. The broadband spectral purity is a function of the number of bits used in the ROM and the DAC, and of the low-pass filter response. If the synthesizer is used to replace a 5- to 6-MHz VFO in a transceiver, a band-pass filter can be used, which will give excellent performance. As the spectral photograph at Fig. 29 shows, the phase noise performance of the synthesizer is outstanding. The speed of changing frequency is limited only by the time it takes to transmit the 24 data bits from the controller to the synthesizer. One nice feature of this approach is that the output is phase-continuous when the change is made. This means the sidebands generated by the change are minimized.

Microprocessor Controller

The frequency must be supplied to the synthesizer as a binary number. Although it is possible to enter the desired frequency directly as a binary number by setting 24 switches, that is rather inconvenient. In addition, most people are not capable of rapidly converting decimal numbers into binary ones — nor should they be. This task is best left to inexpensive microprocessors. As a bonus, the computing power inherent in these chips allows extra

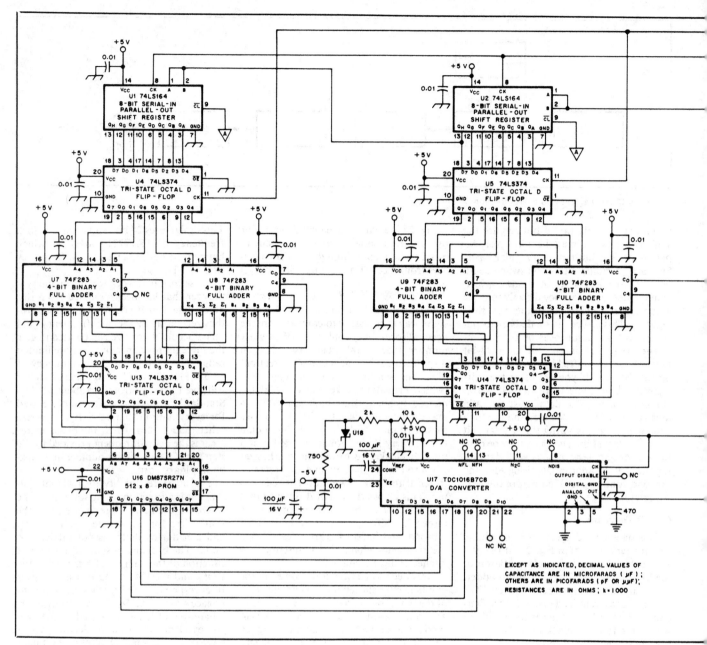

Fig. 27 — Digital frequency synthesizer schematic diagram.

convenience features to be added at no extra parts cost, just the programming effort to tell the unit what to do. A controller based on the Zilog Z80™ microprocessor is described here. It provides keypad entry of the desired frequency, for use with the digital frequency synthesizer.

A Quick Review of Microprocessors

Any computer system uses a limited number of basic component sections. These are:

1) A central processing unit (CPU), which performs all of the required operations on the information it is presented with.

2) A memory section that stores temporary and permanent information, and also stores the sequence of instructions that tells the processor what to do and when to do it.

3) Some way of communicating with the "outside" world (called input and output, or I/O for short).

Since microprocessors are inexpensive, the CPU of most small systems is already designed. For most control purposes, there is virtually no difference in final results between the major microprocessors.

Once the CPU is specified, all that remains to build a unit that uses a microprocessor is to design a memory section, the I/O circuits and the interconnections between these sections. The design philosophy on this project was that simple parts should be used wherever possible, to reduce the cost and difficulty of debugging the controller.

Before beginning a design, you must know what you want the unit to do. Since you are going to be the operator, it is a good idea to start by describing how it should appear to you as a user. In other words, describe how it communicates with the world outside the equipment that you're designing, since you're part of that outside world!

This project includes keypad entry of the frequency and control information, with a digital readout of the frequency as it is being entered. Since the synthesizer is a wide-range device, covering from low audio frequencies to around 6.5 MHz, the frequency readout will range over several decades. To get around this problem, an ENTER key is used. This key tells the synthesizer to change to the frequency just entered. The keypad and display work somewhat like a pocket calculator.

The second set of outputs are the ones that tell the synthesizer what frequency to

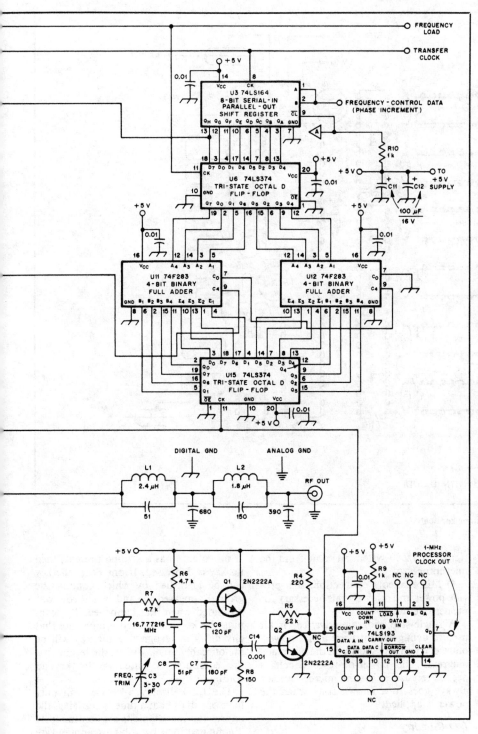

the oscillator portion of Fig. 27), and then using the driver circuit that appears in the Zilog data sheet, as shown in the clock-input portion of Fig. 31.

Memory Circuit

Two kinds of information are stored in the memory section. One kind is temporary data. Examples of temporary data held in memory are the present operating frequency, recorded in both decimal and binary forms, and intermediate results of calculations. The other kind is permanent data, such as the sequence of instructions that tell the controller what to do, or the information that specifies which segments of the LEDs to light for each display digit. It isn't surprising to find that different kinds of memory chips are often used for different kinds of data. The temporary data is stored in memory that can be changed. This kind of memory is referred to as read/write memory (RWM) or random access memory (RAM). Despite the name being less descriptive, the term RAM is normally used. The permanent data (which includes the program or sequence of instructions to tell the processor what to do) is stored in memory that cannot be changed. One kind of memory which does this is called erasable, programmable, read-only memory (EPROM). This is the type of program memory that is used in the controller. The information stored in an EPROM does not "go away" when the power is turned off, like the information in RAM memory does.

For convenience, chips which store data in 8-bit "bytes" can be used to simplify design, because they match the 8-bit groups that the microprocessor handles. The cost of these chips is so low that a designer of a simple, not mass-produced device like this controller can make the task much easier just by using chips that are larger than needed. Each of the chips used in this controller is capable of storing 2048 eight-bit numbers. The single RAM chip (U2) is a 6116-type static memory, but a 2016-type memory IC would also work. The EPROM (U3) is a 2716 IC. Each of these memory ICs almost forms a complete memory subsystem in itself. The CPU needs to have some way of distinguishing between these memories, which it does by using different addresses for the two memory chips, and different I/O addresses for the I/O chips. A simple gate circuit selects which memory is used, as shown in the memory portion of the Fig. 31 schematic diagram.

Interconnection Circuitry

Interconnections between the three different sections (CPU, memory, and I/O) fall into four different categories:

1) Connections that carry information to and from each section. A set of wires which carries information is called a data bus. The Z80 uses a set of 8 lines to carry data, because it handles information in

provide. Three signals are required to drive the synthesizer; one line for the data bits, another to tell the synthesizer when a data bit is valid and can be loaded into the shift register, and one to tell the synthesizer that all data bits have been transferred from the controller to the synthesizer. The data bits must be sent with the most significant bit first. The load-data line must go from a logic LOW value to a HIGH one after all 24 bits have been transferred to the synthesizer. Fig. 30 shows how these waveforms should look.

Circuit Design

Microprocessor Requirements

In order for the Z80 chip to work properly, it needs a signal to tell it when to perform each operation. This signal is called a clock. There are specifications both on the frequency (minimum 100 kHz, maximum 4 MHz for the standard part) and voltage of the clock signal. To avoid the cost and trouble of providing a separate oscillator, the clock signal is obtained by taking the 16.777216-MHz synthesizer clock signal, using a 74LS193-counter IC to divide this frequency by 16 (shown on

Spread Spectrum Theory and Projects 8-15

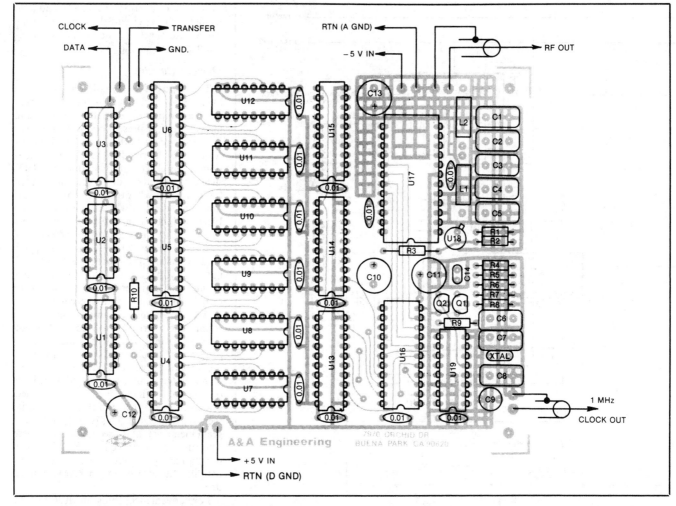

Fig. 28 — A parts-placement diagram for the synthesizer board.

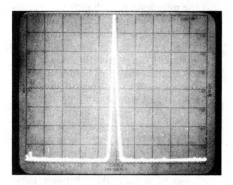

Fig. 29 — Spectral display of the digital frequency synthesizer output.

8-bit bytes. Because this bus carries instructions to the microprocessor from the memory, all 8 lines must be used.

2) Connections that tell a section where to find or put information. This set of lines is called an address bus. The Z80 provides a set of 16 lines to carry the address, but not all of these lines need to be used.

3) Connections that tell a section what to do. This set of lines is called a control bus. This bus tells each section whether it will receive or send information, and what it should do with the information. Control signals in a Z80 system all come from the microprocessor.

4) Miscellaneous connections that provide power and other signals necessary for proper operation.

A single-step program-advance circuit used to debug the software and a power-on-reset circuit are shown on the schematic diagram. The power-on-reset circuit is needed to ensure that the microprocessor always goes to a known state when the power is applied.

I/O Circuitry

At specified times, the processor has to communicate with a keypad, a display, and the synthesizer itself. The Z80 has special instructions for performing I/O operations, but these instructions still must be matched to the circuitry that surrounds the processor.

The keypad shown in the title photo was salvaged from a pushbutton telephone. A calculator-style keypad, available from many surplus dealers, will also do fine. The keypad was wired up as a set of SPST switches in a rectangular array, as shown in Fig. 31. The keypad is scanned by grounding one row at a time and seeing which (if any) column has a ground on it. If no column has a ground present, then no key is depressed. If one does, the key can be identified by which column the ground appeared in, and which row was grounded when that happened. By scanning the keypad at such a high rate that any delay in recognizing the key will be unnoticeably small, you get the feeling that the computer is watching the keypad continuously.

The LED display, likewise, will only have one digit lit at a time, but it gives the impression that all digits are continuously illuminated by using a high scanning rate to take advantage of the persistence of human vision. Both the keypad-scanning technique and the display-scanning technique are widely used in pocket calculators.

To communicate with the synthesizer, the three output lines must each be able to change independently of the times that the other lines change. This function is handled quite simply by three flip-flops that hold the data-bus contents when an output occurs. This is shown in the synthesizer-output section of Fig. 31. The synthesizer circuit itself does not provide any information back to the controller, so the only input comes from the keypad.

There are four I/O chips in the con-

8-16 Chapter 8

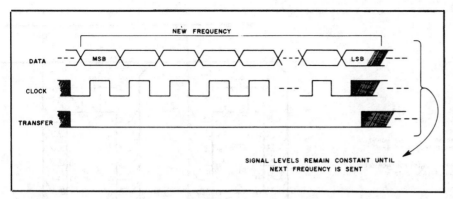

Fig. 30 — Controller-output signal waveforms on the lines to the synthesizer.

Table 9
Controller-Program Memory Map

Address Range	Contents
$0000-$07FF	ROM (program and tables)
$0000-$0037	Main program
$0100-$010D	Memory-clear subroutine
$0200-$0218	Display-scan subroutine
$0300-$033B	Keyboard-scan subroutine
$0380-$03A7	Subroutine to convert keystroke to decimal
$0400-$0412	Subroutine to convert decimal to seven-segment display
$0480-$04AA	Subroutine to shift display digits during entry
$0500-$0535	Subroutine to send binary frequency data to synthesizer
$0600-$060E	Delay subroutine
$0620-$0687	BCD-to-binary conversion subroutine
$0700-$07AF	Unused ROM space
$07B0-$07C4	Decimal-to-binary conversion-factors table
$07D1-$07F0	Keycode translation table
$07F1-$07FF	Seven-segment display translation table
$0800-$0FFF	RAM
$0800-$0807	Display memory
$0808-$080F	BCD frequency memory
$0810-$0816	Intermediate results
$0817-$0819	Binary frequency data
$0820-$0FFF	Unused RAM space

troller. The CPU needs to have some way of distinguishing each one. It does this by using different I/O addresses for each of these chips. Two 74LS138 3- to 8-line decoder chips are used to direct the chip-enable commands to the right IC.

To aid in troubleshooting the controller, a small circuit has been included to perform only one instruction, then wait for a push-button to be pressed and released before performing the next instruction. This single-step circuit is shown connected to pin 24 of the microprocessor in Fig. 31.

Program Design

A program is a list or sequence of instructions that the computer follows to perform a specific task. Unlike humans, microprocessors can only do one thing at a time, but they can do each operation extremely fast.

The best programs, like the best circuits, are put together from sections that do not have a lot of connections. In transceiver design, for example, different circuits are often placed in different metal boxes, with only the inputs and outputs connected, so that interaction between sections can be minimized. Likewise, programs are best designed with simple modules, which can then be strung together to perform the desired function. Instead of wires or coaxial cable to connect different modules or sections together, program sections communicate by placing numbers in certain memory locations, somewhat like communicating with your neighbors by leaving messages at their doors.

There are several major program modules used in this controller:

1) Main program — selects which of the other modules is used.
2) Memory-clear module — makes sure that we start with zeros in every RAM location that is used.
3) Display-driver module — sends one digit to the LED display.
4) Keyboard-scan module — checks to see if a key is pressed; if so, it reports which one.
5) Decimal-to-binary conversion subroutine.
6) Module to send frequency data to the synthesizer in binary form.

Each of these modules is composed of smaller sections. For example, the module that sends the frequency to the synthesizer uses one section to send a "zero" to the synthesizer, and another to send a "one" to the synthesizer. In turn, each function is built up out of individual instructions.

With this idea in mind, look at Fig. 32. This diagram is the programming equivalent of a block diagram. It's called a flow chart. Unlike a block diagram for a piece of electronic equipment, only one block can be working at any instant. Since only one block can be working, the computer has to store the results at each block, so that they will be available when needed. Where is each number stored? The diagram that gives this information is called a "memory map," and the map for this controller is shown in Table 9. Much of this program is "table driven" so changes will be relatively easy to implement if you want. ("Table driven" means that all the information that the program uses for a particular function is stored in a single, unbroken area of memory).[2]

Construction

Fig. 33 is a parts-placement diagram for the controller PC board. Double-sided boards with plated-through holes are available from A & A Engineering. (See note 1.) The keypad and displays were built on separate boards, to make it easier to mount them in a chassis. Preassembled flat cables with dual-in-line-package (DIP) plugs connect the keypad and display to the main computer board. The supply voltage for each IC is bypassed with a 0.01-μF capacitor. No special construction techniques are required.

Access to an EPROM programmer is necessary if you plan to program the EPROM yourself. Many distributors of electronic components offer programming as a service, and computer stores will also often perform programming services for a fee. You might even find a local computer hobbyist who would be willing to program the EPROM for you. Alternatively, you can purchase a preprogrammed EPROM from A & A Engineering. (See note 1.)

One possible problem area is the keypad. If your keypad has different connections for the various keys, the table in EPROM that tells which key was depressed for each possible code will have to be corrected to match your keypad. Since there is no standard, you will either have to write your own conversion table to replace KCTBL, or use individual keys and wire them according to the diagram shown in Fig. 30.

Debugging

Ideally, the synthesizer and controller will work perfectly the first time you turn them on. The real world seldom turns out

---

[2] A commented assembly-code listing is available from ARRL for $2.50. Send your request to ARRL Technical Department, 225 Main St., Newington, CT 06111. Mark the outside of the envelope: Williams Synthesizer. Please print your name and address clearly on your request.

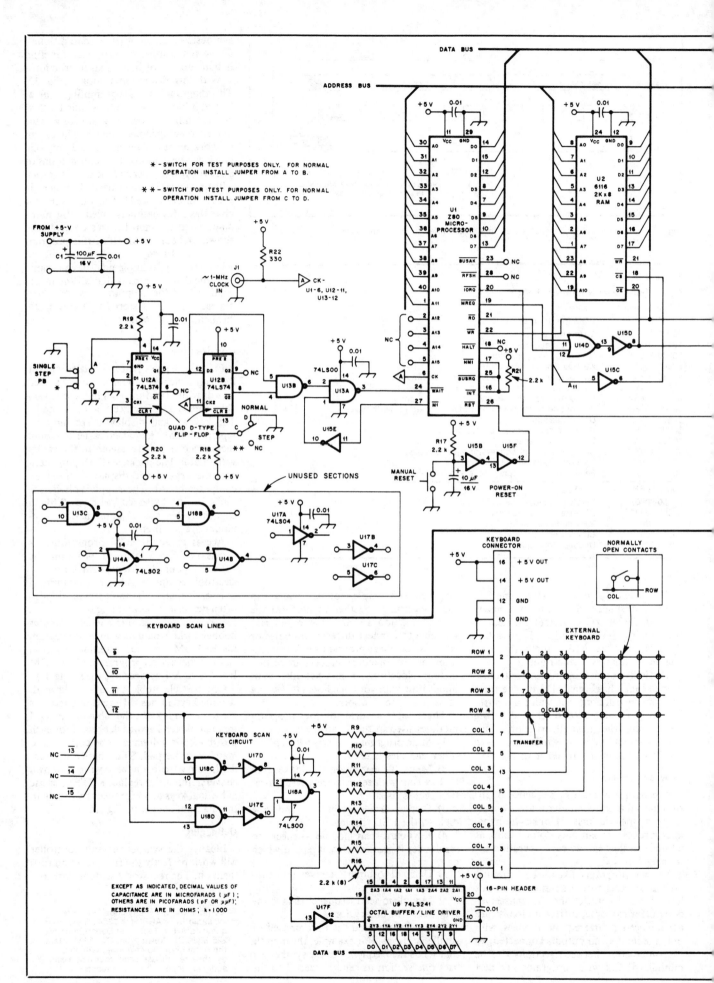

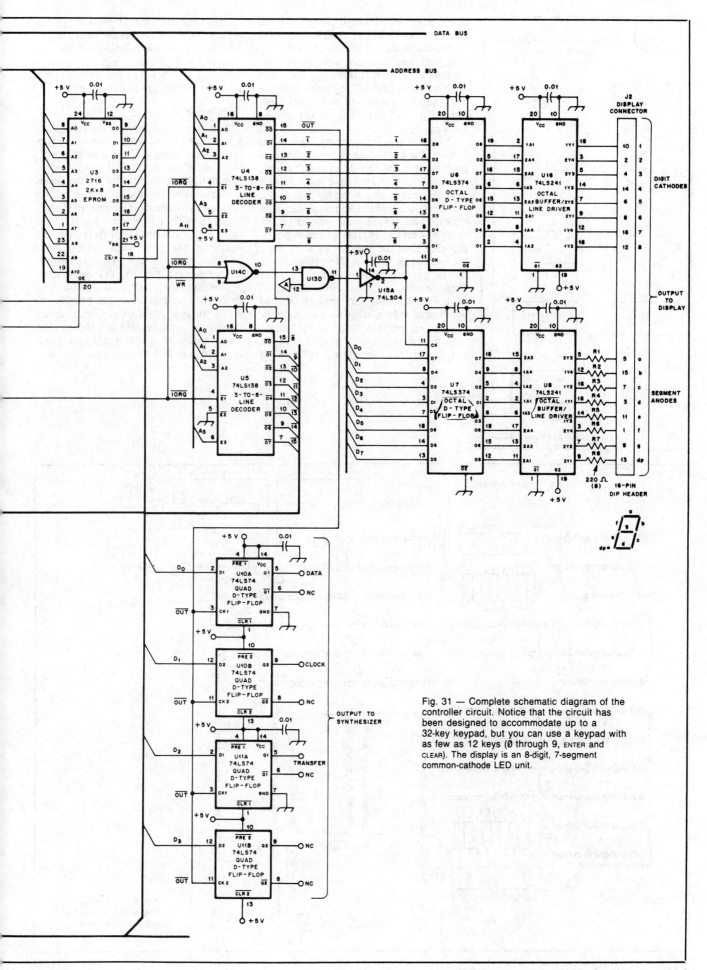

Fig. 31 — Complete schematic diagram of the controller circuit. Notice that the circuit has been designed to accommodate up to a 32-key keypad, but you can use a keypad with as few as 12 keys (0 through 9, ENTER and CLEAR). The display is an 8-digit, 7-segment common-cathode LED unit.

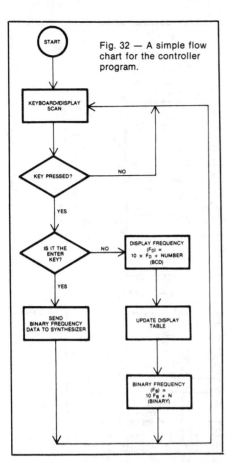

Fig. 32 — A simple flow chart for the controller program.

that way, however! A logic probe and pulse generator, and a triggered oscilloscope will be helpful in debugging the projects.

If your unit does not work when first turned on, verify that there is +5 V at the supply pin of each IC. Next be sure the clock waveform is present on the Z80 clock-input pin, and that it is of the proper voltage.

Once the required driving signals are present, the proper operation of the CPU must be checked. To do this, use a logic probe or scope to verify the presence of a negative-going pulse at the keypad. If this is missing, use the reset and single-step controls to check for the presence of the proper signals on the address and data buses. If these are not the same at each chip, check the bus connections to those chips. Once the keypad strobe is present, verify the presence of strobe pulses at the display. If the keypad can enter numbers correctly into the display, then synthesizer output can be checked either on a scope or by using a logic probe to check the signals at the output of the serial-to-parallel shift registers in the digital frequency synthesizer. A reminder, so you aren't scared away by all this talk of debugging: knowing that you have a functioning program in EPROM to begin with provides an immense advantage in debugging!

Conclusion

There are unused keys on the keypad and a lot of unused memory space in the program memory chip (U3), so several features could be added at only the cost of reprogramming the 2716 EPROM. These include a scan feature, the ability to store a large number of frequencies (for nets, skeds, and the like), or the ability to use a dial similar to that of traditional rigs. Only your imagination and programming skills need limit what you can do with the synthesizer. This project is a good way to improve both.

The combination of the digital frequency synthesizer and controller provide the ability for an amateur who enjoys construction to obtain the benefits of a high-quality, low-phase-noise, stable signal source that is convenient and easy to use.

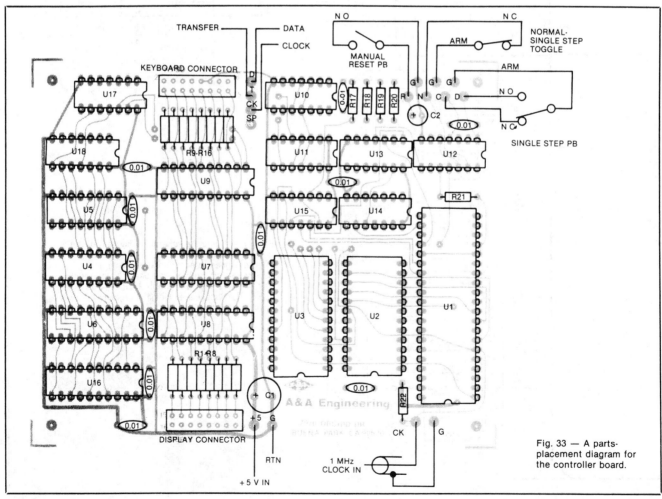

Fig. 33 — A parts-placement diagram for the controller board.

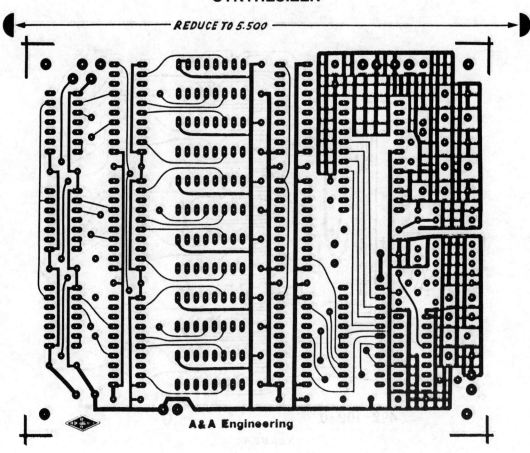

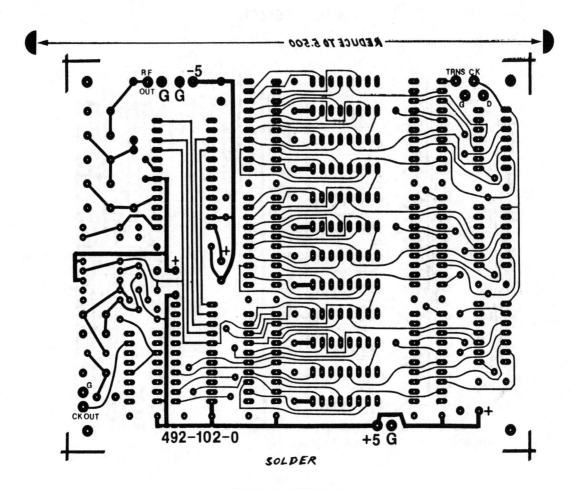

SYNTHESIZER

CONTROLLER

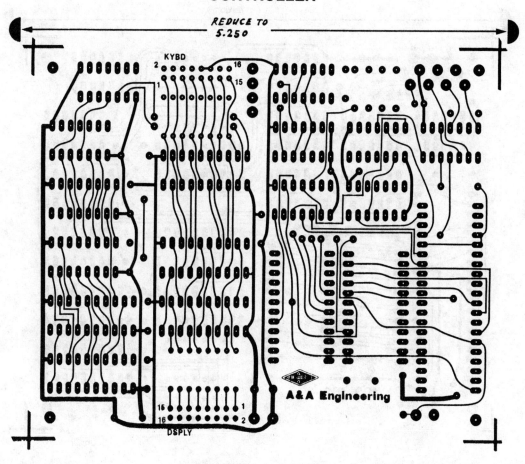

CONTROLLER

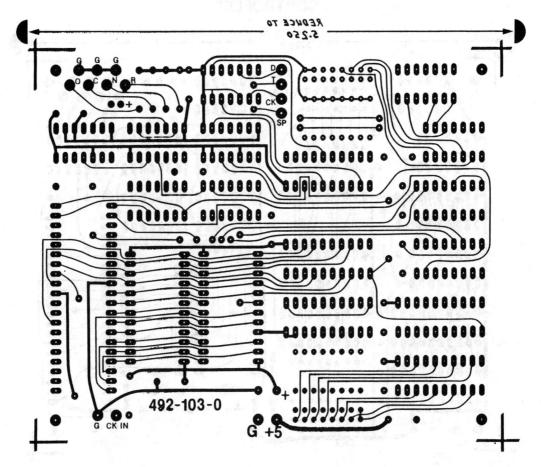

SOLDER

Spread Spectrum: Frequency Hopping, Direct Sequence and You

Are you ready for this month's spread-spectrum rule implementation? How do you operate in spread mode? Here's the answer.

By Hal Feinstein, WB3KDU
Member, ARRL Ad Hoc Spread-Spectrum Committee

As regular readers of *QST* know, after studying the issue since June 1981, the FCC passed a delayed rule allowing radio amateurs to use spread-spectrum frequency-hopping and direct-sequence systems, effective June 1, 1986.[1] However, readers may not know about experimentation that has been ongoing in the spread-spectrum area and how Amateur Radio interoperability will work. This article covers those topics.

What Is a Spread-Spectrum System?

In the 1986 *ARRL Handbook,* Chapter 21 contains a description of spread-spectrum communications. That section of the *Handbook* should be consulted for an extensive discussion of the subject. The basic explanation is that spread spectrum is a modulation scheme whereby the signal is spread over a very wide bandwidth. This results in a dilution of the signal energy such that the power density present at any point within the spread signal is slight. Beyond a certain distance from the transmitter, the spread signal can be below the noise level yet still be recovered with the proper spread-spectrum receiver. Only the intended receiver (or receivers in a net operation) can recover the signal, as both sender and receivers hold copies of the binary sequence that is used to spread the signal and know when it was started in time. Interference to other users of the same spectrum is slight or nil (unless they are close to the transmitter).

There are two spread-spectrum modes authorized to the radio amateur. Frequency hopping is a mode in which the operating frequency is changed rapidly over the spread bandwidth. Both the transmitter and receiver visit the same frequencies at the same time, and must stay in exact synchronization. Each holds the same list of pseudorandom-ordered frequencies, and the transmitter and receiver start hopping together using the same starting point on the list.

In direct sequence (the other authorized

[1]Notes appear at end of article.

This is the station used in the FCC direction-finding test. An ICOM IC-2A was used as a frequency-hopping spread-spectrum beacon. The hop rate was 80 hops/s.

mode), a high-speed pseudorandom binary data stream is used to shift the carrier phase between 0 and 180 degrees. The phase shifting is normally done in a balanced mixer, and the information being transmitted is normally added to the high-speed code sequence.

Why Use Spread-Spectrum Systems?

Effective spectrum management allows the greatest use of a band of frequencies by the largest number of possible users. A large number of spread-spectrum systems can occupy the same band and not interfere with each other. Spread spectrum can make use of unused portions of a frequency band—such as between repeater channels. Only by experimentation can radio amateurs learn the true potential of this new mode. The military has been using these systems for years for "antijam" communication, and the radio amateur can benefit from this experience. Computer-assisted power control can be used effectively in this mode to meet the FCC requirement that amateurs run only the minimum power needed for communication. This has not been done to date, but it is a feature for advanced experimenters to work on.

What Has Been Done Up to Now?

The Amateur Radio Research and Development Corporation (AMRAD), a nonprofit club composed of radio and computer amateurs, has been involved with a series of investigative experiments with spread-spectrum systems since September 1980. The investigations fall into five general experiments authorized by a series of FCC Special Temporary Authorities (STAs).

1) Commercial/military frequency-hopping radios were borrowed and used to test on amateur frequencies in the HF bands. The hop rate was slow—only 5 hops

The HF spread-spectrum station at K8MMO. A Xerox 820 computer was used for control. The transmit position is on the right; receive is on the left.

The heart of the HF spread-spectrum station is the frequency synthesizer designed by Fred Williams. This version of "Fred," as the synthesizer is affectionately called by AMRAD members, came from A & A Engineering.[2]

per second (hops/s). The system worked well using voice while hopping in a nonvoice portion of a ham band. Little, if any, interference was noted in a nearby receiver. When the hopper came by, an "aup" or "thu" noise resulted. This experiment was lots of fun. However, the greatest benefit was to get everyone fired up to go build some equipment for the ham bands.

2) Chuck Phillips, N4EZV, constructed hand-held, frequency-hopping FM transceivers. These units hopped at 10-80 hops/s over 3.2 MHz. If the receiver or transmitter lost synchronization, the lost unit would return to a "homing frequency" and "scream" for help. Periodically, the transmitting station would stop and listen on the homing frequency for screams. If one was heard, a new start-up would be initiated. Further details of this system are not available.

3) ICOM IC-2A transceivers were modified to allow a computer (Commodore C64) to set the frequency instead of using the radio's thumbwheels. A beacon was set up to test for interference caused in the 2-meter repeater band. Receiver synchronization schemes were tested. This project was abandoned after spectral analysis of the transmitted signal revealed that the synthesizer was never "locking up." For that reason, fast-hopping operation was not feasible. Modifications of the phase-locked loop filter improved the lockup time, but produced some "strange audio." The FCC located this unit in a direction-finding test in only 25 minutes, proving that the agency is able to regulate this mode of operation using current equipment.

4) Fred Williams of the TRW LSI Products Division designed a direct-synthesis oscillator that is described in February 1985 *QST* and Chapter 29 of the 1986 *ARRL Handbook*. AMRAD has used this synthesizer with the Yaesu FT-7 and FT-101ZD and Xerox 820 computers to frequency hop cleanly on the HF bands. Experiments with these units continued as synchronization software was being perfected. This experiment came to a standstill when the FCC denied AMRAD's latest STA request for permission for further HF experimentation. The net effect of this denial is to force all experiments to 420 MHz and above. Andre Kestleloot, N4ICK, is building a mixer to convert signals from HF to UHF. This mixer will allow these experiments to resume on the air at UHF. Conventional UHF transverters could also be used, but are costly—the AMRAD gang prefers to build their own and save money.

5) Dick Bingham, W7WKR, has produced a single PC board add-on for a 440-MHz hand-held transceiver that allows direct-sequence transmission or reception. It appears that this unit, while allowed under the AMRAD STA, would not be allowed under the new Part 97 rules, because of the method used to add information to the basic spreading sequence. This requires further study.

What Is Interoperability?

The ARRL Board of Directors has authorized, and President Price has appointed, an ad hoc Spread-Spectrum Committee to study interoperability. When the rules go into effect, two radio amateurs who have been working together closely can communicate using spread-spectrum techniques. This is possible because both operators understand how both systems operate and how to begin their spread-spectrum communication and maintain synchronization. But how does an amateur call CQ using spread spectrum, and expect to receive a reply? How does the amateur communicate the spreading sequence in use, starting signal, hopping frequencies, hopping rate and the like to another amateur listening for a spread-spectrum CQ call? How does the amateur identify spread-spectrum transmissions? The rules require conventional identification.

In packet radio, we have a protocol called AX.25 to answer similar questions. It is too early to agree on a protocol for this new mode of communication, but some basics need to be established, such as a home or calling frequency on each band from 420 MHz and above.

The Committee recommends that you use the national FM calling frequency (or another FM simplex frequency if the national frequency is in frequent use in your area) for calling CQ and for announcing spread-spectrum operating parameters. A further recommendation is that you let others in your area know of your experiments through clubs, over repeaters or by any means that seems appropriate. The committee will publish more on these matters as we gain on-the-air experience.

Your suggestions and comments concerning these matters are welcome. Please remember that it is not our goal to stifle experimentation by needless regulation or standardization. Send your comments and suggestions to the committee secretary: Chuck Hutchinson, K8CH, ARRL Headquarters, 225 Main St, Newington, CT 06111.

How Can an Amateur Keep Current on Spread-Spectrum Progress?

Read *QST* and *QEX* to keep current on fast-breaking spread-spectrum news updates. Take advantage of this unique form of Amateur Radio communication. Start by learning the basics that are found in the 1986 *ARRL Handbook*.

Notes
[1] D. Newkirk, "Our new Spread-Spectrum Rules," *QST*, Apr 1986, p. 45.
[2] A & A Engineering, 2521 W LaPalma Ave, Unit K, Anaheim, CA 92801, tel 714-952-2114.

Practical Spread-Spectrum: A Simple Clock Synchronization Scheme

By André Kesteloot N4ICK

On June 1, 1986, US Amateur Radio operators were given authorization to use a transmission method known as spread spectrum.[1] For a good theoretical coverage of the subject, refer to chapter 21 of *The 1986 ARRL Handbook*.[2] In addition, a condensed report on spread-spectrum experiments conducted by the Amateur Radio Research and Development Corporation (AMRAD) has been published in a recent issue of *QST*.[3]

In Amateur Radio spread-spectrum transmissions, in addition to the application of some traditional method of modulation to an RF-carrier, a pseudo-random code is used either to (a) alter the phase of the carrier (direct sequence), or (b) force the carrier to hop from one frequency to another (frequency hopping).

[1]Notes appear at end of article.

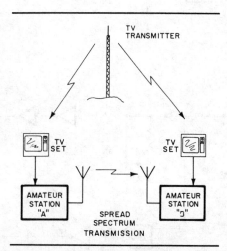

Fig 1—All US television stations transmit horizontal and vertical synchronization pulses that amateurs can use as clock pulses for spread-spectrum applications.

This pseudo-random code is generated in the transmitter at a certain clock frequency. If that clock and the original code are available at the receiving site, decoding (despreading) can be achieved.

The code can be easily dismissed from the picture. The only pseudo-random sequences presently authorized in Part 97 of the latest *FCC Rule Book* are those that can be obtained from the output of one binary linear-feedback shift register either 7, 13 or 19 stages long, and using only certain feedback taps (see note 1).[4] Thus, both transmitting and receiving parties can agree in advance on a particular pseudo-random code and set up their equipment accordingly.

The clock poses a much more intricate problem. In 1983, William Sabin, W0IYH, suggested that one solution is for both parties to be synchronized to the same external reference, such as WWV.[5]

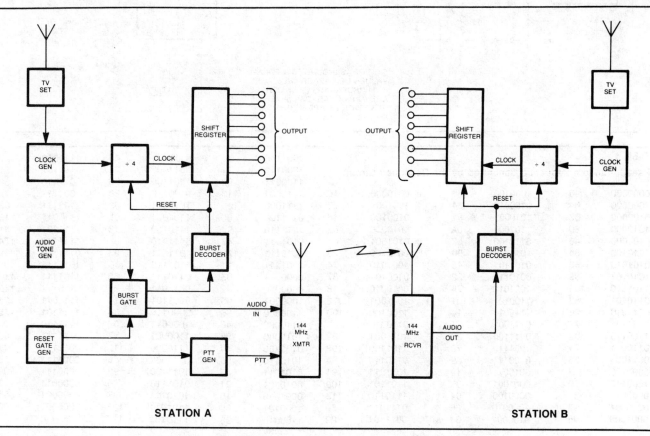

Fig 2—A block diagram of the AMRAD station arrangements used in their June 1986 demonstration.

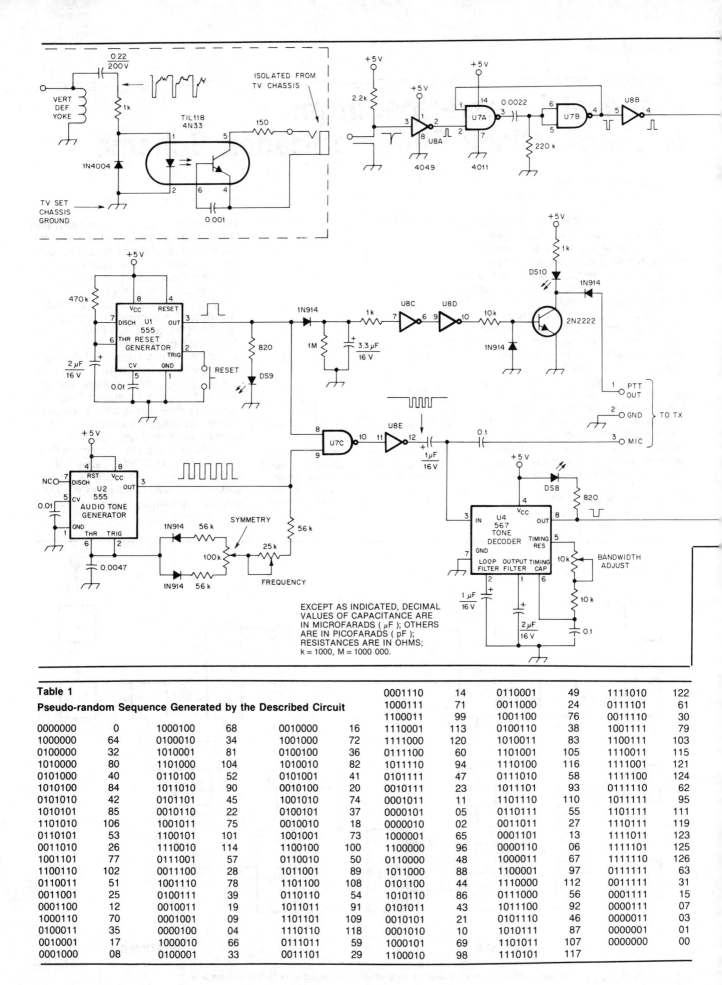

Table 1
Pseudo-random Sequence Generated by the Described Circuit

						0001110	14	0110001	49	1111010	122
						1000111	71	0011000	24	0111101	61
						1100011	99	1001100	76	0011110	30
0000000	0	1000100	68	0010000	16	1110001	113	0100110	38	1001111	79
1000000	64	0100010	34	1001000	72	1111000	120	1010011	83	1100111	103
0100000	32	1010001	81	0100100	36	0111100	60	1101001	105	1110011	115
1010000	80	1101000	104	1010010	82	1011110	94	1110100	116	1111001	121
0101000	40	0110100	52	0101001	41	0101111	47	0111010	58	1111100	124
1010100	84	1011010	90	0010100	20	0010111	23	1011101	93	0111110	62
0101010	42	0101101	45	1001010	74	0001011	11	1101110	110	1011111	95
1010101	85	0010110	22	0100101	37	0000101	05	0110111	55	1101111	111
1101010	106	1001011	75	0010010	18	0000010	02	0011011	27	1110111	119
0110101	53	1100101	101	1001001	73	1000001	65	0001101	13	1111011	123
0011010	26	1110010	114	1100100	100	1100000	96	0000110	06	1111101	125
1001101	77	0111001	57	0110010	50	0110000	48	1000011	67	1111110	126
1100110	102	0011100	28	1011001	89	1011000	88	1100001	97	0111111	63
0110011	51	1001110	78	1101100	108	0101100	44	1110000	112	0011111	31
0011001	25	0100111	39	0110110	54	1010110	86	0111000	56	0001111	15
0001100	12	0010011	19	1011011	91	0101011	43	1011100	92	0000111	07
1000110	70	0001001	09	1101101	109	0010101	21	0101110	46	0000011	03
0100011	35	0000100	04	1110110	118	0001010	10	1010111	87	0000001	01
0010001	17	1000010	66	0111011	59	1000101	69	1101011	107	0000000	00
0001000	08	0100001	33	0011101	29	1100010	98	1110101	117		

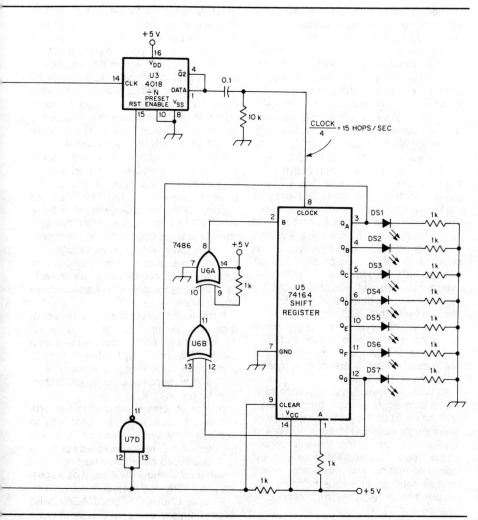

Fig 3—The schematic diagram of demo station A. Station B is similar except that the 555 reset generator, 555 audio pulse generator, audio-burst gate and the PTT generator are not needed. Audio output from the 144-MHz receiver is connected to pin 3 of the 567 audio-burst detector.

Choosing WWV as a source was plausible at the time because amateur HF spread spectrum was still being considered as a possibility. The new FCC ruling, however, allows for spread-spectrum transmissions to take place only above 420 MHz. This effectively restricts the coverage of any given spread-spectrum transmission to a relatively small geographical area, possibly that of a given community (at least until spread-spectrum repeaters are deployed)! If that is the case, then the local TV station's signals can be used as a source of accurate clock pulses (see Fig 1).

All US TV stations transmit horizontal synchronization pulses (H sync) at approximately 15,750 Hz and vertical synchronization pulses (V sync) at approximately 60 Hz. These sync pulses can be used as clock pulses for amateur spread-spectrum applications. At this time, AMRAD members are interested in frequency hopping at a fairly slow rate (between 10 and 60 hops per second) and choosing the V sync as a clock signal seemed logical for a first attempt.

The Arrangement

This article describes a practical frequency-hopping scheme using the discussed approach. A model for spread-spectrum transmissions was built for, and demonstrated at, the June 2, 1986 AMRAD meeting in Vienna, VA. A block diagram of the general arrangement used for the demonstration is shown in Fig 2.

Refer to Fig 3. Vertical sync pulses are readily available at the deflection yokes of any TV set. Since most TV sets have one side of the mains connected to the chassis, the use of an optoisolator stage is strongly recommended as an interface between the TV set and the ham gear. The 60-Hz V sync pulses are fed to a shaping stage and then used to feed a divide-by-N stage, the output of which is the hopping clock frequency. (To make the hopping frequency 15 hops/second, make N = 4, etc.) If both parties use the same arrangement, they will obtain clock pulses of the same frequency. To ensure that these clock pulses are actually in phase, we need now only reset both dividers simultaneously. This is achieved by one station issuing an audio tone burst, used as the reset signal at both transmitting and receiving sites. This burst needs to be sent only once, in clear, before spread-spectrum transmissions may commence. (The audio burst can be transmitted on the spread-spectrum service frequency, or over a local FM repeater to allow for an Amateur Radio net.) From this point on, the two clocks will be in sync unless the TV signal to either ham station is momentarily lost (fading, and the like).

The Circuit

Refer to Fig 3: A 555 (U2) is connected as an astable pulse generator to oscillate continuously at about 2 kHz. The exact frequency is determined by the setting of the 25-kilohm potentiometer, while the symmetry (duty-cycle) of the waveform is adjusted by means of the 100-kilohm potentiometer. Another 555 (U1) is connected as a monostable generator and produces one pulse (of about a one-second duration) every time pin 2 is grounded (depress the reset button). This pulse opens a 4011 gate and sends a tone burst to the transmitter's microphone input. The same pulse is fed to a pulse stretcher consisting of a 1-Megohm/3.3-μF/ 4049 combination, the output of which is used as the PTT signal. (Pulse stretching ensures that the transmitter stays on a little longer than the duration of the tone burst.)

Each station uses the same burst-decoding circuit to allow for an easier decoder adjustment. The audio tone burst, when applied to pin 3 of the 567 tone decoder, produces a negative-going output at pin 8. The negative pulse resets the 4018 divide-by-N and clears the 74164 shift register.

The 4018 divide-by-N stage is fed 60-Hz (V) pulses. Its output is the clock signal used to drive a pseudo-random generator of the type allowed by the FCC (see note 1). For our model, a simple 7-stage serial-in, parallel-out shift register was chosen with feedback taps from stages 1 and 7. A 74164 and a 7486 were used. (The 4015 and 4070-CMOS chips could have been used instead to maintain an all-CMOS design, but I had the TTL chips in my junkbox!) The 74164 is an 8-stage shift register and the 7486 is a quad XOR gate. One section of the 7486 is used to combine the feedback from taps 1 and 7, and another section is used as an inverter. The purpose of the inverter can be understood by considering the operation of the shift-register stage: When the 74164 is reset, all register-stage outputs are set to 0. Thus, feedback taps 1 and 7 are 0s, and the output of the XOR stage is also 0, feeding a new 0 to the input of the shift register.

Spread Spectrum Theory and Projects 8-29

This arrangement would normally "hang up" the shift-register in a nonallowable state. By adding the inverter stage (one section of the 7486), the 0 output of the XOR stage, obtained on reset, is changed to a 1 and the counting sequence can start over again.

The outputs of the shift-register stage, at pins 3, 4, 5, 6, 10, 11 and 12 are now the complements of the original sequence. But since common emitter transistor (or inverting buffer) stages are used to interface with the ham gear, the original sequence, as specified by the FCC, is restored. Note that the tone burst detector's output resets (clears) the shift register and the divider stage simultaneously.

The pseudo-random sequence generated by this described arrangement is shown in Table 1. The numbers listed are the decimal equivalents of the binary states appearing at the 74164's output pins.

The Demonstration

During the AMRAD demonstration, two TV sets were used to simulate a real-life situation. Each set was modified to produce V sync pulses. Both TV sets were tuned to the same local channel. (As luck would have it, the location used for the AMRAD meeting was a genuine RF hole and TV reception was poor at best. Nevertheless, acceptable sync pulses were derived from each set.) The output of each TV set was connected to a separate pulse-shaping stage and a shift register. Each station also had its own separate audio burst detector. An audio tone burst used to synchronize both boards was sent from one station to the other. This was done using 144-MHz transceivers. Both stations remained in phase for the duration of the lecture (about 90 minutes).

Comments

A few words of explanation and some suggestions may be in order

a) A 7-stage shift register was chosen because its parallel output interfaces easily with any parallel scheme, and specifically with the spread-spectrum sets designed several years ago by Chuck Phillips, N4EZV. (At the time, Chuck used extremely stable crystal oscillators in each set, from which he derived clock signals. Clocks were reset whenever sync was lost.)

b) Although a simple serial-in, parallel-out shift register was used for the demonstration, a more elaborate parallel loading scheme could be envisaged. This would allow for different "words" to be loaded initially in the shift registers, thereby permitting different shift sequences and, hence, operation of several different stations on the same frequency band with a minimum of interference.

c) Other external references are available: In addition to WWV, already suggested by William Sabin, Loran C, Satnav satellites and our ubiquitous 60-Hz mains readily come to mind. Each has advantages; each has drawbacks. The TV sync pulses, on the other hand, are easy to obtain. In practice, a portable amateur station would not require a complete TV set, but only a TV tuner, an IF stage and an AM detector followed by a sync separator to produce acceptable V sync pulses. Inexpensive all-band receivers (which have an AM detector) could be modified to cover part of the TV channels. Modern portable FM/VHF TV-sound receivers, however, are often only equipped with a combined IF amplifier and FM detector single-chip arrangement which does not easily lend itself to modification for AM detection.

d) Whichever external reference system is favored, remember that any such scheme could be particularly attractive in future single-sideband frequency hopping applications, as no continuous carrier would need be transmitted to convey clock information. The only output would then be the instantaneous hopping frequency, further reducing the likelihood of interference.

e) The 555/567 audio-burst generator and decoder combination I have used (for the sake of expediency) could profitably be replaced by a more reliable (and easier to adjust) dual-tone, multi-frequency (DTMF) decoder. This would allow for a standard transceiver's DTMF pad to be used as the source of reset pulses. (Radio-Shack sells a DTMF decoder chip requiring no adjustment.)

f) In the arrangement used for the AMRAD demonstration, (Fig 3), the trailing edge of the push-to-talk pulse was controlled by a half monostable to keep the transmitter on slightly longer than the audio burst. It also ensured that the receiver's squelch tail did not interfere with the operation of the burst detector.

g) A maximal-length 7-stage shift register requires 127 clock pulses to complete one sequence. For high-speed spread spectrum, the H sync pulses could be used as a clock, as follows: There are 262.5 TV lines (H sync) per field, two fields (V sync) per frame and 30 frames per second. By feeding the H sync pulses to a divide-by-2, 131 clock pulses per V sync pulse could be obtained. (A "flywheel" circuit would be required to oversee that the equalization pulses do not create false clock signals.) The V sync pulse could be used to trigger a monostable of 4H duration which would be the shift-register reset pulse. Thus, there would be exactly 127 clock pulses until the next reset pulse. The audio burst scheme could then be eliminated and the frequency-hopping rate, or the carrier phase-reversal rate, would be 7650 per second. With this arrangement, if there were to be a temporary loss of clock synchronization between the two amateur stations, sync would automatically be restored after a maximum of 1/60th of a second. This could be a particularly useful feature for mobile spread-spectrum applications.

If the audio burst scheme is retained, then any submultiple of H (between 15,750 Hz and 60 Hz) can be used as a clock pulse, simply by feeding H sync to a divide-by-N. Hence, by making N = 150, 105 hops per second is obtainable, and so on.

h) The scheme described in this article is not necessarily the best way to achieve clock synchronization. It is only my version 1.0, ie, a simple way of doing it. It is offered here with the hope that other radio amateurs will use it as a starting point to design their own equipment.

I thank Chuck Phillips, N4EZV, with whom I am exploring several other practical frequency-hopping methods. My thanks are also extended to Terry Fox, WB4JFI, David Borden, K8MMO, and Hal Feinstein, WB3KDU, three other AMRAD members whose encouragements helped me to complete the project.

Notes

[1]D. Newkirk, "Our New Spread Spectrum Rules" QST, Apr 1986, p 45.

[2]M. Wilson, ed., The 1986 ARRL Handbook, (Newington: ARRL, 1986) pp 21-7 to 21-14.

[3]H. Feinstein, "Spread Spectrum: Frequency Hopping, Direct Sequence and You," QST, Jun 1986, p 42.

[4]R. Palm, ed., The FCC Rule Book, (Newington: ARRL, 1986).

[5]W. Sabin, "Spread-Spectrum Applications in Amateur Radio," QST, Jul 1983, p 16.

Experimenting With Direct-Sequence Spread Spectrum

By André Kesteloot N4ICK

Introduction

During the past several years, AMRAD (Amateur Radio Research and Development Corp) has been conducting experimentation with spread spectrum technology.[1,2] As a member of the group, I decided to build a board that would demonstrate some practical aspects of direct-sequence spread spectrum (DSSS). I chose a carrier frequency of 2 MHz so that the phenomena could easily be displayed on any low-frequency (5-MHz bandwidth) oscilloscope likely to be available to the average radio amateur. The equipment, which can transmit and receive data, test square waves and Morse code signals, was demonstrated at the September 1986 AMRAD meeting in Vienna, VA.

Description

Referring to Fig 1, it is fairly easy to conceive of an amateur spread-spectrum communication link if an existing external reference, such as a local TV or radio transmitter, is used.[3] In this experiment, the external reference is a 4-MHz oscillator. A 2-MHz carrier is obtained by dividing the reference frequency by 2; this carrier is also used as clock for a pseudo-random generator. Data is XORed with the output of the pseudo-random generator and the result is XORed with the carrier. This produces a binary phase shift keyed direct sequence spread spectrum (BPSK DSSS) signal. At the receiving site, a pseudo-noise sequence similar to that created at the transmitter, is generated and synchronized to the original one. By mixing this sequence with the reference carrier, it is possible to extract the original data.

In my demonstration model, points T and R were connected to each other with a jumper. In a practical application, point T would be connected to an antenna via an amplifier; point R would similarly be connected to a receiving antenna, AGC amplifier and the like.

Circuit Operation

Reviewing the transmitter stage in

[1]Notes appear at end of article.

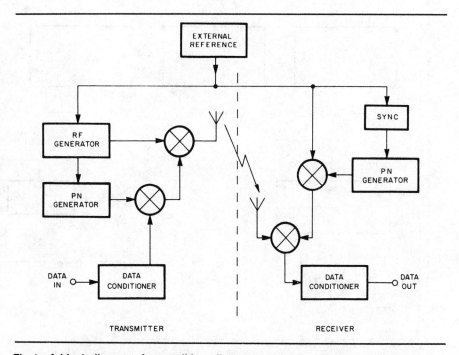

Fig 1—A block diagram of a possible radio amateur's spread-spectrum communications link. The external reference signal can be received from a TV or radio transmitter.

Fig 2, U1 is a 4-MHz DIP oscillator that is used as the reference frequency. U2A, a 4013 D flip-flop divides this frequency by 2 to produce a 2-MHz carrier (referred to as CA). The 2-MHz frequency is further divided by 1,000 (through three consecutive 4017 divide-by-10 stages, U4, U5 and U6) to produce a 2-kHz square wave. This signal is used as a test square wave (Figs 3, 4 and 5) and also serves as clock for another 4013 D flip-flop, U2B.

Under normal conditions, the Morse key is open, the D input of U2B is low and its $\overline{Q}$ output is low. When the key is pressed, the D input goes high. At the next clock pulse $\overline{Q}$ goes to +5 V. When the key is released, the D input goes low again, and at the next clock pulse, $\overline{Q}$ also goes low. Thus, the data (in this case the Morse sequence) is now synchronized to the 2-kHz clock, itself a submultiple of the 2-MHz carrier. This synchronized operation ensures phase-coherent switching.

The $\overline{Q}$ output of U2B is our data, or DA.

The 2-MHz carrier (CA) is also used to clock U7, a 74164 7-stage linear-feedback shift register connected as a pseudo-random noise generator (PN), with feedback taps at outputs 1 and 7. Feedback is applied using an XOR stage U8B, while U8A is connected as an inverter. (This is one of the FCC-authorized PN sequences. For a more complete study of this circuit, see note 3.)

The PN sequence is available at pin 2 of U7 and is called the original code or OCO. It is modulo-2 added to the data output of U2B in an XOR stage, U8C, the output of which is OCO × DA. This signal is XORed with the 2-MHz carrier (CA in U8D) to produce the output signal OCO × DA × CA, a BPSK DSSS signal. (In reality, this signal would be applied to an antenna using a power amplifier. On my experimental board, it is simply jumpered to the input of the receiver.)

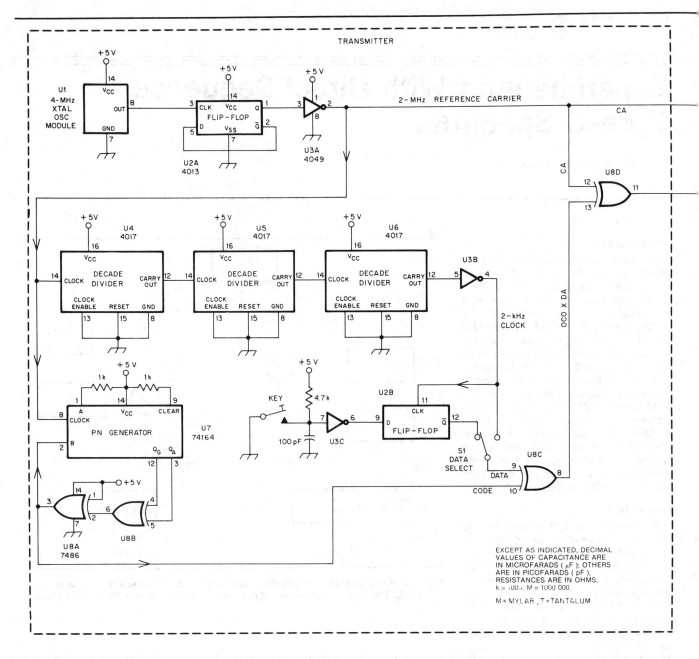

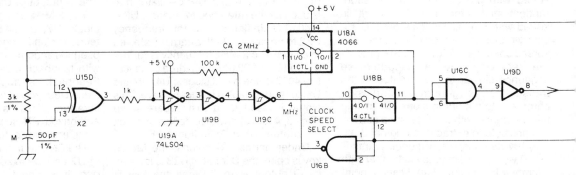

Fig 2—A complete schematic digram of the DSSS transmitter and receiver stages. Operation of the circuit is explained in the text.

Q1—2N4402, PNP transistor.
S1—SPDT switch.
S2—Normally open push-button switch.
U1—4-MHz DIP crystal oscillator.

U2—CD4013BC dual D flip-flop.
U3, U11—CD4049UBC hex inverting buffer.

U4, U5, U6—CD4017BC decade counter/divider with 10 decoded outputs.
U7, U13, U20—74164 8-bit serial-in parallel-out shift register.

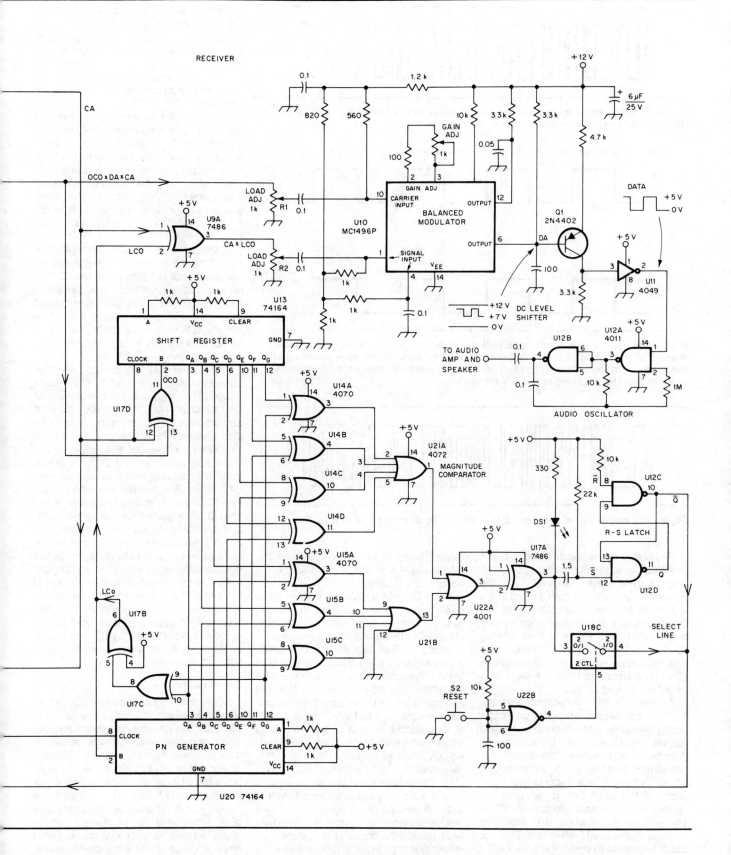

- U8, U9, U17—7486 quad 2-input exclusive OR gate.
- U10—MC1496 balanced demodulator.
- U12, U16—CD4011 quad 2-input NAND B series gate.
- U14, U15—CD4070BC quad 2-input exclusive OR gate.
- U18—CD4066BC quad bilateral switch.
- U19—74LS04 hex inverter.
- U21—CD4072BC dual 4-input OR gate.
- U22—CD4001BC quad 2-input NOR buffered B series gate.

Spread Spectrum Theory and Projects 8-33

Fig 3—A DSSS carrier is shown at the jumper between the transmitter (output of U8D) and receiver. The scanning rate is 1 µs/div.

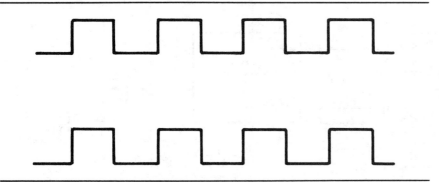

Fig 4—The upper oscilloscope trace shows a 2-kHz test square wave before transmission at the output of U2B. The lower trace is the same signal after decoding at the receiver, at the output of U11A. There is no discernible difference between the two signals. The scanning rate is 0.2 ms/div.

Fig 5—The output of PN generator U7, at pin 2. The scanning rate is 5 µs/div.

Reviewing the receiver stage, we can see that the input signal (OCO × DA × CA) and CA (the carrier derived from the external reference) are both applied to the inputs of an XOR stage, U17D. At the beginning of a transmission, no data is transmitted and the result of XORing (OCO × CA) and CA is thus OCO, the original code sequence. This sequence is serial-loaded into U13, a 7-stage shift register clocked by CA.

Another 74164 shift-register stage, U20, is connected as a PN generator using the same pseudo-noise sequence as the one used in the transmitter. (In Amateur Radio spread spectrum, it is mandatory to announce *in clear* the kind of register used. This enables both operators of the radio link to set their PN generators to the same sequence.) When the receiver is turned on, U20 is clocked at twice the normal 2-MHz rate. This 2× clock is generated by applying the 2-MHz CA to U15D, a multiply-by-two stage, the output of which is shaped by a Schmitt trigger (U19A, U19B). U18A and U18B (part of a 4066-quad switch) are connected as a SPDT switch. U18B is normally closed, U18A is normally open and the 4-MHz clock signal is applied to the PN generator U20.

Since U8 and U20 use the same feedback taps, they produce the same random sequences. Our goal is to synchronize both signals so that they will be in phase. As was discussed earlier, when no data is transmitted, the output of U13 is OCO, the original code sequence. The words produced by U13 and by U20 are compared, bit for bit, in a magnitude comparator comprised of XOR stages U14A, B, C, D, U15A, B, and C, quad input OR gates U21A and B and NAND gate U16A. As U20 is clocked at twice the rate of U13, its output tries to "catch up" with that of U13. There will be one instant when both output words are equal bit for bit. At that time, all outputs of the seven XOR stages will be low and the output of U16A goes high, driving the output of U17A low. This illuminates the "lock" LED and drives the Q output of U12C low. In turn, U18B opens and U18A closes, thereby switching the clock rate of U20 from 4 MHz to 2 MHz. From then on, both U7 and U20 will produce the same PN sequences, will be clocked at the same frequency and will be in phase (OCO = LCO). The edge-triggered RS latch (U12C, D) prevents further changes at the output of the magnitude comparator from influencing the clock speed, something that would otherwise happen whenever data is received.

Loss of sync between OCO and LCO can be simulated by momentarily grounding pin 9 of U7. OCO and LCO will now be out of phase. When this occurs, press the reset button—in the absence of data, or when data is low, the magnitude comparator output is compared to the clock speed selector (by bypassing the RS latch through U18C) to resynchronize LCO to OCO.

If Morse code is sent, for instance, notice that the lock LED extinguishes while the data is being transmitted. This occurs because the code now extracted from the DSSS transmission is no longer in phase with either OCO or LCO. The latter two, however, are still in phase, and as soon as the key is released, the LED lights again.

The carrier (CA) and the LCO are now XORed in U9A, the output of which is CA × LCO. This signal is applied to one of U10's inputs, a MC1496P/1596 balanced demodulator.[4] The other input is fed with the received DSSS signal (OCO × DA × CA). Since OCO = LCO, the output of the demodulator is the data DA. (To adjust the three potentiometers associated with U10, an oscilloscope should be connected to U10, pin 6. Both input potentiometers should be adjusted for no overloading, and the 1-kΩ potentiometer between pins 2 and 3 can usually be left in the minimum resistance position to provide maximum gain for this stage.) As U10 operates from +12 V, a dc voltage shifter composed of Q1 and U11A is used to shift the data to TTL levels. In my experimental board, this data is then used to gate ON and OFF an audio oscillator (U12A and B) that is connected to a small outboard audio amplifier and speaker.

Conclusions

Using components and test equipment readily available to the average radio amateur, it is possible to experiment with DSSS and gain hands-on experience and a better understanding of the fundamentals of this newly authorized technology. No attempt was made to optimize this demonstration board. The use of some circuitry was dictated by the availability of a particular IC, or because a portion of a quad gate was yet unused. No doubt several gates could have been saved, other more elegant ways of achieving the same results could have been developed and additional circuit simplifications could have been carried out. However, this was never meant to be a finished product, but simply a test bed for ideas. The reader is invited to dream up their own refinements.

If some quad-gate ICs were shared between the transmitter and receiver sections on the prototype board, they were renumbered (for clarity) and shown as separate ICs in Fig 2. (Incidentally, the 555 timer, located at the bottom left of the board [cover photo], is an adjustable clock generator only used in some early experiments. It is not shown on the schematic.) The three square potentiometers near the center of the board are associated with demodulator U10. The reset switch is mounted on the rear of the board and is not shown in the photograph.

To build a radio link for 440 MHz, XOR stages U8C, U8D, U9A and U17D would have to be replaced with doubly-balanced mixers such as the SBL-1 from Micro-Circuits, or the M53T manufactured by Magnum Microwave.[5,6] You'll need a reference carrier higher than 440 MHz, and for that purpose a local UHF-TV transmitter carrier comes to mind. At both amateur sites, a free-running oscillator could be used at approximately the reference frequency and a phase-lock loop (PLL) to lock this oscillator to the external reference TV transmitter would be necessary. This arrangement is equivalent to the CA on my demonstration board.

In the experiment described earlier, no preamble signal was sent for synchronization purposes. Sync was acquired before data was transmitted. On 440 MHz, it may be advisable to send a short preamble to synchronize the divider chain from which the 440-MHz carrier will be derived.

Finally, to transmit audio sine waves rather than square waves, it should be relatively easy, after raising the low clock rate from 2 kHz to 20 kHz, to apply that new clock and the audio sine waves to a conventional 555 pulse-width modulator, for example. To raise the clock rate, lower the divider ratio from 1/1,000 to 1/100 by eliminating one of the 4017 divide-by-ten stages.

Further Reading

Radio amateurs who reference readily available literature on spread spectrum will find ample mathematical treatment, plenty of block diagrams and a remarkable paucity of detailed schematic diagrams or specific information on hardware. Chapter 21 of *The ARRL 1986 Handbook for the Radio Amateur*[7] offers a good theoretical background on the subject. The articles by Scholtz[8] and Pickholtz et al[9] should also be referenced. Details of prototype realization are available in a paper by Van Der Gracht[10] and to a lesser extent in an article by Maskara and Das.[11] Finally, anyone contemplating involvement in spread spectrum should have access to the fundamental book by Dixon.[12]

Acknowledgements

I am grateful to AMRAD members Hal Feinstein, WB3KDU, and Mike O'Dell, N4NLN, for their suggestions and encouragements.

Notes

[1] H. Feinstein, "Spread Spectrum, Frequency Hopping, Direct Sequence and You," *QST*, Jun 1986, p 42.
[2] A. Kesteloot, "Practical Spread Spectrum: A Simple Clock Synchronization Scheme," *QEX*, Oct 1986, p 4.
[3] See note 2.
[4] R. Hejhall, "MC1596 Balanced Modulator, Application Note AN-531," Motorola Semiconductor Products, Inc, Phoenix, AZ 1982.
[5] Micro-Circuits Co, Inc, RR 1, PO Box 518, New Buffalo, MI 49117, tel 616-469-2727.
[6] Magnum Microwave Corp, 365 Ravendale Dr, Mtn View, CA 94043, tel 415-968-9281.
[7] M. Wilson, ed., *The ARRL 1986 Handbook for the Radio Amateur*, Newington: ARRL, 1985.
[8] R. Scholtz, "The Spread Spectrum Concept," *IEEE Transactions on Communications*, Vol Com-25, No. 8, Aug 1977.
[9] R. Pickholtz, D. Schilling and L. Milstein, "Theory of Spread-Spectrum Communications—A Tutorial," *IEEE Transactions on Communications*, Vol Com-30, pp 855-884, May 1982.
[10] P. Van Der Gracht and R. Donaldson, "Communication Using Pseudo-Noise Modulation on Electric Power Distribution Circuits," *IEEE Transactions on Communications*, Vol Com-33, No. 9, Sep 1985.
[11] S. L. Maskara and J. Das, "Concatenated Sequences for Spread Spectrum Systems," *IEEE Transactions on Aerospace and Electronic Systems*, Vol AES-17, No. 3, May 1981.
[12] R. Dixon, *Spread Spectrum Systems*, New York; Wiley, 1976.

Extracting Stable Clock Signals From AM Broadcast Carriers for Amateur Spread-Spectrum Applications

By André Kesteloot N4ICK

This circuit provides jitter-free clock pulses by locking onto a readily available external reference signal source. Although it was designed primarily to provide clock pulses for Amateur Radio spread-spectrum transmissions, this circuit can also be used as a stable reference for frequency-calibration purposes.

Introduction

To recover data from spread-spectrum transmissions, it is necessary to reintroduce, at the receiver end, locally generated signals locked in frequency and phase to the clock used in the original transmissions. At carrier frequencies of up to 50 MHz, the extraction of clock pulses from direct-sequence (DS) spread-spectrum transmissions can be achieved fairly easily, using readily available ICs.[1] The 1986 FCC decision to relegate Amateur Radio spread-spectrum transmissions to frequencies above 400 MHz, however, singularly complicated the design and realization of this type of equipment.[2] Although it will be eventually possible to find a way to extract the clock signal at those frequencies, it is a major stumbling block to the development of uncomplicated amateur spread-spectrum equipment.

Another way to obtain a clock is to use an external reference signal readily available at both the transmitting and the receiving sites.[3] Since our amateur transmissions must take place at UHF, communications will usually occur when the two parties are within a fairly short distance of each other. Locally available reference signals are generated by TV, FM and AM radio stations. I have already explored the recovery of such signals, and have demonstrated the possibility of using vertical synchronization signals from local TV stations to generate reliable clock pulses for slow frequency-hopping spread-spectrum experiments.[4]

DS spread-spectrum on the other hand, requires that clock signals be in the megahertz region.[5,6] In DS, the clock drives a pseudorandom generator, which is modulo-2 added to the carrier, to vary the phase of the UHF carrier by 180 degrees. The carrier is cancelled and replaced by the familiar $[\sin x/x]^2$ spectrum (Fig 1). The ratio of F_c/F_k, where F_c is the RF carrier frequency (in our case $F_c > 400$ MHz), and F_k is the clock, is important. If F_k is low, the ratio is too high, and spreading of the signal does not take place. If F_k is too high, the RF signal is spread over too wide a bandwidth, and the spectrum used exceeds the limits of our amateur bands. In practice, for $F_c = 440$ MHz, an F_k between 500 kHz and 2 MHz appears reasonable. (In Fig 1, $F_c = 146$ MHz and $F_k = 1$ MHz. Nulls are clearly visible at $F_c \pm F_k$, $2F_k$, and so on, ie, 144, 145, 147, and 148 MHz. This experiment was conducted in a test load, and at power levels lower than 0 dBm.)

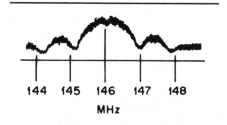

Fig 1—Spectrum-analyzer frequency-domain display of a direct-sequence spread-spectrum signal. $F_{carrier} = 146$ MHz and $F_{clock} = 1$ MHz.

Problems

There are many ways to generate clock signals. You could lock onto a TV station's video AM carrier, and divide down to obtain the required clock. The division process, however, introduces an ambiguity that has to be resolved at the receiving end, thus further complicating receiver design.

You can also extract horizontal TV sync pulses (at 15,750 Hz) and multiply that signal by 100 for a clock at 1.575 MHz. The problem here is that the multiplication process (using, for example, a 4046 phase-locked loop oscillating at 1.575 MHz and two cascaded 4017 divide-by-ten stages between the oscillator and the phase comparator) also multiplies the original jitter by 100! Some jitter is always present on the H sync pulse, and a jitter of 10 ns (10 ns = 0.01 μs), perfectly negligible at 15 kHz, becomes 1 μs at 1.5 MHz. This is clearly an unacceptable solution since the period of a 1.5-MHz signal is only 0.66 μs!

I followed a similar approach using the 19-kHz stereo subcarrier available from all FM stereo stations. Here again, jitter (principally because of incidental amplitude modulation and lack of operating-point stability in simple zero-crossing detectors) does not allow for a reasonably stable signal once the 19-kHz signal is multiplied by 100 to yield a 1.9-MHz clock.

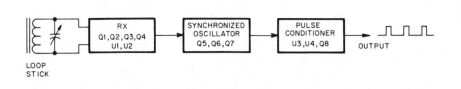

Fig 2—The circuit used to extract stable clock signals from AM broadcast carriers is made up of three parts. The first part consists of a ferrite loop and amplifier that processes the incoming AM signal. The synchronized oscillator locks onto the incoming carrier to produce a jitter-free output. Then, the pulse conditioner converts that signal into a clock pulse with an adjustable phase delay.

[1]Notes appear at end of article.

I eventually decided to lock onto the carrier of an AM broadcast station, and after experimenting with different phase-locked loops, I settled for this design which uses readily available components. The circuit features a stable clock pulse, adjustable in phase to compensate for propagation delays, and it produces *no* measurable jitter.

The Circuit

Fig 2 shows the three parts of the circuit. The first part includes a ferrite loop and a clipping amplifier. The second part consists of a synchronized oscillator that locks onto the incoming carrier to produce a jitter-free output. Finally, the third part converts that signal into a clock pulse with adjustable phase delay.

Fig 3 shows the complete circuit diagram. The ferrite loop receives the signal and U1, a TBA120S, amplifies it. U1 is an FM/IF 6-stage differential amplifier with good limiting and AM rejection properties.[7] Its output is amplified by U2A, a section of a CD4001 operated in its linear mode. The output is subsequently fed to U2B, connected as a Schmitt trigger. A 1-V P-P output is available at the emitter of Q4 (shown in the upper trace of Fig 7). This part of the circuit is the RF head (Fig 4). It can be positioned and oriented for best reception.

Fig 5 is the remote unit. It uses the 1-V P-P square wave from the RF head to lock the "synchronized oscillator."[8] The remote unit can be placed several hundred feet away from the RF head. This circuit, used for clock recovery in satellite installations, is yet generally unknown in Amateur Radio circles. Q5, Q6, and Q7 comprise the synchronized oscillator. Q5 functions as a modified Colpitts oscillator. It has two positive feedback paths, one from the common point between the two capacitors in the collector tank to its own emitter, and the other path from the junction of the 220-Ω resistor and the collector tank to the base of Q5, via C2. C2 is large and represents a very low impedance at the operating frequency. Q6 can be thought of as a dynamic emitter resistor for Q5. Since Q5 operates in class C, the conduction angle is very small. Each time conduction occurs in Q5, a voltage develops across Q6, and amplification of whatever signal is present at that time at the base of Q6 takes place. Conduction in Q6 is similar to the opening of a very brief "time window" during which synchronization to the input signal occurs. Because there is a tuned circuit in the collector of Q5, in the event of a temporary absence of sync pulses, the tank (functioning as a flywheel) continues to produce sine waves at a frequency close to the frequency of interest.

An AM input signal consists of a carrier (F_c = 1,390 kHz in our case), plus two sidebands ($F_c + F_{mod}$) and ($F_c - F_{mod}$). Assuming a single modulating frequency of 2 kHz, the input signal consists of three discrete RF frequencies—the carrier at 1,390 kHz and the two sidebands at 1,388 and 1,392 kHz (Fig 6). In practice, the instantaneous frequencies and amplitudes of the two sidebands depend on the audio input and depth of modulation, respectively. These discrete frequencies in the RF spectrum are visible as jitter on the upper trace of Fig 7.

Although the frequency and amplitude of the sidebands vary continuously, the mentioned attributes of the carrier are constant. Because of the flywheel effect, the synchronized oscillator, operating as a sort of coherent amplifier, tends to accept the carrier as the sync information. Jitter on the input signal tends to be perceived as an aberration, and is essentially ignored. Hence, the output sine wave at the collector of Q5 does not exhibit input jitter.

Finally, Q5's output is buffered by Q7, an FET stage. With the output of a synchronized oscillator being constant in amplitude throughout the synchronized range, it is acceptable to feed that output to U3A, a 7414 Schmitt trigger to obtain a stable trigger pulse. That pulse is then fed to the first of two monostable oscillators connected in series. The first portion of U4, a 74123, introduces a variable delay, adjustable by means of R2, over a range of approximately 270 degrees. U4B, the second monostable, outputs a short positive-going clock pulse, with Q8 connected as a 50-Ω line driver.

Fig 7's upper trace is the leading edge of the input signal to the synchronized oscillator, available at the top of R1. Peak jitter covers about one division, or approximately 0.02 μs. (In practice, average jitter is about 0.01 μs. This is consistent with an upper audio modulation frequency of 10 kHz. The peak jitter displayed in Fig 7 is probably the resultant of several causes, including some incident phase modulation.) The lower trace shows the signal at the emitter of Q8. There is *no* visible jitter at the output, so we know that the synchronized oscillator is operating properly. (Jitter on the upper trace is essentially caused by residual amplitude modulation.)

Construction

Observe good RF construction practices when building the synchronized oscillator, particularly with respect to ground returns and shielding. The prototype of my oscillator is built on fiberglass circuit board, using self-adhesive silvered circuit decals for connection points.

Because of U1's sensitivity, I recommend that it be housed in a separate enclosure from that of the synchronized oscillator. The synchronized oscillator produces several volts of RF at the same frequency as that of the input of U1.

The portion of the circuit comprising Q1, Q2, Q3 and U1 is housed in a small (¼ × ½ × 3½ inches) aluminum die-cast box. U2 and Q4 are mounted in a larger (2¼ × 3½ × 4½ inches) box used for the base of the receiving head. The loopstick is mounted in a plastic box that swivels on a wooden dowel and is positioned for best reception. In my present installation, the receiving head is located near a window so I can orient the ferrite loop for maximum signal reception. The receiving head connects to the synchronized oscillator with 25 feet of RG58 coax cable, with no visible degradation. You could easily use 100 feet of cable, if required. The larger aluminum box also contains a 12 V dc power supply (not shown on the schematic) for the receiving head. A separate aluminum box (2 × 5 × 9½ inches) houses the synchronized oscillator, the pulse conditioner elements, and a 5 V dc power supply (see Fig 5).

Adjustments

An oscilloscope and a frequency counter are required to adjust this unit. (I locked onto WMZQ, a Northern Virginia radio station that uses a solid-state transmitter to broadcast on 1,390 kHz.) Connect the oscilloscope probe at pin 14 of U1, and the frequency counter probe at the emitter of Q4. The counter will indicate the carrier frequency of the AM broadcast transmitter you are receiving. The counter will probably jump ± 100 Hz around the carrier frequency, representing modulation peaks (this reading depends on the integration time of your counter). On the scope screen, adjust C1, the trimmer across the loopstick coil, for maximum signal amplitude of the carrier you are trying to lock onto. By moving the scope probe to the emitter of Q3, you should see a fairly clean looking 150 mV P-P square wave. Look for a similar signal, 4 V P-P in amplitude, at the emitter of Q4.

In the remote unit, connect the oscilloscope probe to the source of Q7 (*not* to the tank of Q5). Adjust the slug in the collector tank of Q5 to produce a free-running frequency close to that of the AM broadcast station you are locking onto. This adjustment must be made with R1's wiper turned to ground potential. If you have a dual-trace triggered sweep oscilloscope, connect the synchronized channel to the source of Q7 (oscillator output), and the other channel to the input pulse. As the input signal applied to Q6 is slowly increased by adjusting R1, you should see the input pulse lock onto the sine wave (it is actually the sine wave that locks onto the input pulse). Jitter should be visible on the input signal only. (If the oscilloscope were synchronized on the

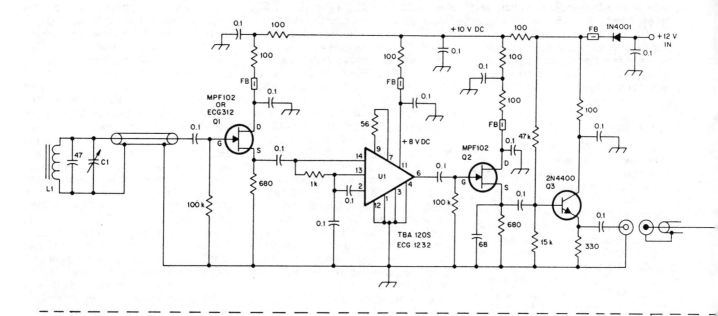

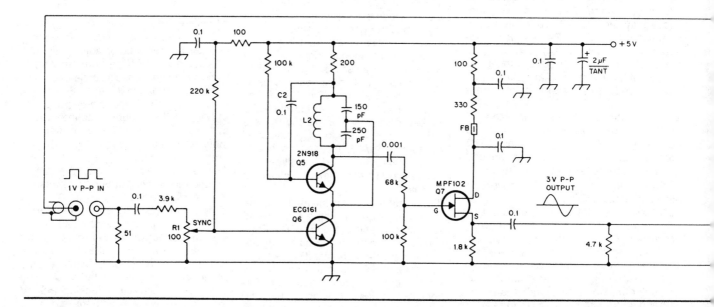

Fig 3—This circuit extracts stable clock signals from AM broadcast carriers for amateur spread-spectrum applications. All 0.1-μF capacitors in coupling and decoupling circuits are miniature 50-V polyester types. All capacitors smaller than 0.1-μF are silver mica.

C1—Trimmer capacitor, 4-64 pF.
FB—Ferrite beads, Amidon FB-73-101.
L1—30 turns of 30-gauge wire, wound on an Amidon ferrite rod R-61-050-750.
L2—JW Miller shielded subminiature adjustable RF coil, catalog no. 9055, 60-120 μH.

Q1, Q2, Q7—MPF102 or ECG312 FET.
Q3—2N4400 (or 2N2222) NPN transistor.
Q4, Q8—2N2222 NPN transistor.
Q5—2N918 NPN transistor.
Q6—ECG161 NPN transistor.
R1—Adjustable potentiometer, 100 Ω.

R2—Adjustable potentiometer, 50 kΩ.
U1—TBA120S (Motorola) or ECG1292 (Sylvania) IF amplifier.
U2—CD4001 Quad NOR gate.
U3—7414 Hex Schmitt trigger.
U4—74123 Dual monostable.

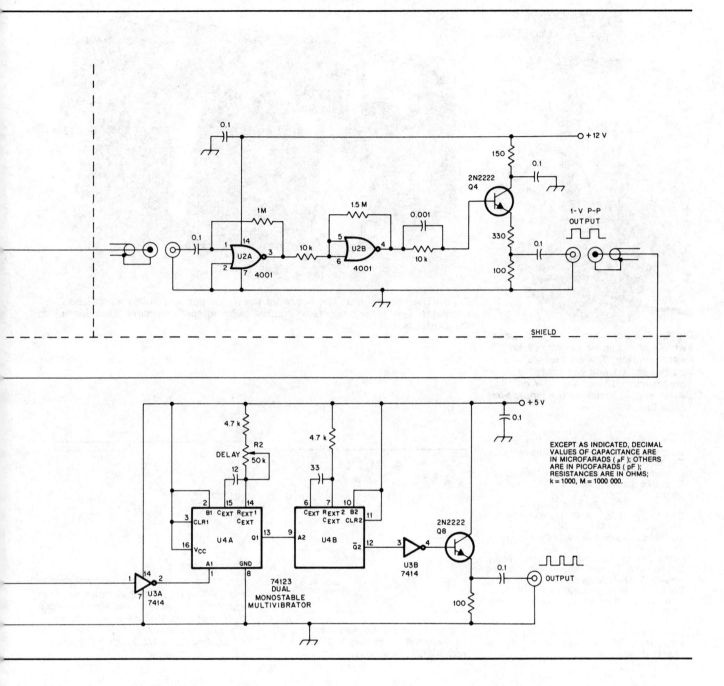

jittery input signal, both the input *and* the output signal would appear to jitter.) Increasing the input level setting can "oversynchronize" the oscillator and possibly drive it into an "injection-oscillator" mode.[9] This results in distortion at the output, and possibly jitter, as the tracking range of the circuit increases.[10] Keep the input signal as low as possible.

Your frequency counter, now connected to the emitter of Q8, should indicate the carrier frequency you are locking onto, with a maximum deviation of ±1 Hz. This deviation represents the least significant digit resolution of your counter, and not a loss of synchronization.

Conclusion

In my synchronized oscillator, a 2N918 transistor was chosen for Q5 because of its good RF properties. An ECG161 was selected for Q6 because of its low noise. If you are only interested in breadboarding a synchronized oscillator for experimental purposes, you may use 2N2222s or 2N4400s in both positions. The result is some degradation of performance. Depending on the type of transistor used, the value of the 220-kΩ bias resistor is adjusted so that Q5's emitter is at about $V_{cc}/2$.

This circuit provides extremely accurate clock signals and uses readily available parts. Whether you are interested in spread-spectrum applications or another phase of Amateur Radio, build a synchronized oscillator, and experiment with this very versatile building block. It can also be used as a divider or multiplier.

Acknowledgements

I thank Mr V. Uzunoglu for devoting his time to discuss with me some of the properties of the synchronized oscillator.

Notes

[1] R. Dixon, *Spread Spectrum Systems*, New York, Wiley, 2nd ed; 1984, pp 215-255.
[2] D. Newkirk, "Our New Spread Spectrum Rules," *QST*, Apr 1986, p 45.
[3] W. Sabin, "Spread Spectrum Applications in Amateur Radio," *QST*, Jul 1983, p 16.

Fig 4—The upper part of the RF head swivels so it can be positioned for best reception. The smaller aluminum box houses U1 and associated components. U2 and the 12-V dc power supply reside in the larger box at the base.

Fig 5—The remote unit. The board on the right supports the synchronized oscillator. The voltage regulator and the pulse-shaping circuitry are mounted on the center board.

[4]A. Kesteloot, "Practical Spread Spectrum: A Simple Clock Synchronization Scheme," *QEX*, Oct 1986, pp 4-7.
[5]A. Kesteloot, "Experimenting With Direct Sequence Spread Spectrum," *QEX*, Dec 1986, pp 5-9.
[6]A. Kesteloot, "First Steps in Direct Sequence Spread Spectrum," *AMRAD Newsletter*, Vol XIV, No. 1, Jan 1987, pp 5-9.
[7]*Linear and Interface Integrated Circuits*, Motorola, Inc, Phoenix, AZ 1985, p 6-142.
[8]V. Uzunoglu and M. White, "The Synchronous Oscillator: A Synchronization and Tracking Network," *IEEE Journal of Solid State Circuits*, Vol SC-30, No. 6, Dec 1985, pp 1214-1224.
[9]V. Uzunoglu, private communication, May 15, 1987.
[10]V. Uzunoglu and M. White, "Some Important Properties of Synchronous Oscillators," *Proceedings of the IEEE*, Vol 74, No. 3, Mar 1986, pp 516-518.

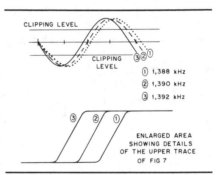

Fig 6—A simplified time-domain representation of amplitude modulation (see text).

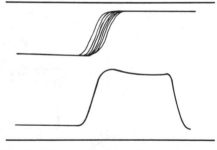

Fig 7—The upper trace shows the input signal going to the synchronized oscillator, and the lower trace is the signal at the emitter of Q8. The vertical scale is 1/div; the horizontal scale is 0.02 μs/div.

Feedback

Please refer to my article, "Extracting Stable Clock Signals From AM Broadcast Carriers For Amateur Spread Spectrum Applications," *QEX*, Oct 1987. There is an error in the schematic on page 7. The numbering of pins 14 and 15 of U4A should be reversed. The junction of R2 and the 12-pF capacitor connect to pin 15 (which should be labeled R_{ext}, C_{ext}), and the other capacitor lead connects to pin 14 (which should be labeled C_{ext}).—*Andre Kesteloot, N4ICK*

Correct Your Schematic

Andre Kesteloot, N4ICK, author of "Extracting Stable Clock Signals From AM Broadcast Carriers for Amateur Spread-spectrum Applications," (Oct 1987 *QEX*, p 5), notes two things missing from Fig 3 on p 6: U1 should be labeled ECG 1292, and insert a 560-pF mica capacitor between the wiper of R1 and the junction of the 220-kΩ resistor and the base of Q6.

Practical Spread Spectrum: Achieving Synchronization with the Slip-Pulse Generator

By André Kesteloot N4ICK

A two-way, spread-spectrum contact was successfully conducted in January 1988 as part of a continuing series of experiments in spread-spectrum communications administered by the Amateur Radio Research and Development Corporation (AMRAD). This article describes the experiment and the equipment used. Before detailing the experiment, a brief introduction to direct-sequence spread-spectrum (DSSS) communications is in order.

Direct-Sequence Spread-Spectrum Communications

The RF bandwidth of conventional Amateur Radio modulations (AM, NBFM, FSK) is usually proportional to the amount of information that is transmitted. The bandwidth of the RF signal is kept as narrow as possible (you generally think of narrow contiguous channels), possibly a few kilohertz wide, with each channel containing separate information. When a conventional channel is in use, it is impossible for other users to transmit on the same channel without creating interference.

Using DSSS techniques, you can have a wide RF bandwidth—possibly several megahertz wide—and different networks can transmit on the same frequencies without interfering with one another. Also, the bandwidth of the signal is not related to the information rate.

At the DSSS transmitting site, a carrier is mixed with the output of a pseudo-random noise (PN) generator to spread the information over a wide bandwidth. The same pseudo-random sequence is reintroduced in the receiver (see Fig 1). If the original carrier is frequency modulated (to transmit voice, for instance), the original modulation information can be extracted at the receiver. If data other than voice is sent, this data can be logically XORed with the PN sequence at the transmitter, and similarly recovered at the receiver.[1] In either case, only the signals corresponding to the original PN sequence are correlated at the receiving

[1]Notes appear at end of article.

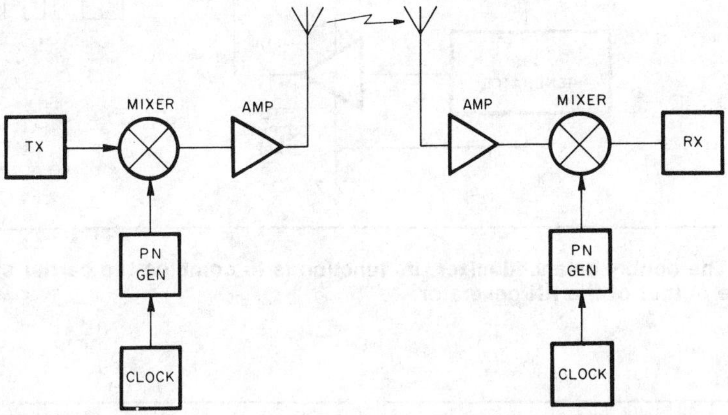

Fig 1—Block diagram of a direct-sequence spread-spectrum communications system.

site. The noise is spread over a wide bandwidth and filtered out. The signal-to-noise improvement thus derived, called *process gain*, is an important derivative of the DSSS process.

The Doubly Balanced Mixer

Fig 2 shows the schematic of a doubly balanced mixer used to mix the carrier signal and the output of the PN generator. Note that D1 through D4 are connected as a ring-modulator. The RF carrier is fed to the input transformer, while the PN signal is fed to the center tap of the output transformer. The PN generator uses TTL ICs, and by connecting it via a capacitor, the center tap of the output transformer swings from +2 V to -2 V, with respect to the center tap of the input transformer. When +2 V is applied, D1 and D3 conduct, and D2 and D4 are reverse-biased, connecting point A to C, and point B to D. When the output of the TTL goes low, the center tap of the output transformer goes to -2 V. At this instant, D2 and D4 conduct, connecting point A to D and point B to C. This produces a succession of 180-degree phase reversals of the RF carrier at the output, effectively canceling the carrier. The PN signal, which is fed in opposing phase to the input and output windings, also cancels. On the other hand, sidebands are created because of the heterodyne process. Since the output of our PN generator is a square wave in the time domain, its Fourier transform in the frequency domain is a (sin x/x) waveform. Because we are dealing with a signal proportional to the output *power* of the mixer stage, what we see on a spectrum analyzer connected to the output is proportional to the *square* of the voltage, or (sin x/x)^2 (see Fig 3). As shown in the waveform of Fig 3, the original 445-MHz signal is spread over almost 10 MHz. In practice, a suitable filter shapes the transmitter output to allow radiation of only the main lobe. (The main lobe contains approximately 95% of the total power.)

At the receiving site, a similar doubly balanced mixer arrangement recreates, or "despreads," the original carrier (Fig 4). To maximize isolation between ports, each port must be properly terminated in 50-ohm loads. The setup used at both the transmitting and receiving ends is shown in Fig 5.

Synchronization

To despread a signal, the PN sequence

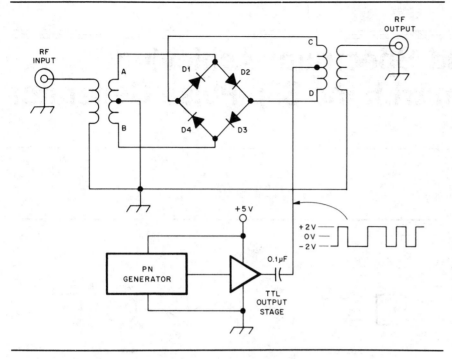

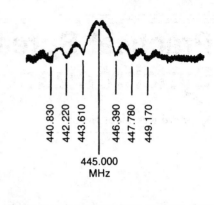

Fig 3—Output of the doubly balanced mixer at the transmitting site. The signal is spread over several megahertz, where $F_{carrier}$ = 445 MHz and F_{clock} = 1.390 MHz.

Fig 2—The doubly balanced mixer. Its function is to combine the carrier signal with the output of the PN generator.

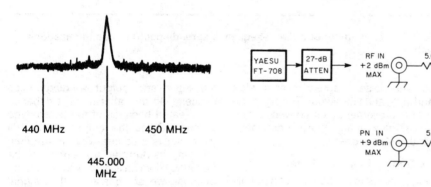

Fig 4—Output of the doubly balanced mixer at the receiving site. The signal is "despread."

Fig 5—Schematic of the doubly balanced mixer port termination at the transmitting site.

at the receiver must be synchronized in both frequency and phase with that of the transmitter. Synchronization is considered to be the most difficult problem to solve in spread-spectrum applications.[2]

There are several ways to achieve synchronization: (1) recover the transmitter clock at the receiving site (a considerable undertaking at 445 MHz), (2) transmit the clock separately, (3) transmit the clock as part of the signal, or (4) synchronize both transmitter and receiver to an external reference, an approach suggested by William Sabin, W0IYH.[3]

The latter technique is used in this experiment, and the equipment required to extract a stable clock pulse from an AM radio station is described in an earlier article.[4] Thus, the PN generator for both the transmitter and receiver are clocked at the same frequency. The remaining problem is for the two PN sequences to operate in phase.

The station setup used for the successful spread-spectrum QSO is shown in Fig 6. Although the actual QSO was conducted at 445 MHz, some of the preliminary work was performed at 146 MHz. At 146 MHz, the output of the transmitting equipment was connected via a cable to the receiving gear. At 445 MHz, actual antennas were used at both transmitting and receiving sites. Since the equipment designed for this experiment is not frequency specific, the operation of the synchronization arrangement was exactly the same in both cases.

Referring to Fig 6, the transmitter used was a Yaesu hand-held FT-208R for 146 MHz and an FT-708R for 445 MHz. At the receiving site, I used a Yaesu FT-23R for 146 MHz and an ICOM IC-4AT for 445 MHz. The output of the transmitter and the output of the PN generator are fed to a doubly balanced mixer, as explained earlier, and the signal at the output of the mixer looks like that of

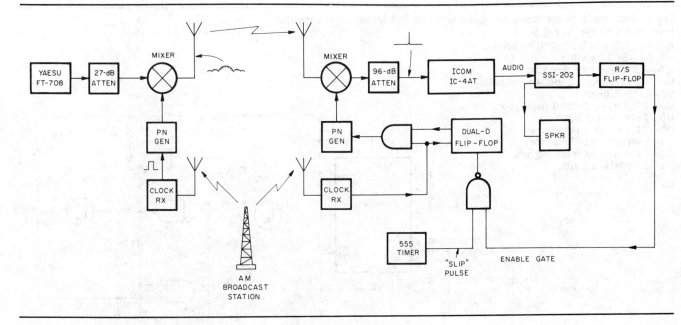

Fig 6—General equipment arrangement for the DSSS experiment.

Fig 3. To decode the original FM modulation at the receiving site, the signal must be despread. (This happens only when the two PN generators are in phase.)

To understand how the phase-synchronization process works, imagine that the two PN generators are receiving the same 1.390-MHz external clock pulse. These pulses, at both sites, are used to clock identical 7-stage PN generators (Fig 7).[5] Although the two PN generators are identical, their sequences are not necessarily in phase because they may have been started or reset at different times. To obtain phase synchronization, it is necessary to shift the phase of the receiver's PN generator, with respect to that of the transmitter's, until the two sequences coincide. To that effect, the *slip-pulse generator* occasionally introduces an extra pulse in the 1.390-MHz clock stream. Thus, the receiver's PN generator slowly "catches up" with the transmitter's until coincidence is achieved. When this happens despreading takes place, the receiver's squelch opens, and the slip-pulse generator inhibits the introduction of additional slip pulses.

Circuit Description

To achieve phase synchronization, I used the built-in DTMF (dual-tone multi-frequency) generator of the Yaesu FT-208R/708R to send a "1." I also prewired U1, an SSI-202 DTMF decoder, in the slip-pulse generator to accept a "1" as a valid output (see Fig 8).

Let's assume that the two PN sequences are not in phase. There is no despreading and thus no output from the receiver. The slip-pulse generator continues to insert pulses at a very slow rate. Since a 7-stage PN register is 127 steps long, after a maximum of 126 additional slip pulses, there will be a moment when the two PN sequences are in phase. At that moment, despreading takes place, the squelch opens, and the receiver generates a "1" that is recognized by U1. The green LED illuminates and the reset-set (R-S) flip-flop (U2A and U2B) latches, turning on the red LED and grounding pin 4 of U4B.

U3, a 555, free-runs at about 200 Hz. If U4B is enabled (ie, when there is an absence of a valid DTMF signal), U3's output pulses are channeled to U5. U5 is a 7474 connected in a one-and-only-one configuration.

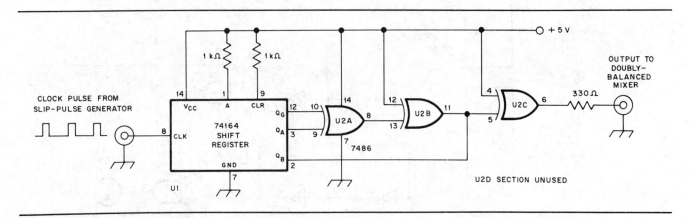

Fig 7—Schematic of the pseudo-random noise generator.
U1—74164 8-bit shift register.
U2—7486 quad 2-input XOR gate.

Fig 8—Schematic of the slip-pulse generator. Circled letters reference the waveforms shown in Fig 9.

Q1, Q2, Q3—2N2222 NPN transistor.
S1, S2—SPST switch.
U1—SSI-202 DTMF decoder.
U2—CD4011 quad 2-input NAND gate.
U3—LM555 timer.
U4—7400 quad 2-input NAND gate.
U5—7474 dual D flip flop.
U6—74123 dual retriggerable monostable multivibrator.
U7—7414 hex Schmitt trigger.
Y1—3.58-MHz crystal.

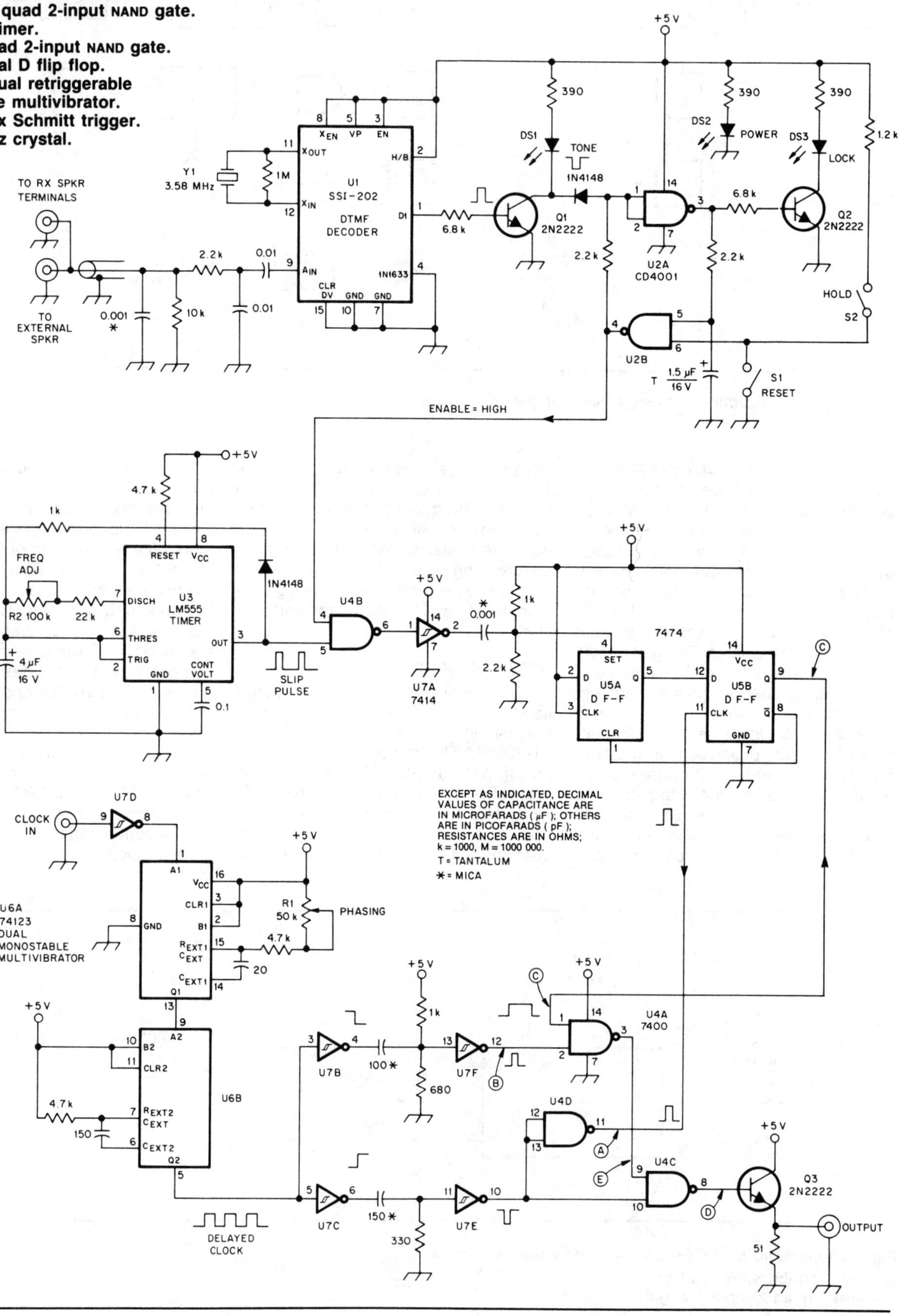

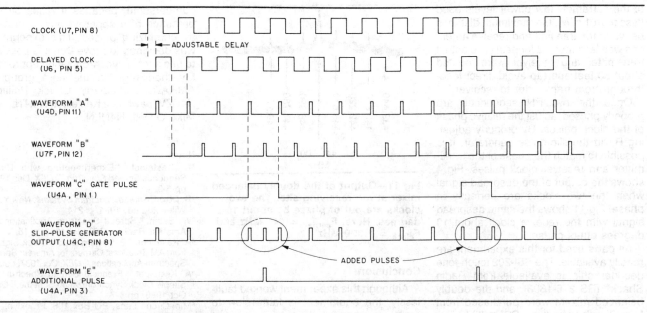

Fig 9—Slip-pulse generator timing sequence.

The external clock signal is applied to U6A and U6B, two monostable multivibrators connected in series. This circuit provides an adjustable delay to compensate for TTL gate propagation delays, different cable lengths, and so on. The delayed clock signal is then buffered by U7B and U7C, and fed to U7F and U7E (connected as two "half-monostable" sections that produce a short pulse on the trailing and leading edges of the clock pulse, respectively). The signal from U7E is also used to clock U5. U5 produces a gating signal, the length of which is the interval between two clock pulses, only after it receives a pulse from U3. Whenever this gating pulse occurs, U4A is enabled and allows a pulse synchronized on the trailing edge of the clock (ie, half-way between two regular clock pulses) to reach U4D, where it is added to the normal stream of 1.390-MHz pulses. (A pulse timing diagram is shown in Fig 9.) We have thus "slipped-in" an extra clock pulse for every pulse created by U3 (hence my name for the slip-pulse generator), and the effective clock pulse at the receiver end is 1,390,200.

This process continues as long as U4B remains enabled. A valid DTMF signal inhibits U4B. To prevent further tones or audio noise from adding extra slip pulses once phase synchronization has been achieved, the output of the DTMF decoder is connected to U2A and U2B, an R-S latch, that can only be reset by grounding pin 6 of U2B via S1. Similarly, S2 can simulate the reception of a valid tone by stopping the slip-pulse generator, a help during testing.

The Experiment

To facilitate experimentation, various parts of the equipment were built in separate shielded boxes and connected as shown in Fig 6. Fig 10 is a photograph of the slip-pulse generator. I found that at slip-pulse frequencies above 250 Hz, the system cannot achieve lock. This upper limit occurs because it takes a finite time for the receiver's squelch to open and the DTMF decoder to recognize a valid tone. By adjusting R2 so that the output frequency of U3 is about 200 Hz, it took a maximum of 42 seconds for locking to occur reliably. (Depending on the respective position of the two PN sequences, it can take less time; 42 seconds was the maximum time required when the two PN sequences were 126 steps apart.) Once synchronization is achieved, however, both transmitter and receiver clocks remain in phase, whether communication takes place or not. In a regular spread-spectrum system, synchronization must be achieved each time the push-to-talk switch is engaged. In this experiment the transmitting site and the receiving site were symmetrical (no amplifier was connected between either antenna and doubly balanced mixer), and it was possible to establish a two-way contact between the two units. Because

Fig 10—The slip-pulse generator.

of the extremely low power levels used (less than 1 mW), the maximum distance between the transmit and receive antennas was less than 12 inches! (The actual transmitter and receiver were located about 50 feet apart to avoid direct feedthrough from transmitter to receiver.)

Once the two PN sequences are properly phased, adjust the relative phase of the clock pulses. By properly adjusting R1 in the slip-pulse generator, it is possible to match the phase of the transmitter and receiver clock pulses. Fig 4 shows the output of the despread signal when the two clocks are perfectly in phase. Fig 11 shows the same despread signal with the receive clock about 10 degrees out-of-phase.

All parts used for this experiment are readily available. The SSI-202 touch-tone decoder chip is available from Radio Shack® (RS 276-1303), and the doubly balanced mixers were purchased from Mini-Circuit Labs (type SBL-1(6)).[6]

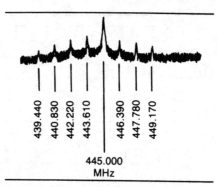

Fig 11—Output of the doubly balanced mixer at the receiving site. The two clocks are out of phase by about 10 degrees. Here, $F_{carrier}$ = 445 MHz and F_{clock} = 1.390 MHz

Conclusion

Although this experiment worked faultlessly, the equipment is fairly slow to reach synchronization and somewhat cumbersome, because it relies on the presence of a separate unit to provide synchronizing pulses. This experiment was designed to prove that it is possible to use such a synchronization approach. My thanks go to the core group of AMRAD, particularly Chuck Phillips, N4EZV, Lawrence Kesteloot, N4NTL, and Mike O'Dell, N4NLN.

Notes

[1] A. Kesteloot, "Experimenting with Direct Sequence Spread Spectrum," *QEX*, Dec 1986, pp 5-9.
[2] R. Dixon, *Spread Spectrum Systems*, New York, Wiley 2nd ed, 1984, p 214.
[3] W. Sabin, "Spread Spectrum Applications in Amateur Radio," *QST*, Jul 1983, p 16.
[4] A. Kesteloot, "Extracting Stable Clock Signals from AM Broadcast Carriers for Amateur Spread Spectrum Applications," *QEX*, Oct 1987, pp 5-9.
[5] A. Kesteloot, "Practical Spread Spectrum: A Simple Clock Synchronization Scheme," *QEX*, Oct 1986, pp 4-7.
[6] Mini-Circuit Labs, PO Box 166, Brooklyn, NY 11235, tel 718-934-4500

A Practical Direct-Sequence Spread-Spectrum UHF Link

In this *QST* exclusive, amateur spread-spectrum communication moves off the drawing board and into practical reality. Read all about it—and warm up your soldering iron!

By André Kesteloot, N4ICK
ARRL Technical Advisor
6915 Chelsea Rd
McLean, VA 22101

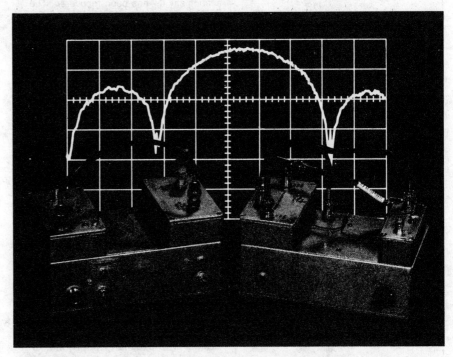

Radio amateurs under FCC jurisdiction have been authorized to use spread-spectrum emission since June 1, 1986,[1] but practical information on how to get an amateur spread-spectrum system up and running has been scarce. This is so for good reason: Military applications of spread spectrum are routinely classified, while space and other civilian applications are usually proprietary. Because of this, the literature abounds with spread-spectrum articles replete with mathematical treatments and block diagrams, but very few articles that give *practical details* of spread-spectrum systems have been published.

This situation has, in effect, forced Amateur Radio experimenters to reinvent parts of the wheel! Notwithstanding the fact that spread spectrum has been around for about 50 years, this article is, to my knowledge, the first to give complete details of the concept, design and realization of a direct-sequence spread-spectrum UHF radio link. Although this article does not present construction information at the component level, experimenters with a good grounding in UHF construction techniques should have no trouble building a functional spread-spectrum system based on the information given here.

Introduction to Direct-Sequence Spread Spectrum

A direct-sequence (DS) spread-spectrum

[1]References appear at end of article.

signal can be obtained by mixing a carrier with the output of a clock-driven pseudo-noise (PN) generator (see Fig 1). This is readily achieved in a doubly balanced mixer (DBM), and results in the suppression of the original carrier and the creation of a new signal that is spread over a wide bandwidth (typically several megahertz). This biphase modulation, described in more detail in the references cited at notes 2 and 3, creates sidebands, or "spectral lines," the envelope of which has a $(\sin x/x)^2$ shape as shown at Fig 2A.

A simplified understanding of this phenomenon can be arrived at by considering the PN sequence as a succession of identi-

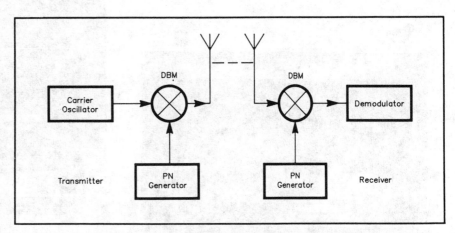

Fig 1—Simplified block diagram of a direct-sequence spread-spectrum communication system. In the transmitter, left, a spread-spectrum signal is generated by mixing carrier oscillator and pseudo-noise (PN) signals in a doubly balanced mixer (DBM). The energy of the resultant *biphase-modulated* suppressed-carrier signal is spread over a wide bandwidth. In the link receiver, right, the spread-spectrum signal is *despread* by mixing it with a PN signal identical to that used in the transmitter. In practice, the most difficult aspect of making this system work is that of synchronizing the PN-generator clocks at the transmitter and receiver sites.

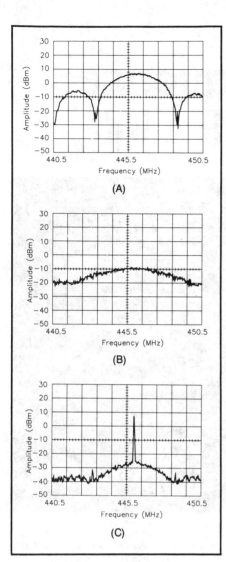

Fig 2—(A) Envelope of the unfiltered biphase-modulated spread-spectrum signal as viewed on a spectrum analyzer. In the practical system described, band-pass filtering is employed to confine the spread-spectrum signal to the amateur band. (B) At the receiver end of the link, the filtered spread-spectrum signal is apparent only as a 10-dB hump in the noise floor. (C) Despread signal at the output of the receiver DBM. The original carrier—and any modulation components that accompany it—has been recovered. The peak carrier is about 45 dB above the noise floor—more than 30 dB above the hump shown at B. (These spectrograms were made at a sweep rate of 0.1 s/div and an analyzer bandwidth of 30 kHz; the horizontal scale is 1 MHz/div.)

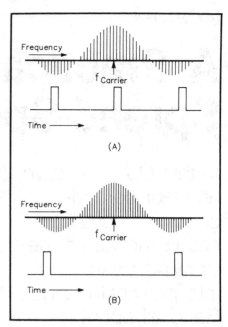

Fig 3—The PN signal (lower trace at A and B) consists of constant-width pulses that vary in repetition rate as the PN-generator shift register advances. Varying the PN-signal pulse rate changes the number and position of the spread signal's characteristic "spectral lines" (upper trace at A and B) without changing the shape of the signal envelope.

cal rectangular pulses of constant width, but variable repetition rate, the latter always being some submultiple of the (constant) clock frequency. Because the modulating signal is a rectangular pulse (rich in harmonics), a multitude of sidebands is created. As the shift register advances, the varying repetition rate changes the *number* and *position* of the spectral lines. Neither the shape of the envelope nor the position of the nulls—which are functions of the pulse width (see Fig 3)—changes.

The envelope is so shaped because a rectangular pulse in the time-domain has a (sin x/x) Fourier transform in the frequency domain. Hence, the sidebands, or spectral lines, created by such a pulse have a (sin x/x) envelope.[4] As we are dealing with a signal proportional to the output *power* (that is, a function of the square of the output voltage), the resulting signal appears as a $(\sin x/x)^2$ spectrum when observed with a spectrum analyzer (Fig 2A). Note that the spectrum-analyzer display shows only the *envelope* of all the sidebands—that is, an *imaginary* curve joining the peaks of all the spectral lines.[5]

At the receiver end of the spread-spectrum link, the output of the DBM

The spread-spectrum transmitter is based on a Hamtronics TA-451 446-MHz transmitter strip (Z1, right). The transmitter PN generator—the two piggybacked boards at left—uses 111.5-MHz energy from the TA-451 as the PN-sequence clock.

The heart of the spread-spectrum receiver is a synchronized oscillator (right) that recovers the transmitter PN-sequence clock from the received signal. The modules at left contain clock divider ICs (U8 and U9), a voltage regulator (U7) and the receiver PN generator (U10 and U11). R1, the OSC FREQ control, is at top right.

shows—in the absence of correlation—a slight rise in the noise floor, as shown in Fig 2B. If we now reintroduce PN identical in sequence, frequency and phase to that used at the transmitter, the received signal will be correlated or "despread," and the output of the DBM will be as shown at Fig 2C.

In Amateur Radio applications, the PN sequence is announced before spread-spectrum transmission begins[6] and can thus be easily duplicated at the receiver. A major problem—one common to all spread-spectrum systems—remains: that of synchronizing the receiver's PN sequence with that used at the transmitter. This can be done by recovering the transmitter clock frequency from the received signal. Because the received signal appears to be noise (Fig 2B), however, this can be a formidable challenge. Commenting on this problem, Robert Dixon writes that "...more time, effort and money has been spent developing and improving synchronizing techniques than any other area of spreadspectrum systems. There is no reason to suspect that this will not continue in the future."[7]

This article describes a simple solution to the synchronization problem. (Because of its simplicity, the solution does not offer all the antijamming properties of more sophisticated systems, but this should not be a concern to Amateur Radio operators.)

Overall Description of the Link

The UHF link described in this article was first demonstrated at the June 1988 AMRAD meeting. (Several members of AMRAD [the Amateur Radio Research and Development Corporation] have been involved in spread-spectrum experimentation since 1980.) Fig 4 shows the general arrangement used. The output of a 446-MHz transmitter is fed to one input port of a doubly balanced mixer. The other input port of the mixer is connected to a PN sequence generator that is clocked at a submultiple of the transmitter's carrier frequency.

At the receiver end of the link, a sliding correlator is used as follows: The output of a free-running "synchronized oscillator"[8] is divided to create a clock signal at a frequency close to that of the transmitter's clock. This locally generated clock is used to drive a PN generator identical to that used at the transmitter site. The resulting PN sequence is mixed with incoming RF in a doubly balanced mixer. The free-running frequency of the synchronized oscillator can be adjusted to run the receiver's PN sequence faster or slower than the transmitter's. The speed difference causes the receiver PN sequence to "slide by" the transmitter sequence (hence the name *sliding correlator*).

Because the transmitter and receiver sequences differ in speed, the PN sequence fed to the receiver's doubly balanced mixer will coincide with the transmitter sequence at some instant. Correlation, or *despreading*, then occurs, and the output of the DBM duplicates the original carrier as shown in Fig 2C. This carrier is fed to the input of the synchronized oscillator, which locks onto the incoming signal. The loop is now closed: The original clock has been recovered, and the system stays locked.

The Direct-Sequence Transmitter

As shown in Fig 5, I used a Hamtronics® model TA-451 transmitter strip (Z1), the output power of which is adjusted to 10 mW (+10 dBm), as an exciter. This transmitter is based on a 12.388-MHz crystal oscillator, the output of which is multiplied by 36 to produce output at 446 MHz. The link clock is based on transmitter energy sampled at one-fourth the output frequency (111.5 MHz, at the base of Q8 in the Hamtronics transmitter). This signal is fed to a divide-by-40 chain consisting of a MC3396P and a 7474. The divider chain produces a series of pulses at 2.7875 MHz. These pulses are used to clock a seven-stage shift register composed of a 74164 and a 7486. (This circuit, described in detail in the reference cited at note 9, is identical to that used in the link receiver [Fig 6].) This shift register, the PN generator, drives one input port of a doubly balanced mixer (U1). Attenuators are used on all ports of the DBM to reduce the effects of impedance mismatch and keep IMD products at a low level.

The output of the DBM, a biphase-modulated signal, is then amplified by an MMIC (U2), and a UHF amplifier module (U3) that produces about 0.39 W in the 440- to 450-MHz range. The output spectrum of the transmitter (ahead of the band-pass filter, Z2) is shown at Fig 2A.

Because about 90% of the output power appears between the two first nulls (located at 443.2125 MHz [446.00 − 2.7875] and 448.7875 MHz [446.00 + 2.7875], respectively) the use of a band-pass filter (Z2) to attenuate the signal below 440 MHz and above 450 MHz does not appreciably affect reception of the radiated signal. (Reliable lock was obtained over a distance of more than a mile of fairly flat terrain using 0.39 W output and a quarter-wave groundplane antenna.)

As Fig 5 shows, the link clock is slightly frequency modulated by the audio signal. This does not introduce jitter or cause false synchronization because a peak deviation of 5 kHz at 446 MHz translates to a shift of only 15 Hz at the clock frequency (2.7875 MHz)—a negligible variation. (As an alternative approach, the clock could be derived directly from the transmitter's crystal oscillator, ahead of the phase modulator. This would, of course, require a different divider chain at the receiver end of the link.)

The Direct-Sequence Receiver and Synchronizer

See Fig 6. Signals from the antenna are fed to a Hamtronics Model LNW-432 preamplifier (Z3) aligned for a flat response from 430 to 460 MHz, an arrangement that provides some limited RF selectivity. The output of Z3 is boosted another 20 dB by a Mini-Circuits MAR-8 MMIC (U4) and then applied to one port of an SBL-1 Mini-Circuits DBM (U5). The output of the DBM is further amplified by a second MAR-8 (U6), the output of which feeds Z4, a 446-MHz narrow-band-FM receiver (in my system, a Yaesu FT-708R transceiver), and the input of a synchronized oscillator.

The synchronized oscillator is a deceptively simple-looking circuit that is unfortunately as yet little-known in amateur circles. Q2 is a modified Colpitts oscillator that, in this application, free-runs at approximately 111.5 MHz, or one-fourth of the expected input frequency. Because Q2 operates in class C, it draws supply current only during a small portion of each cycle of its output sine wave. The resulting pulsating emitter current develops a pulsating voltage across Q1. Because there is no voltage drop across Q1 when Q2 is cut off, Q1 operates as an amplifier only on the peaks of Q2's output sine wave. Thus, an RF synchronizing signal applied to Q1's base will be allowed to steer Q2's oscillation frequency only for the brief periods during which Q2 conducts. R1, OSC FREQ (labeled TUNING in the title photo), allows adjustment of Q1's base current and, hence, Q2's free-running frequency.

In practice, the synchronized oscillator provides the function of a phase-locked loop (PLL) while offering several advantages over a PLL. The synchronized oscillator, which uses only two transistors, is simpler to implement. Unlike a PLL, which depends on a multistage feedback loop (phase detector, loop filter and so on) to achieve and hold lock, the synchronized oscillator locks onto the input signal directly. Further to its advantage, the synchronized oscillator can operate with very noisy input signals.[10,11,12]

The output of Q2 is thus a sine wave at 111.50 MHz. It is divided by 20 in U8 (a MC3396P prescaler) and then again by 2 in U9 (a 7474 flip flop), the output of which—a 2.7875-MHz square wave—feeds a seven-stage shift register (U10, a 74164), the application of which is described in more detail in the work cited at note 9. This arrangement exactly duplicates that used at the transmitter end of the link.

The output of the PN generator is connected to the IF port of the DBM (U5). This constitutes a sliding correlator in the sense that the transmitter and receiver PN sequences—identical, but running at slightly different speeds—slide by each other within the DBM. At the instant the receiver PN sequence coincides with that of the transmitted signal, correlation takes place in the DBM, and the DBM delivers a despread signal (Fig 2C). This despread signal, recognized by the synchronized oscillator as a valid input, forces the oscillator into lock and keeps it there.

Spread Spectrum Theory and Projects 8-49

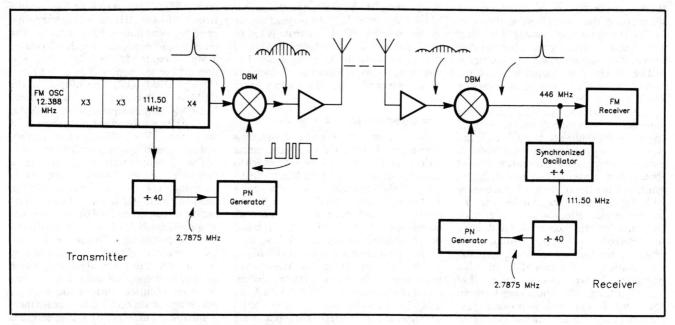

Fig 4—Block diagram of the practical spread-spectrum link. The success of this arrangement lies in the use of a synchronized oscillator (right) to recover the transmitter clock signal at the receiving site.

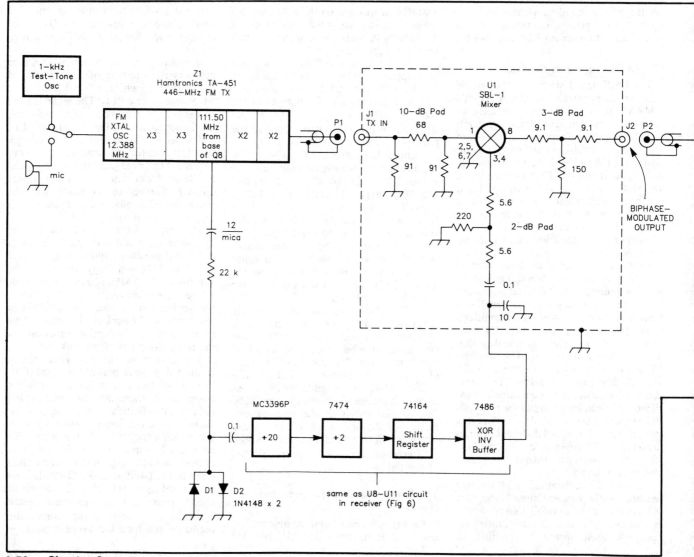

8-50   Chapter 8

Construction of the Link Prototype

As shown in the photographs, the various elements of the circuit were built in individual die-cast aluminum boxes. This provides ample interstage shielding; it also allowed flexibility during development of the link. A second version could no doubt be made much simpler mechanically.

Adjustments

The Hamtronics transmitter (Z1 in Fig 5) requires only one adjustment specific to its use in this application: *Its output must be reduced to 10 mW.* If you do not own an RF-power meter capable of measuring this level accurately, you can easily construct one as follows. Build a dummy load by connecting a 1.5-V, 25-mA lamp (one of the two identical lamps available as Radio Shack® no. 272-1139) in series with a 1- to 8-pF variable capacitor. Connect this dummy load to the Hamtronics transmitter via a sensitive SWR indicator. Turn on the transmitter and adjust the capacitor to minimize the SWR presented by the dummy load to the transmitter. Connect the second 272-1139 lamp across a 1.5-V cell in series with a 50-Ω adjustable resistor, and place this lamp so that you can simultaneously view it and the dummy-load lamp. Adjust the current through the dc-fed lamp so that the lamp dissipates 10 mW. Turn on the transmitter and adjust R37 (on the transmitter) so that the RF-fed lamp glows at the same brightness as the dc-fed lamp. This simple comparison method yields surprisingly accurate results. No other adjustments are required at the transmitter site.

At the receiver site, the only adjustments necessary concern the synchronized oscillator (Fig 6). Using a dip meter coupled to L5 as a resonance indicator, adjust C1, OSC INPUT TUNING, for resonance at 111.5 MHz. (Because R2 loads the tuned circuit, you may have to temporarily disconnect the 68-pF capacitor from the R2 wiper to obtain a discernible dip.) Connect a frequency counter to the output of the divide-by-40 chain via J11, FREQ COUNTER. Adjust R1, OSC FREQ, to the center of its range. Adjust the wiper of R2, GAIN SET, to the ground end of its range. Set C2, OSC TANK TUNING, to the center of its range, and adjust L6 for a reading of approximately 2.78 MHz on the frequency counter. Note that this reading can be varied by adjusting the OSC FREQ control.

Operation

After identifying your station on the link carrier frequency in accordance with §97.84(g)(5) of the FCC rules (identification by means of a narrow-band emission—AFSK or voice—is the easier option to implement) and stating the characteristics of your PN sequence, you may turn on the spread-spectrum transmitter.[13]

At the receiver end of the link, adjust the wiper of R2 (Fig 6) to about 30° from the ground end of its range. (Further advancing this control only "oversynchronizes" the oscillator and produces distortion at its output, and could lead to false triggering of the divide-by-20 stage.) Connect a frequency counter to J11 and adjust R1 for a counter indication of about 2.7875 MHz. Assuming that the received signal is sufficiently strong, you should observe that the counter suddenly displays exactly

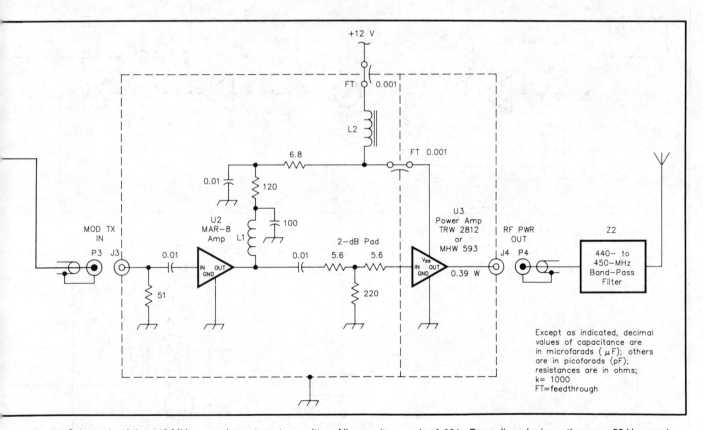

Fig 5—Schematic of the 440-MHz spread-spectrum transmitter. All capacitors under 0.001 µF are dipped mica; others are 50-V ceramic (0.1 µF units: monolithic ceramic) unless marked otherwise. Resistors are 5% tolerance, ¼- or ½-W carbon film or composition. Q8 (in the Z1 box) is a Hamtronics part designator. See Fig 6 for details on the PN generator. The 1-kHz tone generator, included to facilitate testing in the author's version of the link, is not described in this article.

D1, D2—1N4148 or equiv.
L1—10 close-wound turns of no. 30 KYNAR®-insulated (wire-wrap) tinned copper wire, 1 mm diam (use a no. 64 [0.036-in. diam] drill as a removable form).
L2—1 turn of no. 30 KYNAR-insulated wire through an Amidon FB-64-101 (Palomar FB-1-64, RADIOKIT FB64-101 also suitable) ferrite bead.
U1—Mini-Circuits SBL-1 mixer (Mini-Circuits, PO Box 350166, Brooklyn, NY 11235-0003, tel 718-934-4500).
U2—Mini-Circuits MAR-8 MMIC.
U3—TRW 2812 or Motorola MHW 593 amplifier module.
Z1—Hamtronics TA-451 446-MHz FM transmitter (Hamtronics, Inc, 65 Moul Rd, Hilton NY 14468-9535, tel 716-392-9430).
Z2—Hamtronics HRF-432 440- to 450-MHz helical-resonator band-pass filter.

Fig 6—Mixer and PN-generator circuitry for the 440-MHz spread-spectrum link. Except for 0.1-µF units, all capacitors are 50-V ceramic unless marked otherwise; 0.1-µF capacitors are 50-V monolithic ceramic. Resistors are 5% tolerance, ¼- or ½-W carbon film or composition.

C1—1- to 10-pF, ceramic-dielectric piston trimmer.
C2—1- to 12-pF, air-dielectric trimmer.
FB1, FB2—Amidon FB-64-101 (Palomar FB-1-64, RADIOKIT FB64-101 also suitable) ferrite bead.
L3, L4—10 close-wound turns of no. 30 KYNAR-insulated (wire-wrap) tinned copper wire, 1 mm diam (use a no. 64 [0.036-in. diam] drill as a removable form).
L5—2½ turns of no. 16 tinned copper wire, ¼ in. diam. Tap ¾ turn from ground end.

Use a ¼-in.-diam drill as a form. After winding the coil, remove the drill and spread the turns to make a coil ½ inch long.
L6—4 space-wound turns of no. 20 enam copper wire on a ¼-in. diam, slug-tuned plastic form. Nominal inductance, approx 0.2 µH.
R1—100-kΩ, linear-taper control.
RFC1—100-µH miniature molded choke.
U4, U6—Mini-Circuits MAR-8 MMIC.
U5—Mini-Circuits SBL-1 doubly balanced mixer.

U7—7805 regulator IC
U8—MC3396P divide-by-20 prescaler IC.
U9—7474 dual-D, positive-edge-triggered flip-flop IC.
U10—74164 8-bit, parallel-output shift register, asynchronous clear IC.
U11—7486 quad two-input exclusive-OR gate IC.
Z3—Hamtronics LNW-432 preamplifier adjusted for an input response flat from 430 to 460 MHz. See text.
Z4—446-MHz FM receiver, or transceiver in receive mode.

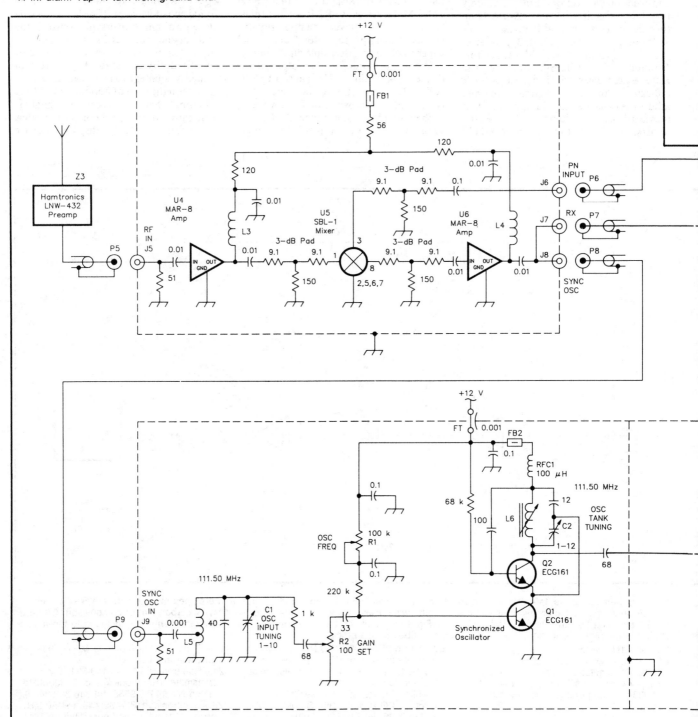

Construction of the transmitter biphase modulator. J1, TX IN, is at left; the SBL-1 DBM, U1, at top center; and J2, BIPHASE-MODULATED OUTPUT, at right. U1 receives PN-generator injection via the wire in the foreground.

The transmitter MMIC (U2) and power (U3) amplifiers. J3, MOD TX IN, is to the lower left; J4, RF POWER OUT, to the right. U3, a TRW 2812 amplifier module, commands the right two-thirds of this view; U2, a Mini-Circuits MAR-8, is the tiny black pill above and to the right of J3.

2.7875 MHz as you adjust R1. When this happens, the receiver PN sequence has locked to the transmitter sequence.

The synchronized oscillator should be able to achieve lock at free-running frequencies from about 2.7860 to 2.7890 MHz. In my version, the receiver stays in lock for hours without needing readjustment of R1 once the synchronized-oscillator enclosure has stabilized (about 30 minutes after turn-on).

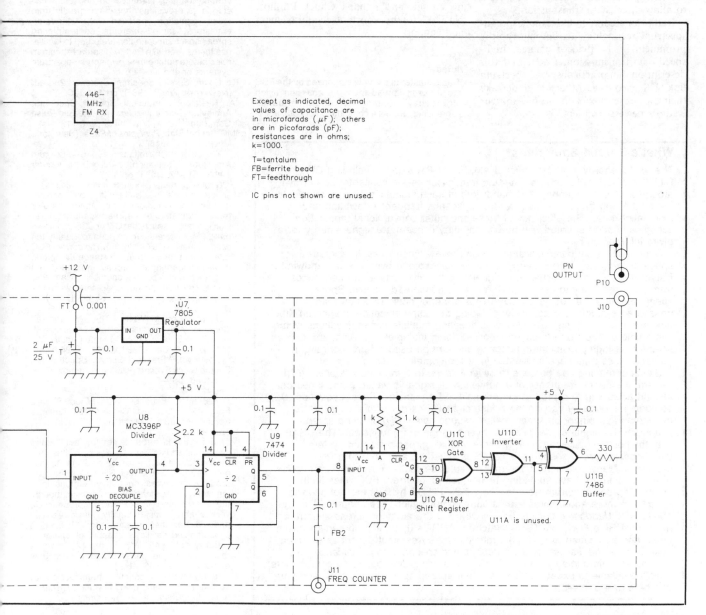

Spread Spectrum Theory and Projects 8-53

The receiver preamp, Z3, a Hamtronics LNW-432 module, fits neatly into its own die-cast box.

The three ICs in the receive-mixer module are contained on a single PC board. U4 is at left (above J5, RF IN); U5 is almost hidden from view by the disc-ceramic capacitor at top, and U5 is at the right (above J8, SYNC OSC). (In this model, a coaxial tee is used at J8 in lieu of mounting J7, RX, on the box.) The jack at center is J6, PN INPUT.

Summary

This article has described a direct-sequence spread-spectrum UHF link that uses readily available components and does not require sophisticated equipment for adjustments and tuning. As it stands, the system transmits and receives voice, or packets by means of AFSK. Work is currently proceeding to modify the system to allow direct data transmission.

Radio amateurs should expect spread-spectrum technology to become rapidly prominent in the field of amateur high-speed data transmission. I hope that this description of a practical spread-spectrum link will encourage others to undertake their own experiments on one of Amateur Radio's newest frontiers.

Acknowledgments

Vasil Uzunoglu, codeveloper of the synchronized oscillator, deserves much credit for clarifying my understanding of his circuit, and I thank Chuck Phillips, N4EZV, for his constant encouragement and support.

Notes

[1] Radio amateurs are now authorized by the FCC to transmit spread-spectrum emission using frequency hopping or direct-sequence spreading, but not a combination of both, at frequencies above 420 MHz. This article describes a direct-sequence 446-MHz link that uses a seven-stage shift register with feedback taps at stages 7 and 1. Because the FCC rules pertaining to amateur spread-spectrum operation contain logging, identification and technical-standards provisions that differ greatly from those regulating operation with "standard" emissions, radio amateurs contemplating spread-spectrum operation are urged to familiarize themselves with Part 97's spread-spectrum rules before putting their own spread-spectrum systems on the air.—*Ed.*

[2] R. Dixon, *Spread Spectrum Systems,* 2nd ed, (New York: Wiley, 1984), pp 114-125.

[3] A. Kesteloot, "Practical Spread Spectrum: Achieving Synchronization with the Slip-Pulse Generator," *QEX,* May 1988, pp 6-11.

[4] MIT School Staff, *Principles of Radar* (New York, McGraw-Hill, 1946), pp 4-12.

[5] A. Kesteloot, "Of Weltanschauung, Jitter and Amplitude Modulation," *AMRAD Newsletter 1987,* Vol XIV, No. 4, pp 4-7.

[6] FCC rules currently require that amateur spread-spectrum emissions be identified by means of narrow-band emissions, or by altering one or more parameters of a spread-spectrum emission in a fashion such that CW or SSB or narrow-band FM receivers can be used to identify the sending station (§97.84 (g) (5)). The current FCC rules do not require that the station ID include details concerning the spreading method and sequence in use, but interoperability of amateur spread-spectrum stations is facilitated when this information is given as part of the ID.—*Ed.*

[7] Dixon, p 214.

[8] V. Uzunoglu and M. White, "The Synchronized Oscillator: A Synchronization and Tracking Network," *IEEE Journal of Solid State Circuits,* Vol SC-30, No 6, Dec 1985, pp 1214-1224.

[9] A. Kesteloot, "Practical Spread Spectrum: A Simple Clock Synchronization Scheme," *QEX,* Oct 1986, pp 4-7.

[10] See note 8.

[11] A. Kesteloot, "Extracting Stable Clock Signals from AM Broadcast Carriers for Amateur Spread-Spectrum Applications," *QEX,* Oct 1987, pp 5-9.

[12] An ambiguity is introduced by the fact that the synchronized oscillator divides by 4. It is thus theoretically possible for the oscillator to lock onto any one of the four RF cycles that can produce the correct frequency relationship at the receiver end of the link. In practice, though, the receiver tends to lock onto the proper cycle very reliably. (This ambiguity will be resolved once it becomes possible to [1] build a synchronized oscillator capable of operating reliably at 446 MHz, or [2] find a reasonably priced phase-locked loop capable of operating at 446 MHz.)

[13] I identify my spread-spectrum transmissions by speaking into a hand-held FM voice transceiver tuned to the link carrier frequency.

What's Spread Spectrum?

The useful energy in a conventional AM, FM or PM signal—including CW, RTTY, AMTOR, SSB and packet-radio transmissions—is concentrated narrowly around a center frequency. The bandwidth of such signals is directly related to the modulating frequency (and, where applicable, frequency deviation and modulation rate). The efficiency of these and other conventional modulation schemes is often equated with how tightly they concentrate signal energy for a given information rate.

Spread-spectrum communication doesn't follow these rules. In spread-spectrum work, the signal energy is *intentionally spread* over a wide bandwidth. Because of this, spread-spectrum signals are largely immune to interference from, and less likely to cause interference to, nonspread signals. Spread spectrum offers the additional advantages of better noise rejection than nonspread systems, the possibility of hiding the communication channel in the ambient noise, and the possibility of conducting multiple communications at the same time on the same frequency (code-division multiplexing). Also, spread spectrum is highly resistant to jamming and can be used for precise ranging—characteristics that make it valuable in space communications.

Signal spreading can be done in several ways. In a *frequency hopping (FH)* system, the center frequency of a conventional signal is varied many times per second according to a predetermined table of frequencies. *Direct sequence (DS)* spreading is done by varying the phase of an RF carrier with a very fast, pseudorandom binary bit stream called pseudo-noise (PN). In *chirp* spread spectrum, the signal carrier is swept over a range of frequencies. (The USAF over-the-horizon-backscatter HF radar is a chirp spread-spectrum system.) In *time hopping* spread spectrum, a carrier is keyed on and off with a PN sequence. Many commercial and military spread-spectrum systems are *hybrids* of two or more of these spreading techniques, but current FCC rules limit amateur spread-spectrum work to FH *or* DS—FH-DS hybrids are not allowed.

To learn more about spread-spectrum communication, see chapter 21 of the 1989 *ARRL Handbook,* back issues of *QEX* and the amateur spread-spectrum rules in ARRL's *FCC Rule Book.* Watch for ARRL's upcoming spread-spectrum book. And stay tuned to *QST*: This month, André Kesteloot describes what we believe to be *the first practical amateur spread-spectrum communication system—an affordable* system that's simple to adjust and based on components readily available to experimenters. Amateur spread spectrum is here—and *you* can be a part of it.—*Ed.*

Practical Spread Spectrum: Clock Recovery With the Synchronous Oscillator

By André Kesteloot N4ICK

Introduction

Clock recovery circuits are usually built around convolvers, such as surface acoustic wave devices, or phase-lock loops such as Costas loops, τ-dither loops etc (see references 1, 2). The formers tend to be extremely expensive and therefore out of the reach of radio amateurs, while the latters can be quite complicated and are inclined to work better when the clock information is present most of the time.

When, however, synchronization information is missing, a large portion of the time (ie, in the presence of heavy interference, or long pseudorandom sequences) most PLLs revert quickly to their free-running frequency, thus producing output jitter.

A simple solution to the above problem is the Synchronous Oscillator (SO), an interesting and simple circuit which can be used for clock recovery. This article describes its principle, as well as a test apparatus designed to evaluate the SO's practicality, ease of use, and performance.

The Basic Circuit

Referring to Fig 1, the Synchronous Oscillator is basically a modified Colpitts oscillator with an extra transistor in its emitter-to-ground path. It has two positive feedback paths: one from the junction of the two capacitors in the collector tank circuit back to the emitter of the upper transistor; the other one from the junction between the tank coil and the RF choke back to the base of the transistor. The operation of this circuit is described in detail in references 3, 4 and 5. Briefly, the upper transistor is a free-running sinusoidal oscillator operating in class C. It thus conducts only during very short periods of time. Whenever it draws current, it develops a voltage across the bottom transistor, and allows the latter to conduct. Therefore, of all the signals applied to the base of the bottom transistor, only those which appear during that very short "time-window" which is the moment of conduction, can be amplified by the bottom transistor and used to synchronize the upper one. This "coherent amplification" arrangement explains the excellent noise rejection characteristics of the SO. (Such a circuit is used for carrier recovery in my 440 MHz direct-sequence spread-spectrum link, published in the May 1989 issue of *QST*.) Note that, contrarily to what happens in a phase-lock loop, in an SO the input signal directly synchronizes the output signal.

The Clock Recovery Circuit

Clock recovery circuits do not generally need the ability to work with noisy signals, a quality usually demanded of carrier-recovery circuits, and it is thus possible to further simplify the SO (see Fig 2). The tuned circuit in the collector introduces a flywheel effect which supplies the stability required to produce steady output clock pulses in the momentary absence of input synchronization pulses.

Practical Application

To put the SO to the test, an apparatus was built which generates an incomplete train of synchronization pulses. (In other words, properly timed synchronization pulses are produced, but some of them are purposely deleted.) The general arrangement is shown in Fig 3, while the actual circuit is shown in Fig 4.

U1, a crystal oscillator operating at 2 MHz is used to drive a seven-stage pseudorandom noise (PN) generator of the kind used in most of my spread-spectrum equipment. The PN generator consists of U2, a 74164 shift register, while two sections of U3, a 7486, are used as an XOR and inverter stage (see reference 6 for a more complete description of the PN generator used). This PN sequence is fed to an "edge-detector" consisting of U4, another 7486 XOR integrated circuit. Three of the four gates are connected in series, and their cumulative propagation delay is put to good use to retard the incoming pulse. This pulse and the original pulse are XORed in the fourth section of U4, and a short output pulse is thus created for each input transition. Since the output of a seven-stage pseudonoise generator presents at times up to six "0" or six "1"

[1]Notes appear at end of article.

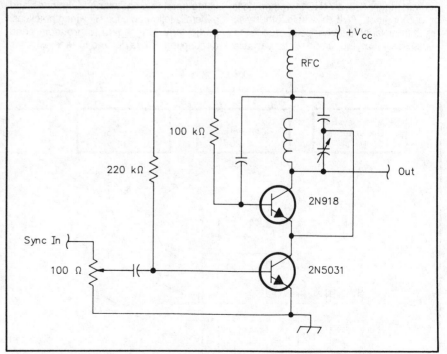

Fig 1—The Synchronous Oscillator. Component values depend on frequency of interest and transistors used.

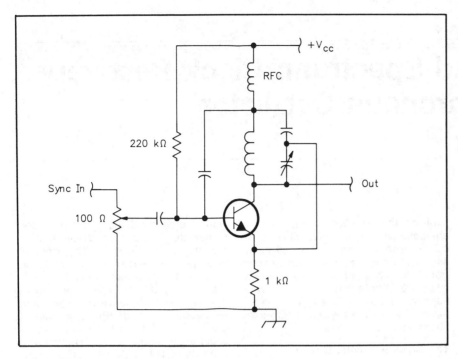

Fig 2—Simplified Synchronous Oscillator for clock recovery.

in a row, the edge-detector will create a pulse train with up to six synchronization pulses missing from time to time.

This latter pulse-train is the synchronization information supplied to our SO. It is displayed on the lower oscilloscope trace of Fig 5. One of the characteristics of the SO, which consists of Q1, is that it is a sine-wave oscillator whose output amplitude is always constant throughout the synchronization range. Since the output amplitude is constant, it is acceptable to feed it to a CMOS Schmitt trigger (U5, a 74HC14) without fear of output jitter. (Which would otherwise result from variable trigger points due to the ceaselessly varying slope of a variable amplitude signal, should such a signal be used.) The output of the Schmitt trigger stage is a 2-MHz square wave shown on the upper oscilloscope trace of Fig 5. Notice that, although 6 synchronization pulses are missing, there is no visible jitter on the traces in the interval between sync pulses.

Construction

The circuit described above was constructed on perforated phenolic board, using point-to-point wiring techniques and generally following good HF wiring practices, particularly with regard to grounding and decoupling. If it is desired to recover clock frequencies much above 50 MHz, the circuit can be scaled up, but it will be necessary to use Schmitt triggers made of discrete components (such as MPF102 FETs). Similarly, VHF transistors (such as the 2N918) should replace the 2N2222 used for Q1.

Adjustments

A digital frequency counter is connected to the output of U4 (pin 11) to measure the frequency of the input signal. The counter probe is then transferred to the output of the synchronous oscillator at U5 pin 10. The wiper of R1, the potentiometer used to adjust the amount of sync signal applied to the SO, is first turned all the way down to ground. The slug of L2 is then adjusted to bring the free-running frequency of the SO near the target frequency (in our case, 2 MHz). The variable capacitor C1 is then adjusted until the output frequency, read on the frequency meter, is as close as possible to the target frequency (say plus or minus 5 kHz, or between 1,995 and 2,005 MHz). The input signal is then slowly increased by means of R1. After less than ¼ turn, the frequency meter will suddenly display the target frequency, as the SO has reached lock.

If a dual-trace oscilloscope is available, connect channel B to the output of the edge-detector U4, connect channel A to the output of U5 and synchronize on channel A. The channel (the upper trace on Fig 5) will display a square wave, while the other trace will be fuzzy. As you slowly turn up R1, the lower trace (channel B) will suddenly synchronize with the bottom trace, as shown on Fig 5.

Tracking Range

To simulate the effects of a Doppler shift in input frequency (satellite operation, for instance), the crystal oscillator was replaced by a variable-frequency oscillator. With the values shown on Fig 4 and the circuit adjusted for a free-running frequency of 2 MHz, the tracking range (ie, the bandwidth within

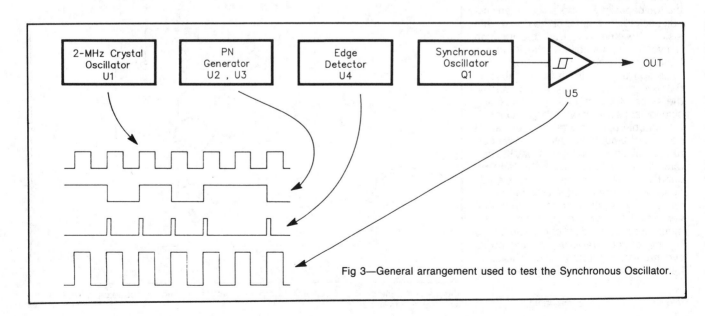

Fig 3—General arrangement used to test the Synchronous Oscillator.

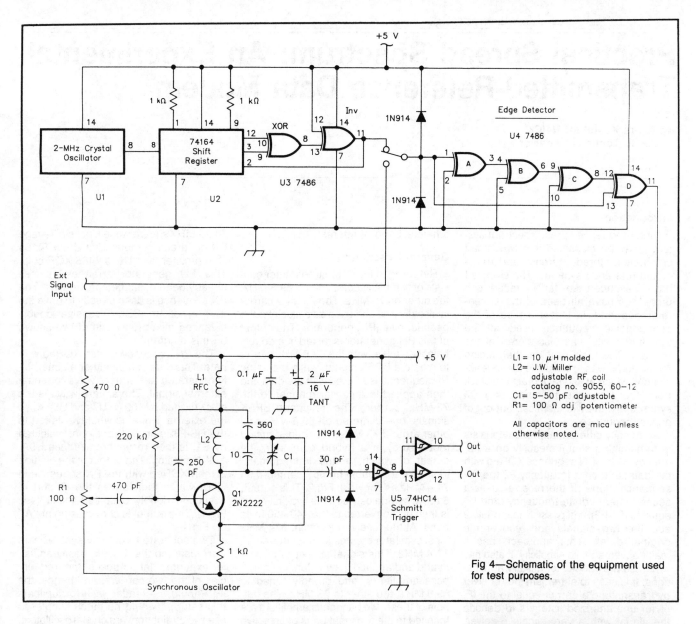

Fig 4—Schematic of the equipment used for test purposes.

which synchronization takes place without returning) extended from 1.986 MHz to 2.015 MHz. This 1.5% tracking range is generally in agreement with measurements reported in references 3 and 4. Several typical synchronization curves have been published in reference 3. It should be kept in mind that the tracking range depends on the level of the synchronization signal at the input, and on the Q of the collector tank circuit. (Incidentally, per reference 7, the maximum doppler shift created by amateur radio satellites is typically less than 0.006% which is but a fraction of the tracking range available with the SO.)

Conclusion

The SO is a circuit remarkable both for its performance and its simplicity of construction. The reader is encouraged to breadboard a SO to appreciate its ease of use. (If a commercial application is contemplated, the reader should note that the SO is covered by several US patents, see note 8.)

I am indebted to Professor Marvin White (of Lehigh University) and Mr. Vasil Uzunoglu (formerly with Fairchild Communications), the co-developers of the Synchronous Oscillator, for their encouragements and stimulating discussions.

References

[1] R. Dixon, *Spread Spectrum Systems*, (New York: Wiley) 2nd Ed, 1984, pp 186-187 and pp 232-234.
[2] R. Best, *Phase-Locked Loops*, (New York: McGraw Hill) 1984, pp 212-217.
[3] V. Uzunoglu and M. White, "The Synchronous Oscillator: A Synchronization and Tracking Network," *IEEE Journal of Solid State Circuits*, Vol SC-30, No. 6, Dec 1985, pp. 1214-1224.
[4] V. Uzunoglu and M. White, "Some Important Properties of Synchronous Oscillators," *Proceedings of the IEEE*, Vol 74, No. 3, Mar 1966, pp 516-518.
[5] A. Kesteloot, "Extracting Stable Clock Signals from AM Broadcast Carriers for Amateur Spread-Spectrum Applications," *QEX*, October 1987, pp 5-9.
[6] A. Kesteloot, "Practical Spread-Spectrum: A Simple Clock Synchronization Scheme," *QEX*, October 1986, pp 4-7.
[7] M. Davidoff, *The Satellite Experimenter's Handbook*, (Newington, CT: ARRL), 1984, pp 10-3 to 10-5.
[8] US Patents 4,274,067; 4,355,404; 4,356,456.

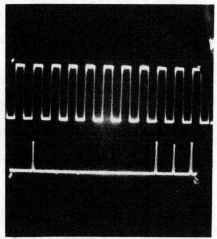

Fig 5—The upper oscilloscope trace shows the output of the Synchronous Oscillator (U5, pin 8) while the lower trace shows the train of synchronization pulses. Note that there is no visible jitter on the output signal in the absence of input signal. Horizontal scale: 0.5 μSec/division.

Practical Spread Spectrum: An Experimental Transmitted-Reference Data Modem

By André Kesteloot N4ICK
ARRL Technical Advisor

Introduction

Direct-sequence spread-spectrum systems may be classified into two broad categories: stored-reference and transmitted-reference systems. The circuits I have described so far in these columns[1,2,3,4] have all been of the stored-reference kind, in that a replica of the pseudonoise sequence used at the transmitter site was also stored at the receiver. The main problem then facing the designer was to build into the receiver a circuit which would (a) extract the clock from the transmitted signal, and (b) synchronize the two identical, but out of phase, sequences.

A radically different approach consists in transmitting simultaneously on a frequency F1 the PN sequence XORed with the data, and on a frequency F2 the PN sequence alone. If the receiver uses a second intermediate frequency (2nd IF) equal to (F1-F2), it is possible to modulo-2 add the two signals and recover the original data[5]. This approach has a major advantage: its simplicity. It also has two main drawbacks for military applications: It is easy to either jam the receiver by transmitting a carrier equal to the IF, or for unauthorized listeners to decode the signal with a very simple receiver. Neither of these problems is of concern to the radio amateur, and it was therefore decided to design a simple data link using the transmitted reference approach. The description of the circuitry used will be brief, as most of its elements have already been described in some of my previous articles. (This paper does not aim to be a "construction article," but to give enough pointers to the interested that they may decide to use this information as a starting point, and try their hands at experimenting with spread spectrum.)

Specifications

The equipment described, an experimental modem operating around 70 MHz, can transmit and receive ASCII data at rates higher than 38 kilobauds. The 70-MHz output, about 7 MHz wide, can be up-converted to 440 MHz or any other amateur band of interest.

General Description

Referring to Fig 1, the transmitter consists of two oscillators, one on 69.8 MHz, the other on 73 MHz. The 73-MHz carrier is also divided and used as a clock for the pseudonoise (PN) generator. The output of this PN generator is mixed in a doubly balanced mixer with the 69.8-MHz carrier to create a BPSK signal, while the same PN sequence is XORed with the data and then applied in a similar manner to the 73-MHz carrier. The resultant BPSK signals, one centered on 69.8 MHz, the other around 73 MHz, are then *summed* (not mixed) in a hybrid combiner, the output of which, in the time domain, looks like a "double hump" centered on 71.4 MHz as shown in Fig 2. This output, 6.85 MHz wide between the two first nulls, is then up-converted to the 440-450 MHz band. At the receiver, a similar 374-MHz local oscillator creates an IF centered on 71.4 MHz. This signal is split through a hybrid and applied to two identical bandpass amplifiers, one broadly tuned to 69.8 MHz, the other to 73 MHz. The outputs of these two bandpass amplifiers are then fed to the two input ports of an active doubly balanced mixer, the output of which is our baseband data. It is then filtered, amplified and regenerated through a Schmitt-trigger stage.

Transmitter Circuit

As shown in Fig 3, the transmitter consists of two identical frequency synthesizers, one functioning on 73 MHz, the other on 69.8 MHz. (Only one, the 73 MHz, is shown for simplicity.) The circuit is comprised of an MC1648 voltage-controlled oscillator tuned to the frequency of interest, an MC3396 divide-by-20 stage, and a MC145151 synthesizer chip. (The operation of this synthesizer was described in more detail in the November 1988 *AMRAD Newsletter*, Vol XV, No. 5.) The output of the MC3396 divider-by-20 stage is a clock at 3.65 MHz. This clock is buffered by four sections of U4, a 4049 hex-inverter, and used for three separate functions. After another division by two in U9B, a 7474 flip-flop, the resulting 1.825-MHz square wave clocks a 7-stage PN generator comprised of U5, a 74164 shift register, and U6, a 7486 XOR chip. This PN generator arrangement has already been described previously.[6] The PN sequence is used directly to drive the IF port of an SBL-1 passive doubly balanced mixer (only one of two shown on this diagram).

The PN sequence is also XORed with data. This data can be either an internally generated test square wave, or some external signal. The square wave is the clock divided by 100 (in U7 and U8), while the external signal is whatever is fed to the RS-232 port. Either signal, selected by S1, is then resynchronized via U9A, a 7474 flip-flop. The output of the flip-flop is then XORed with the PN sequence in U6D, which then drives the IF port of a second doubly balanced mixer (see Fig 1). This output is shown as point "A" on Fig 1.

Two separate frequency synthesizers were used so that I could easily adjust, for experimental purposes, the frequencies of the two carriers, and hence, the IF. Should the reader wish to duplicate this setup, it would be much simpler to use two straightforward crystal oscillators.

The outputs of the two doubly balanced mixers are then added in a summing network, here a TV antenna coupler/hybrid combiner (Radio Shack® part #15-1141). One could also manufacture one's own by winding a few turns of trifilar wire on a ferrite core, but the Radio Shack part is perfectly satisfactory, and comes already shielded and fitted with F connectors.

The output of the summing hybrid is centered on 71.4 MHz, and occupies a minimum bandwidth of about 6.9 MHz. As shown in Fig 1, the output of the summing hybrid feeds yet another doubly balanced mixer, wherein is mixed the 374-MHz local oscillator signal, thus heterodyning the whole band, from 67.975 to 74.825 MHz to the 440-MHz band.

Although the heterodyning and amplification process are by no means trivial, they are not novel and have already been covered in the literature. This article concentrates on the spread-

[1]References appear at end of article.

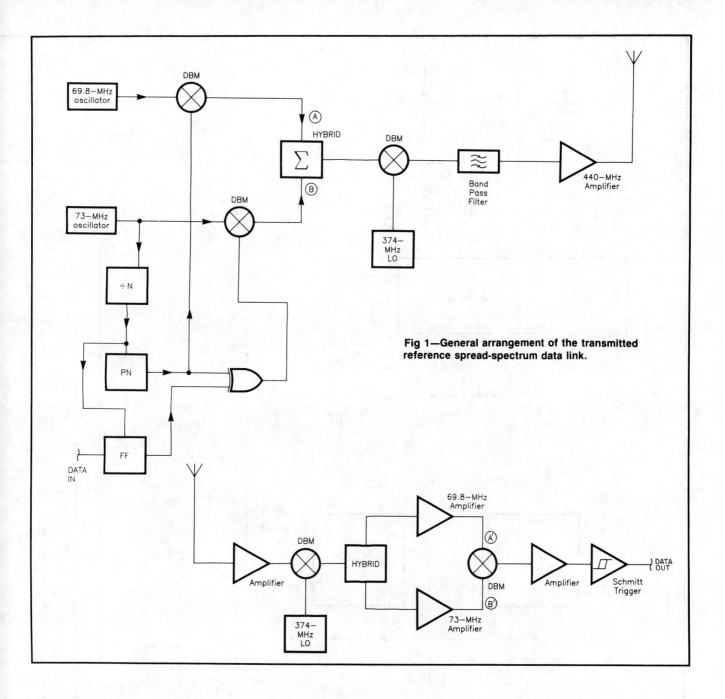

Fig 1—General arrangement of the transmitted reference spread-spectrum data link.

spectrum aspect of the project.

The Receiver

Fig 1 shows the general arrangement, while Fig 4 shows the details. After mixing with a local 374-MHz oscillator, the incoming signal is once again centered on a 71.4-MHz intermediate frequency. This signal is then split into two paths, each fed to a band pass amplifier, one centered on 69.8 MHz, the other around 73 MHz. The outputs of those two amplifiers are shown as A and B in Figs 1 and 4. These outputs are now fed to the two input ports of U101, an MC1496P active doubly balanced mixer. In the absence of an input signal at the transmitter, the two PN streams are in phase, but if data is applied, that portion of the PN sequence which corresponds to a "data high" will be inverted. (See reference 1 for additional details of a similar data recovery scheme.)

A replica of the data is thus available at the output of the MC1496P DBM. After passing through a lowpass filter, the data is amplified through U102 a CMOS op-amp, and finally regenerated through U103, a 4093 Schmitt trigger. RS-232 level conversion and buffering is then accomplished with U104, an MC1488 IC.

Operation

Testing was conducted as shown in Fig 5, with the data used being ASCII characters generated by the computer. A test program was written by Lawrence Kesteloot, N4NTL, and is available on the

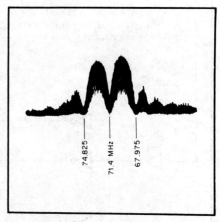

Fig 2—Output of transmitter summing hybrid.

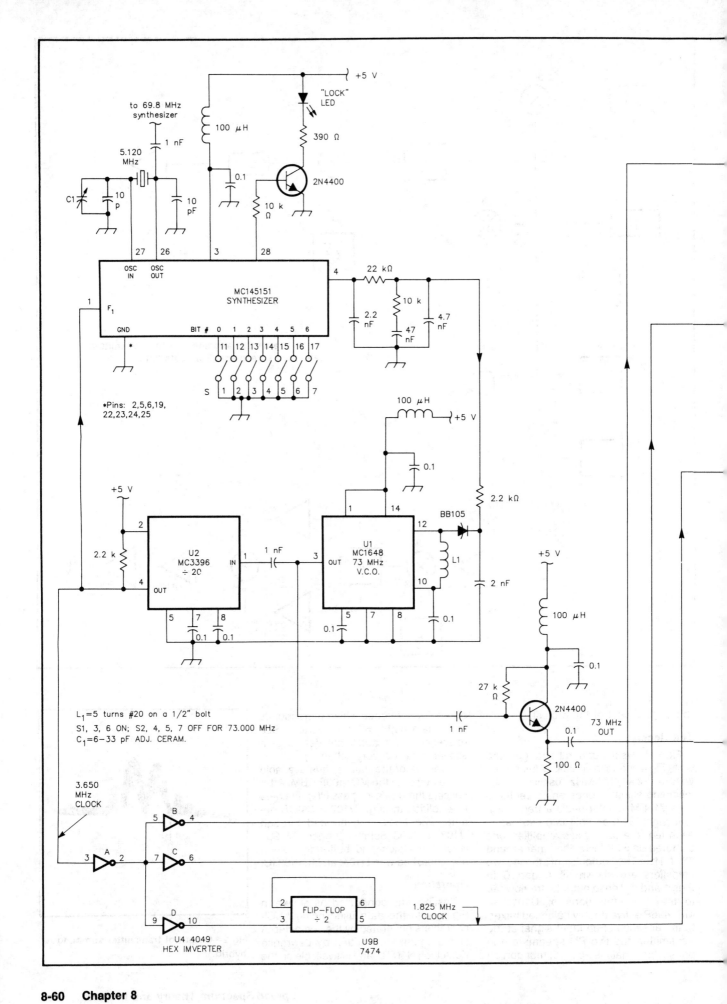

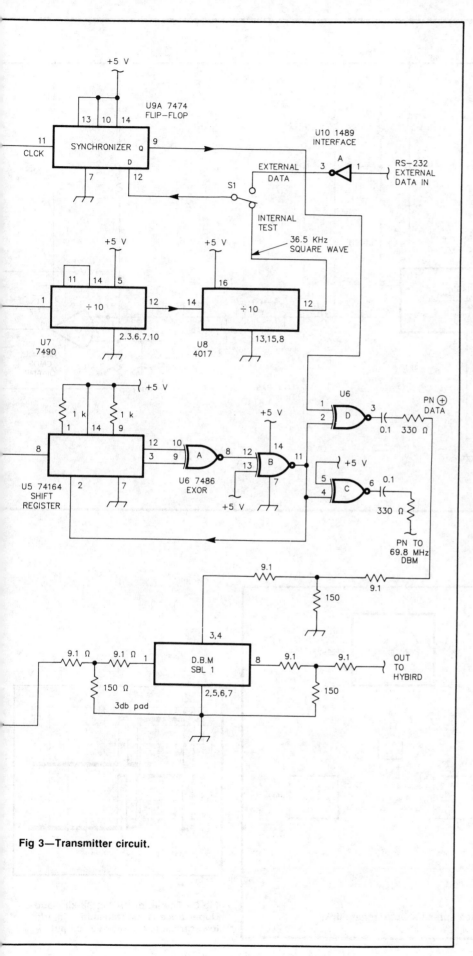

Fig 3—Transmitter circuit.

AMRAD BBS in the spread-spectrum area[7]. The program continuously sends, in ASCII format, all the letters of the alphabet, and after each letter, checks whether the letter coming back from the receiver is the same as the one just sent. The highest speed (easily) available at the RS-232 port of my IBM® clone is 38 kilobauds, and Fig 6 shows on the upper trace, the signal at the input of the transmitter, while the lower trace shows the receiver output. The two signals are essentially indistinguishable from each other, and we may infer from these preliminary results that much higher speeds are attainable. (The final version of the hardware will probably include a more sophisticated data interface, such as NRZ, etc.) At 38 kilobauds, during tests repeated over several days, the error rate was always 0 (zero) error for over 2,000,000 (two million) letters of transmitted data. This kind of error rate is due, at least in part, to the fact that this was an ideal bench setup without any interference at the IF. In a practical realization, one would obviously benefit from the addition of an LC circuit tuned to the IF of interest, at the output of U101 (between pin 12 and ground). Another circuit which would need to be added is an automatic gain control (AGC) for the IF strip, something not exactly trivial for direct sequence.

Acknowledgements

Some of the ideas implemented in this design were originally discussed with two other members of the AMRAD core group: Glenn Baumgartner, KA0ESA, and Chuck Phillips, N4EZV. Their support is gratefully acknowledged.

Figures 4-6 appear on the next page.

References

[1] A. Kesteloot, "Experimenting with Direct Sequence Spread Spectrum," *QEX* Dec 1966, pp 5-9.
[2] A. Kesteloot, "'Practical Spread Spectrum: A Simple Clock Synchronization Scheme," *QEX* Oct 1986, pp 4-7.
[3] A. Kesteloot, "Practical Spread Spectrum: Achieving Synchronization with the Slip-Pulse Generator," *QEX* May 1988, pp 6-11.
[4] A. Kesteloot, "A Direct Sequence Spread Spectrum UHF Link," *QST* May 1989.
[5] R. Dixon, *Spread Spectrum Systems*, (New York, Wiley) 2nd ed, 1984 pp 224-225.
[6] *The 1989 ARRL Handbook*, Bruce Hale, Ed, (ARRL, Newington, CT), 1988, Chap 21, p 12.
[7] The AMRAD BBS: 703-734-1387. (300/1200/ 2400 baud, 8 bits, 1 stop bit, no parity.)

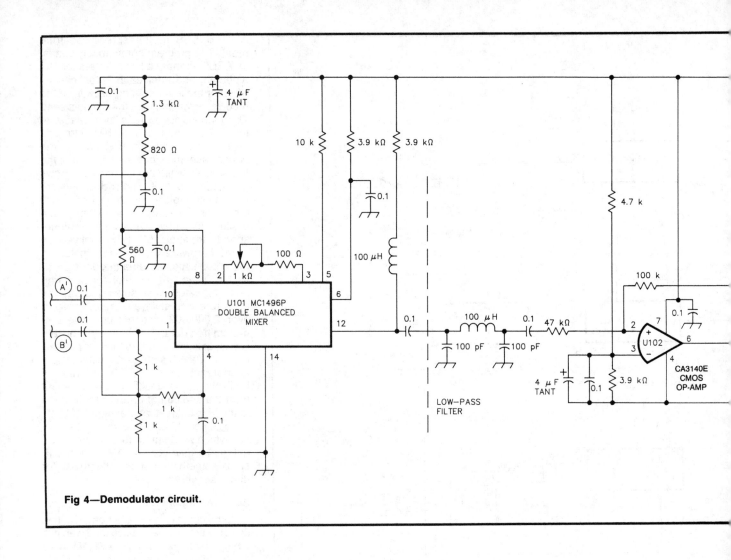

Fig 4—Demodulator circuit.

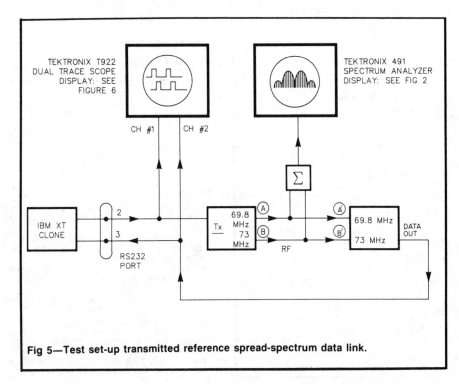

Fig 5—Test set-up transmitted reference spread-spectrum data link.

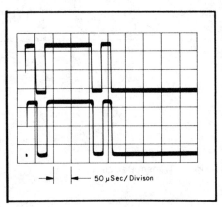

Fig 6—The letter "A" at 38 kilobauds. Upper trace is the transmitter input, lower trace is the receiver output.

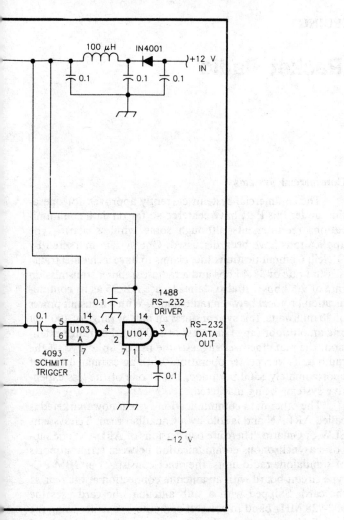

License-Free Spread Spectrum Packet Radio

By Albert G. Broscius, N3FCT
   Distributed Systems Lab, CIS Dept
   University of Pennsylvania
   200 S 33rd St
   Philadelphia PA 19104-6389

ABSTRACT

Under Part 15.126 of CFR Title 47 the FCC has authorized the use of unlicensed spread spectrum transmitters with output power not to exceed one watt in the 902-928 MHz, 2400 MHz, and 5800 MHz bands shared by the amateur service. Several manufacturers have already offered products which have been approved by the FCC for this class of operation. Although this is not ham technology authorized by Part 97 of the FCC rules, it may be advantageous for amateurs to be aware of the use and possibilities of such equipment to augment our regulated packet communications. Conversely, the proliferation of unlicensed transmitters in these amateur shared bands could spell trouble for weak signal work in densely populated areas.

Introduction

Amateur experiments with spread spectrum techniques under an STA have provided a basis for FCC rule changes in 1986 allowing restricted forms of spread spectrum modulation on the amateur bands. Only frequency hopped (FH) and direct sequence (DS) spreading has been authorized, though, prohibiting hybrid spreading techniques which are often preferred in modern designs. The pseudo-noise (PN) code sequences available for DS are also restricted by the amateur rules to a set of three linear feedback shift register (LFSR) sequences. Further, transmitted power is restricted to one-hundred watts and a technical log of all spread spectrum activity must be kept. Unfortunately, work in this area has not yet yielded commercial or even kit-form amateur spread spectrum transceivers in spite of the fine engineering effort spent on this technique by several groups around the country.

Deciding to bring industrial resources to bear on the technology transfer from the military to consumer electronics, the FCC added Part 15.126 of its Rules in June, 1985 in order to speed along the development efforts already authorized for several amateur groups under a 1981 STA. This new section of the low-power communication devices regulations mandates only the power distribution uniformity, the occupied bandwidth for DS transmitters, and the bandwidth and number of hopped channels for FH transmitters. The FCC's attitude toward this technology appears to foster commercial applications which are capable of taking advantage of "wasteland" properties used by Industrial, Scientific, and Medical (ISM) equipment for non-communication purposes and by high EIRP radiolocation systems[4] without undue interference to amateur transmissions on these shared bands. The ability of spread spectrum systems to improve signal-to-noise ratio should enable communication transceivers to overcome the high noise floor resulting from RF sources operating on these bands. Narrowband interference sources such as amateur transmissions on these shared bands would also be combated by a properly designed spread spectrum system.

Commercial Systems

The commercial systems currently approved for operation under this Part have centered so far on data communications requirements although some wireless alarm type applications have been discussed. One system marketed by O'Neill Communications Inc. claims to have a channel transmission rate of 38.4 kbps and a radio-computer transmission rate of 9.6 kbps[1]. It also claims to use AX.25 as its communication protocol between radio nodes. With a transmit power of 20 milliwatts, this system states an indoor range of 100 feet and an outdoor range in excess of 500 feet. For range extension, the manufacturer suggests the use of up to two of the radio units in repeater operation. With an estimated cost of approximately $500 per node, this is one of the less expensive systems being marketed.

The other data communication system now marketed is called ARLAN and is sold by a Canadian firm, Telesystems SLW of Ontario. There are two versions of ARLAN: one supports asynchronous communication between terminal ports of standalone radio units, the other consists of an IBM PC-type circuit board with an antenna connection at the rear of the card. Shipped with a stub antenna, the card uses the 902-928 MHz band to connect computers together using the Novell Netware protocols at 200 kbps and up to 1 watt of transmitted power. Interestingly, the company claims to have tested a pair of the IBM-PC-type machines together with beam antennas successfully at a distance of six miles. The price on these units, however, is approximately $1500 per node which may be prohibitive for use by individuals.

Implications for the Radio Amateur

While it may be disheartening that commercial systems have become available before their amateur counterparts, it should be mentioned that these license-free systems may be used to augment or supplement our communications abilities even though they are not regulated under Part 97 of the Rules. It is also possible that a system which qualifies under 15.126 could be modified to be pursuant to Part 97 spread spectrum rules and thus allowed to operate at the higher power limit, one hundred watts, available for amateur spread spectrum as long as the control operator satisfied all appropriate requirements of the Rules. And of course, placing a 15.126 unit on a Microsat-class vehicle[5] could pave the way for license-free space operation although there may be other restrictions which come into play in that situation.

The design of a power-limited spread spectrum network with realistic inter-node distances would require substantial antenna engineering skills which could be provided by amateur operators familiar with propagation conditions on these bands. However, the resulting network would be free of Part 97 restrictions in the spirit of the pre-Commission Ham activities. Realistically, a Wild West scenario of competing BBS networks and CB-style chaos could make this non-Ham world an unpleasant environment. Unfortunately, unless a pro-active position on this technology is taken, we may see a digital CB world forming around our shared allocations.

Neglecting intentional interference to amateur transmissions and power-limit abuses, there is still the issue of a high noise floor on the weak signal portions of the shared bands. Although these bands now suffer from their shared status, some feel that an influx of consumer electronics items which may each transmit up to one watt will cause unacceptable degradation on the "quiet regions" of the band plan. Considering the possible density to be tens of radiators per city block, the argument of RF pollution seems credible.

Recommendations

To responsibly address this technology, we feel amateur operators should experiment with the commercial systems now available in establishing long distance communication paths using high-gain antenna systems coupled with the maximum legal power of one watt, determining interference levels seen by weak signal receivers attributable to spread spectrum transmissions, and carefully introducing this technology to computer bulletin board operators who could financially support development of an unlicensed computer internet.

References

[1] Information from *Computer Shopper*, September 1989, pp 448-450, relayed to the tcp-group@ucsd.edu Internet mailing list by N6PLO

[2] *Spread Spectrum Communications*, Vol 1, Simon, Omura, Sholtz, and Levitt, Computer Science Press, 1985

[3] *The ARRL 1986 Handbook for the Radio Amateur*, Edited by Mark Wilson, ARRL, 1985

[4] See "NOTE" in Appendix concerning high EIRP radiolocation; See Chapter 38 of [3] concerning the ISM status of these bands.

[5] "AMSAT's MICROSAT/PACSAT Program," Tom Clark, W3IWI, *ARRL Seventh Computer Networking Conference Proceedings*, October 1988

Appendix:
Code of Federal Regulations Title 47: Part 15.126
Operation of spread spectrum systems.

Spread spectrum systems may be operated in the 902-928 MHz, 2400-2483.5 MHz, and 5725-5850 MHz frequency bands subject to the following conditions:

(a) They may transmit within these bands with a maximum peak output power of 1 watt.

(b) RF output power outside these bands over any 100 kHz bandwidth must be 20 dB below that in any 100 kHz bandwidth within the band which contains the highest level of the desired power. The range of frequency measurements shall extend from the lowest frequency generated in the device (or 100 MHz whichever is lower) up to a frequency which is 5 times the center frequency of the band in which the device is operating.

(c) They will be operated on a noninterference basis to any other operations which are authorized the use of these bands under other Parts of the Rules. They must not cause harmful interference to these operations and must accept any interference which these systems may cause to their own operations.

NOTE: Spread spectrum systems using the 902-928 MHz, 2400-2500 MHz and 5725-5850 MHz bands should be cautioned that they are sharing these bands on a noninterference basis with systems supporting critical government requirements that have been allocated the usage of these bands on a primary basis. Many of these systems are airborne radiolocation systems that emit a high EIRP which can cause harmful interference to other users. For further information about these systems, write to: Director, Office of Plans and Policy, U.S. Department of Commerce, National Telecommunications and Information Administration, Room 4096, Washington, D.C. 20230.

Also, future investigations of the effect of spread spectrum interference to Government operations in the 902-928 MHz band may require a future decrease in the power limits.

(d) For frequency hopping systems, at least 75 hopping frequencies, separated by at least 25 kHz, shall be used, and the average time of occupancy on any frequency shall not be greater than four-tenths of one second within a 30-second period. The maximum bandwidth of the hopping channel is 25 kHz. For direct sequence systems, the 6 dB bandwidth must be at least 500 kHz.

(e) If the device is to be operated from public utility lines, the potential of the RF signal fed back into the power lines shall not exceed 250 microvolts at any frequency between 450 kHz and 30 MHz.

[50 FR 25239, June 18, 1985]

Chapter 9

Potential Use of Spread Spectrum Techniques in Non-Government Applications

Walter C. Scales

December 1980

MTR-80W335

CONTRACT SPONSOR:
Federal Communications Commission
CONTRACT: FCC-0320
PROJECT: 14570
DEPT.: W-47

The MITRE Corporation
Metrek Division
1820 Dolley Madison Boulevard
McLean, Virginia 22102

TABLE OF CONTENTS

		PAGE
1.	EXECUTIVE SUMMARY	9-7
1.1	What Are Spread Spectrum Techniques?	9-7
1.2	What Are The Advantages Of Spread Spectrum And How Is It Used?	9-7
1.3	Why Would Spread Spectrum Techniques Be Considered For Non-Government Use?	9-7
1.4	Are Spread Spectrum Systems Efficient In Their Use Of Spectrum Resources?	9-7
1.5	Can The Benefits Of Spread Spectrum Systems Be Achieved With Conventional Technologies?	9-7
1.6	What Non-Government Applications Can Be Expected?	9-7
1.7	What Economic Factors Are Involved?	9-8
1.8	What Are The Risks Of Increased Interference?	9-8
1.9	What Are the Risks With Respect To FCC Monitoring And Enforcement?	9-8
1.10	What Additional Information Is Available On Spread Spectrum?	9-8
2.	INTRODUCTION	9-9
2.1	Study Background and Objectives	9-9
2.2	Report Organization	9-9
2.3	Uses of Spread Spectrum	9-10
2.3.1	Functions	9-10
2.3.2	Ongoing Programs	9-12
2.4	Spread Spectrum Technology: A Brief Overview	9-12
2.4.1	Basic Techniques	9-12
2.4.1.1	Frequency Hopping	9-12
2.4.1.2	Time Hopping	9-16
2.4.1.3	Direct Sequence	9-16
2.4.1.4	Linear FM ('Chirp')	9-16
2.4.2	Discussion	9-16
2.4.3	Implementation Constraints	9-18
2.4.3.1	Synchronization	9-18
2.4.3.2	Matched Filter vs. Correlation Receivers	9-18
2.4.4	Spread Spectrum and Error-Correcting Codes	9-18
3.	THE USE OF SPREAD SPECTRUM IN NON-GOVERNMENT SERVICES	9-20
3.1	Potential Benefits	9-20
3.2	Spread Spectrum in Dedicated Bands	9-20
3.2.1	Efficiency of Spectrum Utilization	9-20
3.2.2	Fast Frequency Hopping in High Capacity Land Mobile Radio Systems	9-23
3.2.2.1	Differential PSK	9-23
3.2.2.2	Multilevel FSK	9-24
3.2.3	Independent Land Mobile Radio Systems	9-24
3.2.4	Spread Spectrum in the Maritime Mobile Service	9-27
3.2.5	Spread Spectrum in Commercial Satellite Communications	9-27
3.3	Spread Spectrum "Overlay" with Conventional Services	9-27
3.3.1	Basic Technical Issues	9-27
3.3.2	Overlay for Land Mobile Service	9-28
3.3.3	Overlay for Maritime Mobile Service	9-28
3.3.4	Overlay with Television	9-29
3.3.5	JTIDS—A Case Study in Band Sharing with Spread Spectrum	9-29
3.3.6	Theoretical Considerations	9-29
3.4	Spread Spectrum in the ISM Bands	9-32
3.4.1	Current Rules and Practices	9-32
3.4.2	Characteristics with Respect to Potential Spread Spectrum Use	9-32
3.5	Alternatives to Spread Spectrum	9-33

4. EXAMPLES OF HYPOTHETICAL IMPLEMENTATIONS ... 9-34
   4.1 Slow Frequency Hopping With FM Voice ... 9-34

   4.1.1 Objectives ... 9-34
   4.1.2 Potential Applications ... 9-34
   4.1.3 Basic Concept ... 9-34
   4.1.4 Equipment Configuration ... 9-34
   4.1.5 Factors Affecting Performance ... 9-39

   4.1.5.1 Noise ... 9-39
   4.1.5.2 Interference ... 9-39
   4.1.5.3 Propagation Factors: Fading and Shadowing ... 9-41
   4.1.5.4 Synchronization Errors ... 9-41
   4.1.5.5 Intrinsic FH Self-Noise ... 9-41
   4.1.5.6 Random FM and Harmonic Distortion ... 9-41

   4.1.6 Address Code Design ... 9-41
   4.1.7 Degree of Privacy ... 9-41
   4.1.8 System Parameters ... 9-43
   4.1.9 Performance and Spectrum Efficiency ... 9-43
   4.1.10 Repeater Operation and Full Duplex Operation ... 9-44

   4.2 Sensor Systems Using Homodyne Detection ... 9-46

5. SPREAD SPECTRUM DEVELOPMENT AND IMPLEMENTATION: COSTS AND RISKS ... 9-49
   5.1 The Risk of Increased Interference ... 9-49

   5.1.1 In-Band Interference ... 9-49
   5.1.2 Out-of-Band Interference ... 9-49

   5.2 Monitoring and Enforcement ... 9-49
   5.3 Development and Implementation Costs ... 9-50

   5.3.1 Fast Frequency Hopping/DPSK ... 9-50
   5.3.2 Fast Frequency Hopping/Multilevel FSK ... 9-51
   5.3.3 Direct Sequence Modems ... 9-51
   5.3.4 SAMSARS ... 9-51
   5.3.5 Slow FH/FM ... 9-52
   5.3.6 Homodyne Detector Applications ... 9-52
   5.3.7 Cost Trends ... 9-52

6. CONCLUSIONS ... 9-53
   6.1 General Comments ... 9-53

   6.1.1 Performance And Efficiency Of Spectrum Utilization ... 9-53

   6.2 Costs and Risks ... 9-54

   6.2.1 Costs ... 9-54
   6.2.2 The Risk Of Increased Interference ... 9-54
   6.2.3 Monitoring And Enforcement ... 9-54

   6.3 Potential Applications ... 9-54
   6.4 Acknowledgments ... 9-54

7. REFERENCES ... 9-56

APPENDIX A: SPECTRUM EFFICIENCY OF SSMA ... 9-59

    A.1 Introduction ... 9-59
    A.2 Direct Sequence ... 9-61

        A.2.1 Synchronous Direct Sequence ... 9-61

            A.2.1.1 Noise-Limited Case ... 9-61
            A.2.1.2 Interference-Limited Case, Restricted Access ... 9-61
            A.2.1.3 Interference-Limited Case, Unrestricted Access ... 9-62

        A.2.2 Asynchronous Direct Sequence ... 9-62

    A.3 Frequency Hopping ... 9-67

        A.3.1 Synchronous FH ... 9-67

A.3.2	Asynchronous FH	9-67
A.4	Performance of Asynchronous DS and FH in Fading	9-70
A.4.1	Direct Sequence	9-70
A.4.2	Frequency Hopping	9-70
A.5	Error-Correction Coding: Fundamental Limitations	9-73
A.6	Discussion	9-74
A.7	References	9-75

APPENDIX B: SPECTRUM EFFICIENCY OF FDMA VS. SSMA NETWORKS: THE IMPACT OF FREQUENCY TOLERANCES AT LOW INFORMATION RATES — 9-76

B.1	Introduction	9-76
B.2	FDMA Notation	9-76
B.3	Asynchronous Direct Sequence SSMA (CDMA)	9-76
B.4	SSMA With Slow Frequency Hopping	9-77
B.5	References	9-78

APPENDIX C: MONTE-CARLO SIMULATION OF LAND MOBILE RADIO WITH SLOW FREQUENCY HOPPING — 9-79

C.1	Objectives	9-79
C.2	Scope	9-79
C.3	Description of Model	9-79
C.3.1	Frequency Hopping	9-79
C.3.2	Sources of Performance Degradation	9-79
C.3.3	Propagation Model	9-79
C.3.4	Antennas and Power Levels	9-80
C.3.5	Basic Simulation Models	9-80
C.3.5.1	Signal-to-Interference Model	9-80
C.3.5.2	Intelligibility Model	9-83
C.3.6	Distribution of Transmitter-to-Receiver Distances	9-83
C.3.7	Output Format	9-83
C.4	Simulation Results	9-87
C.4.1	Signal-to-Interference Simulation Model	9-87
C.4.2	Intelligibility Model	9-101
C.4.2.1	Mobile-to-Base Operation	9-101
C.4.2.2	Mobile-to-Mobile Operation (Without Repeaters)	9-109
C.4.2.3	Base-to-Mobile Operation	9-109
C.4.3	The Impact of Background Noise and Fading	9-109
C.4.4	Summary of Simulation Results	9-109
C.5	References	9-123

LIST OF ILLUSTRATIONS

		PAGE
TABLE 1:	ISM Band Allocations And Uses	9-31
TABLE 2:	Typical System Parameters	9-43
TABLE 3:	Estimated Maximum Number Of Users Per Channel For Interference-Limited Operation	9-44
TABLE 4:	Estimated Relative Costs of Six Spread Spectrum Configurations	9-51
TABLE C-1:	Propagation Parameters	9-79
TABLE C-2:	List Of Simulation Runs (Intelligibility Model)	9-101
TABLE C-3:	Simplified Mobile-To-Base Link Budget For Quiet Receiver Location	9-109
TABLE C-4:	Estimated Maximum Number of Users Per Channel For Interference-Limited Operation	9-123
FIGURE 1:	Response of Idealized Multipath Channe	9-11
FIGURE 2:	Simplified Frequency Vs. Time Trajectory Of A Frequency Hopping Carrier	9-13
FIGURE 3:	Basic Frequency Hopping System With Waveforms	9-14
FIGURE 4:	Effect of Narrowband Interference On A Frequency Hopping System (Simplified)	9-15
FIGURE 5:	Overall Direct Sequence System Showing Waveforms	9-17
FIGURE 6:	Typical Matched Filter (Passive Correlator) For Direct Sequence Spread Spectrum Reception	9-19
FIGURE 7:	Spectrum Efficiency of Asynchronous Direct Sequence SSMA	9-22
FIGURE 8:	FH/FSK Transmitter	9-25
FIGURE 9:	FSK/FH Receiver	9-26
FIGURE 10:	Composite JTIDS Spectrum	9-30
FIGURE 11:	Hypothetical Frequency Plans For Frequency Hopping Mobile Communication	9-35
FIGURE 12:	Simplified Block Diagram Of A Conventional Half-Duplex Synthesizer-Driven Transceiver	9-36
FIGURE 13:	Basic Design Modification For Frequency Hopping	9-37
FIGURE 14:	FH Controller	9-38
FIGURE 15:	Simplified State Diagram For FH Transceiver	9-39
FIGURE 16:	Interference In A FH System	9-40
FIGURE 17:	Simplified Timing Diagram Showing The Impact Of Synchronization Errors	9-42
FIGURE 18:	Maximum Spectrum Utilization As A Function Of Communication Distance For Interference-Limited Systems	9-45
FIGURE 19:	Conventional Homodyne System	9-47
FIGURE 20:	Range-Limited, Interference-Resistant Homodyne System (FH Version)	9-48
FIGURE A1:	Spectrum Efficiency Of Synchronous Direct Sequence SSMA: Number Of Assigned Codes Greater Than Number Of Simultaneous Users	9-63
FIGURE A2:	Spectrum Efficiency Of Synchronous Direct Sequence SSMA: Number Of Codes = Number Of Users	9-64
FIGURE A3:	Spectrum Efficiency Of Asynchronous Direct Sequence SSMA	9-66
FIGURE A4:	Typical Frequency Hopping Sequence	9-68
FIGURE A5:	Spectrum Efficiency Of Frequency Hopping SSMA Spectrum Efficiency	9-69
FIGURE A6:	Spectrum Efficiency Of Asynchronous Direct Sequence SSMA With Fading	9-71
FIGURE A7:	Spectrum Efficiency Asynchronous Frequency Hopping In Rayleigh Fading	9-72
FIGURE C1:	Signal-To-Interference Simulation: Simplified Flow Chart	9-81
FIGURE C2:	Simplified Flow Chart Of Intelligibility Simulation	9-82
FIGURE C3:	Empirical Basis Of Intelligibility Model	9-84
FIGURE C4:	Radial Density For Uniform Deployment Of Interferers	9-85
FIGURE C5:	Density Of Distance From Receiver To Intended Transmitter	9-86
FIGURE C6:	Cumulative Distribution Of R.F. Signal-To-Interference Ratio: 1 Channel, 1 Interferer	9-88
FIGURE C7:	Cumulative Distribution Of R.F. Signal-To-Interference Ratio: 1 Channel, 2 Interferer	9-89
FIGURE C8:	Cumulative Distribution Of R.F. Signal-To-Interference Ratio: 1 Channel, 4 Interferers	9-90
FIGURE C9:	Cumulative Distribution Of R.F. Signal-To-Interference Ratio: 100 Channels, 5 Interferers	9-91
FIGURE C10:	Cumulative Distribution Of R.F. Signal-To-Interference Ratio: 100 Channels, 10 Interferers	9-92
FIGURE C11:	Cumulative Distribution Of R.F. Signal-To-Interference Ratio: 100 Channels, 20 Interferers	9-93
FIGURE C12:	Cumulative Distribution Of R.F. Signal-To-Interference Ratio: 100 Channels, 40 Interferers	9-94
FIGURE C13:	Cumulative Distribution Of R.F. Signal-To-Interference Ratio: 100 Channels, 80 Interferers	9-95
FIGURE C14:	Cumulative Distribution Of R.F. Signal-To-Interference Ratio: 100 Channels, 160 Interferers	9-96
FIGURE C15:	Cumulative Distribution Of R.F. Signal-To-Interference Ratio: 100 Channels, 200 Interferers	9-97

FIGURE C16:	Probability Of Achieving A 1- dB Carrier-To-Interference Ratio With M Interferers	9-98
FIGURE C17:	Probability Of Achieving A 10-dB C/I As A Function Of Traffic Intensity, 1 Channel	9-99
FIGURE C18:	Probability Of Achieving A 10-dB C/I As A Function Of Traffic Intensity, 100 Channels	9-100
FIGURE C19:	Cumulative Distribution Of PB Word Articulation: 5 Interferers, 10/30 Splatter, 8 KM Mean Comm. Dist., 1 HOPS/S, 200 Trials, Mobile-To-Base	9-102
FIGURE C20:	Cumulative Distribution Of PB Word Articulation: 5 Interferers, 10/30 Splatter, 8 KM Mean Comm. Dist., 20 HOPS/S, 200 Trials, Mobile-To-Base	9-103
FIGURE C21:	Cumulative Distribution Of PB Word Articulation: 20 Interferers, 10/30 Splatter, 8 KM Mean Comm. Dist., 20 HOPS/S, 200 Trials, Mobile-To-Base	9-104
FIGURE C22:	Cumulative Distribution Of PB Word Articulation: 40 Interferers, 10/30 Splatter, 8 KM Mean Comm. Dist., 20 HOPS/S, 200 Trials, Mobile-To-Base	9-105
FIGURE C23:	Cumulative Distribution Of PB Word Articulation: 20 Interferers, 10/30 Splatter, 4 KM Mean Comm. Dist., 20 HOPS/S, 200 Trials, Mobile-To-Base	9-106
FIGURE C24:	Cumulative Distribution Of PB Word Articulation: 40 Interferers, 10/30 Splatter, 4 KM Mean Comm. Dist., 20 HOPS/S, 50 Trials, Mobile-To-Base	9-107
FIGURE C25:	Cumulative Distribution Of PB Word Articulation: 40 Interferers, 200/200 Splatter, 4 KM Mean Comm. Dist., 20 HOPS/S, 50 Trials, Mobile-To-Base	9-108
FIGURE C26:	Cumulative Distribution Of PB Word Articulation: 80 Interferers, 200/200 Splatter, 4 KM Mean Comm. Dist., 20 HOPS/S, 50 Trials, Mobile-To-Base	9-110
FIGURE C27:	Cumulative Distribution Of PB Word Articulation: 120 Interferers, 200/200 Splatter, 4 KM Mean Comm. Dist., 20 HOPS/S, 50 Trials, Mobile-To-Base	9-111
FIGURE C28:	Cumulative Distribution Of PB Word Articulation: 80 Interferers, 60/80 Splatter, 8 KM Mean Comm. Dist., 20 HOPS/S, 50 Trials, Mobile-To-Base	9-112
FIGURE C29:	Cumulative Distribution Of PB Word Articulation: 80 Interferers, 60/80 Splatter, 4 KM Mean Comm. Dist., 20 HOPS/S, 50 Trials, Mobile-To-Base	9-113
FIGURE C30:	Cumulative Distribution Of PB Word Articulation: 100 Interferers, 60/80 Splatter, 2 KM Mean Comm. Dist., 20 HOPS/S, 50 Trials, Mobile-To-Base	9-114
FIGURE C31:	Cumulative Distribution Of PB Word Articulation: 130 Interferers, 60/80 Splatter, 2 KM Mean Comm. Dist., 20 HOPS/S, 50 Trials, Mobile-To-Base	9-115
FIGURE C32:	Cumulative Distribution Of PB Word Articulation: 160 Interferers, 60/80 Splatter, 2 KM Mean Comm. Dist., 20 HOPS/S, 50 Trials, Mobile-To-Base	9-116
FIGURE C33:	Cumulative Distribution Of PB Word Articulation: 20 Interferers, 60/80 Splatter, 16 KM Mean Comm. Dist., 20 HOPS/S, 50 Trials, Mobile-To-Base	9-117
FIGURE C34:	Cumulative Distribution Of PB Word Articulation: 40 Interferers, 60/80 Splatter, 16 KM Mean Comm. Dist., 20 HOPS/S, 50 Trials, Mobile-To-Base	9-118
FIGURE C35:	Cumulative Distribution Of PB Word Articulation: 60 Interferers, 60/80 Splatter, 16 KM Mean Comm. Dist., 20 HOPS/S, 50 Trials, Mobile-To-Base	9-119
FIGURE C36:	Cumulative Distribution Of PB Word Articulation: 60 Interferers, 60/80 Splatter, 8 KM Mean Comm. Dist., 20 HOPS/S, 50 Trials, Mobile-To-Base	9-120
FIGURE C37:	Cumulative Distribution Of PB Word Articulation: 80 Interferers, 60/80 Splatter, 4 KM Mean Comm. Dist., 20 HOPS/S, 50 Trials, Mobile-To-Base	9-121
FIGURE C38:	Cumulative Distribution Of PB Word Articulation: 80 Interferers, 60/80 Splatter, 2 KM Mean Comm. Dist., 20 HOPS/S, 50 Trials, Mobile-To-Base	9-122

1. EXECUTIVE SUMMARY

This summary presents the important features of the report in concise, nontechnical terms. In order to provide for quick, easy reference, the individual paragraphs of the Executive Summary are written in question-and-answer form. The reader who is interested in detailed technical results should turn to Section 2.2, *Report Organization*, for guidance.

1.1 What are Spread Spectrum Techniques?

The term "spread spectrum" has been applied to a wide variety of electronic systems and techniques. The common thread uniting these diverse areas of technology is that each uses signals requiring significantly more radio frequency bandwidth than a conventional signal would require. The expanded bandwidths provide certain features and characteristics that would otherwise be difficult or expensive to attain. Spread spectrum techniques are not new; they have been evolving since the late 1940s.

1.2 What are the Advantages of Spread Spectrum and How is it Used?

Spread spectrum techniques have evolved primarily in military environments, where they have been used to provide one or more of the following features:
- Resistance to Jamming
- Resistance to Unintentional Interference
- Resistance to Unauthorized Interception
- Sharing of a Common Radio Frequency Band by Multiple Users
- Discrete Addressing
- Accurate Distance or Location Measurements
- Pulse Compression

A more complete explanation of these features can be found in Section 2.3.1.

1.3 Why would Spread Spectrum Techniques be Considered for Non-Government Use?

Aside from the threat of intentional jamming, many of the features and characteristics of spread spectrum have found potential applications outside of the military environment. These include nonmilitary aerospace applications now in the planning stages and several non-Government applications that have been proposed, but not implemented. (The term "non-Government," as used in this report, refers to applications outside the Federal Government for which FCC authorization would be required.) Examples of potential non-Government applications are presented in Sections 3 and 4.

1.4 Are Spread Spectrum Systems Efficient in Their Use of Spectrum Resources?

Since bandwidth is an important measure of the spectrum resources used by a signal, the idea that a given signal should occupy more bandwidth, rather than less, runs against intuition and existing regulatory policy. Indeed, if each of many spread spectrum signals were allowed to occupy a separate, dedicated band, the resulting waste of spectrum resources would be unthinkable. However, because spread spectrum systems are resistant to interference, several spread spectrum signals can co-exist simultaneously in a single, common band.

The question of whether this mode of spectrum-sharing is more effective than the existing frequency-channelized approach is not simple. Certainly, it is easy to produce examples in which spread spectrum systems make poor use of the spectrum, even when as many signals as possible share a common bandwidth. In some applications, spread spectrum signals suffer a fundamental, theoretical disadvantage when compared to frequency-channelized signals on the basis of spectrum utilization.

But the usual mode of sharing the radio spectrum on the basis of frequency channelization and geographic separation has practical limitations that prevent it from achieving the degree of efficient spectrum utilization predicted by simple theoretical models. These limitations include the need for guard bands between channels, the difficulty of maintaining a uniform demand or traffic loading on all channels, and the need for geographic gaps between regions in which a particular channel is re-used. For this reason, it is possible to produce examples in which spread spectrum techniques make more effective use of the spectrum than the conventional frequency-channelized approach. Naturally, it is also possible to improve the spectrum efficiency of the existing frequency-channelized approach without resorting to spread spectrum techniques.

When comparing practical implementations of spread spectrum and conventional narrowband technologies, it becomes evident that both approaches require a compromise between communication performance (or radiolocation performance, in the case of radio-location systems) and efficiency of spectrum utilization. Increasing the number of users in a given band increase the utilization of the spectrum, but also increases interference, thus degrading performance. For this reason, such comparisons must be carefully formulated if the results are to be meaningful.

1.5 Can the Benefits of Spread Spectrum Systems be Achieved With Conventional Technologies?

In some cases, some of the advantages of spread spectrum systems can be achieved by other techniques. One example is resistance to unauthorized interception, which is readily achievable with conventional secure voice or data encryption techniques. On the other hand, spread spectrum techniques offer a unique method of sharing a common band between multiple users without requiring the users to coordinate their transmissions in any way. Spread spectrum systems have features and characteristics that differ from those of conventional narrowband systems both qualitatively and quantitatively. These distinct features and characteristics are naturally better-suited to some applications than to others.

1.6 What Non-Government Applications Can Be Expected?

In the near-term, the applications that are most likely to engender practical implementations are those that require little development and offer attractive costs and performance characteristics. Most existing spread spectrum equipment was developed for military and aerospace applications. The functional and technical requirements for such applications are usually more severe than the corresponding requirements for commercial equipment. As a result, much of the existing spread spectrum equipment will be beyond the economic reach of commercial users, even if the equipment characteristics happen to be compatible with the application. On the other hand, commercial versions of existing designs could appear if the manufacturers perceive substantial markets.

In recent years, significant attention has been given to spread spectrum implementations of land mobile radio systems. However, the designs that have been proposed would require very substantial development efforts, and are not likely to be implemented in the foreseeable future. This report contains, in Section 4.1, a brief description of a simple spread spectrum technique for land mobile radio that could be implemented with significantly less development effort. This approach may entail certain performance compromises as a result of its simplicity, and thus may not be suitable for some applications. The performance and spectrum utilization aspects of this simplified approach are addressed in Section 4.1 and Appendix C.

A second application that has received considerable attention is the use of spread spectrum signals for distress alerting via satellite in the maritime mobile service. Although an experimental system has been developed and partially tested, implementation before the late 1980s is considered unlikely. A brief description of the proposed system is presented in Section 3.2.4.

There are probably a number of potential non-Government applications of spread spectrum technology that have not received serious attention simply because designers of commercial equipment are generally not well-versed in this area. A hypothetical example of such an application is described in Section 4.2.

1.7 What Economic Factors are Involved?

Most spread spectrum systems can be viewed as conventional narrowband systems to which special features have been added. This added level of complexity yields new performance characteristics at an increased cost. Clearly, the costs and benefits can be evaluated only on a case-by-case basis. However, the overall trend for the technologies used in spread spectrum systems has been one of decreasing costs during recent years. Non-Government spread spectrum systems, if developed, could be significantly less expensive than their military counterparts. Section 5.3 provides a preliminary assessment of the relative costs of several spread spectrum techniques that have been proposed for non-Government applications.

1.8 What Are the Risks of Increased Interference?

Before addressing this question directly, it is important to realize that there are several ways in which spread spectrum frequency assignments might be made. In the simplest case, a particular band might be set aside for exclusive use by spread spectrum signals in a particular geographic area. This approach might be used, for example, when moderate bandwidths are required at microwave frequencies. The alternative is to "overlay" spread spectrum signals on bands that are already in use. This approach would rely on the interference resistance and low spectral density of spread spectrum systems to minimize mutual interference.

Current users of the spectrum are likely to be concerned about the risk of interference from spread spectrum signals, especially if the "overlay" approach to spectrum management is adopted. On the other hand, the ever-increasing demand for spectrum resources will make it increasingly difficult to find suitable "dedicated" bands for exclusive use by spread spectrum systems. One potential solution is to provide frequency allocations for spread spectrum systems in the Industrial, Scientific, and Medical (ISM) bands. With one exception, current users of these bands are not protected from unintentional interference. The technical aspects of potential spread spectrum operation in the ISM bands are presented in Section 3.4.

Regardless of the approach to spectrum management, spread spectrum systems are likely to suffer interference from other spread spectrum systems. The inherent interference resistance of spread spectrum techniques will make this situation acceptable in many cases. The performance of spread spectrum systems in the presence of various types of interference has been extensively studied and can be reliably predicted given the system parameters and propagation characteristics.

1.9 What Are the Risks With Respect to FCC Monitoring and Enforcement?

This issue encompasses several facets, including the ability to monitor message content and operational procedures, the ability to measure and monitor the technical characteristics of signals, the ability to collect statistical data on spectrum utilization, and the ability to identify a particular station as the source of a particular signal.

Individual spread spectrum signals are normally difficult to detect with conventional receivers, except within a relatively short distance of the transmitting antenna. Even if they are detectable, the spread spectrum signals will probably not be intelligible on a conventional receiver. Thus, the question of monitoring and enforcement is not trivial.

Nevertheless, the FCC has recourse to regulatory measures that could eliminate or substantially mitigate this potential problem. Some of these measures are listed in Section 5.2. Within the proper regulatory framework, monitoring and enforcement should be no more difficult with spread spectrum signals than with conventional secure voice or encrypted data signals.

1.10 What Additional Information Is Available on Spread Spectrum?

The body of technical literature on spread spectrum techniques is extensive. Of several tutorial papers that have been published, the most recent can be found in the September 1978 issue of the IEEE Communications Society Magazine. (See Reference 9 in Section 7 of this report.) A textbook entitled *Spread Spectrum Systems* is also available (Reference 1).

2. INTRODUCTION

2.1 Study Background and Objectives

The demand for radio spectrum from virtually all sectors of society has grown steadily and is expected to continue to grow in the future. Against a background of limited spectrum resources, the established method of dealing with this steady pressure of increasing demand involves a carefully formulated combination of regulatory and technical measures. Such measures have included, among others:

- shifts to higher frequency bands,
- use of closer channel spacings (often coupled with more stringent frequency tolerance or narrower limits on occupied bandwidth, or both), and
- the use of more efficient modulation techniques.

Such measures have been introduced carefully with the intent of minimizing the economic burden on the user groups. Although most efforts at improving spectrum utilization have been directed at increasing the efficiency of frequency division channelization, other channelization techniques have been successfully implemented using time division or other methods. The efficient sharing of a given bandwidth for communication on a time-division basis requires that transmissions from the various users be synchronized. This approach has been used, for example, in sharing a satellite repeater. A common bandwidth can also be shared on a time basis without a network synchronization protocol (i.e., by random access), but this results in interference between users and is generally effective only at low values of channel utilization, that is, when the common bandwidth is unused most of the time. Sharing on the basis of both time and frequency has been demonstrated in "trunked" land mobile systems. Again, coordination between users is required.

Another method of sharing a common bandwidth between multiple users is code division multiple access (CDMA), which has also been known as spread spectrum multiple access (SSMA). In this approach, as in time division multiple access (TDMA), users occupy a common bandwidth. In CDMA, however, multiple users may occupy the common bandwidth simultaneously. The signals from the various users are separable because each is modulated with a unique underlying code, and the various user codes are very nearly orthogonal to one another. Synchronization, in a network sense, is not required as in TDMA.* In fact, the only coordination that is required in such a system is a simple protocol between pairs of users who are (or want to be) in mutual communication.

The idea that such techniques could be used to improve the efficiency of spectrum utilization and to provide new capabilities is not original. The basis for spread spectrum communication dates back to the late 1940s, and civilian applications were discussed during the 1950s. The body of technical literature on the subject is extensive, and much of it is unclassified, despite the fact that spread spectrum technology evolved in a military environment. But high equipment cost and a conservative regulatory environment for years prevented the serious consideration of non-Government spread spectrum applications. Both of these factors are changing. The advent of low-cost large scale integrated (LSI) circuits has driven the cost of communication equipment down to the point where frequency synthesizers are commonly used in citizens band equipment, including inexpensive hand-held portable transceivers. This would have been economically unthinkable twenty years ago. Entire receivers (less tuned circuits and electromechanical components) can be built on a single "chip." Whether or not this trend will continue at anything like its previous pace is beside the point. Short of an economic catastrophe, the trend is not likely to be significantly reversed. Furthermore, the current regulatory climate is one in which innovation is encouraged, subject to common-sense caveats.

It is clear, then, that the time has come for a fresh look at spread spectrum technology as it might be applied under the FCC's regulatory domain. A certain amount of audacity would be needed in order to imagine that all such potential applications could be examined in a single study. The purpose here is to present, in concise form, a number of potential applications that have been described prior to this study and to suggest a few new potential applications that have not been previously considered for non-Government use. The benefits, costs, and risks associated with these various potential applications are also presented.

In any effort of this sort, it is inevitable that a focus will develop on a limited area, to the detriment or exclusion of other areas. Much of the work that has been performed on military and aerospace applications of spread spectrum technology will go unmentioned, simply because there is so much of it. The interested reader can pursue these areas with the aid of the list of references presented in Section 7 of this report, or through the bibliography presented in Dixon's textbook on spread spectrum systems (1).

2.2 Report Organization

This report is organized into six major sections, plus references and appendixes. The *Executive Summary* (Section 1) is a primarily nontechnical presentation designed to convey the results of this study in concise form. The *Introduction* (Section 2) is intended to provide the reader with a working vocabulary of terms that are normally associated with spread spectrum technology, as well as a general feeling for the contexts within which that technology normally arises. *Section 3* describes the potential motivations for using spread spectrum techniques under the FCC's regulatory domain and summarizes various technical approaches that have been proposed or studied for such applications. *Section 4* provides two hypothetical examples of non-Government spread spectrum implementations which have not been previously studied. *Section 5* describes the potential costs and risks of FCC-authorized spread spectrum implementations. The study conclusions are summarized in *Section 6*. References are presented as *Section 7*. Finally, the purely technical and analytical aspects of this study are presented in three appendixes.

Appendix A describes the theoretical spectrum efficiencies attainable with spread spectrum multiple access under various conditions. In *Appendix B*, the impact of frequency tolerances on the spectrum efficiency of frequency-channelized systems is examined as a basis of comparison for spread spectrum multiple access systems. *Appendix C* presents the development and results of a Monte-Carlo simulation of a hypothetical spread-spectrum implementation described in Section 4.

---

* However, synchronization on individual links is necessary whenever information is actually being communicated on those links.

2.3 Uses of Spread Spectrum

2.3.1 Functions

The term "spread spectrum" has been used to describe a variety of techniques and applications. The common thread that unites these diverse areas of technology is the use of signals having bandwidths that are far in excess of the bandwidth of the underlying information. The signals have a "fine structure" that is used by the receivers to separate them from interference, jamming, and other undesired components that may be present at the receiver input. Spread spectrum techniques have evolved primarily in military environments, where they have been used to provide one or more of the following features.

Resistance to Intentional Jamming. Potential jammers span the range from crude radio frequency generators to sophisticated devices that imitate the desired signals, either by amplifying and retransmitting them (repeater jamming) or by discovering and using their "fine structure." Although no system is completely immune to jamming, a properly designed spread spectrum system can make it so expensive for the enemy to jam effectively that he will consider conventional military alternatives (e.g., destroying the transmitter or the receiver, assuming that he can find them).

Resistance to Unintentional Interference. Clearly, any system that is resistant to intentional jamming will also be resistant to unintentional interference from other communication signals, radar signals, spurious emissions, ignition noise, and other sources. Again, the resistance to unintentional interference is not absolute, but is typically high compared with conventional narrowband signaling.

Resistance to Interception. In some military applications, it is important that the existence (and location) of a transmitter not be discovered. Conventional narrowband signals can be detected, identified, and triangulated upon, even by an enemy having access to only crude and inexpensive equipment. Spread spectrum signals, on the other hand, can occupy such a large bandwidth that the average signal-to-noise ratio (measured in the signal bandwidth) is significantly less than unity, except within a small area in the immediate vicinity of the transmitting antenna. Under these conditions, the difficulty of detecting the signal is greatly increased.

Even in cases where it is possible to detect the presence of a spread spectrum signal, some degree of security can be provided by designing the signal so that it cannot be "decoded" by the casual listener. The possibility that messages will be recorded and played back with the intent of creating confusion or deception is also minimized.

Low Spectral Density. Interference to narrowband receivers from co-channel signals is minimized if the interference power per unit bandwidth is made small. This end can be achieved by spreading the fixed (spread spectrum) interference power over the widest possible bandwidth. The resulting interference to a narrowband receiver may still be greater than that caused by an adjacent-channel narrowband signal, but will be significantly lower than that caused by a co-channel narrowband signal of equal power.

Multipath Resistance. "Multipath" is a term that is used to refer to the existence of more than one propagation path between a transmitter and a receiver. Signals traversing the various paths arrive at the receiving antenna with random phase angles, so that at any moment these signal components may add constructively or they may cancel one another, leaving the total received signal far below its "normal" or "average" level. Such fading effects can degrade the performance of narrowband communication systems. In addition, the existence of multipath propagation can significantly degrade the performance of narrowband ringing or radio-location systems.

If the various propagation paths are all of different lengths, they can be separated by using signals of sufficiently large bandwidth. As a trivial example, consider a system in which information is transmitted in discrete pulses, each of which is much shorter than the difference between the propagation delays associated with any two paths. In this case, the receiver will "see" a sequence of pulses (one for each propagation path) in response to every transmitted pulse (see Figure 1). If the interval between transmitted pulses is sufficiently long, no two received pulses will overlap, so the problem of multipath fading will be eliminated. The receiver may use the first received pulse in each sequence, or the strongest pulse, or it may attempt to combine the pulses in such a way that fading is mitigated or eliminated. Spread spectrum techniques can also be effective for combating diffuse multipath, in which individual paths are not separable.

Of course, narrowband techniques have also been developed to mitigate multipath fading. The most familiar of these are "diversity" techniques, which include frequency diversity (where two or more narrowband signals are transmitted on separate channels and recombined at the receiver) and space diversity (in which two or more separate receiving antennas are used). Time diversity and polarization diversity techniques have also been used.

Multiple Access. This term traditionally refers to the use of a satellite repeater by multiple signals originating from transmitters that are spatially dispersed. In a more general context, it applies when access to a channel (or group of channels) is involved, whether or not there is a repeater. The various signals may be separated in frequency (frequency division multiple access or FDMA) or time (time division multiple access or TDMA). A third multiple access technique involves the use of the entire repeater bandwidth continuously by each signal. Because the various signals overlap completely in both time and frequency, the receiver can separate the "desired" signal from the other signals only if each signal is carried by an underlying waveform that is very nearly orthogonal to all of the other signal waveforms. (Two waveforms are said to be orthogonal if their product, integrated or averaged over some fixed time period, is zero). In general, this arrangement is possible only when the signal bandwidth greatly exceeds the information bandwidth for each signal. Thus, the term "spread spectrum multiple access" (SSMA) is often used to describe this approach. Because digital codes are usually employed to form the underlying quasi-orthogonal waveforms, the term "code division multiple access" (CDMA) is also used. Such a system has the characteristic that it displays no distinct "saturation point" as the number of users is increased. Rather, the signal-to-interference ratio experienced by each user continuously declines as more and more users enter the system.

Another advantage is that the various users need not coordinate their transmissions, but simply transmit at will. SSMA techniques have the disadvantage that they often require greater transmitter power and greater bandwidth per user (for a given information rate) than FDMA or TDMA techniques.

Discrete Addressing. For some applications, it is desirable to code individual signals in such a way that they can

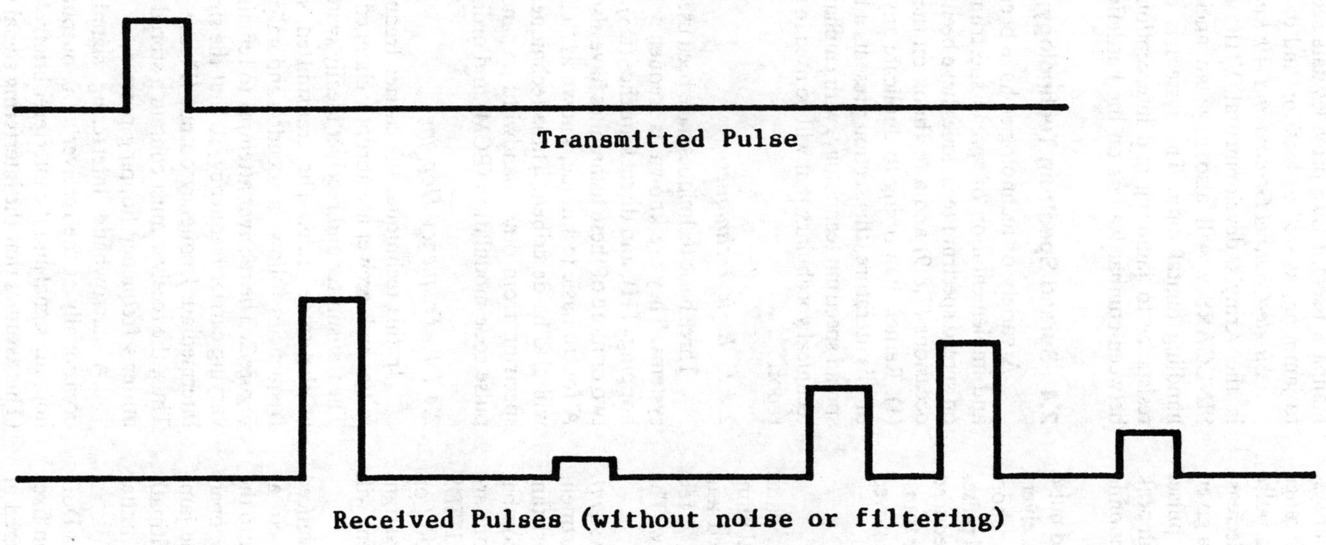

**FIGURE 1
RESPONSE OF IDEALIZED MULTIPATH CHANNEL**

Transmitted Pulse

Received Pulses (without noise or filtering)

be received only at their intended destinations. A spread spectrum system will provide this feature if each user is assigned a unique code that is employed to produce the underlying "fine structure" of his particular signal. The level of security provided by this approach is not normally comparable with cryptography, but protection from the casual listener can be reasonably assured.

Pulse Compression and Ranging. Although this application is not necessarily associated with "spread spectrum," the same underlying techniques are involved. Many radar systems transmit single, unmodulated pulses. The intrinsic range resolution in such systems improves with decreasing pulse width, but the signal-to-noise ratio also decreases with decreasing pulse width, since decreasing the pulse width, for a given peak pulse power, decreases the total energy of the pulse. Good range accuracies can be attained by using very high peak pulse power, but this imposes a corresponding economic burden.

An alternative approach is to design the transmitted pulse so that its width can be significantly reduced at the receiver. (Hence the term "pulse compression.") The resulting short pulse contains all of the energy of the longer transmitted pulse, but has the intrinsic range resolution implicit in the reduced pulse width. Again, the basic approach entails giving the transmitted pulse a "fine structure" that can be efficiently removed at the receiver.

2.3.2 Ongoing Programs

The spread spectrum concept, in its basic form, dates back to the late 1940s (1). Since that time, a substantial number of spread spectrum systems have been developed and implemented. Others are currently under development. A few of the more well-publicized applications are listed below. The list is by no means exhaustive.

JTIDS (Joint Tactical Information Distribution System). This system is currently being developed by the Department of Defense (DoD) to provide secure, jamming-resistant communications, navigation and identification for combat elements such as aircraft. The communication functions are digital, but a provision is made for digitized voice (2). JTIDS is of particular interest because its spectrum is "overlaid" on existing (60-1215 MHz aeronautical radionavigation assignments (3). The details of the band-sharing arrangement will be described in Section 3.3.5.

Packet Radio is a project of the Defense Advanced Research Projects Agency (DARPA) that is aimed at extending a technology known as "packet switching" into the area of radio communications. Spread spectrum signals, operating in the 1710-1850 MHz band, are used to provide jamming resistance, security, multiple access, and multipath protection (4). An experimental system has been under evaluation for several years.

Global Positioning System (NAVSTAR/GPS) is a DoD effort that will provide accurate satellite-derived location fixes anywhere in the world. Spread spectrum signals centered at 1227 MHz and 1575 MHz provide the basis for range measurements and data transfer. The features provided by the spread spectrum signal format include resistance to jamming and interference, multiple access, good range resolution, multipath rejection, and security (5). The system is currently under development.

Tracking and Data Relay Satellite System (TDRSS) is a NASA program aimed at providing improved control, communication, and ranging for low-orbiting satellites.

Spread spectrum signals from the user satellites (including the space shuttle) are relayed to ground stations via TDRS. The spread spectrum signal format provides resistance to multipath and unintentional interference, multiple access, and good range resolution (6). TDRSS is scheduled to become operational in 1981.

Position Location Reporting System (PLRS) is an Army/Marine Corps UHF system designed to provide digital data and position location for land vehicles, aircraft, and manpack users. PLRS provides data security and resistance to jamming, as well as location and data communications.

Single-Channel Ground and Air Radio (SINCGARS-V) is the Army's developmental VHF voice radio system. SINCGARS-V will also have an ancillary capability for handling digital data. The system is designed to provide resistance to jamming and interception in communication between combat leaders on the battlefield.

2.4 Spread Spectrum Technology: A Brief Overview

A variety of technologies have been used in the design and implementation of spread spectrum systems. Tutorials on spread spectrum techniques have been published on several occasions (7,8,9) and a textbook on the subject is available (1). Rather than trying to duplicate any significant amount of this literature, this section presents a brief introduction to spread spectrum technology, with emphasis on developing the technical vocabulary that will be used in the remainder of the report.

2.4.1 Basic Techniques

Three basic techniques have been used in spread spectrum systems. They are known as frequency hopping (FH), time hopping (TH), and direct sequence (DS). Hybrids employing two or more of these techniques have also been implemented. A fourth basic technique, known as "chirp," or linear FM, will also be described. These techniques distinguish spread spectrum from other bandwidth-expanding techniques like pulse code modulation (PCM) and ordinary wideband FM.

2.4.1.1 Frequency Hopping

In this technique, the carrier frequencies of the transmitter and receiver are abruptly changed at regular intervals. The transmitter and receiver ideally switch carrier frequency at the same time. The transmitted sequence of carrier frequencies follows a pseudo-random pattern, as shown in Figure 2. The receiver attempts to remain tuned to this time-varying carrier frequency, so that the signal at the receiver's intermediate frequency output (Figure 3) is "de-hopped." Thus, the receiver must contain a stored replica of the transmitter's frequency hopping pattern.

A narrowband interfering signal will appear only occasionally at the receiver's i.f. output, as the frequency-hopping signal and the interferer briefly "collide" (Figure 4). (This assumes that the interfering signal is not so strong that it is still measurable after being attenuated in the stop-band of the receiver's i.f. filter.) If there are N separate hopping frequencies, each being used for an equal fraction of the transmission time, then the interference power, averaged over all hops, is reduced by a factor of N.

In principle, information can be modulated onto the transmitted carrier in any convenient form, since the carrier is de-hopped at the receiver. In practice, however, it is not always possible to maintain a constant or predictable carrier

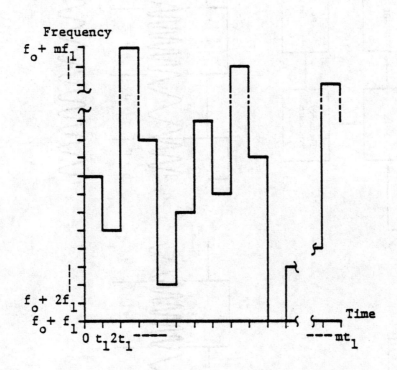

**FIGURE 2
SIMPLIFIED FREQUENCY VS. TIME TRAJECTORY
OF A FREQUENCY HOPPING CARRIER**

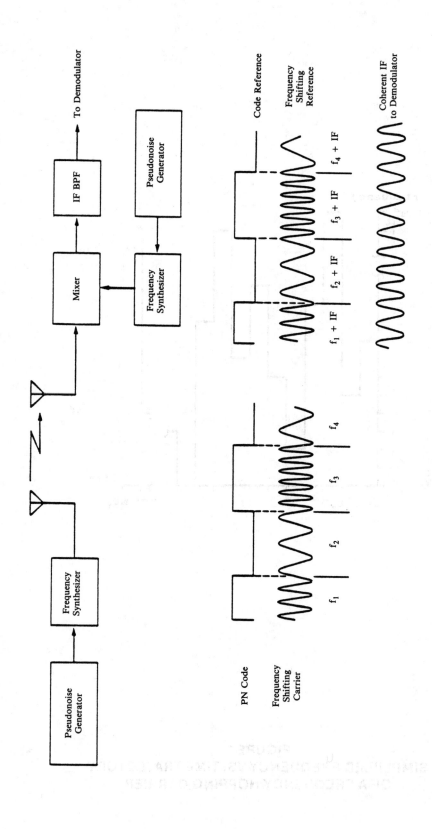

**FIGURE 3
BASIC FREQUENCY HOPPING SYSTEM WITH WAVEFORMS**

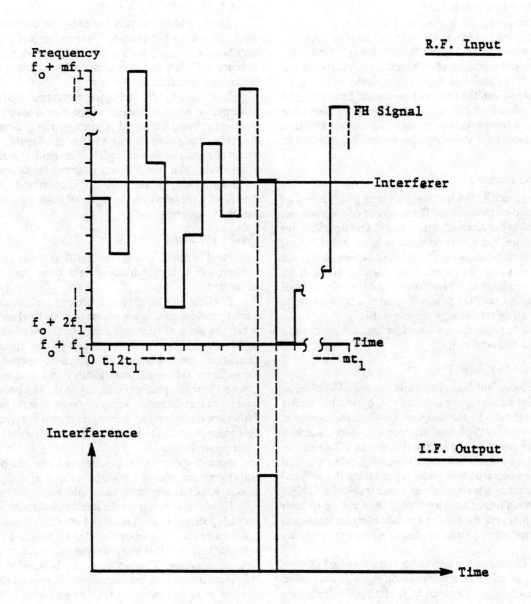

**FIGURE 4
EFFECT OF NARROWBAND INTERFERENCE ON A FREQUENCY
HOPPING SYSTEM (SIMPLIFIED)**

9-15

phase between frequency hops. For this reason, coherent demodulation techniques are seldom used. In analog FM transmission systems using frequency or phase modulation, the phase discontinuities between frequency hops produce background noise ("clicks") that cannot be mitigated by increasing the signal power (10).

For some purposes, it may be convenient to classify frequency hopping systems as fast hopping or slow hopping. This report will use the term "fast-hopping" when the hopping rate significantly exceeds the information rate. In systems of this type, multiple hops are typically added (i.e., integrated) to re-form the information signal. Examples will be given in Section 3.2.2. In slow-hopping systems, the hopping rate is comparable to or less than the information rate. An example of a slow-hopping system is presented in Section 4.1.

2.4.1.2 Time Hopping

In this approach, the transmitter emits short pulses or bursts at pseudo-random times. The sequence of transmission times is stored in the receiver, which tracks and demodulates the transmissions but otherwise ignores the channel. Thus, a jammer must either discover the hopping pattern, or spread its power over a high duty cycle (thus wasting much of its energy) or be content with randomly jamming only a small fraction of the transmitted pulses. Repeater jamming can be thwarted by using sufficiently narrow pulses.

Pure time-hopping has not found widespread application outside of the military sector.

2.4.1.3 Direct Sequence

This method, which is also known as pseudo-noise, or PN, consists of switching the phase of the transmitted carrier at regular intervals. In the simplest version, the carrier is switched between two phases that are 180 degrees apart, according to a pseudo-random binary pattern. The receiver tracks these pseudo-random phase inversions using a stored replica of the binary pattern, thus reproducing the original carrier. A period of constant carrier phase is called a "chip" in order to avoid use of the term "bit" in reference to both information bits and the smaller spread spectrum elements. The term "chip" has also been applied to a frequency hopping interval.

Figure 5 illustrates the operation of a simplified DS system. At the transmitter, the information signal is multiplied by a rapidly-switched sequence of chips. Its bandwidth is thus significantly expanded beyond the information bandwidth. At the receiver, the signal is restored to the information bandwidth by re-multiplying with the pseudo-random chip sequence. Interference, on the other hand, is subjected to only one multiplication process (this occurs in the receiver) so that its bandwidth is increased to at least the bandwidth of the chip waveform. Filtering the receiver output effectively removes interference energy beyond the information bandwidth.

Thus, the signal-to-interference ratio at the receiver output is higher than the channel signal-to-interference ratio by a factor that is approximately equal to the ratio of the spread spectrum bandwidth to the information bandwidth. This factor is known as "process gain" or "processing gain."

Other versions of DS spread spectrum involve the use of four or more carrier phases, or the use of two carrier phases that are separated by an angle other than 180 degrees.

In principle, virtually any conventional technique can be used to modulate the carrier with the information, either before or after spreading. In practice, most DS systems have been used for digital data transmission, although analog modulation systems have also been implemented.

2.4.1.4 Linear FM ("Chirp")

Linear FM techniques are based on constant envelope, swept-frequency waveforms. These techniques have traditionally been associated with pulse compression radar systems, and are not necessarily associated with the usual connotation of spread spectrum.

In the basic chirp technique, the carrier frequency of the transmitted pulse is linearly swept over a given bandwidth. At the receiver, the signal is processed in a dispersive delay line that compresses the pulse to a width roughly equal to the reciprocal of swept bandwidth. The gain in signal-to-noise ratio during the compression is approximately equal to the time-bandwidth product of the transmitted pulse. Time-bandwidth products on the order of 1000 have been demonstrated in chirp systems.

2.4.2 Discussion

At this point, it may be useful to make several observations about the techniques that have just been briefly described.

I. None of the spread spectrum techniques described above provides any improvement over narrowband techniques when the noise or unintentional interference has a constant, uniform spectrum over the entire spread spectrum band. (However, forcing a willful jammer to spread his limited power over this wider bandwidth is advantageous.) Thus, "pure" spread spectrum techniques do not provide performance improvements against background noise alone. However, digital spread spectrum systems are frequently used in conjunction with error-correcting codes that do provide such improvements.

Spread spectrum techniques can provide performance improvements against multipath-related effects, but only if they are specifically designed to do so.

II. Multiple access can be implemented with virtually any spread spectrum technique. In the usual implementation, each transmitter (or group of transmitters) is assigned a unique digital code that underlies either its hopping sequence or its phase-keying sequence. The set of codes is usually designed to minimize interaction between transmitters and receivers that are not in communication with one another.

III. It has been a common practice to combine two or more spread spectrum techniques to produce hybrid systems. Thus, an early concept known as RADA—Random Access Discrete Address—combined frequency hopping and time hopping (11). JTIDS and PLRS uses a combination of frequency hopping and DS techniques. TDRSS and GPS, on the other hand, use pure direct sequence techniques (5,6).

IV. An important feature of slow frequency hopping is that advantage can be taken of the selectivity of inexpensive crystal filters. Unless a narrowband interferer is extraordinarily strong, it will produce interference only 1/N of the time, where N is the number of FH channels or frequency slots. Subject to the constraint just stated, the amount of interference produced is limited to a fraction that is independent of the signal-to-interference ratio at the receiver input.

In order to see the significance of this, consider a multiple access application in which a receiver must contend with a single interferer that is 40 dB stronger than the desired sig-

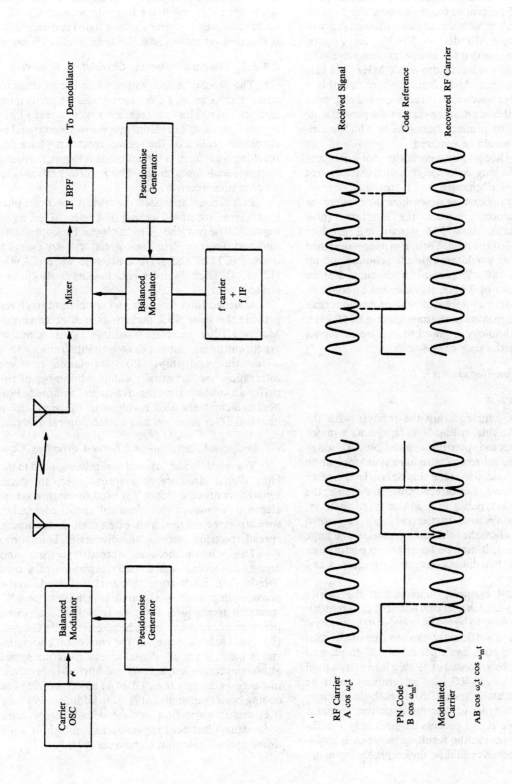

FIGURE 5
OVERALL DIRECT SEQUENCE SYSTEM SHOWING WAVEFORMS

nal. (This is the so-called "near-far" problem. For this example, non-fading conditions will be assumed.) Suppose, furthermore, that a 10 dB output signal-to-interference ratio is required at least 95 percent of the time under these conditions. A direct sequence system would then require a processing gain of 40 dB + 10 dB = 50 dB. As a good approximation, this means that for an information bandwidth of 3 kHz a spread spectrum bandwidth of 300 MHz (= 3 kHz $\times 10^5$) would be required. Aside from the fact that this is an extraordinarily large bandwidth, direct sequence chip rates on the order of 300 MHz are near the limit of the present state of the art. It is easy to produce examples in which multi-gigahertz chip rates would be required.

With a frequency hopping approach, the problem is substantially alleviated. As long as the design is such that adjacent channels do not "splatter" through the i.f. passband, or overcome the i.f. filter's stopband, or desensitize the receiver, or produce intermodulation products, the interferer (now assumed to be another slow FH signal) will produce interference only 1/N of the time. Thus, a properly-designed twenty-channel system would provide the required output signal-to-interference ratio 95 percent of the time. For an information bandwidth of 3 kHz, this number of channels would typically represent a total bandwidth of no more than 1.0 MHz, assuming a maximum channel spacing of 50 kHz. Furthermore, the technology required to implement such a system is well within the state-of-the-art.

2.4.3 Implementation Constraints

2.4.3.1 Synchronization

The problem of synchronizing the receiver with the incoming signal at the chip or hop level represents a major design issue in most spread spectrum applications. Although most conventional digital receivers require synchronization at the bit level, synchronization at the chip or hop level is often more difficult to achieve. For direct sequence systems, the SNR for a single chip is typically less than unity. In addition, the frequency of the received carrier may not be known accurately enough to allow the coherent integration of a large number of chips. Thus, it may be necessary to perform an exhaustive search over two dimensions: carrier frequency and code chip epoch.

As a hypothetical example, suppose that the carrier frequency uncertainty is such that one hundred different trial carrier frequencies need to be tested in order to assure that the signal will be found sufficiently close to one of the trial frequencies. If the DS code has a period of 127 chips, and the possible code epochs are tested at ½ chip intervals, a total of 100 × 127 × 2 = 25,400 trial integrations must be performed in an exhaustive search. After each integration, a test is performed in order to determine whether or not a signal is present. Even if this process requires only 1 millisecond per trial integration, the resulting worst-case acquisition time of 25.4 seconds could be unacceptable for many applications.

Slow FH systems can be easier to synchronize because the SNR for a single hop is typically sufficient to make reliable decisions on the presence or absence of the signal. Since the receiver "knows" the hopping sequence, it can simply observe a single frequency until a signal is detected. If energy is detected on subsequent hops, acquisition is assumed.

2.4.3.2 Matched Filter vs. Correlation Receivers

The process of integrating a number of chips to reconstruct a data bit at a DS receiver can be performed in two distinct ways. The first technique, which was illustrated in Figure 5, consists of multiplying the received signal by a synchronized replica of the spread spectrum phase code. The resulting waveform is integrated in a low-pass filter or in an integrate-and-dump circuit. This implementation is known as a correlation receiver.

In a second approach the spread spectrum phase code coefficients are stored as part of a linear filter, as shown in Figure 6. The two leading technologies for implementing such matched filters or "passive correlators" are charge transfer devices (CTDs) and surface acoustic wave (SAW) devices (12,13,14). Digital implementations have also been demonstrated.

Matched filter receivers and correlation receivers ideally provide the same SNR performance after synchronization. Matched filter receivers facilitate rapid acquisition and synchronization, but are limited to integration over at most a few thousand chips (15). Correlation receivers allow integration over a virtually unlimited number of chips, but suffer acquisition-time disadvantages, as noted above. Economic factors are also involved in the trade-off between matched filter receivers and correlation receivers.

2.4.4 Spread Spectrum and Error-Correcting Codes

The word "code" as used earlier has applied to the underlying digital structure of a spread spectrum signal. In a separate context, a "code" is used to mean a set of digital elements employed for forward error correction. Since forward error correction is often used in conjunction with spread spectrum systems, the distinction is important.

The relation between spread spectrum and error correction coding was recently explained in a tutorial by Viterbi (16). Error correction uses increased bandwidth to provide improved performance. Unlike "pure" spread spectrum techniques, error correction coding can provide performance improvements in additive white Gaussian noise (i.e., background noise). Error correction coding is beneficial in most, if not all, digital spread spectrum applications, although economic tradeoffs are obviously involved. Multiple access concepts (i.e., CDMA) based on error correction coding have been analyzed (17). In addition, low rate coding (i.e., error correction coding with a large amount of redundancy) has been suggested as a substitute for conventional spread spectrum techniques (87).

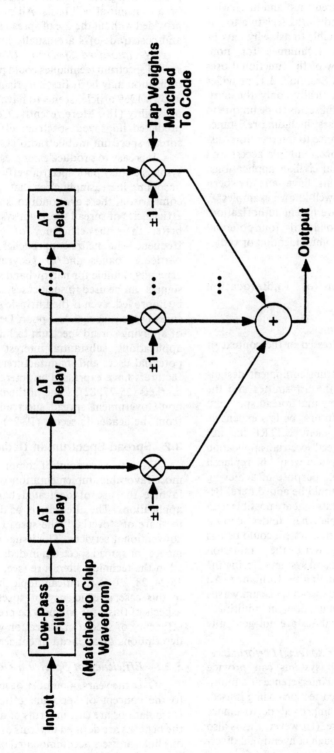

**FIGURE 6
TYPICAL MATCHED FILTER (PASSIVE CORRELATOR)
FOR DIRECT SEQUENCE SPREAD-SPECTRUM RECEPTION**

3. THE USE OF SPREAD SPECTRUM IN NON-GOVERNMENT SERVICES

3.1 Potential Benefits

Given that spread spectrum techniques have evolved largely in response to military requirements, and in view of the fact that they require large bandwidths (relative to the information bandwidth), it is reasonable to ask why anyone would consider spread spectrum techniques for non-Government* applications. A review of the functional uses of spread spectrum, as presented in Section 2.3.1, provides a background for this issue. Specifically, only the anti-jamming property of spread spectrum seems to be unique to military environments. The other uses, including resistance to unintentional interference, resistance to interception, discrete addressing, multipath resistance, multiple access and pulse compression all have potential civilian applications. Indeed, certain civilian applications have already been exploited; TDRSS and GPS provide well-known examples**, although both will operate under Government authorization.

In general terms, it is possible to identify four potential motivations for introducing a new communication or radiolocation technology:
- Reduced Cost
- Improved Communication or Radiolocation Performance
- Expanded Capabilities
- Improved Spectrum Utilization

Each of these will now be addressed in the context of spread spectrum techniques.

I. *Reduced Cost*: Spread spectrum equipment designs require added complexity in both the transmitter and the receiver. Thus, it seems most unlikely that spread spectrum equipment will, in the foreseeable future, be less expensive than comparable narrowband equipment. MITRE has not been able to identify any potential spread spectrum application in which an existing conventional system would be replaced by a spread spectrum system for the purpose of achieving reduced costs. However, a caveat should be added here. Because of the performance improvements that are possible with spread spectrum, it is conceivable that under certain conditions, a particular spread spectrum system could be less costly—due to reduced transmitter power or the elimination of ancillary circuits—than a narrowband system offering the same level of communication or ranging performance. An example of such a trade-off might be spread spectrum versus polarization diversity for communication in multipath environments. Comparative cost analyses are not generally available in this area.

II. *Improved Communication (or Ranging) Performance*: As noted earlier, spread spectrum systems can provide significant resistance to unintentional interference and multipath fading. To the extent that error correction coding is used, spread spectrum systems provide improved performance against additive white Gaussian noise. (However, this is also true of conventional systems.) Although conventional diversity techniques also provide fading resistance in communication systems, high-resolution ranging systems frequently depend on wide bandwidths for good performance.

III. *Expanded Capabilities*: Spread spectrum systems can provide user privacy, discrete addressing, and multiple access on a transmit-at-will basis. All of these features can also be provided without the use of spread spectrum, but performance and cost trade-offs are usually involved.

IV. *Improved Spectrum Utilization*: The notion that spread spectrum techniques could provide improved spectrum utilization may be at first surprising. J. P. Costas, in a well-known 1959 article, seems to have been the first to raise this possibility (18). More recently, Cooper and Nettleton have predicted improved spectrum efficiency for high-capacity spread spectrum mobile radio systems (19-24).

It is easy to produce examples in which spread spectrum systems display poor spectrum efficiencies. However, it is also becoming increasingly clear that, depending on the basis of comparison, there are conditions under which the spectrum efficiencies of spread spectrum systems are comparable to or better than the efficiency of conventional narrowband frequency-channelized approaches. In addition, there may be particular bands and/or geographic areas that are not generally suitable for narrowband signals. If spread spectrum signals can be used in such cases, spectrum utilization might be improved, even if the intrinsic spectrum efficiency of the spread spectrum signals is poor. Despite the potential benefits of applying spread spectrum techniques in non-Government applications, substantial interest has yet to be shown by potential users and manufacturers. Although a few manufacturers have expressed an interest in pursuing specialized markets (25-27, 82), some of the most visible proponents of non-Government spread spectrum applications have been from the academic sector (19-24).

3.2 Spread Spectrum in Dedicated Bands

From the viewpoint of communication performance, the most favorable implementation of spread spectrum would involve the use of dedicated bands for spread spectrum applications. The alternative, as discussed in Section 3.3, is to share of "overlay" the spread spectrum allocation with conventional services. The issue of communication performance for spread spectrum in dedicated bands has been treated in the technical literature (see, for example, 1, 5, 10-12, 15-24, 28, 29). A significant topic that remains to be addressed in this case is efficiency of spectrum utilization. Various aspects of this issue will now be presented. The discussion on efficiency of spectrum utilization will be followed by several descriptions of application concepts.

3.2.1 Efficiency of Spectrum Utilization

Over the years, a number of measures have been applied to the concept of spectrum efficiency (30-38). The usual formulations are cast in terms of a benefit/cost ratio, where the benefits are defined in terms of the number of simultaneous links or users accommodated or the network information throughput; the costs are normally defined in terms of the amount of spectrum resources occupied. Typical examples are number of users per unit bandwidth and number of users per unit bandwidth per unit area (for a given information rate or information bandwidth). Comparisons of various modulation techniques and system concepts may be sensitive to the measure of spectrum efficiency upon which the comparison is based.

---

* The term "non-Government," as used in this report, coincides with the regulatory use of the term. That is, non-Government means non-Federal-Government, but includes state and local governments.

** Although GPS is a DoD program, civilian applications are anticipated.

One of the most straightforward measures of spectrum efficiency for a network or an associated group of users is total information throughput per unit bandwidth. Under this formulation, the rates of information transfer for all user links are summed and the total is divided by the total bandwidth required to accommodate all the links. All links are assumed to be in continuous use, and no reference to physical area or volume is made. This definition can be used to measure the spectrum efficiencies of satellite links, for example.

On this basis, the spectrum efficiency of spread spectrum signaling can be quite low. Detailed analyses supporting this conclusion are presented in Appendix A and in References 17, 29, and 39-41. Basically, the reason for this result is that under idealized conditions, complete orthogonality between the signals of various user links is more closely approachable with TDMA or FDMA than with SSMA, unless the various SSMA user links can be mutually synchronized. In designing a set of signals for an asynchronous* multiple access application, it is found that the best achievable worst-case cross-correlation between signals degrades (i.e., increases) if each signal is required to be spread uniformly across the available bandwidth (76). If the cross-correlation is minimized, then signals with highly nonuniform spectra result, as in FDMA. More will be said about this in Section 3.3.6.

Example 1

As an example, consider an asynchronous direct sequence SSMA system in which the various user links employ bi-phase keying to embed the information in the spread spectrum signal. All links operate at the same data rate, but each link uses a separate spread spectrum code. The codes are assumed to be quasi-orthogonal. Equal power levels and a non-fading channel are assumed. In addition, the number of active users is assumed to be sufficiently large that, for the purpose of computing the bit error rate, the interference from other links that is experienced by any given receiver can be modeled as a Gaussian random process. Coherent detection (without error-correction coding) is assumed.

Under these conditions, the spectrum efficiency, as a function of the required bit error rate, is shown in Figure 7 (from Appendix A, Figure A3). The lower curve applies when the ratio of received energy per bit to background noise power density (E_b/N_o) is 10. It can be seen that the spectrum efficiency is quite low (less than 0.2) for bit error rates below 0.01. As shown by the upper curve, this conclusion is not substantially changed if background noise is eliminated completely, that is, if $E_b/N_o = \infty$. (The term "background noise", as used above, refers to the noise level in the absence of interference from other spread spectrum links.) If asynchronous frequency hopping with binary frequency shift keying and noncoherent detection are used instead of direct sequence signaling, spectrum efficiency is degraded even further (see Figure A5, Appendix A). The existence of fading can also degrade spectrum efficiency relative to Figure 7. (See Appendix A, Figures A6 and A7.)

As a basis of comparison, an FDMA system requiring 2.5 Hz of r.f. bandwidth for each bit-per-second of transmitted information has an efficiency of 1/2.5 = 0.4 at low error rates, using the definition set forth above. In the absence of frequency errors, this level of performance appears to be attainable with existing narrowband modulation techniques (42, 43) as long as the received power levels in adjacent channels are not greatly different. Thus, at low error rates FDMA can provide at least twice the spectrum efficiency of SSMA for the conditions of this example.

Of course, there is more to the trade-off between FDMA and SSMA than spectrum efficiency. In addition, there are several factors that can mitigate the relatively negative view of SSMA spectrum efficiency presented above. One of these is the impact of guard bands in FDMA, which are necessary to provide for frequency tolerances, Doppler shifts, modulation sidebands (i.e., "splatter"), finite filter roll-offs, or a combination of these factors. As shown in Appendix B, the existence of frequency tolerances can significantly reduce the spectrum efficiency of FDMA, particularly when the data rate per channel is low and the assigned frequency is high. The following is presented as an example of how this factor can influence the comparison of SSMA and FDMA in terms of spectrum efficiency.

Example 2

The basic assumptions for direct sequence SSMA are as in Example 1, except that a frequency tolerance of 2 parts per million is assumed (see Appendix B, Example 1). In addition, the following parameters apply:

R = information rate = 100 Hz
M = Maximum number of users = 100
A_1 = predetection SNR in the absence of interference = 30 dB
A_m = required predetection SNR at full capacity = 10 dB

From Appendix B or reference (44), the required number of code chips per information bit is approximately

$n = (1/3)(M - 1)(1/A_m - 1/A_1)^{-1} = 333$

The required r.f. bandwidth is approximately twice the chip rate, or $2 \times 333 \times 100$ Hz = 66.6 kHz.* In addition, we add an equal bandwidth to provide for guard bands. The total SSMA channel bandwidth is then 133.2 kHz, plus twice the frequency tolerance. (The frequency tolerance is 2×10^{-6} times the center frequency.) At a center frequency of 1.0 GHz, the spectrum efficiency is $(100 \text{Hz})(100 \text{ users})/(133.2 \times 10^3 + (10^9)(2)(2 \times 10^{-6})) = 0.073$.

As a comparison, consider an FDMA system having the same number of users, operating at the same data rate, with the same frequency tolerance (i.e., 100 users, 100 Hz per user, and 2 parts per million). Let the bandwidth expansion factor for the FDMA users be 2.0. The total required r.f. signal bandwidth is then (100 Hz)(100 users)(2.0) = 20 kHz. If 100 Hz guard bands are added between channels, with 50 Hz guard bands at each of the band edges, the total required bandwidth is approximately 20 kHz + (100 Hz)(100 channels) + $2(2 \times 10^{-6})(100)f_c$ = 30 kHz + $4 \times 10^{-4} f_c$, where f_c is the center frequency of the band. The spectrum efficiency at $f_c = 1.0$ GHz is (100 Hz)(100 users)/(30 $\times 10^3$ + (4 $\times 10^{-4})(10^9)$) = 0.023, or less than half the efficiency of SSMA. As shown in Appendix B, the break-even frequency at which the spectrum efficiencies of FDMA and direct

---

* In the context of multiple access, "asynchronous" means that the various links need not be mutually synchronized.

* In a practical design, the number of code chips per information bit might be increased to the next-highest value of $2^L - 1$ in order to permit the use of L-bit linear shift register generators. In this case, a 511 chip code would allow the use of 9-bit linear shift-register generators. However, the number of chips per bit is generally not constrained to values of $2^L - 1$.

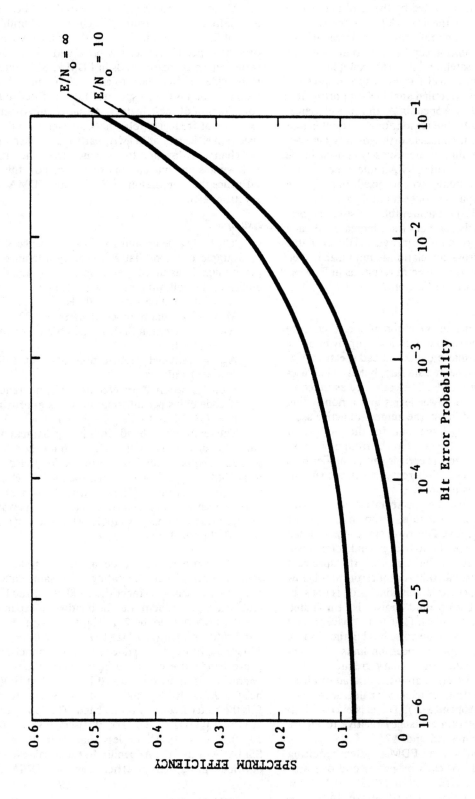

**FIGURE 7
SPECTRUM EFFICIENCY OF ASYNCHRONOUS DIRECT SEQUENCE SSMA**

sequence SSMA are equal (for this example) is 261 MHz. At higher frequencies, SSMA can provide better spectrum efficiency than FDMA.

Similar examples can be produced using frequency-hopping SSMA.

A second factor that alters the trade-off between SSMA and FDMA is the inclusion of geographic coverage in the definition of spectrum efficiency. For example, the spectrum efficiency of a land mobile (voice) communication system might be defined in terms of number of users per unit bandwidth per unit of coverage area. Clearly, if SSMA provides better geographic reuse of the spectrum than FDMA, the trade-off will be shifted in favor of SSMA. The advantage of SSMA in this context is that each user has access to the entire system bandwidth regardless of his geographic location. In conventional frequency-channelized systems, each channel may be available over as little as one-seventh of the coverage area*. In small cell systems, the number of user links per channel per cell is typically on the order of 0.045 to 0.055 (see Reference 55). For example, in a 667-channel system, 30 to 37 channels per cell would be available. This apparent disadvantage is overcome by providing a large number of cells in each system.

Unfortunately, a general in-depth analysis of SSMA spectrum efficiency in land mobile radio (LMR) systems has not yet been performed, although some steps in this direction have been taken (19, 29, 45). One of the basic problems is that, particularly in urban areas, there is a great deal of variability in transmission loss over various paths of equal distance for given antenna heights, within a single geographic area (46). There is also variation in path loss characteristics from one city to another (47-49), so the incorporation of simple path loss models (e.g., $1/R^N$ law) in spectrum efficiency calculations is at best an expedient approximation. In addition, the wide variety of possible spread spectrum implementations (including hybrid techniques and code variations) and possible geographic configurations make it unlikely that a general analysis will be performed.

Another dimension affecting spectrum utilization is time. In many services, the average utilization of channels (with respect to time) is low. Nevertheless, the assignment of channels to users on a fixed basis makes it more likely that a particular user will find his channel blocked when it is needed. A spread spectrum multiple access system averts this problem by making a large bandwidth continuously available to each user. Performance degrades gradually as the number of active users increases.

Trunked narrowband systems also provide a solution to this problem, by dynamically assigning channels to users only when the channels are actually needed. Trunked systems assign channels to users on a demand basis until all available channels are in use. At this point, new calls cannot be placed until one or more ongoing calls is terminated. Thus, trunked systems are capable, in principle, of attaining high efficiencies of spectrum utilization. However, trunked systems require a central control of channel assignments, and exhibit a "hard" saturation characteristic when the number of users in the system becomes equal to the number of available channels. It seems likely that there will always be applications in which, for one reason or another, trunking is not practical.

In summary, the question of efficiency of spectrum utilization, as it applies to spread spectrum systems, should be viewed in the context of the major qualitative differences that exist between conventional frequency channelized or TDMA approaches and random access spread spectrum. Spread spectrum may be an attractive alternative where uncoordinated use of the spectrum over one or more geographic areas is an important requirement. In making comparisons of spectrum efficiency, it becomes clear that the actual capacity of practical FDMA approaches may differ significantly from their theoretical capacities computed on the basis of uniform instantaneous channel loading. The need for geographic frequency reuse intervals and guard bands also limit the spectrum efficiency that can be attained by conventional frequency channelization in practical systems.

3.2.2 Fast Frequency Hopping in High Capacity Land Mobile Radio Systems

3.2.2.1 Differential PSK

In 1977, Cooper and Nettleton (50-52) proposed a unique approach to high-capacity land mobile communications. The following salient features were incorporated in their technique:
- Digitized voice with differential PSK modulation
- Fast frequency-hopping
- A small-cell network implementation
- A unique receiver design based on the use of tapped delay lines and multiple bandpass filters
- Dynamic control of transmitter power levels

The following benefits have been claimed for this approach (23):

1. Resistance to fading is provided by the frequency diversity that is inherent in the fast frequency-hopping technique.

2. Each user has access to the entire system bandwidth. Thus, there are no "clocked calls."

3. There is no hard limit on the maximum number of simultaneously-active users. Communication performance degrades gradually as more and more users enter the system.

4. The system offers privacy from casual listeners.

5. Since it is not necessary for users to switch channels when they cross a boundary between cells, "forced termination" of calls cannot occur.

6. The hardware employed by each (mobile) user is identical, except for the filters associated with that user's unique signal set.

7. Priority messages can be accommodated even when the system is heavily loaded.

8. Under conditions of reduced capacity, the system may be able to co-exist with conventional narrowband services.

The disadvantages of this approach, as recognized by its advocates, include (23):

1. The need for dynamic control of mobile transmitter power levels in order to mitigate the "near-far" problem (i.e., the problem of having a mobile located near the base station interfere with a mobile far from the base station).

2. The requirement for digitized voice, together with the spread spectrum signal format, will increase the complexity of transmitters and receivers.

3. A vehicle location technique of modest accuracy is necessary in order to monitor the vehicles as they move from cell to cell. (This would also be required for conventional small-cell techniques—see Reference 53).

---

* This is based on a cellular approach with one ring of hexagonal cells between base stations using the same frequency.

4. Fully coherent detection is not possible in a rapid-fading LMR environment.

5. The proposed approach is not economically attractive for large-cell systems.

In addition, Mikulsky (54) has pointed out that, because of the receiver implementation technique advocated by Cooper and Nettleton, realization of the "no calls blocked" and "no forced terminations" claims would require that each base station accommodate all possible user codes. Since each code requires a unique delay-line/filter configuration in the receiver, the economic impracticability of this is evident. Thus, blocked and terminated calls would occur as in any cellular system.

In References 19 and 23, Cooper and Nettleton present comparisons of their proposed technique with conventional cellular systems in terms of spectrum efficiency (defined in terms of calls per unit bandwidth per cell). These comparisons show estimated spectrum efficiencies for fast frequency-hopping of 2.0 to 4.7 times the spectrum efficiency of conventional narrowband systems, depending on the specific configurations being compared. These estimates were based on a baseband data rate of 30 kbits/s for digitized voice.

In a subsequent independent analysis, Henry (55) found the DPSK/fast-FH technique to be inferior to conventional narrowband FM techniques on the basis of the average number of usable channels per cell. In his example, narrowband FM was superior by a factor that ranges from 1.2 to 2.8, based on a 32 kbits/s digitized voice rate for the fast frequency hopping system. (If a 30 kbit/s rate is used as in the earlier analysis by Copper and Nettleton, Henry's spectrum efficiency penalty (for fast FH) is 1.12 to 2.62.)

Further improvements in the spectrum efficiency of this spread spectrum approach can be attained by lowering the digitized voice rate. Digital voice has been used successfully at rates as low as 2.4 kbits/s. However, low data rates usually result in increased equipment cost or degraded voice quality, or both. If degradations in voice quality are allowed, then comparable narrowband systems can also increase spectrum efficiency by using closer channel spacings (with or without corresponding reductions in the r.f. bandwidth). On the other hand, digital voice technology is advancing rapidly, and it is possible that low cost systems for high quality voice transmission at data rates on the order of 10 kbits/s (with an acceptable tolerance for transmission errors) will be available in the future. (See Flanagan, Reference 100, for an overview of digital voice technology.) If so, then the fast FH/DPSK technique will exhibit superior spectrum efficiency even in Henry's example.

It has also been suggested that the spectrum efficiency of the Cooper-Nettleton approach might be further improved by controlling the mobile transmitter power to optimize the signal-to-interference ratios at all base stations, rather than trying to maintain a constant mean received signal power at the base station with which the mobile is communicating (89, 90).

3.2.2.2 Multilevel FSK

In a recent analysis of a concept originally described by Viterbi (101), Goodman and others have evaluated the performance of a fast frequency hopping multiple access technique that uses multi-level FSK as a basis for signaling (102-105). The approach is quite similar to the one described by Cooper and Nettleton, except for the method in which the transmitted information is embedded in and recovered from FH waveforms.

Figure 8 illustrates the operation of the FH/FSK transmitter. The digital data stream representing the voice input is divided into blocks of k bits (typically, k = 8). Each k-bit block is used to select one of 2^k time-frequency sequences. Each such sequence has a length of L time chips, or hops (typically, L = 19). To provide for discrete addressing, the basic time-frequency sequence used by each transmitter is different. Information is imposed on this basic time-frequency sequence by selecting one of 2^k possible cyclic shifts of the basic sequence in the frequency domain for each k-bit input data block. Thus, 2^k channels are required.

The operation of the receiver is illustrated in Figure 9. The transmitter's address code (i.e., its basic time-frequency sequence) is used to de-hop the intended signal, yielding a matrix of 2^k channels by L time chips. A hard decision is made on the presence or absence of a signal in each of the $L \times 2^k$ matrix elements. In the absence of noise, fading, and interference, the de-hopped signal will occupy a single row of the matrix; all of the other $2^k - 1$ rows will be vacant, indicating the absence of signal energy. With noise, fading, and interference, the row corresponding to the correct signal can contain deletions; other rows can contain insertions. Interference in the row containing the correct signal nominally has no effect. The simplest decoding algorithm involves selecting the row containing the largest number of signal "hits."

For 32 kbit/s digitized speech, k = 8 and L = 19, about 170 users can be supported at an error rate of 10^{-3}. This estimate includes the effects of multipath and noise, as well as mutual interference in an isolated system, but it does not include the effects of urban shadowing, synchronization errors, "splatter" between adjacent channels, or interference from nearby cells in a cellular system. Nevertheless, it appears that the number of users per unit bandwidth under this approach is several times the corresponding number for FH/DPSK, as described in the previous section, and is at least comparable to the number of users per unit bandwidth that can be supported in a conventional narrowband FM system in which frequencies must be re-used over a particular service area.

A major disadvantage of this approach is that it requires that a real-time spectrum analysis be performed in the receiver. For k = 8 and L = 19, and a 32 kbit/s data rate, this analysis must be performed every 13.2 microseconds and it must resolve $2^8 = 256$ separate channels over a total bandwidth of about 20 MHz. While this requirement is entirely within the state-of-the-art, the implementation costs could be excessive.

3.2.3 Independent Land Mobile Radio Systems

In a recent study performed by NTIA for the FCC's UHF Task Force, Berry and Haakinson (45) analyzed multiple independent land mobile radio (LMR) systems using spread spectrum in a common band. It was found that the spectrum efficiency of spread spectrum in this case was low compared with conventional techniques. The analyses apply to direct sequence techniques but do not explicitly treat the case of frequency-hopping.

The study also contains a section on spread spectrum "overlay" with conventional services (see Section 3.3.2 below) as well as tutorials on spread spectrum and land mobile radio.

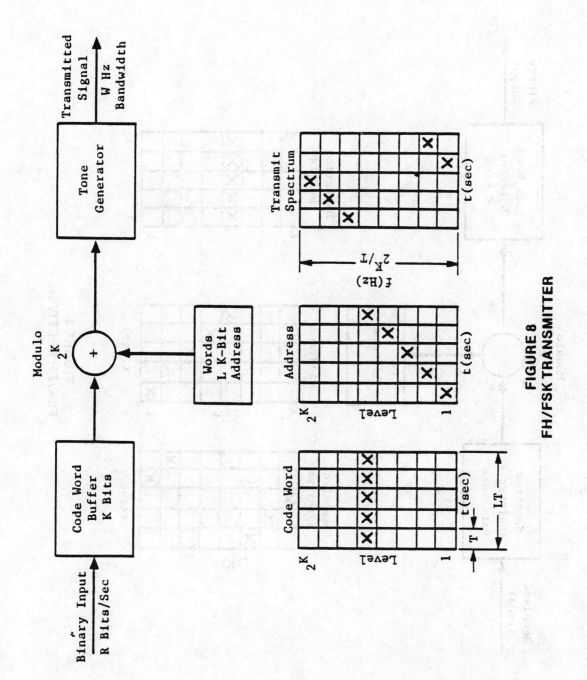

**FIGURE 8
FH/FSK TRANSMITTER**

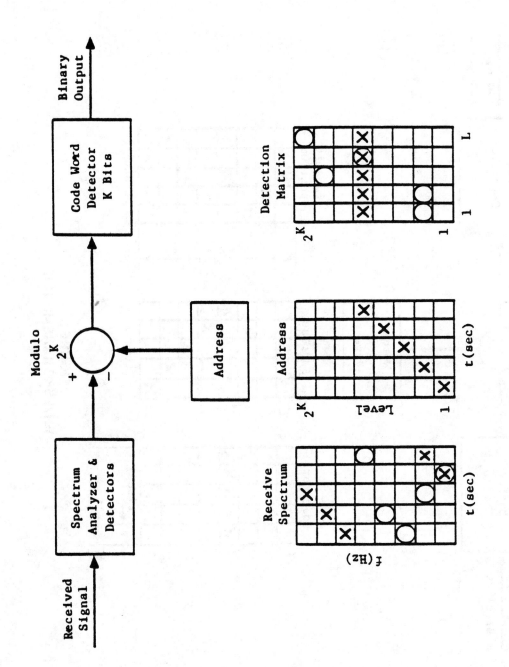

**FIGURE 9
FSK/FH RECEIVER**

3.2.4 Spread Spectrum in the Maritime Mobile Service

Experimental use of spread spectrum for maritime mobile applications dates back to 1974, when NASA and the Maritime Administration performed tests via NASA's ATS-5 and ATS-6 satellites to evaluate direct sequence ranging for maritime navigation purposes (88). Later efforts to use spread spectrum signaling for maritime satellite navigation have focused on GPS.

More recently, spread spectrum has been suggested for two other applications in the maritime mobile service. They are: emergency communications via satellite and overlaid communications in the terrestrial VHF marine band. The latter will be discussed in Section 3.3.3. This section will provide brief descriptions of proposed spread spectrum signaling for distress alerting via satellite.

SAMSARS (Satellite Aided Maritime Search and Rescue System) is a recent concept for the application of spread spectrum techniques to maritime distress alerting (56-60).* Under the current version of this concept, a commercial ship in distress would activate a low-power (about 10 watts), battery operated transmitter operating at L-land in a channel reserved for distress alerting. These signals would be received and translated to C-band by a commercial maritime communication satellite. At the satellite ground station, the distress message (which would typically include vessel identification, location coordinates, and a distress code) would be decoded and relayed to a rescue coordination center by conventional communication links. In order to provide multiple access and resistance to unintentional interference, a spread spectrum signal format is used.

In this application, conventional frequency channelization is not practical for the following reasons:

1. Since it is, by nature, a one-way signaling application, the user cannot listen for a clear channel before transmitting nor can he be instructed to switch to a clear channel by the ground station.

2. If only 1000 units were built and deployed, the bandwidth required to give a separate frequency channel to each would be about 3.2 MHz ($= 2 \times 1600$ MHz $\times 10^{-6} \times 1000$) for a frequency tolerance of 1 part per million. (This bandwidth could be reduced by assigning several transmitters to one channel, at the expense of occasional interference.) Less severe frequency tolerances would result in even larger bandwidths.

SAMSARS would use a 200 kHz-wide dedicated channel to accommodate its 128 kHz chip rate. For reasons of economy, each transmitter would employ the same direct sequence code. Multiple signals are separated by code epoch, random carrier frequency offset (resulting from a 10 ppm tolerance) and transmitter duty cycle. At least 58 simultaneously active transmitters can be accommodated within the field of view of any satellite, with a 95 percent message detection probability and a bit error rate of 10^{-5} (Reference 60).

A similar direct-sequence spread spectrum technique for maritime distress signaling has been proposed by Japan (67). The Japanese system would operate with a 406 MHz up-link.

Although narrowband alternatives to SAMSARS have been proposed (61, 62), they would not use frequency channelization in the usual sense. Instead, they would rely on transmitter duty cycle and random frequency offsets to provide multiple access, with occasional interference.

3.2.5 Spread Spectrum in Commercial Satellite Communications

The use of spread spectrum for multiple access in satellite communication systems provides for privacy and resistance to interference. However, as noted earlier, spread spectrum satellite networks tend to require more power and more bandwidth per user than equivalent FDMA or TDMA networks. These characteristics run directly against the current trend of optimizing the use of spacecraft power and bandwidth in commercial satellite communication systems. For this reason, it seems unlikely that spread spectrum or CDMA techniques will come into widespread use in commercial satellite systems within the foreseeable future. Within this overall trend, spread spectrum techniques may find specialized applications, such as the emergency alerting system described in Section 3.2.4.

In a 1977 study funded by NASA, various multiple access techniques were evaluated for satellite systems providing service to non-Government mobile users (40). FDMA was selected on the basis of its low implementation cost. The low spectrum efficiency of CDMA in this application was also noted.

3.3 Spread Spectrum "Overlay" with Conventional Services

Despite the ever-growing demand for spectrum resources in most non-Government services, the average utilization of the spectrum, even at peak hours in urban areas, is probably low. A recent FCC report (63) indicates that as few as 48 percent of the 25-470 MHz land-mobile channels monitored in Los Angeles had "very high occupancy" (defined as at least 60 percent peak hour message occupancy). The corresponding rate of "very high occupancy" channels in San Diego was 25 percent. Substantial numbers of channels were found to have zero occupancy. Analogous results were obtained in a 1977 study of VHF Maritime Mobile channel occupancy in the New Orleans area (64).

Such observations, however well-founded, are of little use to someone who is trying to get a new frequency assignment in a congested area (or to someone who is trying to use an occupied channel for which he is already licensed). The user of a busy channel (or set of channels) has only one legitimate way to "average out" the variations in channel occupancy—he must wait for a clear channel. He cannot switch to a temporarily-vacant channel that is allocated to a service in which he is not authorized to operate. Indeed, he will only be aware that the channel or channels on which he is authorized to operate are busy.

If spread spectrum signals were "overlaid" on existing allocations, it might be possible to improve the overall efficiency of spectrum utilization, even if the spectrum efficiency of the individual spread spectrum signals were quite poor. This prospect has given rise to a number of studies, which will be summarized below. First, however, a brief review of the basic technical issues will be presented.

3.3.1 Basic Technical Issues

In considering spread spectrum band-sharing with conventional services, three distinct interference modes must be considered:

---

* This concept was developed and analyzed independently by the MITRE Corporation. Subsequent development and test phases were funded by the Maritime Administration and the Coast Guard.

1. Interference to the conventional services from spread spectrum signals,
2. Interference to spread spectrum signals from conventional services, and,
3. Interference to spread spectrum signals from other spread spectrum signals.

The third interference mode—multiple access interference to spread spectrum signals has been treated in the technical literature (6, 11, 17-24, 28, 29, 40, 44, 45, 50-52) and in Appendix A. In general, meaningful analyses of spread spectrum self-interference require at least partial definitions of the application and the specific spread spectrum techniques to be used.

Spread spectrum interference to conventional receivers can take two forms. In the most serious form, the spread spectrum signal produces interference when the intended conventional signal is present at the receiver input. The second form occurs when the intended conventional signal is absent, but the spread spectrum signal or signals become noticeable above the level of the background noise. In either form, the effective interference is normally reduced relative to the interference that would have resulted from a co-channel narrowband signal. The amount of interference reduction depends on the specific spread spectrum technique and the type of conventional receiver to which the spread spectrum signal is applied.

As an example, consider a direct sequence signal that has an approximately uniform power spectrum over a bandwidth of 1.0 MHz. If the interference at the input of the conventional receiver is not so great as to cause nonlinear operation of a receiver stage "upstream" from the I.f. filter, and if the i.f. filter has a bandwidth of 14 kHz, for example, then the effective interference power is reduced by approximately 10 log (14 kHz/1000 kHz) = -18.5 dB, due to the "mismatch" between the i.f. bandwidth and the spread spectrum bandwidth.

A frequency-hopping signal can produce short bursts of relatively unattenuated interference in a narrowband receiver. A compendium of methods for analyzing spread spectrum interference to conventional receivers is available in Reference 65.

The recoverability of spread spectrum signals in a band occupied by conventional signals depends on the number, power, and frequency spacing of the conventional signals and on the particular spread spectrum implementation. In direct sequence systems, the interfering (conventional) signals are attenuated by an amount equal to the receiver's processing gain (approximately the ratio of the r.f. bandwidth to the information bandwidth). Slow frequency-hopping systems can be more severely degraded by a number of interferers at widely separated frequencies than by a single high-power narrowband interferer (66). As in conventional systems, interference to a spread spectrum receiver can degrade the intended signal or it can create "nuisance" effects when the intended signal is not present.

In bands that have been channelized at intervals of 25 to 50 kHz, a substantial fraction of the spectrum is committed to guard bands. In these cases, it may be possible to insert frequency-hopping signals between the existing channel assignments. Channels that are especially vulnerable to interference (public safety, for example) could be "notched out" of the FH spectrum by simple logic changes. (See Section 3.3.5.)

3.3.2 Overlay for Land Mobile Service

The prospect for overlaying spread spectrum signals in a common band with conventional narrowband land mobile communications was discussed by Cooper (68) and by Berry and Haakinson (45) in 1978. Both studies presented encouraging preliminary results, although neither explicitly addressed the "near-far" problem. Later, more detailed analyses by Dvorak (69) and Juroshek (70) yielded substantially more negative results. Specifically, Dvorak concluded that the overlay strategy would be practical only for small cell spread spectrum networks. In examples, he demonstrated that, for the 150 MHz land mobile band:

1. Mobile-to-mobile (SS to FM) interference could be unacceptable if the distance between mobiles were less than 24 meters.
2. Mobile-to-base interference (from a single spread spectrum transmitter to an FM receiver) could cause a 3 dB degradation in receiver noise at distances of 15-20 km.

(Part of this analysis was implicitly restricted to direct sequence or fast frequency-hopping signals.)

Juroshek does not treat the small-cell case, but assumes that the spread spectrum transmitters and the FM transmitters have comparable power levels. His analysis is applicable to fast frequency-hopping and direct sequence techniques. The major conclusion of his study is that, over the 150-900 MHz frequency range, spread spectrum and conventional FM land mobile systems cannot share a common band without significant interference to both types of systems. Again, the required separation distances (between the interfering transmitter and the receiver) were found to be typically on the order of 10 km or more.

3.3.3 Overlay for Maritime Mobile Service

In a study funded by the Maritime Administration, NTIA is assessing the potential for overlaying spread spectrum signals on the VHF maritime mobile band (156-162 MHz). An analysis was performed, with emphasis on direct sequence techniques. The results were applied in a hypothetical case study for the New Orleans area, which is known to suffer from radio frequency congestion in the VHF maritime mobile band. Although the results of this analysis have not been published at the date of this writing, a number of tentative conclusions have evolved. These are summarized as follows:

1. The "near-far" problem for the direct sequence spread spectrum, even without narrowband interference, requires an equalization of received signal power if two or more mobile stations transmit at once to a single base station. This could be done, in principle, by restricting transmitter/receiver distances or by controlling the transmitter power levels. It may be possible to use a direct sequence spread spectrum system to simultaneously communicate with a number of ships in the New Orleans area with acceptably low mutual interference between the spread spectrum signals if a single base station can be located 30 km or more north or south of New Orleans to provide nearly equal mobile-to-base distances, thereby insuring nearly equal received spread spectrum signal powers at the base station.
2. A practical spread spectrum system requires one frequency band for base-to-mobile communications and a second non-overlapping frequency band for mobile-to-base communications.
3. When a direct sequence spread spectrum system with a single base station and multiple mobile stations operate in

the same frequency band and geographic area with one or more conventional narrowband FM land mobile systems, mutual interference severely reduces the ranges of all systems.

4. The use of a small cell configuration for the spread spectrum stations can reduce the total spread spectrum interference to FM systems sharing the same frequency band and geographic area. However, this approach will not reduce the FM interference to the spread spectrum system.

In order to verify the preliminary conclusions, tests were performed in the New Orleans area. To avert the necessity of actually radiating a spread spectrum signal, conventional signals in the 156-162 MHz band were picked up by a test antenna, downconverted to an intermediate frequency, amplified, and combined with a direct sequence spread spectrum signal. The composite waveform was used as the input to a direct sequence demodulator.

Although the results of these tests have not yet been published, it is understood that they are consistent with the conclusions of the earlier NTIA analysis. Again, the negative results apply only to direct-sequence systems.

3.3.4 Overlay with Television

More than 400 MHz of bandwidth has been allocated to broadcast television in the United States. Of this, only a fraction of the potential channel assignments are actually used in any given city. This condition has already resulted in the allocation of 470-512 MHz band (channels 14-20) on a shared basis between television broadcast service and land mobile service. It is logical to ask whether the television bands might be shared with one or more spread spectrum allocations.

Preliminary studies on band-sharing between spread spectrum and television signals were published in 1978 by Juroshek (71) and Ormondroyd (72). Both studies included laboratory experiments involving the use of standard TV receivers and spread spectrum signals. Using a black-and-white TV receiver, Ormondroyd found that spread spectrum signals generated by modulating an AM signal with a direct sequence pseudonoise code caused much less degradation than conventional narrowband AM signals. On the other hand, Juroshek found that direct sequence spread spectrum signals produce about the same amount of interference to color TV as narrowband FM signals of the same level, as long as the spread spectrum bandwidth is about 2.0 MHz or less. For spread spectrum bandwidths greater than 6 MHz, spread spectrum should cause reduced interference relative to narrowband FM signals, simply because of the TV receiver's ability to attenuate interference outside of its nominal 6 MHz passband.

In 1979, Coll and Zervos proposed the use of spread spectrum techniques for multiplexing digital data with video in television-based information distribution systems (73). A laboratory experiment was performed in which the direct sequence chip rate was 63 times the horizontal scan rate (63 × 15,734 Hz = 991.25 kHz). A spread spectrum level 40 dB below the peak-to-peak video amplitude was used. Reliable operation of both the television and the data signal was reported.

3.3.5 JTIDS—A Case Study in Band-Sharing with Spread Spectrum

The U. S. table of frequency allocations was recently amended to permit the operation of JTIDS* in the 960-1215 MHz band. Although JTIDS is a Government system, the process by which its allocation was obtained, as well as the band-sharing characteristics of the system, are of interest because they demonstrate the use of a common frequency band by several independent wideband systems.

The design of JTIDS led to the early definition of a requirement for 150 MHz of bandwidth and in the range of 200-2000 MHz (3), based on technical and operational considerations. It was clear that obtaining this much bandwidth below 2 GHz on a dedicated basis was out of the question.

JTIDS shares the 960-1215 MHz aeronautical radionavigation band with TACAN/DME and the Air Traffic Control Radar Beacon System (ATCRBS), both of which use pulse signals. JTIDS uses a frequency-hopped signal format that is designed to minimize emissions in the 1030 MHz and 1090 MHz ATCRBS bands. An illustration of the composite JTIDS spectrum is shown in Figure 10. The 1030/1090 MHz "notches" in the spectrum are achieved by the elimination of the corresponding frequency slots from the hopping sequences, in conjunction with controlled pulse rise and fall times, continuous phase shift modulation, and sidelobe filtering of the modulating waveform (3).

JTIDS was found to be compatible with all present and planned uses of the 960-1215 MHz band, including ATCRBS monopulse, Discrete Address Beacon System (DABS), Beacon Collision Avoidance System (BCAS) and the DME portion of the Microwave Landing System (MLS/DME), as well as TACAN/DME and ATCRBS. However, the frequency coordination process, if it can be called that, was unquestionably one of the most expensive ever undertaken. Assuring compatibility between JTIDS and all present and planned uses of the 960-1215 MHz band entailed approximately 1500 hours of bench tests, 1000 hours of flight tests and eight man-years of analysis (3).

The resulting data were compiled in a 2000 page report (74). The magnitude of this effort was related partly to the fact that services protecting the safety of human life were involved, and partly because JTIDS is one of the first spread spectrum systems to share a band with such services.

3.3.6 Theoretical Considerations

In designing a SSMA system to operate in a band with narrowband signals, it is desirable to provide a set of spread spectrum user codes with the following properties:

1. The interference between spread spectrum users should be minimal.

2. The interference produced by a narrowband signal at the worst-case frequency should be minimal.

3. The interference suffered by a narrowband receiver at the worst-case frequency should be minimal.

These goals place conflicting requirements on the spread spectrum code set. A practical design thus entails a trade-off. Consider first a direct sequence system.

Property 1 implies a uniformly low cross-correlation between different user codes. Properties 2 and 3 imply a "maximally flat" power spectrum for each code, which, for rectangular chips, means, in effect, a $(\sin kf)^2/(kf)^2$ power spectrum of the type produced when the periodic autocorrelation function has no sidelobes.

That a trade-off is necessary in direct sequence systems is implied by the Welch bound (75). In any phase-coded sig-

---

* Joint Tactical Information Distribution System—See Section 2.3.2.

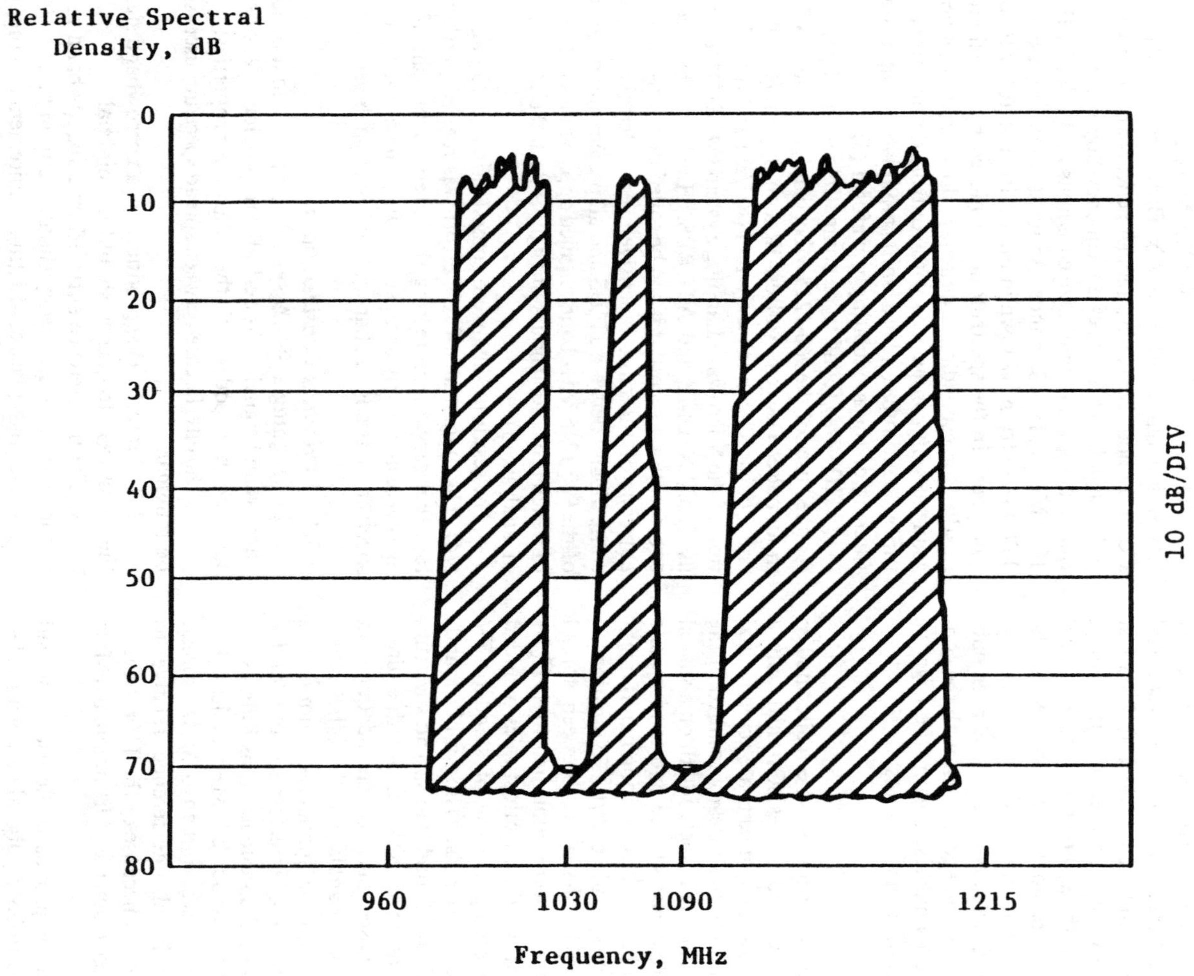

**FIGURE 10
COMPOSITE JTIDS SPECTRUM**

9-30

nal set, the largest of the cross-correlations or auto-correlation sidelobes has a minimum attainable value. Specifically, if C_1 is the largest cross-correlation in a signal set and C_2 is the largest auto-correlation sidelobe in the signal set, then

$$C_{max}^2 \geq f(M,L)$$

where

C_{max}^2 = Max (C_1^2, C_2^2)

M = number of codes in the code set
L = number of elements (chips) in each code

The particular form of f(M,L) depends on whether periodic or aperiodic correlations are being considered. Both forms result in a positive value of f(M,L) for values of M greater than 1. If uniform spectra are required, then

$C_2 = 0$ and $C_1^2 \geq f(M,L)$.

If low cross-correlations are required, then

$C_1^2 < C_2^2$ and $C_2^2 > f(M,L)$.

Thus, it is impossible to achieve uniform spectra ($C_2 = 0$) and arbitrarily low cross-correlations.

A second derivation of the trade-off between low auto-correlation sidelobes and low cross-correlation in a signal set was developed by Sarwate (76). This bound has the form

$$C_1^2 + g(M,L) > C_2^2 L$$

Sarwate also shows that in any set of L signals, each containing M = L elements, a requirement for zero cross-correlations implies that the magnitude of the periodic auto-correlation takes on its maximum value for every possible shift of every signal. That is, $C_1 = 0$ implies $R_k(j) = R_k(0) = L$, for all j and all k. (Here, $R_k(j)$ is the periodic auto-correlation function of the kth sequence.) What can be said about the spectra of such signals is that they are highly non-uniform, since uniform spectra imply $R(j) = 0$ for $j \neq 0$.

Although such bounds have not been explicitly formulated for frequency-hopping systems, it seems likely that a similar trade-off exists. For example, the so-called "one-coincidence" codes* (77) provide low mutual interference between FH users while producing a relatively uniform power spectrum. However, the number of such codes is only equal to the number of FH channels (i.e., frequency slots). In order to increase the number of user codes, the original one-coincidence code set can be modified in such a way that low mutual interference between FH users is preserved, but the individual modified codes no longer use all available frequency

---

* So-called because any two codes or their time shifted replicas occupy the same frequency slot during no more than one hop per code cycle.

TABLE 1
ISM BAND ALLOCATIONS AND USES

CENTER FREQUENCY, MHz	TOLERANCE	DESIGNATION*	TYPICAL ISM USES (78)	TYPICAL NON-ISM USES
13.56	±6.78 kHz			• Aeronautical Fixed Service (where landline comm. is not available at en-route stations)
27.120	±160 kHz		• Commercial Food Processing • Medical Diathermy • Land Transportation Radio Serv. • Low Power Comm. (under Part 15)	• Class C and D Citizens Band • Public Safety Radio Serv. • Industrial Radio Serv.
40.680	±20 kHz	G	• Commercial Food Processing	• Telemetry from Ocean Buoys and Wildlife (Secondary)
915	±13.0 MHz	G	• Commercial Food Processing • Medical Diathermy • Home Microwave Ovens • Tracking, Telemetry, Control and Comm. (Government)	• Automatic Vehicle Monitoring Systems (Secondary) • Field Disturbance Sensors (Intrusion & Theft Alarms)
2450	±50.0 MHz	G, NG	• Commercial Food Processing • Medical Diathermy • Medical Research • Plasma Research • Microwave Power Transmission Experiments • Home Microwave Ovens	• Amateur • Fixed • Mobile • Radiolocation • Field Disturbance Sensors
5800	±75 MHz	G, NG	• Radiolocation • Amateur • Field Disturbance Sensors	
24,125	±125 MHz	G, NG	• Radiolocation • Amateur • Airport Surface Detection Equipment (Primary) • Field Disturbance Sensors	

• G = (Federal) Government
NG = Non (Federal) Government

slots (23). Interference to and from conventional narrowband signals at the worst-case frequency is therefore increased. Alternately, the number of user codes could be increased while using every frequency slot once (or N times) in every code, but mutual interference between FH users would be increased with respect to the one-coincidence codes.

3.4 Spread Spectrum in the ISM Bands

Frequencies allocated for Industrial, Scientific, and Medical (ISM) purposes provide for the unlicensed operation of devices which use radio waves for the purposes other than communication. Such devices include, for example, medical diathermy equipment, industrial heating equipment, and microwave ovens. Because of the inherent interference resistance of spread spectrum receivers, it is logical to examine the possibility of operating spread spectrum systems in these bands.

3.4.1 Current Rules and Practices

There are currently seven ISM bands below 25 GHz, as shown in Table 1. These allocations encompass a total of more than 526 MHz. They range in width from 13.56 kHz (at 13.56 MHz) to 250 MHz (at 24.125 GHz). Besides accommodating ISM equipment operated under Part 18 of the FCC Rules and Regulations, these bands provide, on a shared basis, communication and radiolocation functions as shown in the last column of Table 1. In addition, field disturbance sensors (e.g., intrusion alarms and theft detectors) and low power communication devices are operated in certain ISM bands under Part 15 of the Rules and Regulations. Note also that most bands are shared with Federal Government stations or are allocated to such stations except through coordination with the FCC.

In general, radiocommunication services and radio frequency devices operated under Part 15 are required to accept interference from ISM devices. In addition, radiocommunication services in the 890-902 MHz and 928-940 MHz bands, which are adjacent to the 902-928 MHz band, are required to accept ISM interference from devices which met existing FCC standards on their date of manufacture. The only operation within the ISM bands that is protected from interference is Airport Surface Detection Equipment (ASDE) operating in the 24.-24.25 GHz band.

3.4.2 Characteristics with Respect to Potential Spread Spectrum Use

The only ISM band below 900 MHz having a significant bandwidth (if 320 kHz can be called a significant bandwidth) is the 26.96-27.28 MHz band. However, this band is already extensively in use as citizens band channels 1-27. As of February 1980, there were nearly fifteen million licensed citizens band stations, authorized to operate the range of 26.965-27.405 MHz. In addition, the upper 50 kHz of the 27.12 MHz ISM band is authorized for use by other services, as indicated in Table 1. A small number of Government stations also operate in this band. The 13.56 MHz and 40.68 MHz bands may find limited spread spectrum applications in special cases where very low information rates are used.

The 915 MHz ISM band encompasses a total bandwidth of 26 MHz that appears to have few non-ISM uses, aside from Government and experimental stations. Although the 902-912 MHz and 918-928 MHz portions of this band are available to automatic vehicle monitoring systems (on the basis of non-interference to Government stations), only one such system has been operationally deployed in the band. A search of the FCC frequency listing for this band produced only 35 assignments, other than experimental, ISM*, and Government listings, as of September 2, 1979. A total of 112 unclassified Government listings (many of which include multiple stations) were on file with IRAC (Interdepartment Radio Advisory Committee) as of February 27, 1980. A small number of classified Government listings were also on record.

The increased path loss and slightly higher equipment costs for the 915 MHz ISM band (relative to lower bands) may be of concern to potential users. However, the increased path loss (for transmit and receive antennas of fixed gain) is primarily relevant to attaining given levels of performance in the presence of natural background noise and receiver noise. For interference-limited systems using omnidirectional antennas, performance is relatively independent of frequency, since the interferer-to-receiver links will exhibit the same frequency dependence as the intended transmitter-to-receiver link. Indeed, frequencies in the vicinity of 1 GHz may be favorable for such systems, because the external background noise is low.

Although equipment costs for the 915 MHz band may be somewhat higher than for lower bands, commercial communication equipment for the 800 MHz land mobile bands has been available for some time.

One of the current uses of the 915 MHz ISM band is for home microwave ovens, although most models operate in the 2450 MHz ISM band. Microwave oven emissions typically occupy several megahertz of bandwidth. Emission levels up to 18 dB above one microvolt per meter have been observed in a 30 kHz bandwidth at a distance of 1000 feet from a single oven, with no intervening obstacles between the oven and the receiving antenna (79). Since most emissions occupy less than 10 MHz, an upper bound on the total field strength can be placed at about 18 + 10 log (10 MHz/30 MHz) = 106 dB above one microvolt per meter at 1000 feet.

The observed interference levels were several dB lower when outdoor measurements of emissions from home-installed ovens were made (79). Building penetration loss in the vicinity of 900 MHz is on the order of 10 to 25 dB (80).

The 2450 MHz ISM band (2400-2500 MHz) is extensively used for home microwave ovens and for commercial, medical, and research purposes. In addition, this band is occasionally used for non-ISM purposes by such diverse groups as industry, local governments, fire, police, railroads, and highway maintenance services. The FCC Master Frequency List contained 319 non-Government assignments (other than experimental and ISM listings) in the 2450 MHz ISM band as of September 2, 1979. Only nine unclassified Government assignments were recorded with IRAC as of February 27, 1980.

The 2450 MHz ISM band has also been proposed for the transfer of microwave power from a solar power satellite to earth (114). If implemented, the total radiated power from such a system would be on the order of several Gigawatts (cw) in a beam having a diameter of several kilometers at the surface of the earth. The beam sidelobes, combined with scattering in the ionosphere, could make this band unusable over wide areas.

---

* ISM equipment must be licensed if and only if it does not comply with the standards set forth in Part 18 of the Rules and Regulations.

The principal use of the 5800 MHz ISM band (5725-5825 MHz) is for Government-operated radar systems. As of February 27, 1980, there were 121 Government assignments in this band, virtually all of them for radar or radar-related systems. Only four non-Government listings were on file with the FCC as of September 2, 1979, exclusive of experimental and ISM uses.

The 24,125 MHz ISM band (24,000-24,250 MHz) is near the 22,235 MHz water-vapor absorption line and is, for that reason, better suited to short and medium range applications than to long range applications. Total atmospheric attenuation, under good meteorological conditions, is typically 0.05 to 0.2 dB per kilometer at the surface of the earth (81). The principal use of this band is for police radar systems. Over 1600 licenses had been granted for non-Government use of the band (exclusive of experimental and ISM licenses) as of September 2, 1979. There were only 31 IRAC listings for government stations as of February 27, 1980. Most of these are for police radar.

3.5 Alternatives to Spread Spectrum

In order to achieve a complete view of the spread spectrum issue, it is necessary to recognize that in some cases, there are narrowband alternatives to spread spectrum that can provide most of the benefits of spread spectrum techniques. It would be impossible to list all of the potential alternatives, since this would require a study of all potential applications, present and future. However, a few examples will not be given. The examples are organized according to the type of benefit that the alternative technique is intended to achieve.

Privacy. Spread spectrum techniques provide a measure of user privacy that falls between encryption and completely uncoded transmission. There are a number of alternatives. The simplest is to substitute use of the public telephone system for radio communications whenever possible. Although this approach is not entirely "secure" in an absolute sense, it provides more privacy than uncoded private land mobile radio. Radio paging is available in most metropolitan areas, and this service can be used to establish telephone communications when a call originating from a fixed location is placed to a mobile user. Since the paging signals are coded, they are not readily intercepted by unsophisticated listeners.

The use of improved mobile telephone service (IMTS) or advanced mobile phone service (AMPS) also provides some privacy. Although the transmissions are uncoded, a listener trying to intercept a call from a particular talker must scan all the IMTS or AMPS channels. The degree of protection is somewhat higher in small-cell systems, because a mobile unit changes channels frequently as it passes from cell to cell.

Finally, it has been possible for some time to purchase voice communication security units ("scramblers") for use with conventional mobile transceivers (83).

Discrete Addressing. The discrete addressing feature of spread spectrum has two benefits. The first of these is privacy, which was just reviewed. The second is selective calling. It is important in many applications to be able to automatically route calls to particular stations. This feature is included in all IMTS and AMPS designs, as well as earlier mobile telephone systems and paging systems. Selective calling equipment is available from the major manufacturers of private land mobile radio equipment. Selective calling is also available for certain conventional maritime mobile equipment (84).

Resistance to Multipath Propagation. A primary manifestation of multipath propagation in narrowband communication systems is fading. As noted in Section 2.3.1, certain spread spectrum implementations (as well as simple, wide-band pulse techniques) can substantially mitigate fading. However, diversity techniques can be used with narrowband signaling to reduce fading (85). Diversity techniques have been only partially successful in mitigating the impact of multipath errors in narrowband ranging and radiolocation systems (86).

Spectrum Efficiency. Examples were given in Section 3.2.1 and References 19 and 23 of cases in which spread spectrum implementations use the spectrum more efficiently than conventional narrowband implementations. However, the outcomes of such examples obviously depend on the particular narrowband implementation with which a spread spectrum approach is compared. The spectrum efficiency of some narrowband systems could potentially be improved by techniques such as trunking, channel-splitting, or the use of single sideband.

In low data rate microwave FDMA systems where frequency tolerances limit spectrum efficiency, a common frequency reference, coupled with a frequency distribution system, may be appropriate. In other applications, the use of ovenized oscillators (instead of temperature-compensated crystal oscillators) may be justified. However, these alternatives are obviously not viable when the frequency tolerances are due primarily to Doppler shifts.

Resistance to Interference. In some cases, spread spectrum techniques may offer the most economical approach to mitigating interference. In other cases, narrowband techniques, combined with the use of directional antennas, special filters or interference blanking may be preferable. In advanced applications, the use of adaptive array antennas may be justified, with or without spread spectrum. It is also possible to envision "smart" narrowband feedback communication systems that would automatically detect the presence of narrowband interference, locate a clear channel, and switch to the new channel, all without human intervention.

Clearly, the choice between the use of spread spectrum techniques and narrowband techniques (where such a choice exists) requires evaluation on a case-by-case basis. The trade-off may affect cost, performance, spectrum efficiency, and the availability of operational features.

4. EXAMPLES OF HYPOTHETICAL IMPLEMENTATIONS

Up to this point, a few Government applications of spread spectrum have been briefly described (Section 2.3.1), and reference has been made to several non-Government applications that have been proposed or analyzed: fast FH/DPSK and FH/FSK for high capacity land mobile service (Section 3.2.2) and SAMSARS, which uses direct sequence signaling for maritime distress alerting via satellite (Section 3.2.4). Naturally, it is impossible to predict what new non-Government applications, if any, will be proposed in the future. However, in order to provide a better grasp of the variety of potential spread spectrum implementations, two hypothetical examples of such implementations are presented in this chapter. These examples are not intended as recommendations, but only as illustrations of potential spread spectrum applications for non-Government services. In selecting these examples, particular emphasis has been placed on approaches that have simple implementations. It is recognized at the outset that the price of simplicity is often reduced performance. Nevertheless, there is a limited payoff in considering potential spread spectrum systems having such complex implementations that the prospects for their practical realization in the foreseeable future are small.

4.1 Slow Frequency Hopping with FM Voice

4.1.1 Objectives

This approach would provide privacy, multiple access without coordination, discrete addressing, and resistance to narrowband interference. It would not provide any resistance to wideband interference.

4.1.2 Potential Applications

In a general context, the applications considered here are mobile applications where the number of potential users in a given area significantly exceeds the number of available narrow-band channels. For the purpose of illustration, a UHF citizens band application has evolved as a principal example. Other mobile applications may be equally appropriate.

4.1.3 Basic Concept

Most late-model citizens band transceivers are based on the use of low-cost phase-lock loop frequency synthesizers for channel selection. Low-cost frequency synthesizers are also used extensively for VHF maritime mobile and aeronautical mobile transceivers, for IMTS transceivers, and for other communication equipment. The concept presented here is based on the premise that such transceivers could be economically modified for slow frequency hopping.

As in conventional signaling, mobiles and base stations might transmit on the same set of frequencies (as in the present 27 MHz citizens band) or in separate sub-bands separated by typically three to five megahertz (as in many land mobile systems). The former option generally results in the least expensive equipment configuration, particularly if direct mobile-to-mobile operation (without repeaters) is required. The latter provides for the possibilities of full duplex and repeater operations.

In a practical implementation, transceivers could provide the option, by switch selection, of operating in a conventional mode or in a frequency-hopped mode. This possibility opens up a second set of system design alternatives, as follows:

1. Provide separate sub-bands for conventional channels and frequency-hopped channels.
2. Allow the same set of channels to be used for conventional communications and frequency-hopping.
3. Allow most channels to be used for either frequency-hopping or conventional communications, but reserve some channels for FH acquisition sequences.
4. Reserve a separate set of channels for conventional communications only, as an adjunct to alternative 2 or 3.

Since this is only a hypothetical example, no attempt is made to analyze the advantages and disadvantages of the various system alternatives. Instead, it has been assumed that base and mobile transmissions occupy separate sub-bands and that conventional and FH transmissions are similarly separated. A hypothetical channelization plan reflecting this strategy is shown in Figure 11a. Although an allocation in the 912-918 MHz band is shown, the basic approach is applicable, in principle, to any available VHF or UHF band. Figure 11b shows an alternate frequency plan in which only the FH channels use separate bands for base and mobile frequencies.

Note that mobile-to-mobile transmissions can occur in two modes: by repeater, or by providing mobile receivers that can be switched to either the uplink or the downlink band.

4.1.4 Equipment Configuration

Figure 12 shows a simplified block diagram of a conventional synthesizer-driven transceiver. This configuration differs from older designs only in that the carrier and local oscillator (L.O.) frequencies, instead of being generated by a separate crystal for each channel, are produced in a low-cost synthesizer using one or two crystals. In some cases, the synthesizer consists of as few as two large scale integrated (LSI) circuits, in addition to a few dozen passive elements. A programmable read only memory (PROM) is sometimes used to restrict the available channels to those for which the unit is licensed*. In transceivers designed for CB or marine use, this function can be accomplished in a permanent fashion by the use of a read only memory (ROM) or by decoding logic between the channel selector and the synthesizer.

Figure 13 shows how a conventional synthesizer-driven transceiver design might be modified for frequency-hopping. It can be seen that this design differs from the conventional design only in the addition of a "Frequency Controller" function and a PROM containing the hopping sequence. The frequency controller provides for switching the synthesizer output frequencies according to a PROM-selected FH pattern. A simplified diagram of the frequency controller is shown in Figure 14. The PROM containing the hopping sequence is cycled through its successive addresses by a binary counter, which is driven at the hopping rate, f_{HOP}. In general, the number of FH channels may be less than the number of available memory locations in the PROM. In this case, the counter must be augmented with a simple circuit to reset the counter at the end of each hopping cycle.

When the transceiver is switched from receive to transmit (typically, by grounding a push-to-talk line), a reset pulse, having a duration of one or more hops, is applied to the counter. This returns the synthesizer to the initial frequency of the hopping sequence. The hopping sequence then proceeds at a rate equal to the quiescent frequency of the voltage-controlled oscillator (VCO). Optionally, a separate crystal-controlled

---

* The PROMs are programmed by the equipment manufacturer, not by the user.

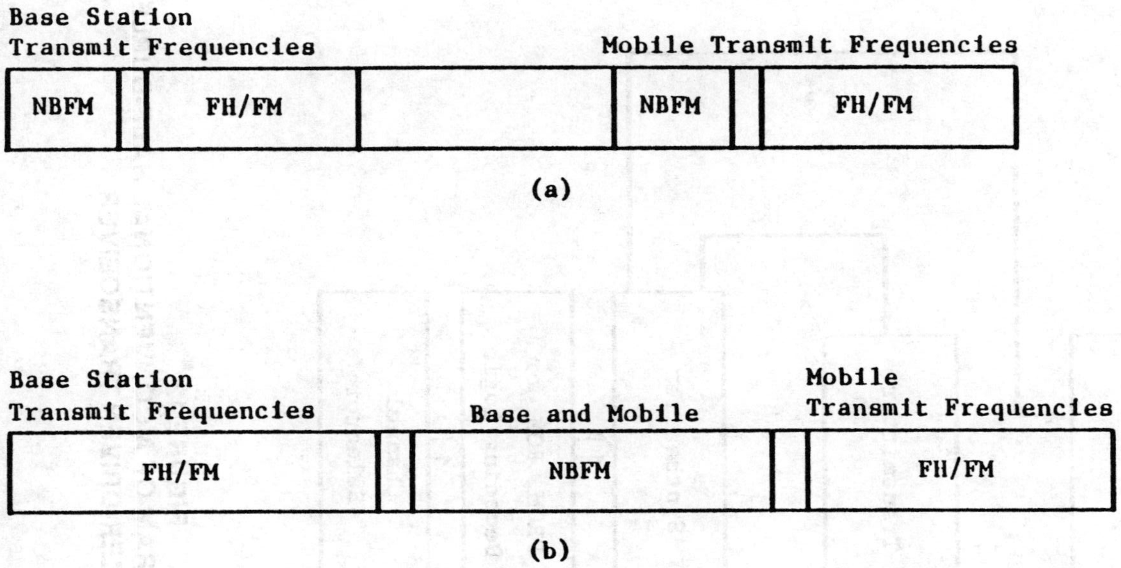

FIGURE 11
HYPOTHETICAL FREQUENCY PLANS FOR FREQUENCY HOPPING MOBILE COMMUNICATION

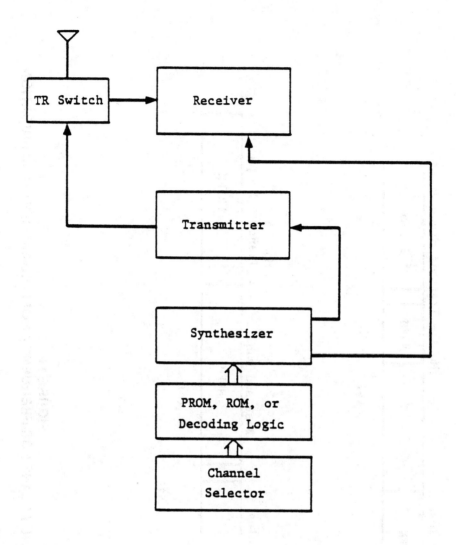

**FIGURE 12
SIMPLIFIED BLOCK DIAGRAM OF A CONVENTIONAL HALF-DUPLEX
SYNTHESIZER-DRIVEN TRANSCEIVER**

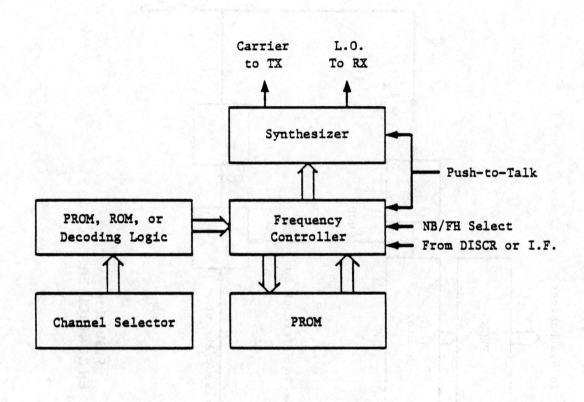

**FIGURE 13
BASIC DESIGN MODIFICATION FOR FREQUENCY HOPPING**

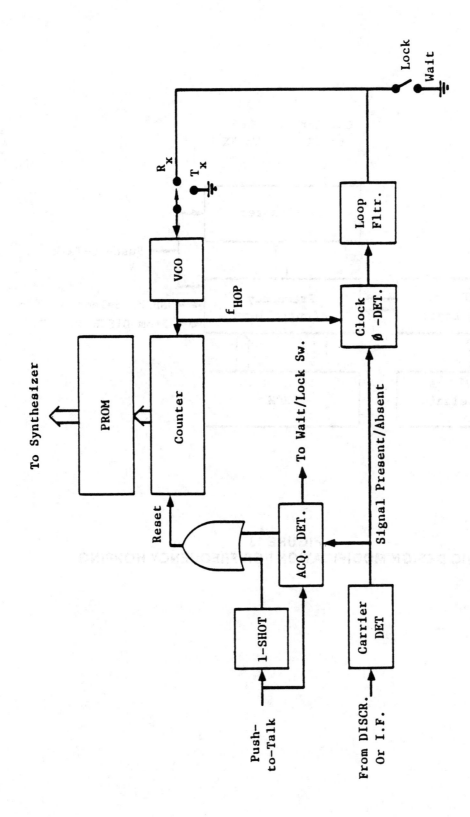

**FIGURE 14
FH CONTROLLER**

clock (not shown in Figure 14) could be used to control the hopping rate during transmissions.

When the transceiver is switched from transmit to receive, the counter is again reset, returning the synthesizer to the initial frequency in the hopping sequence. However, this time the counter is held in the reset condition until the presence of a signal is detected on that frequency. At this point, the counter is enabled and the hopping sequence proceeds. A phase detector measures the phase error of the VCO output with respect to the hops of the received signal. This error signal is used to correct the VCO frequency in a conventional tracking loop. If a signal is detected on each of several subsequent hops, acquisition is declared. If the signal is absent on subsequent hops, or if it disappears for a significant interval after acquisition has been declared, the counter is reset and the receiver returns to the initial frequency in the hopping sequence to wait for a signal.

A simplified state diagram of the process described above is shown in Figure 15. In operation, the receiver may experience several false starts if it encounters interference on the initial hopping frequency, f_o. After each false start, it returns to f_o in search of a valid FH signal.

Note that operating the transceiver in a conventional (non-FH) mode merely entails controlling the synthesizer from the channel selector rather than from the FH PROM. In the FH mode, the channel selector is disabled.

This section has provided a brief description of how a slow-FH, half-duplex transceiver might be easily implemented. Alternate implementations are obviously possible. For example, it might be desirable to use a pseudo-random code generator instead of a PROM to generate the hopping sequence. Such an approach would reduce the number of possible hopping sequences, but would facilitate "tuning" the receiver to any permissible hopping sequence. A very small amount of added logic could be used to restrict the transmitter to a particular hopping sequence, if this is deemed necessary from a regulatory viewpoint.

4.1.5 Factors Affecting Performance

In conventional FM mobile radio systems, performance is limited primarily by a combination of noise, man-made interference and fading. In a slow FH system, some of these factors have different manifestations. In addition, other factors, such as synchronization errors, can degrade performance. The following is a general description of these factors. Specific performance predictions will be presented in Section 4.1.9.

4.1.5.1 Noise

For the purpose of this report, the term "noise" includes receiver front-end noise, natural background noise, and certain types of incidental man-made noise, such as automotive ignition noise. Manufacturers of land mobile communication equipment commonly specify receiver noise in terms of sensitivity rather than noise figure or noise temperature. For receivers operating in the 800-900 MHz region, typical sensitivities range from 0.35 to 0.5 microvolt (across 50 ohms) by the 12 dB SINAD* method. These sensitivities correspond roughly to noise figures in the range of 9 to 14 dB (see Reference 91). Lower noise figures are obviously feasible but probably at increased cost (92). Even at the sensitivi-

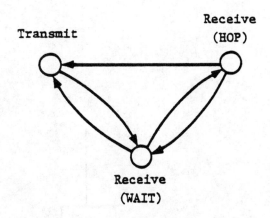

Figure 15—Simplified state diagram for FH transceiver.

ties of current equipment, however, the overall system noise is often dominated by sources external to the receiver. Man-made background noise in the vicinity of 1 GHz is typically on the order of 10 microvolts per meter in a 10 kHz bandwidth (93,94). At lower frequencies, the level of man-made noise increases.

The impact of wideband background noise, whether man-made or natural, will be approximately the same for slow FH and for conventional narrowband FM systems having comparable parameters.

4.1.5.2 Interference

Interference from narrowband communication signals can include co-channel and adjacent channel interference, intermodulation products, harmonics, spurious emissions and interference due to spurious receiver responses. In addition, the hypothetical frequency plan (Figure 11) shows an allocation in the 915 MHz ISM band; in this case interference from ISM devices can be anticipated (see Section 3.4.2). Based on measured data (79,80) applied with an assumed propagation loss of $1/R^3$ to $1/R^4$ and a 15 kHz i.f. bandwidth, it appears that a single 915 MHz microwave oven can produce a 3 dB degradation in receiver sensitivity over a maximum range of about four to twelve miles. This estimate is based on a typical receiver sensitivity of 0.35 microvolts (EIA SINAD).

Finally, the FH system will suffer self-interference from an aggregate of FH signals. Figure 16 illustrates the impact of two types of interference on a slow FH system. The horizontal axis represents time; the vertical axis represents the interference level of the output of the i.f. filter. In Figure 16a, a hypothetical single narrowband interferer is received near the center of one of the FH channels. Interference pulses, having a duration of one hop, are produced whenever the receiver hops to the channel occupied by the narrowband interferer. It is assumed that the interferer is not so strong that it is visible through the stopband of the i.f. filter. Figure 16 b shows the analogous result for interference from a single FH signal. In this case, the interference pulses are of random duration (with a maximum duration of one hop and an average duration of ½ hop) because the clocks that control hopping in the receiver and in the interfering transmitter are not synchronized.

---

* SINAD = (Signal + Noise + Distortion)/(Noise + Distortion)

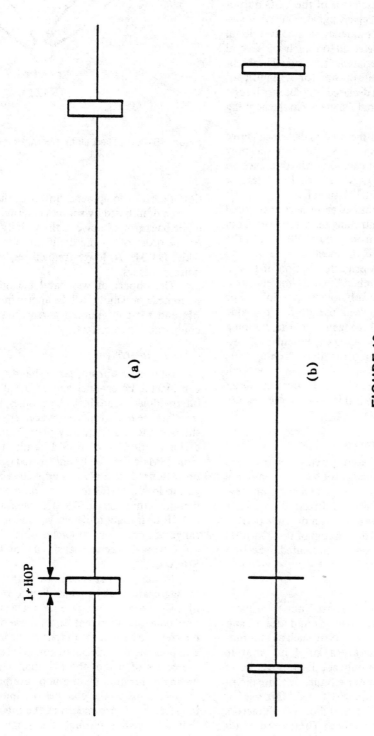

**FIGURE 16
INTERFERENCE IN A FH SYSTEM**

9-40

The types of interference depicted by Figure 16 can be described as co-channel interference. As in conventional narrowband systems, adjacent-channel interference will also occur whenever the interference in one or more of the (hopping) adjacent channels is sufficiently strong. Interference that is two or more channels removed from the intended signal can also occur. The approach described here provides no protection against wideband interference, since, in this case, each FH channel is subject to interference.

4.1.5.3 Propagation Factors: Fading and Shadowing

The characteristics of land mobile propagation have been extensively studied and documented (see, for example, References 46-49 and 85). One of the outstanding characteristics of this type of channel is the usual lack of a direct line-of-sight between the base station and the mobile with which it is communicating. Under this condition, the link is characterized by multiple reflected or diffracted paths. The transmission coefficient of the channel is modeled as a narrowband Gaussian stochastic process (95). As a result, the envelope of a CW signal transmitted over this type of channel is Rayleigh-distributed at the receiver input. The time variations of the fading envelope are a function of the vehicle speed, measured in wavelengths per unit time. For example, at 30 miles per hour a 900 MHz carrier experiences 20 dB fades (relative to the mean-square level) at an average rate of about 11 per second. The average (20 dB) fade duration is about 1.0 ms.

This model may be excessively pessimistic when a direct path exists between transmitter and receiver. Under such conditions, a Ricean fading model (which provides for one specular component plus a diffuse component) may be more appropriate.

A particular manifestation of the fading environment as it relates to slow frequency hopping is that the de-hopped signal level will abruptly change whenever the frequency difference between successive hops exceeds the coherence bandwidth of the channel. The coherence bandwidths of land mobile radio channels vary from 25 kHz to 500 kHz (95), so this can be expected to happen often in a FH system occupying several megahertz of bandwidth. This feature may actually be desirable, since it limits the length of any fade, even at low vehicle speeds. Thus, a measure of frequency diversity is provided.

4.1.5.4 Synchronization Errors

Synchronization errors can degrade the performance of this FH system in two ways. Figure 17 illustrates the effect of minor synchronization timing errors. Since the signal will not be "seen" when the transmitter and receiver are tuned to different channels, this type of timing error causes the level of the de-hopped i.f. signal to drop to zero periodically. These "glitches" occur at the hopping rate. Neglecting the rise and fall times of the transient wave form associated with the i.f. filter, the fraction of the signal that is lost in this way is equal to the timing error expressed as a fraction of the hop duration. Unless the system is provided with a very fast carrier squelch, the receiver's audio output will contain noise or interference during these "glitches." Depending on the hopping rate, the subjective effect of these periodic audio noise bursts can be quite distracting. For this reason, some type of fast audio blanking or squelch may be required.

A more serious type of degration will occur when the receiver experiences interference over such extended periods of time that it gives up on trying to maintain synchronization and goes into the "wait" mode, listening for a signal on the first frequency of the hopping sequence. The receiver may also simply fail to acquire the intended signal when it appears. However, by the time the interference reaches such a severe condition, the audio output of the receiver will probably have become unintelligible anyway. Further investigation is needed to confirm this.

4.1.5.5 Intrinsic FH Self-Noise

The approach described here does not require that either the transmitter or receiver maintain phase coherence between hops. Thus, even in the absence of synchronization errors, external noise, and interference, the audio output of the receiver will contain periodic "clicks" that correspond to the phase discontinuities between hops. Schreiber (10) has shown that the signal-to-noise power ratio attributable to this phenomenon is approximately $3\beta^2 (f_m\tau/2)$ for low hopping rates. In this expression, which is in good agreement with the exact result for $f_m\tau > 2$, the following notation has been used.

β = modulation index
f_m = information bandwidth
τ = interval between hops = 1/(hopping rate)

For the type of system described here audio SNRs on the order of 25-35 dB are attainable. However, there is currently no information available on the effect of this type of noise on intelligibility or subjective voice quality.

4.1.5.6 Random FM and Harmonic Distortion

Random RM is an effect caused by random variations in the complex channel transmission coefficient. It produces an irreducible amount of low frequency audio noise at the discriminator output (85). For the type of system under consideration here, random FM limits the audio SNR to about 26 dB, for vehicle speeds of 55 miles per hour. At lower speeds, the achievable SNR increases.

Harmonic distortion results primarily from nonlinearities in the transmitter and receiver audio sections and from bandwidth limitations in the receiver's i.f. filter. In conventional systems, total harmonic distortion (measured from transmitter input to receiver output) is less than 15 percent (97). The effects of random FM and harmonic distortion should be about the same for FH systems and conventional systems.

4.1.6 Address Code Design

The problem of constructing frequency-hopping multiple access codes having low mutual interference properties has been considered by Yates (98) and by Einarsson (99). No attempt will be made to incorporate specific codes into this example. However, the problem of address code design may be constrained by implementation considerations. Specifically, if the receiver must be able to select any valid user address code (by key pad command, for example) then it may be desirable to use linear feedback shift register generators to control the hopping sequences. If the transmitters and receivers all use fixed codes, such constraints do not exist for hopping sequences of reasonable length.

4.1.7 Degree of Privacy

One of the potential motivations for considering this approach is the provision of privacy from unauthorized listeners. The protection thus provided is not absolute, since slow FH transmissions can be de-hopped by using a sophisticated, frequency-agile receiver or by using a very wideband

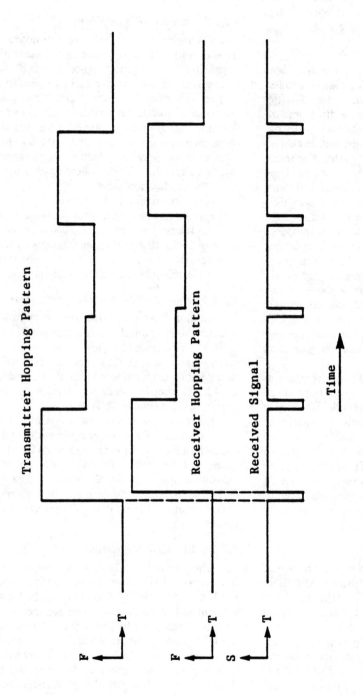

**FIGURE 17
SIMPLIFIED TIMING DIAGRAM SHOWING THE IMPACT OF SYNCHRONIZATION ERRORS**

receiver at close range to the transmitting antenna. Both of these interception techniques can be made more difficult by increasing the number of channels or the hopping rate, or both. However, for the listener with a conventional narrowband FM receiver (or a frequency-hopping receiver with a different code) the degree of privacy should be high. Miller and Licklider (96) have demonstrated that, when speech is interrupted on an average of at least once per second, and the interruptions have a duty factor of at least 90 percent, the result is virtually unintelligible. This implies that a slow FH system having a hopping rate greater than one per second and at least 10 channels should provide privacy from unsophisticated listeners.

4.1.8 System Parameters

Typical parameters for the type of slow frequency hopping system described here are presented in Table 2. Obviously, these figures do not reflect the result of any in-depth trade-off study. They are presented for the purpose of illustration only.

One important pair of parameters not listed in Table 2 is the attenuation of adjacent channel and next-adjacent channel interference relative to co-channel interference. These parameters depend on the modulation index, the channel spacing, the frequency tolerances, the design of the receiver's i.f. filter and the amount of "splatter" generated by the hopping process. In the initial simulation runs (Appendix C), 10 dB adjacent channel attenuation and 30 dB next-adjacent channel attenuation were assumed. (Signals more than two channels away on a particular hop are assumed to produce no interference on that hop.) This is representative of what can be achieved using the least expensive monolithic crystal or ceramic filters, with channel spacing of 15 to 20 kHz. The parameters were later changed to 60 and 80 dB, respectively, to show the performance of more sophisticated crystal filters and 25 kHz channel spacing. A small number of runs were performed with 200 dB attenuation of adjacent and next-adjacent channel interference to slow an upper limit on the gains that could be achieved in this area.

TABLE 2
TYPICAL SYSTEM PARAMETERS

Transmitters

Modulation:	Narrowband FM, with or without FH
Peak Deviation:	5.0 kHz
Audio Passband:	300-3000 Hz
Output Power:	35 Watts
Channel Spacing:	25 kHz
Spurious Outputs:	−75 dB
Hopping Rate:	20/s

Antennas

Base: Omnidirectional, 9 dB gain, vertical polarization
Mobile: Omnidirectional, 6 dB gain, vertical polarization

Receiver

Sensitivity:	0.35 microvolts (EIA SINAD)
Selectivity:	−80 dB (EIA SINAD)
Intermodulation:	−75 dB (EIA SINAD)
Spurious and Image Rejection:	−100 dB

4.1.9 Performance and Spectrum Efficiency

In virtually all spread spectrum multiple access systems, there is a trade-off between communication performance and spectrum efficiency. Indeed, this trade-off is not limited to spread spectrum multiple access, but is inherent in virtually any approach to sharing the spectrum in which the possibility of mutual interference exists. If high levels of interference are acceptable, then many users can operate simultaneously. If the tolerable level of interference is reduced to improve performance, then, for a given system configuration, the number of simultaneous users must be reduced. The issue of cost is also involved, since increased system complexity can allow improvements in communication performance for a given level of interference, or improvements in spectrum efficiency for a given level of communication performance.

In order to facilitate the exploration of these trade-offs for the slow FH technique, two Monte Carlo simulations were developed. The first provides estimates of the statistics of the r.f. signal-to-interference ratio. The second provides estimates of the intelligibility of speech resulting from a slow FH system. Both simulations are described in detail in Appendix C.

The signal-to-interference simulation was a developmental effort that was later supplanted by the intelligibility simulation. Both simulations model the service area as a circular region having a radius of 30 km. In all cases, the computed levels of performance refer to a receiver located in the center of this area. The transmitter with which this receiver is in communication (or with which communication is being attempted) is separated from the receiver by a randomly-selected distance. The distribution of distances has a density $p(X) = X/X_o^2 \exp(-X/X_o)$, where $2X_o$ is the mean communication distance. Interfering transmitters have randomly-selected distances (to the receiver) that correspond to a uniform average number of interferers per unit area. Both of these distributions are obviously simplifications. The signal-to-interference simulation measures the probability of attaining a given r.f. signal-to-interference ratio. The intelligibility simulation measures the probability of attaining a given articulation score. In both simulations, the Longley-Rice propagation model (46-48) is used to estimate the mean and variance of the short-term-average signal and interference levels at the receiver input.

The simulations indicate that it is probably feasible to provide adequate communication performance with slow frequency-hopping FM. The spectrum efficiency of this approach, in terms of the maximum number of active users per channel over a given geographic area, depends on several factors, including the mean communication distance and the degree of adjacent channel interference. Under favorable conditions, the slow FH technique may be able to accommodate at least as many users per channel as conventional systems having uniform channel loadings. With less favorable conditions, conventional systems having uniform channel loadings will provide better spectrum efficiency than a slow frequency hopping for a given level of performance.

The estimated maximum number of active users per channel (in a geographic area having a radius of 30 km), is presented in Table 3 for various operating conditions, based on the results of the intelligibility simulation. The first column represents the attenuation of adjacent and next-adjacent channel interference relative to co-channel interference. The second column is the mean distance over which communication is attempted. The third column shows whether the link is mobile-to-base (MB) or mobile-to-mobile without repeaters (MM). The final column represents the estimated maximum number of active users per channel, based on the assumed requirement for a 90 percent chance of achieving a word articulation score of 0.75.

TABLE 3

ESTIMATED MAXIMUM NUMBER OF USERS PER CHANNEL FOR INTERFERENCE-LIMITED OPERATION

Splatter, dB Attenuation	Mean Communication Distance, km	Type of Link	Estimated Maximum Number of Users Per Channel
10/30	8.0	MB	0.3
10/30	4.0	MB	0.3
200/200	4.0	MB	1.0
60/80	16.0	MB	0.5
60/80	8.0	MB	0.6
60/80	4.0	MB	0.9
60/80	2.0	MB	1.4
60/80	4.0	MM	0.9
60/80	2.0	MM	1.4

Two major results are evident:

(1) The degree of adjacent channel and next-adjacent channel interference can have a substantial effect on performance and spectrum efficiency. The simulation runs that are representative of the least expensive implementations show relatively poor results. Significant improvements in spectrum utilization could accrue if these minimal implementations are averted in favor of commercial-quality communication equipment with 25 kHz channel spacings.

(2) The maximum number of users per channel is a function of the mean communication distance. This relationship is shown in Figure 18 for mobile-to-base operation and 60/80 dB adjacent/next-adjacent channel attenuation. Obviously, short communication distances correspond to high signal-to-interference ratios. Thus, more users can be accommodated per channel for short range communication than for long range communication. This is true for conventional narrowband communication systems as well as spread spectrum communication systems. For conventional narrowband systems, however, the maximum number of users per channel at any given time is limited to integer values. (If two conventional users can be accommodated on one channel over a given geographic area, than two hundred conventional users can be accommodated on one hundred channels over the same area, assuming that all channels are equally loaded. The number of conventional users on a particular channel at a particular time can only be 1, 2, 3, . . .). Simulation runs with one channel indicate that if two conventional users are to share a single channel, the mean communication distance must be less than about 2 km for the parameters used here. This is based on an assumed requirement that the signal-to-co-channel-interference ratio be above 6 dB for at least 90 percent of the transmitter locations. Adjacent channel interference is not included. The relatively low degree of frequency reuse in this case is related to the fact that the transmitters are deployed randomly throughout the service area. (Higher degrees of frequency reuse are possible in cellular systems, but in such systems each channel is available over as little as 14 percent of the service area.)

It can be seen that, for the set of simulation parameters used here, the spectrum efficiency of the slow frequency hopping approach, in terms of the maximum number of simultaneous calls per channel, is comparable to or above the corresponding level for conventional narrowband systems for mean communication ranges of about 2 to 4 km. At shorter ranges, two or more calls per channel can be accommodated with conventional narrowband systems. At longer ranges, the degree of interference suffered by a slow frequency hopping system limits the number of users for a given level of performance.

Some caution needs to be exercised in the interpretation of Figure 18 and the results on which it is based. Specifically, the implicit comparison of the slow frequency hopping approach to conventional FM techniques in terms of spectrum utilization is credible only if both techniques have the same channel spacing and the same voice quality. It is always possible to trade voice quality for improved spectrum utilization in both techniques by simply reducing the channel spacing (with or without a concomitant reduction in the modulation index). This issue can probably be resolved only by experiments which would be beyond the scope of this study.

In the absence of interference, the communication range will be limited by Rayleigh fading and background noise. As shown in Appendix C, this range will be limited to about 19.3 km (12.1 miles) for 50 percent coverage or 8.3 km (5.2 miles) for 90 percent coverage, even under favorable conditions. In Urban areas, the maximum range will be substantially below these values. In the presence of combined interference, fading, and background noise, the maximum range will be further reduced.

In summary, it is clear that the slow frequency hopping approach and the conventional narrowband approach have different characteristics and may be suited to different applications. Whether the predicted performance and spectrum efficiency of the slow frequency hopping approach are compatible with the needs of various user groups remains to be seen.

4.1.10 Repeater Operation and Full Duplex Operation

The feasibility of a slow FH voice communication system does not depend on the use of separate bands for mobile-to-base and base-to-mobile transmissions. Half-duplex operation can be achieved using a common band for both link directions, as in the current 27-MHz citizens band. However, if the uplink and downlink bands are split, the possibility of repeater operation and full duplex operation is preserved. Each of these functions may be beyond the economic reach of the average CB hobbyist, but they may find other applications. Indeed, some types of service may not be practical without the use of repeaters.

Repeaters receive on the mobile-to-base frequency and simultaneously transmit on the base-to-mobile frequency, thus providing for extended-range mobile-to-mobile communication. In a frequency hopping system, it is envisioned that a repeater would consist of one or more FH receivers and FH transmitters connected back-to-back and sharing a common antenna through a duplexer*. Other arrangements (such as retransmitting the entire mobile-to-base band with a wideband linear repeater) are obviously possible, but are probably not as efficient. For any repeater arrangement, it may be necessary to include a digital code in the beginning of each mobile transmission addressed to the repeater. This would provide for immediate activation of the repeater in response to mobile transmissions addressed to it, while allowing the repeater to ignore other transmission.

---

* The ability of a duplexer to accommodate frequency hopping will depend on the separation between the transmit and receive bands and on the bandwidth over which the transmitter is hopped. The bandwidths of current commercially available duplexers range from a few hundred kilohertz or less to a few megahertz.

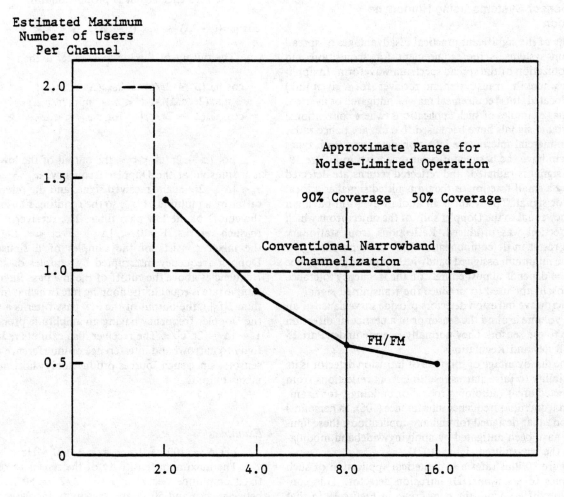

**FIGURE 18
MAXIMUM SPECTRUM UTILIZATION AS A FUNCTION OF COMMUNICATION
DISTANCE FOR INTERFERENCE-LIMITED SYSTEMS**

The use of repeaters to extend communication performance would degrade spectrum efficiency, since a single mobile user occupies an uplink and a downlink simultaneously. This is also true for conventional systems.

Full duplex operation would allow two users in mutual communication to talk simultaneously, telephone-style, at the expense of increased equipment cost and decreased spectrum efficiency. Again, each full-duplex link requires two simultaneously-active transmitters instead of one.

4.2 Sensor Systems Using Homodyne Detection

One of the significant practical disadvantages of spread spectrum systems is the requirement for acquisition and synchronization of the spread spectrum waveform. In applications where the transmitter and receiver (for a given link) are co-located, this requirement may be mitigated or averted. Particular examples of such applications (where conventional narrowband signals have been used to date) are police radar and commercial microwave intrusion detectors. Both types of system have the basic configuration shown in Figure 19. A CW signal is radiated, and reflected returns are detected by using a small fraction of the transmitted signal as a local oscillator signal. The result is an audio waveform having a frequency equal to the Doppler shift of the object from which the reflection was produced. Reflections from stationary objects result in dc components, which are ignored.

The minimum assigned bandwidth required by this type of system depends almost entirely on the frequency tolerance of the oscillator used to produce the transmitted signal.

Microwave intrusion detectors provide surveillance of an area or volume around the sensor or in a particular direction relative to the sensor. They normally operate under Part 15 of the Rules and Regulations.

One disadvantage of this type of intrusion detector is its susceptibility to false alarms resulting from reflections from unprotected areas (adjoining rooms or buildings, for example) and from radio frequency interference (106). In personnel detection radar designed for military applications, these limitations have been mitigated by applying wideband modulation to the transmitted signal (107).

Figure 20 illustrates a hypothetical application of such techniques to a commercial intrusion detector. This configuration differs from the one shown in Figure 19 in that provision has been made for frequency-hopping the transmitted signal at a fast rate. In addition, the reference signal has been delayed by an amount Δ. The frequency difference between adjacent hops is assumed to be large with respect to the maximum Doppler shift, but small with respect to the transmitter's center frequency.

On any particular hop, the transmitted signal is:

$$\cos(\omega_0 t + k\omega_m t) \quad 0 < t < \tau$$
$$k = -n, -n+1, \ldots 0.1, \ldots n$$

where τ is the duration of one hop and ω_m is the angular frequency difference between adjacent hopping "channels." For a specular reflector at a range r, the received signal is of the form:

$$\cos[\omega_0(t - 2r/c) + k\omega_m(t - 2r/c)]$$
$$= \cos[\omega_0 t + k\omega_m t - (2r/c)(\omega_0 + k\omega_m)],$$
$$2r/c < t < \tau + 2r/c$$

The reference signal is proportional to

$$\cos[\omega_0(t - \Delta) + k\omega_m(t - \Delta)], \quad \Delta < t < \tau + \Delta$$

The output of the low pass filter is then

$$\cos[\omega_0(\Delta - 2r/c) + k\omega_m(\Delta - 2r/c)],$$
$$\max(2r/c, \Delta) < t < \tau + \min(2r/c, \Delta)$$
$$= \cos[\omega_0(\Delta - 2r/c)] \quad \text{for } n\omega_m \ll \omega_0$$

For $|\Delta - 2r/c| \ll \tau$, the output of the low pass filter is a sinusoid at the Doppler frequency, $\omega_d = 2r\omega_0/c$. For $\tau < |\Delta - 2r/c|$, the received signal and the reference signal differ by a multiple of ω_m, so the resulting sinusoid is above the cutoff of the low pass filter. The receiver is insensitive to such returns. For $0 < |\Delta - 2r/c| < \tau$, the output of the mixer consists of the samples of a sinusoid at the Doppler frequency interrupted by samples of sinusoids at frequencies above the cutoff of the low pass filter. Since the sample rate is equal to the hopping rate (which is much greater than $2|f_d|$), the output of the low pass filter is a sinusoid at the Doppler frequency having an amplitude proportional to $1 - |\Delta - 2r/c|/\tau$. The receiver also exhibits reduced sensitivity to narrowband interference or interference from other sensors, since such sources produce only short pulses at the mixer output.

Example

Let $\Delta = 100$ ns, $f_H = 1/\tau = 5.0$ MHz.

The maximum sensitivity of the sensor to returns of a fixed amplitude occurs at $r = c\Delta/2 = 50$ ft. At ranges between zero and 50 ft, the sensitivity increases monotonically with increasing range for returns of a given amplitude. For ranges greater than 50 ft, the sensitivity monotonically decreases out to a maximum range of $c(\tau + \Delta)/2 = 150$ ft.

Analogous examples can be produced for intrusion detectors based on direct sequence techniques.

The extension to police radar is obvious. The FCC has received numerous complaints about jamming of these devices by motorists (113). The addition of spread spectrum capabilities would appear to be an economically viable long-term approach to this problem.

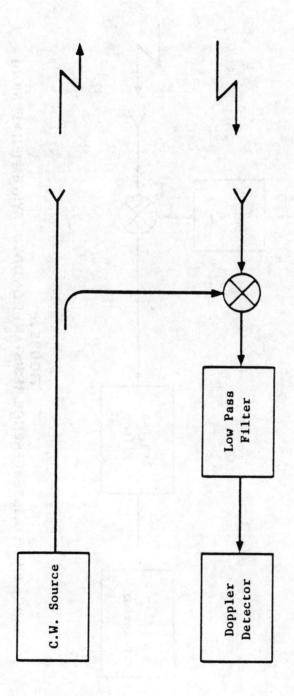

**FIGURE 19
CONVENTIONAL HOMODYNE SYSTEM**

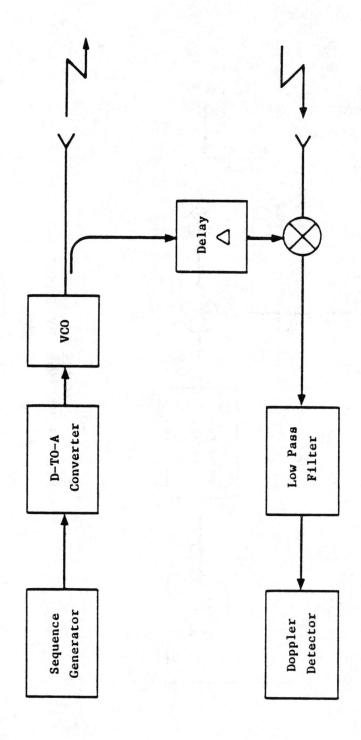

**FIGURE 20
RANGE-LIMITED, INTERFERENCE-RESISTANT HOMODYNE SYSTEM (FH VERSION)**

5. SPREAD SPECTRUM DEVELOPMENT AND IMPLEMENTATION: COSTS AND RISKS

In this section, a number of potential negative implications of spread spectrum technology will be addressed. These include: the risk of increased interference to conventional users, the difficulty of monitoring and the economic burden on the spread spectrum user.

5.1 The Risk of Increased Interference

Current users of the spectrum are likely to be concerned about potential interference from spread spectrum signals. Obviously, it is no more possible to give general a priori assurances on this issue with spread spectrum than with conventional modulation techniques. Proposed modes of operation will have to be evaluated individually to assess their potential for interference. However, it is reasonable to expect that well-conceived spread spectrum applications will be able to evolve with acceptably low levels of degradation to existing services.

5.1.1 In-Band Interference

The most severe type of interference will occur when conventional receivers share a band with spread spectrum transmitters. For continuous direct sequence emissions, the fraction of the received spread spectrum power falling into the receiver passband may be as high as B_{IF}/R_c, where B_{IF} is the IF bandwidth and R_c is the spread spectrum chip rate. For example, if $B_{IF} = 10$ kHz and $R_c = 10$ MHz, the isolation between the interferer and the receiver is only 30 dB. This assumes that the receiver is tuned to the center frequency of the spread spectrum signal, that B_{IF} is large with respect to the code repetition rate (but small with respect to the chip rate, R_c) and that the interfering signal is not so strong as to cause nonlinear amplification prior to the IF filter. Such a low degree of isolation would clearly be unacceptable for some applications. On the other hand, it might be acceptable for satellite applications in which the received power levels are constrained to narrow range or for terrestrial applications involving very low spread spectrum power levels or very low spread spectrum duty cycles, for example.

In-band interference from frequency hopping signals can take several forms, depending on the system parameters and the ratio of signal power to interference power (65). When the FH interference power at the receiver input is not greatly larger than the power of the intended signal (so that interference occurs only when the spread spectrum "hops into" the IF passband of the victim receiver) and the hopping rate is low compared to the IF bandwidth, interference will take the form of short pulses. During each pulse, the signal-to-interference ratio at the IF output will be equal to the signal-to-interference ratio at the receiver input. However, for a single FH interferer using N channels equally, this interference will occur for only a fraction of the time equal to 1/N. (A particular FH signal hopping over 100 channels will cause interference to a particular narrowband receiver only one percent of the time under such conditions.)

When the interfering FH signal power greatly exceeds the power of the intended signal, or when fast hopping rates are used, interference may occur even when the FH signal is not centered within the IF passband. The interference can result from sidelobes (i.e., "splatter") produced by the hopping process or from sidelobes of the modulation process used to impose the transmitted information on the FH signal. In addition, intermittent interference can occur as a result of intermodulation products generated by the interaction of the FH signal with other signals in the early receiver stages.

5.1.2 Out-of-Band Interference

Like conventional signals, spread spectrum signals can cause interference to services operating in adjacent bands. For direct sequence and fast frequency hopping signals, the major risk is likely to be from the sidelobes of the spread spectrum modulation. Sidelobes can be minimized by shaping the transmitted pulses and by using continuous-phase modulation. However, these measures are not without economic consequences and they may, in some cases, degrade the communication performance of the spread spectrum system.

For well-designed slow FH systems, the hopping sidelobes can be low with respect to the sidelobes produced by the information process, particularly when FM is used. Such systems may produce no more out-of-band interference than conventional signals.

5.2 Monitoring and Enforcement

Since the Commission is charged with the enforcement of its Rules and Regulations, the issue of whether to allow the use of emissions that are difficult to monitor must be addressed. This issue appears to have a number of major facets, including the ability to monitor message content and operational procedures, the ability to monitor technical characteristics such as radiated power, modulation parameters, and frequency stability, and the ability to collect statistical data on spectrum occupancy.

To the extent that any radio system provides privacy from unauthorized listeners, that system will be more difficult to monitor for message content and operational procedures than "clear voice" systems. This does not mean that such monitoring will be impossible; only that it will be more costly, more time-consuming, or less effective than it otherwise would be. The problem is not new, nor is it in any way unique to spread spectrum. (See, for example, FCC Dockets 18108, 18261, and 21142, which all relate to digital communication.) Neither is it limited to systems designed to provide privacy. Clear text digital transmissions are generally more expensive to monitor than voice transmissions simply because of the cost of the terminal equipment.

The problem of monitoring may be exacerbated, however, for spread spectrum systems in which the number of possible user codes is large. As a hypothetical example, consider a 1.2-MHz band that is channelized at 25-kHz increments. There are 1200/25 = 48 conventional channels to be monitored. However, if this bandwidth is used by a frequency hopping system having the same channel spacing, the number of possible user codes is $48! = 1.24 \times 10^{61}$, assuming that each code (i.e., hopping sequence) uses every channel once and only once. (The number of possible codes increases if this assumption is relaxed or removed.) Although it can be argued that constraints on mutual interference would prevent all of these codes from being available for use, the point is that the number of possible user codes is so large that the probability of finding an active user on a randomly-selected code is negligibly small. Similar examples can be produced for direct sequence signals.

The situation just described could make any type of monitoring activity impractical if no remedial measures were available. Rules and regulations would be simply unenforceable. Unlicensed operations could grow out of control. Fortunately, however, the Commission has a number of options that could be used to significantly mitigate or eliminate this problem. For example, the Commission could:

1. Use its regulatory powers to allow the sale of equipment having a capability for multiple user codes only in cases where a demonstrated need for such a capability exists. Restrict the number of codes according to demonstrated need.

2. Require that user codes be assigned on a permanent basis and not be subject to modification by users.

3. Require the registration of user code assignments with the FCC.

4. Authorize spread spectrum emissions only for services in which enforcement problems are manageable without extensive monitoring.

5. Authorize only spread spectrum techniques that can be "decoded" using a conventional wideband receiver at short range, such as frequency hopping with AM*, frequency-hopped on-off keying, or chirped pulse modulation.

6. Employ specialized monitoring equipment that would allow a user code to be identified by observing the signal over a sufficient period of time, along with spread spectrum receivers having selectable user codes. Allow only techniques for which such decoding is practical.

In identifying these options, no recommendation is made as to their suitability for specific cases. The intent is simply to list a number of rather obvious regulatory approaches that should be considered. Various combinations of these measures may be appropriate for particular applications.

5.3 Development and Implementation Costs

The ultimate utility of spread spectrum techniques in non-Government applications will depend strongly on economic factors. The majority of all spread spectrum equipment that has been designed and produced to date has been intended for military and aerospace applications. The relatively high cost of this equipment (typically at least ten to twenty thousand dollars per user terminal) often reflects requirements for features such as:

- resistance to intentional jamming and "spoofing,"
- a high degree of information security,
- interoperability with other equipment,
- survivability and operability in extreme environments,
- power, weight, and volume restrictions, and
- very high reliability.

Although some non-Government applications may require one or more of these features, the standards for commercial equipment are not usually as severe as the standards for military and aerospace equipment. For this reason, it is likely that commercial spread spectrum equipment, if produced in sufficient quantities, could be sold for prices significantly below the price of comparable military or aerospace equipment.

Because potential non-Government markets for spread spectrum communication equipment tend to be specialized, poorly defined, or poorly recognized, efforts to design and develop such equipment on a commercial basis have been limited.

The development of reliable, absolute cost estimates is usually an expensive, high-risk endeavor. Prices ultimately reflect such factors as the market size and market share as perceived by individual producers, the degree of competition in the market, unforeseen development or production problems, inflation, and other relatively unpredictable factors. The detailed development of such absolute price or cost estimates would be beyond the scope of this study. Nevertheless, it is possible to provide a coarse ranking of the likely relative costs of several of the spread spectrum concepts that have been described in this report.

The results of this ranking are summarized in Table 4. The basis of this table will be discussed in the following paragraphs. The six techniques or system configurations that are included in the table have been grouped into categories of high, moderate, and low cost. The order of the entries within any of these categories is arbitrary except as noted below. Each of the six techniques was placed in its respective category on the basis of the likely increase in commercial price relative to the price of comparable conventional narrowband equipment, not on the basis of absolute price level. This increase includes the impact of development costs, where applicable. Production within the next five years is assumed.

The "high" cost category includes configurations that seem likely to sell for at least 1.8 times the price of comparable commercial narrowband equipment. The corresponding factor for "moderate cost" configurations is 1.2 to 1.8, and for "low cost" configurations, 1.2 or less. The definition of "comparable" commercial narrowband equipment clearly requires some subjective judgment. Also, "low cost" configurations could become more expensive over the long run due to unforeseen requirements; "high cost" configurations could become less expensive over the long run in the event of unforeseen technological breakthroughs. Before taking even the coarse categorization of Table 4 as a solid ranking, such risks need to be taken into account.

The basis for the cost/price ranking of each of the six configurations included in Table 4 will now be discussed.

5.3.1 Fast Frequency Hopping/DPSK

The specific FFH/DPSK technique under consideration here is described in Section 3.2.2.1. It is potentially one of the most expensive spread spectrum techniques ever suggested for use under the regulatory aegis of the FCC. The factors contributing to this conclusion include the following.

- *State of Development.* Although a significant number of analyses and simulations have been performed, no equipment appears to have been developed, even on an experimental basis. In view of the uniqueness of the configuration, development costs could be significant.

- *The Need for Digitized Speech.* Digital speech techniques vary in cost, in voice quality and in susceptibility to transmission errors. However, the least expensive digital speech techniques providing reasonable voice quality at 30 kbits/s are likely to be more expensive than conventional analog transmission techniques.

- *The Need for Coherent Frequency Synthesis.* The frequency synthesizer (or other waveform generator) incorporated in the transmitter must be capable of hopping rates on the order of 10^5/s. Moreover, the phase of the transmitted waveform must remain coherent not just from one hop to another, but over at least one entire FH code cycle consisting, typically, of 16 to 32 hops. This is because the demodulation technique requires a phase comparison between chips on successive code cycles.

- *The Need for Dynamic Control of the Transmitter Power.* This feature is needed to combat the "near-far" problem.

---

* i.e., double sideband AM with carrier: A3 emission with frequency hopping.

TABLE 4

ESTIMATED RELATIVE COSTS OF SIX SPREAD SPECTRUM CONFIGURATIONS

SYSTEM CONFIGURATION	POTENTIAL APPLICATIONS	STATE OF DEVELOPMENT
High Cost		
Fast Frequency Hopping/DPSK	High Capacity Land Mobile	Concept
Fast Frequency Hopping/FSK	High Capacity Land Mobile	Concept
Moderate Cost		
Direct Sequence Modems	Voice/Data Communications	Fully Developed and Tested for Satellite Applications
SAMSARS	Maritime Distress Alerting	Engineering Models have been Developed and partially tested
Low Cost		
Slow FH/NBFM	Personal Radio Service / Private Land Mobile Radio	Commercial applications untested
Homodyne Detectors	Intrusion Alarms / Police Radar	Tested for military applications

- *The Number and Type of Components Required for the Demodulator*. Several potential demodulator designs have been suggested (24). Each of these designs requires, at a minimum, multiple bandpass filters and multiple product detectors. In typical designs, the number of each of these components might be on the order of 16 to 32. In addition, the demodulator requires this number of analog delay lines or, alternatively, a single tapped delay line having a delay-bandwidth product that is currently near or beyond the state of the art for charge-coupled devices (CCDs) and surface acoustic wave (SAW) devices. Furthermore, a yet-to-be-developed technique for rapidly and economically changing the FH address code is required if the equipment is to be used in cellular mobile radio systems.

5.3.2 Fast Frequency Hopping/Multilevel FSK

The specific FFH/FSK technique considered here was described in Section 3.2.2.2. The potential cost implications are similar to those just described for FFH/DPSK, except that coherent frequency synthesis is not required, and the demodulator configuration is significantly different. The first of these factors represents a cost advantage for FFH/FSK. The second probably does not, since the FFH/FSK demodulation technique described earlier would require, in effect, that a spectrum analysis of the transmission band be performed on each hop. It has been suggested that this function could be performed with charge coupled devices (CCDs) using the chirp-Z transform algorithm (101) but the details of the demodulator design do not appear to have been investigated.

It is not clear whether dynamic control of the transmitter power would be required with this approach. The obviation of this requirement, together with a successful CCD demodulator implementation, could bring this technique into a lower cost category.

5.3.3 Direct Sequence Modems

The type of equipment configuration to which this section refers is not as general as the title suggests. Examples of the specific types of equipment that belong in this category are described in detail in References 108 and 109. The use of suppressed clock pulse duration modulation (SCPDM) provides for voice transmission without digitization. The processing gain for voice is on the order of 21 dB. Digital data can be accommodated at rates up to 2.4 kbits/s. This configuration provides privacy, multiple access and discrete addressing (up to 2048 user codes). Similar equipment was used for the maritime VHF tests described in Section 3.3.3.

The current prices of these modems*, in small quantities, could put them in the "high cost" category. However, the relatively advanced state of development and the basic design both suggest that there is significant potential for price reductions in a commercial market environment, should one develop.

5.3.4 SAMSARS

The SAMSARS (Satellite Aided Maritime Search and Rescue System) concept was briefly described in Section 3.2.4. Direct sequence techniques are applied to low-rate digital data in order to provide for multiple access and resistance to unintentional interference. The SAMSARS concept has been tested via satellite, although not from a floating buoy, as an operational deployment would require (59).

The total cost of SAMSARS will include the cost of special-purpose receivers at each of at least three existing satellite ground stations. The way in which the ground station costs will be passed on to users has not yet been determined. Indeed, no reliable estimates of the ground station costs has been developed, although the cost of each receiver (including commercial development, testing, and installation) is believed to be on the order of $250,000. There are similar uncertainties

---

* About $20,000, depending on the features required.

in the ground station costs for alternative narrowband satellite distress signaling techniques, although the ground station costs for the narrowband techniques should be lower, all other things being equal.

In 1977, the Intergovernmental Maritime Consultative Organization (IMCO) Subcommittee on Radiocommunications estimated the cost of various Emergency Position-Indicating Radio Beacons (EPIRBs) that would signal via geosynchronous maritime communication satellite (110). The estimate was $1500 for a 1.6-GHz spread spectrum EPIRB and $1275 for a 1.6-GHz narrowband EPIRB. Although the basis of these estimates is not clear, their ratio (1.18) would place the spread spectrum EPIRB in the "low cost" category. If the cost of the EPIRB alone is considered, this is probably a legitimate conclusion. However, on the basis of overall system costs, the SAMSARS concept can reasonably be placed in the "moderate cost" category.

5.3.5 Slow FH/FM

This concept, as it might apply to non-Government mobile radio applications, was described in detail in Section 4.1. It could be implemented as a relatively simple modification of existing transceiver design.

Slow frequency hopping has been applied in a number of military designs, including SINCGARS-V (see Section 2.3.2). Target prices for the SINCGARS-V transceivers are comparable to the current price range for commercial VHF transceivers for private land mobile applications (111, 112).

5.3.6 Homodyne Detector Applications

As described in Section 4.2, this concept is also intended as a simple modification of existing designs. The current prices for commercial microwave intrusion sensors (in small quantities) range from several hundred dollars to a thousand dollars or more. In order to qualify for the "low-cost" category, the cost increment for the improved versions must be no more than 50 to 200 dollars. It seems quite reasonable to expect that this target could be attained, considering the small amount of circuitry that must be added.

5.3.7 Cost Trends

For much of the technology that is used in the implementation of spread spectrum systems, costs have been decreasing due to the use of new technology and concomitant increases in productivity. Particular examples are frequency synthesizers (now used in citizens band "walkie-talkies"), microprocessors (used in a wide range of consumer applications) and SAW devices (now used in consumer-grade television receivers).

6. CONCLUSIONS

6.1 General Comments

Spread spectrum techniques cover a wide variety of potential Government and non-Government applications. These techniques have been in use long enough to produce some advocates and some opponents. In any particular application, arguments can be made for and against the use of spread spectrum. Advocates and opponents alike will be able to find information in this report to support their respective cases, since the purpose of this report is not to defend individual viewpoints—whether general or specific—but to present a reasonably balanced view of the potential benefits, costs, and risks of spread spectrum communications as they might be applied under the FCC's regulatory domain.

The preparation of this report covered a time span of roughly nine months. The amount of original work that can be performed in such an interval is obviously limited. In addition, common sense dictates that the length of the final report not be expanded to an unreasonable size by the inclusion of large quantities of material that are already available in open technical literature. For these reasons, it was necessary to focus on limited subject areas, possibly to the detriment of others. A certain amount judgment is involved in this process, and the particular areas covered in this report are likely to interest some readers more than others.

Spread spectrum systems have characteristics that differ both qualitatively and quantitatively from the characteristics of conventional narrowband systems. The costs, risks, and benefits of particular applications can be intelligently assessed only on a case-by-case basis. Provision for such assessments already exist within the FCC's rulemaking process. Under this process, decisions are affected by evaluations performed by the FCC staff and by comments from users of the spectrum who have an interest in such decisions. The structure of the rulemaking process provides for the introduction of new technologies, but tends to discourage or inhibit reckless use of the spectrum under ill-conceived implementations.

Many potential spread spectrum applications are likely to be economically unattractive. This factor seems to have been ignored or minimized in previous studies of non-Government spread spectrum applications. Economically unattractive technologies are not likely to have significant constituencies among potential user groups and are thus unlikely to become the subject of FCC regulatory proceedings.

Other potential spread spectrum applications may be economically feasible, but may make poor use of the spectrum resources that they would require. The commission should then be prepared to determine, on a case-by-case basis, whether the benefits provided by such applications justify their inefficient use of spectrum resources. Many of the potential benefits of spread spectrum technology can be achieved with more spectrum-efficient narrowband technologies, but a cost trade-off is likely to be involved. In some cases, it is possible that particular services, functions, benefits, or levels of performance can be provided by spread spectrum technology at a lower cost than with conventional narrowband technology. However, examples of such cases are not easy to find. In other cases, it is possible that particular services, functions, benefits, or levels of performance can only be provided by spread spectrum or other wideband techniques.

In certain applications, spread spectrum techniques can make more efficient use of the spectrum than the usual implementations of narrowband techniques. Such applications tend to be easier to identify when the information bandwidth per user is low and the operating frequency is high. They also tend to result when constraints are placed on the achievable spectrum efficiency of the narrowband technique with which spread spectrum is being compared. Examples of such constraints include the need for guard bands, the existence of interference, and the use of specific modulation techniques. Comparisons of alternative modulation techniques or multiple access techniques must be carefully defined if the results as to be meaningful.

With these introductory comments, the following conclusions are submitted on the use of spread spectrum techniques for non-Government applications.

6.1.1 Performance and Efficiency of Spectrum Utilization

A trade-off exists between communication performance (or radio-location performance) and efficiency of spectrum utilization in virtually all spread spectrum systems. In the absence of interference and multipath, spread spectrum systems can achieve about the same levels of performance attainable with corresponding narrowband systems. However, such single-link spread spectrum systems make very inefficient use of the spectrum. As additional spread spectrum links are added in the same bandwidth, spectrum utilization improves, but higher signal energy is required to maintain a given performance level. The continued addition of user links will eventually increase mutual interference enough to prevent the attainability of any fixed level of performance, even with infinite signal energy. Very high levels of spectrum utilization can be achieved, but typically at very low levels of performance. This general concept is presented in quantitative form in Appendix A.

A second aspect of the trade-off between performance and spectrum efficiency is that communication techniques yielding high performance in background noise alone exhibit higher levels of spectrum utilization (for a given performance level) than do less-efficient techniques, when applied in a multiple-access environment. Thus, digital CDMA systems using powerful forward error correction techniques can provide better spectrum efficiencies than corresponding systems with only simple coding.

In multiple access systems where all signals are received with equal average power levels, spread spectrum multiple access techniques suffer a theoretical disadvantage in their maximum attainable spectrum efficiency relative to CDMA or TDMA. The disadvantage can be severe when high levels of performance are required. This statement holds even if error correction coding of M-ary signaling are allowed. Again, Appendix A provides a quantitative background for this conclusion.

Common sense dictates that the utmost care be used in drawing conclusions about practical systems based on idealized theoretical models. In particular, the work presented in Appendix A may be directly applicable only to certain geosynchronous satellite links and similar applications where the received signals have relatively uniform power levels and the frequency tolerances (including Doppler shifts) are narrow. In such applications, spread spectrum multiple access techniques typically require more bandwidth and higher power per user than do TDMA or FDMA techniques.

In applications where received signal powers vary over a wide range, or where frequency tolerances are significant (relative to the information rate per user), the trade-off between CDMA and FDMA must be viewed in a different light. Comparisons of this type must be formulated carefully if the results are to be meaningful. It is often (if not always) possible to produce comparative examples in which practical implementations of spread spectrum systems exhibit better spectrum utilization than the conventional frequency-channelized approach if the performance levels of the two systems being compared are not carefully defined.

A number of previous studies have compared the spectrum efficiency of spread spectrum with the spectrum efficiency of conventional FM systems for land mobile communication systems. In some of these comparisons, the spread spectrum approach appears to be equivalent to or somewhat superior to conventional FM systems on the basis of efficiency of spectrum utilization. In other comparisons, conventional FM techniques exhibit somewhat better performance. Such exercises call into question not only the voice quality for the approaches being compared but also the cost of the equipment, since the higher cost of spread spectrum equipment opens the door to other narrowband approaches (like amplitude-companded SSB) that may be more efficient in their use of the spectrum than narrowband FM.

One area in which spread spectrum techniques probably make better use of the spectrum than FDMA techniques is in the transmission of low-rate data at microwave frequencies when substantial frequency tolerances (due to oscillator tolerances or Doppler shifts, or both) are involved. This topic is explored in further detail in Appendix B. But again, economic factors are involved. In some of these cases TDMA may prove to be the most effective approach from the viewpoint of spectrum utilization, if its typically higher cost (with respect to FDMA) is acceptable. In the case of simple homodyne detector applications (see Section 4.2) spread spectrum techniques may be able to provide very high spectrum efficiencies at low cost.

Another area in which spread spectrum techniques may be able to improve the utilization of the spectrum is in cases where use can be made of ISM bands that are relatively unsuitable for applications requiring guaranteed high levels of performance. Indeed, since users of the ISM bands are not nominally protected from interference, it can be argued that any productive use of these bands frees other spectrum resources that are needed by applications requiring protection from interference.

6.2 Costs and Risks

6.2.1 Costs

Most spread spectrum systems can be viewed as conventional narrowband systems to which a higher level of complexity has been added for the purpose of achieving particular design characteristics. This added level of complexity naturally entails increased cost. In some cases, the cost increase may be relatively small. A preliminary assessment of the relative costs of various spread spectrum applications was provided in Section 5.3.

6.2.2 The Risk of Increased Interference

Current users of the spectrum are likely to be concerned about the risk of interference from new spread spectrum systems. This will be particularly true if attempts are made to "overlay" spread spectrum signals on bands that are now used for conventional signals in the same geographic area. Such "overlays" may be more acceptable in certain ISM bands than in bands where users are now protected from interference.

Even if "dedicated" bands are set aside for spread spectrum users, these users will suffer interference from one another. But the impact of such interference is minimized by the inherent interference resistance of the spread spectrum approach. The interference potential of any implementation must be assessed on the basis of the proposed system parameters.

6.2.3 Monitoring and Enforcement

Spread spectrum techniques can be used to produce systems in which each user has a unique code that underlies his transmitted signal. It can be difficult or impossible to demodulate a particular signal without knowledge of its "user code," and the number of possible user codes can be astronomical. If the FCC licenses such systems, it will face roughly the same type and degree of risk that it faces with conventional secure voice and secure data systems. Some possible measures for mitigating these risks are listed in Section 5.2.

6.3 Potential Applications

Although it is obviously not possible to predict specific applications that the FCC will be asked to license, it may be useful to identify the potential applications that have been proposed to date. Of these, potential land mobile applications using fast frequency hopping have dominated in the area subject to FCC licensing. Although a significant amount of analysis has been performed in this case, little or no experimental activity is evident, and implementations in the near-term seem unlikely, except possibly on an experimental basis. The prospects for the realization of such systems is weakened by their potentially high cost, although future technological breakthroughs could negate this factor.

In the maritime mobile area, direct sequence spread spectrum signaling is being proposed for emergency signaling via satellite. This concept has been partially field-tested, and could be implemented in the late 1980's. Direct sequence signaling has also been examined for possible use in terrestrial VHF maritime mobile applications, but the well-known vulnerability of direct sequence techniques to nonuniform signal levels (i.e., the "near-far" problem) will work against their realization in any mobile application that does not involve relay from geosynchronous satellites or dynamic control of transmitter powers.*

Aside from these better-known potential applications, the use of spread spectrum may be proposed in areas that have not previously received such attention. Near-term applications are likely to involve low-cost techniques like slow frequency hopping, the modification of homodyne sensors to provide a spread spectrum capability, chirp radar or chirp radiolocation systems.

6.4 Acknowledgments

The number of people who contributed to this study is so large that a complete list would be impractically long. Preliminary discussions on a possible study of non-Government uses of spread spectrum techniques were first held between Mr. Jack Robinson of the FCC's Plans and Policies and Dr. Feisal Keblawi of MITRE, after they met at a short course on the subject. A proposal was subsequently prepared and, after coordination with the FCC's Chief Scientist, Dr.

---

* However, hybrid techniques involving direct sequence techniques are entirely possible in such applications.

Stephen Lukasik, a contract was awarded. Dr. Michael Marcus of the FCC's Office of Science and Technology was named as COTR.

Besides the individuals named above, the following people contributed to the study or provided useful cooperation in one form or another.

Yarsolav (Bob) Kaminsky	MITRE
Joe Fee	MITRE (currently with FAA)
Jim Morrell	MITRE
George Steelman	MITRE
Art Smith	MITRE
Mary Ann Davis	MITRE
Bill Zeiner	MITRE
Bob Noyer	MITRE
Richard Horn	MITRE
Paul Ebert	COMSAT Corporation
Aaron Weinberg	Stanford Telecommunications
Richard Tell	EPA/Las Vegas
George Cooper	Purdue University
Ray W. Nettleton	Michigan State University
Jerry Silverman	Magnavox
Charles Cahn	Magnavox
Paul Goodman	Bell Laboratories/Holmdel
Thijs de Haas	DoC, NTIA/Boulder
Carl S. Mathews	DoC, Maritime Administration
John Juroshek	DoC, NTIA/Boulder
Samuel Musa	DoD
Don J. Torrieri	DoD

7. REFERENCES

1. R. C. Dixon, *Spread Spectrum Systems*, Wiley, 1976.
2. Donald D. Neuman, "JTIDS—An Integrated Navigation and Identification System—and Its Potential for Air Traffic Control."
3. A. L. Covitt and D. D. Neuman, "Band Sharing—A Case Study," National Telecommunications Conference, Washington, DC, November 1979.
4. F. H. Dickson, "Packet Radio Communications," Collins Radio Co., AD-786155, 30 September 1974.
5. J. Inst. of Navigation, Vol. 25, No. 2, Summer 1978, Special Issue on GPS.
6. Final Report, Multipath/Modulation Study for the Tracking and Data Relay Satellite System, Contract No. NAS5-10744, The Magnavox Company, Government and Industrial Systems Division.
7. B. G. Glazer, "Spread Spectrum Concepts—A Tutorial," Proceedings of the 1973 Symposium on Spread Spectrum Communications," Volume 1, pp. 5-8, March 1973.
8. R. L. Harris, "Introduction to Spread Spectrum Techniques," in *Spread Spectrum Communications*, Advisory Group for Aerospace Research and Development, AD-766914, July 1973.
9. William F. Utlaut, "Spread Spectrum Principles and Possible Application to Spectrum Utilization and Allocation," IEEE Communications Society Magazine, September 1978.
10. Heinz H. Schreiber, "Self-Noise of Frequency-Hopping Signals," IEEE Trans. Comm. Tech., Vol. COM-17, No. 5, October 1969.
11. D. H. Hamsher, *Communication System Engineering Handbook*, Chapter 18, McGraw-Hill, 1967.
12. Edward G. Magill, et al., "Charge-Coupled Device Pseudo-Noise Matched Filter Design," Proc. IEEE, Vol. 67, No. 1, January 1979.
13. Ronald D. Haggarty, el at., "SAW Processors vs. Competing Technologies for Spread Spectrum Data Link Applications," IEEE Ultrasonics Symposium, 26 December 1978.
14. Carlo H. Sequin and Michael F. Tompsett, *Charge Transfer Devices*, Academic Press, 1975.
15. Donald M. Grieco, "Inherent Signal-to-Noise Ratio Limitations of Charge-Coupled Device Pseudo-Noise Matched Filters," IEEE Trans. Comm., Vol. COM-28, No. 5, May 1980.
16. Andrew J. Viterbi, "Spread Spectrum Communications—Myths and Realities," IEEE Communications Magazine, May 1979.
17. Andrew R. Cohen, et al., "A New Coding Technique for Asynchronous Multiple Access Communication," IEEE Trans. Comm. Tech., Vol. COM-19, No. 5, October 1971.
18. J. P. Costas, "Poisson, Shannon, and the Radio Amateur," Proc. IRE, December 1959.
19. George R. Cooper and Ray W. Nettleton, "Spectral Efficiency in Cellular Land-Mobile Communications—A Spread-Spectrum Approach," PB293264, 1978.
20. G. R. Cooper and R. W. Nettleton, "Spread Spectrum in a Fading, Interference-Limited Environment," International Communications Conference, June 10-13, 1979.
21. D. P. Grybos, et al., "Probability of Error Performance of the Spread Spectrum Mobile Communications Receiver in a Non-Rayleigh Fading Environment," 29th IEEE Vehicular Technology Conference, March 27-30, 1979.
22. G. R. Cooper, et al., "Cellular Land Mobile Radio: Why Spread Spectrum," IEEE Communications Magazine, Vol. 17, No. 2, March 1979.
23. G. R. Cooper and R. W. Nettleton, "A Spread-Spectrum Technique for High-Capacity Mobile Communications," IEEE Trans. Vehicular Technology, Vol. VT-27, No. 4, November 1978.
24. D. P. Grybos and G. R. Cooper, "A Receiver Feasibility Study for the Spread Spectrum High Capacity Mobile Radio Systems," 28th IEEE Vehicular Technology Conference, March 22-24, 1978.
25. Charles Cahn (Magnavox), Private Communication, December 10, 1979.
26. Jerry Silverman (Magnavox), Private Communication, December 10, 1979.
27. Del Norte Technology, Inc., Comments on FCC Notice of Proposed Rulemaking PR Docket 80-9, "Police Surveillance."
28. IEEE Trans. Comm., Vol. COM-25, No. 8, August 1977 (Special Issue on Spread Spectrum).
29. Samuel A. Musa and Wasyl Wasylkiwskyj, "Co-Channel Interference of Spread Spectrum Systems in a Multiple User Environment," IEEE Trans. Comm., Vol. COM-26, No. 10, October 1978.
30. Leslie A. Berry, "Spectrum Metrics and Spectrum Efficiency: Proposed Definitions," IEEE Trans. Electromagnetic Compat., Vol. EMC-19, No. 3, August 1977.
31. R. P. Gifford, "EMC Revisited—1966," IEEE Trans. Electromagnetic Compat., Vol. EMC-8, No. 3, September 1966.
32. K. Powers, "Diversity of Broadcasting," Science and Technology, April 1968.
33. L. C. Tillotson, et al. "Efficient Use of the Radio Spectrum and Bandwidth Expansion," Proc. IEEE, Vol. 61, No. 4, April 1973.
34. D. R. Ewing and L. A. Berry, "Metrics for Spectrum-Space Usage," Office of Telecommunications Report 73-24, U.S. Department of Commerce, Nov. 1973.
35. O. L. Luk'yanova, "Characteristics of the Utilization of the Radio-Frequency Spectrum by a Particular System," Telecommunications and Radio Engineering, Vol. 28/29, No. 4, April 1974.
36. C. Colavito, "On the Efficiency of the Radio Frequency Spectrum Utilization in Fixed and Mobile Communication Systems," Alta Frequenza, Vol. XLIII, No. 9, September 1974.
37. "The Effect of Modulation Characteristics on the Efficiency of Use of the Geostationary Orbit," Documents of the XIIIth Plenary Assembly, ITV, CCIR Report 559, Geneva, 1974.
38. D. N. Harfield, "Measures of Spectral Efficiency in Land Mobile Radio," IEEE Vehicular Technology Conference, 1975.
39. Scott H. Cameron, "Efficiency of Spread Spectrum Signaling with Respect to Spectrum Utilization," 19th IEEE Electromagnetic Compatibility Symposium, August 1977.
40. "Mobile Multiple Access Study, Final Report," TRW Defense and Space Systems Group, NASA Contract No. NAS5-23454, August 1977.
41. R. C. Sommer, "Asynchronously Multiplexed Channel Capacity," Proc. IEEE, Vol. 54, January 1966.
42. Lucky, Salz, and Weldon, *Principles of Data Communication*, Ch. 8 and 9, McGraw-Hill, 1968.

43. Frank de Jager and Cornelis B. Dekker, "Tamed Frequency Modulation, A Novel Method to Achieve Spectrum Efficiency in Digital Transmission," IEEE Trans. Comm., Vol. COM-26, No. 5, May 1978.

44. Michael B. Pursley, "Performance Evaluation for Phase-Coded Spread-Spectrum Multiple-Access Communication—Part 1: System Analysis," IEEE Trans. Comm., Vol. COM-25, No. 8, August 1977.

45. Leslie A. Berry and E. J. Haakinson, "Spectrum Efficiency for Multiple Independent Spread Spectrum Land Mobile Radio Systems," NTIA Report 78/11, PB-291539, November 1978.

46. Anita G. Longley, "Location Variability of Transmission Loss—Land Mobile and Broadcast Systems," Office of Telecommunications, Report No. 76-87, May 1976.

47. A. G. Longley and P. L. Rice," Prediction of Tropospheric Radio Transmission Loss Over Irregular Terrain, A Computer Method—1968," ESSA Technical Report ERL 79-ITS67, AD-676874, July 1968.

48. A. G. Longley and P. L. Rice, "Comparison of Propagation Measurements with Predicted Values in the 20 to 10,000 MHz Range," ESSA Technical Report ERL 148-ITS 97, January 1970.

49 Anita G. Longley, "Radio Propagation in Urban Areas," 28th IEEE Vehicular Technology Conference, March 22-24, 1978.

50. George R. Cooper and Raymond W. Nettleton, "A Spread Spectrum Technique for High Capacity Mobile Communications," 27th IEEE Vehicular Technology Conference, March 16-18, 1977.

51. Raymond W. Nettleton and George R. Cooper, "Error Performance of a Spread-Spectrum Mobile Communications System in a Rapidly-Fading Environment," National Telecommunications Conference, 1977.

52. Raymond W. Nettleton and George R. Cooper, "Mutual Interference in Cellular LMR Systems: Narrowband and Broadband Techniques Compared," Midcon 1977 Proceedings, November 8-10, 1977.

53. Gary D. Ott, "Vehicle Location in Cellular Mobile Radio Systems," IEEE Trans. Veh. Technol., Vol. VT-26, No. 1, February 1977.

54. James J. Mikulsky, "Mobile Radio: To Spread or Not to Spread," IEEE Communications Magazine, Vol. 17, No. 4, July 1979.

55. Paul S. Henry, "Spectrum Efficiency of a Frequency-Hopped DPSK Mobile Radio System," 29th IEEE Vehicular Technology Conference, 1979.

56. A. Weinberg, et al., "A Novel Concept for a Satellite-Based Maritime Search and Rescue System," IEEE Trans. Veh. Technol., Vol. VT-26, No. 3 (Special Joint Issue on Maritime Communications), August 1977.

57. A. Weinberg, "Characteristics and Performance Aspects of a Concept for Satellite Based Maritime Search and Rescue," The MITRE Corporation, Report No. M77-101, September 1977.

58. W. Scales, "Development of an Experimental Spread Spectrum Maritime Communications System: Phase 1," The MITRE Corporation, Report No. MTR-79W00217, July 1979.

59. J. J. Fee, et al., "Development of an Experimental Spread Spectrum Maritime Communications System: Phase II," The MITRE Corporation, Report No. MTR-80W00052, November 1980.

60. "Operational Concept, Technical Characteristics and Test Results of the Satellite-Aided Maritime Search and Rescue System (SAMSARS)," U.S. CCIR Document USSG 8/12C, 21 May 1980.

61. SARSAT System Summary, April 1980.

62. "European Communications Experiments in L Band with ATS-6," European Space Research Organization, Volumes 6 and 11.

63. Larry D. Reed, "Land Mobile Spectrum Utilization, Los Angeles and San Diego, CA," FCC Report No. PRB/RDL 79-02, November 1979.

64. "Spectrum Requirements for the Civil Maritime and Aeronautical Services," Report to the FCC by the UHF Task Force, Appendix A, FCC Order No. 1820-2 (30E25.13), November 1978.

65. Paul Newhouse, "Procedures for Analyzing Interference Caused by Spread-Spectrum Signals," ITT Research Institute Report No. ESD-TR-77-033, AD A056911, February 1978.

66. Don J. Torrieri, "Frequency-Hopping in a Jamming Environment," AD A067299, December 1978.

67. "Japanese EPIRB Experiment Plan," CCIR XVth Plenary Assembly, Japan No. 8/7/5, June 1980.

68. George R. Cooper, "Multi-Service Aspects of Spread-Spectrum Mobile Communication Systems," MIDCON '78, Paper No. 28.2, Dallas, Tx, December 12-14, 1978.

69. T. Dvorak, "Compatibility of Spread-Spectrum Signals with Narrow-band-FM Receivers in VHF Mobile Networks," IEEE Electromagnetic Compatibility Symposium, p. 19-23, June 20-27, 1978.

70. J. R. Juroshek, "A Compatibility Analysis of Spread-Spectrum and FM Land Mobile Radio Systems," NTIA Report 79-23, August 1979.

71. John R. Juroshek, "A Preliminary Estimate of the Effects of Spread-Spectrum Interference on TV," NTIA Report 76-8, PB-286623, June 1978.

72. R. F. Ormondroyd, "Spread Spectrum Communication in the Land Mobile Service," IEE Conference on Communication Equipment and Systems, Birmingham, England, April 4-7, 1978.

73. D. C. Coll and N. Zervos, "The Use of Spread Spectrum Modulation for the Co-Channel Transmission of Data and Television," National Telecommunications Conference, December 27-29, 1979.

74. "EMC Analysis of JTIDS in the 960-1215 MHz Band," Report No. OT 78-140, March 1978.

75. L. R. Welch, "Lower Bounds on the Minimum Cross Correlation of Signals," IEEE Trans. Inf. Theory, May 1974.

76. Dilip V. Sarwate, "Bounds on Cross Correlation and Autocorrelation of Sequences," IEEE Trans. Inf. Theory, Vol. IT-25, No. 6, November 1979.

77. R. D. Yates and G. R. Cooper, "Design of Large Signal Sets with Good Aperiodic Correlation Properties," Technical Report TR-EE 66-13, Purdue University, September 1966.

78. IEEE Proceedings, Vol. 62, No. 1, (Special Issue on Industrial, Scientific, and Medical Applications of Microwaves), January 1974.

79. Richard A. Tell, "Field-Strength Measurements of Microwave-Oven Leakage at 915 MHz," IEEE Trans. Electromagnetic Compat., Vol. EMC-20, No. 2, May 1978.

80. J. M. Durante, "Building Penetration Loss at 900 MHz," 1973 IEEE Vehicular Technology Conference, Cleveland, Ohio, December 3-5, 1973.

81. M. I. Skolnik, *Radar Handbook*, McGraw-Hill, p. 24-17, 1970.

82. Del Norte Technology, Inc., Petition for Rulemaking, General Docket 80-135 (non-Government radiolocation in the 420-450 MHz band).

83. A. M. McCalmont, "Communications Security Devices (CSD)—Techniques, Constraints, and Selection," 28th IEEE Vehicular Technology Conference, Denver, Colorado, March 22-24, 1978.

84. Thijs De Haas, et al., "Maritime Digital Selective Calling System Technical Description," The MITRE Corporation, Report No. MTR-7217, Rev. 1, June 1976.

85. William C. Jakes, Jr., (Ed.), *Microwave Mobile Communications*, Wiley, 1974.

86. W. Scales, "Urban Field Tests of Four Vehicle Location Techniques," The MITRE Corporation, Report No. MTR-6399, Rev. 1, April 1973.

87. H. Helgert and R. L. Pickholtz, "Low Rate Coding for Spread Spectrum," National Telecommunications Conference, Washington, DC, November 27-29, 1979.

88. Bruce P. Gibbs, "Results of the NASA/MARAD L-Band Satellite Navigation Experiment," EASCON-77 (also reprinted in AESS Newsletter, July 1978).

89. Ray W. Nettleton, "Traffic Loads in a Spread-Spectrum Cellular Land-Mobile System," 1979 MIDCON, November 6-8, 1979.

90. Ray W. Nettleton, "Traffic Theory and Interference Management for a Spread Spectrum Cellular Mobile Radio System," International Conference on Communications, Seattle, June 8-12, 1980.

91. Donald H. Hamsher (Ed.), *Communication System Engineering Handbook*, p. 17-14, McGraw-Hill, 1967.

92. Merril I. Skolnik, *Radar Handbook*, p. 5-10 and 5-11, McGraw-Hill, 1970.

93. Edward N. Skomal, "The Range and Frequency Dependence of VHF-UHF Man-Made Radio Noise in and Above Metropolitan Areas," IEEE Trans. Vehicular Technol., Vol. VT-19, No. 2, May 1970.

94. *Reference Data for Radio Engineers*, Fourth edition, p. 763, ITT, 1967.

95. Michael J. Gans, "A Power-Spectral Theory of Propagation in the Mobile-Radio Environment," IEEE Trans. Vehicular Technol., Vol. VT-21, No. 1, February 1972.

96. George A. Miller and J. C. R. Licklider, "The Intelligibility of Interrupted Speech," J. Acoustical Soc. of America, Vol. 22, No. 2, March 1950.

97. "Minimum Standard for Land-Mobile Communication Systems Using FM or PM in the 25-470 MC Frequency Spectrum," EIA Std. RS-237, August 1960.

98. R. P. Yates, "Design of Large Signal Sets with Good Aperiodic Correlation Properties," Ph.D. Thesis, Purdue University, 1966.

99. Goran Einarsson, "Address Assignment for a Time-Frequency-Coded Spread Spectrum System," Bell Laboratories, Crawford Hill Laboratory, Holmdel, NJ, 07733.

100. James L. Flanagan, et al., "Speech Coding," IEEE Trans. Comm., Vol. COM-27, No. 4, April 1979.

101. A. J. Viterbi, "A Processing Satellite Transponder for Multiple Access by Low-Rate Users," Digital Satellite Communications Conference, Montreal, October 1978.

102. David J. Goodman, et al., "Frequency Hopped Multilevel FSK for Mobile Radio," 1980 International Zurich Seminar on Digital Communications.

103. Uzi Timor, "Improved Decoding Scheme for Frequency Hopped Multilevel FSK System," Bell Laboratories, Crawford Hill Laboratory, Holmdel, NJ.

104. Goran Einarsson, "Address Assignment for a Time-Frequency-Coded Spread Spectrum System," Bell Laboratories, Crawford Hill Laboratory, Holmdel, NJ.

105. B. G. Haskell, "Computer Simulation Results on Frequency Hopped MFSK Mobile Radio—Noiseless Case," Bell Laboratories, Crawford Hill Laboratory, Holmdel, NJ.

106. "Selection and Application Guide to Commercial Intrusion Alarm Systems," NBS Special Publication 480-14, August 1979.

107. U. A. Frank, et al., "The Two-Pound Radar," RCA Eng., Vol. 13, No. 2, pp. 52-54, August-September 1967.

108. Leonard Jacobson, et al., "Highly Efficient Voice Modulation for Low C/No Communications Channels with Hard Limiting Repeaters," IEEE Trans. Comm., Vol. COM-21, No. 2, February 1973.

109. "MX-170C for Satellite and Radio Communications," Report No. R-5362, Magnavox, Advanced Products Division, Torrance, CA, May 1976.

110. "Operational Requirements for EPIRBs and Survival Craft Equipment," COMXVIII/WP.8, IMCO Subcommittee on Radiocommunications, 1977.

111. R. D. Boyle, "SINCGARS V—Secure Battlefield Communications for the 1980s," International Defense Review, pp. 1155-1157, 6/1977.

112. "JTIDS/TIES Consolidate Tactical Communications," EW Review, September/October 1977.

113. "More Complaints Received About Speed-Radar Jammers," Signal, p. 52, July 1980.

114. William E. Gordon and Lewis M. Duncan, "SPS Impacts on the Upper Atmosphere," Aeronautics and Astronautics, July/August 1980.

APPENDIX A:

SPECTRUM EFFICIENCY OF SSMA

Summary

The spectrum efficiencies of several spread spectrum multiple access (SSMA) techniques are examined. Spectrum efficiency, for the purpose of this paper, is defined as total network information throughput per unit bandwidth. Results are presented for fading and non-fading channels. Under conditions of equal received power levels, direct sequence techniques can out perform frequency-hopping techniques. This conclusion is reversed when one of the interferers is much more powerful than the desired signal. The spectrum efficiencies attainable with SSMA are generally low, although SSMA may provide better spectrum efficiency than FDMA or TDMA under certain specialized conditions.

A.1 INTRODUCTION

The spectrum efficiency of particular Spread Spectrum Multiple Access (SSMA) networks has been the subject of a number of recent papers. For example, Cooper and Nettleton have analyzed the performance and spectrum efficiency of a unique frequency-hopping system using DPSK (A1-A5). Henry (A6) has performed an independent analysis of the same basic approach (although with different parameters) that includes spectrum efficiency computations and comparisons. Frequency hopping with multilevel FSK was analyzed by Goodman and others (A7).

At a more general level, numerous analyses of SSMA performance have been published, but usually without reference to spectrum efficiency (A8-A13). Two notable exceptions are Costas (A14) and Cameron (A15). In some cases, it may not be possible to derive a simple, closed-form expression for SSMA spectrum efficiency in terms of the network performance parameters. In others, such a relationship follows indirectly from the results of the performance analysis.

Spread spectrum systems are of several basic types. In direct sequence (DS) systems, each information symbol is encoded as a pseudo-random sequence that is superimposed on the carrier phase. In frequency-hopping (FH) systems, the transmit and receiver carrier frequencies are switched at regular intervals according to a pseudo-random pattern. Each method has advantages and disadvantages, independent of the multiple access application. Other relative advantages and disadvantages relate directly to multiple access capabilities.

The purpose of this appendix is to relate SSMA communication performance (in terms of error rate) to spectrum efficiency. It can be argued that such relations should exist for any SSMA system. First, let spectrum efficiency be initially defined as

$$\eta = \frac{MR}{B} = \frac{Mq}{BT}$$

where
M = number of simultaneous SSMA links
R = information rate per link
B = total system RF bandwidth
q = information content of one symbol (bits)
T = symbol duration

Although other measures of spectrum efficiency have been suggested (A16), this one has been used frequently (A5,A6). By this definition, η is approximately the network information throughput per unit bandwidth for systems with low error rates. (The definition will later be modified to reflect the loss of information due to transmission errors.)

As an elementary observation, it is clear that if M exceeds the number of the dimensions in the signal space, 2BT, then there will necessarily be mutual interference between the various links, even if all of the links have the same symbol timing and the relative carrier phases can be perfectly controlled. Furthermore, as $M/2BT = \eta/2q$ increases, the average distance between members of any signal set in the space defined by 2BT dimensions must decrease. Thus, a relationship between spectrum efficiency (as defined above) and communication performance would be expected in every case.

As a first step in established relations between performance and spectrum efficiency, consider the general case in which each of M distinct DS signals occupies an RF bandwidth equal to twice the one-sided bandwidth, W, of its baseband code. In the absence of phase synchronization between the M carriers, the effective dimensionality of the signal space is then 2WT = BT. Assume further that the M − 1 unwanted signals can be represented as a composite waveform characterized as an independent noise process, x(t). Then, for equal signal energies, the one-sided power density of x(t) is approximately

$$S_o = \frac{(M-1)E}{BT} \qquad (1)$$

where E is the energy of a transmitted symbol. The effective total noise power density is

$$N_o' = N_o + S_o \qquad (2)$$

where N_o is the background noise power density. Solving equation (2) for S_o and substituting into equation (1) yields

$$N_o' - N_o = \frac{(M-1)E}{BT}$$

which implies

$$\eta = q\left[\frac{N_o'}{E} - \frac{N_o}{E} + \frac{1}{BT}\right] \qquad (3)$$

The required error rate, for any given data modulation technique, determines the required E/N_0. Also, E/N_0 is just the signal-to-noise ratio without interference, so both of these parameters are presumably known. Thus, equation (3) relates spectrum efficiency to communication performance through E/N_0. Note that for large WT, η can approach one only if E/N_0 is large.

As Mazo recently pointed out (A8), this model must be used carefully, because it embodies the tacit assumption that the M codes are more or less uniformly distributed over the available signal space. In particular, if the M codes in use at any given time are only a subset of a larger number, L, of assigned user codes, then the error rate computed from E/N_0 (whether using a Gaussain noise model or not) will not be a guaranteed error rate for any M users, but rather an average error rate over the ensemble of all possible subsets of M codes out of the total population L.

Note also that, while equation (3) implies that η is a discrete function of the variables on the right, η can, in fact, only take on discrete values, corresponding to particular values of M and N_0. However, it will later become evident that, for large M, the values of η are so close together that equation (3) can, for some purposes, be considered to be continuous.

There are several other noteworthy aspects to equation (3). Since it contains a term inversely proportional to BT, a disadvantage is implied for spread spectrum systems. For narrowband systems, BT $\approx$ 2 (WT $\approx$ 1) and $\eta = 0.5$ for binary signals in the absence of interference. (Note that η can exceed unity if M-ary narrowband signals are used.) However, narrowband systems are not resistant to co-channel interference and must generally be multiplexed by frequency. The requirement for guard bands limits the spectrum efficiency of Frequency Division Multiple Access (FDMA) to values typically smaller than unity, although FDMA may, in many cases, offer spectrum efficiency superior to that of SSMA. For large SSMA systems, BT must be large also, and in this case η is seen to be relatively independent of BT. Also, the use of error-correcting codes to increase BT can substantially lower the required E/N_0 for a given error rate, thus yielding a net increase in spectrum efficiency.

A.2 DIRECT SEQUENCE

A.2.1 Synchronous Direct Sequence

Since it is usually impractical to synchronize transmitters to within one code "chip" (spread spectrum code element) this case is primarily of theoretical interest. It provides an upper bound on uncoded binary systems, and was treated by Mazo (A8) in a recent analysis.

The case to be considered here involves the use of M binary phase-shift keyed (PSK) signals of the form

$$x_i(t) = b_i c_i(t) \cos(\omega_c t + \phi_i) \quad 0 \le t \le T$$

where $b_i = \pm 1$ and all variables are real. The code sequence is

$$c_i(t) = \sum_{j=0}^{K-1} a_j S(t - j\Delta)$$

where again $a_j = \pm 1$. Also, $s(t)$ is the basic "chip" waveform of length Δ. The carrier phase of the desired signal is taken to be zero; the phases of the other signals are assumed to be uniformly distributed over $(0, 2\pi)$. The effective dimensionality of the baseband signal space is $K = T/\Delta$. Note that all signals are assumed to be mutually synchronized except for carrier phase.

A.2.1.1 Noise-Limited Case

First, let us examine the case where the maximum number of simultaneous links, M, is equal to the number of codes that have been assigned. That is, all possible users are allowed to access the network simultaneously. Clearly, if $M \le K$, the M codes can be chosen to be orthogonal, so interference does not result. The maximum spectrum efficiency is then

$$\eta = \frac{M}{2WT} = \frac{M}{2K} \le 0.5$$

where the equality holds for the maximum loading without interference, i.e., $M = K$.

A.2.1.2 Interference-Limited Case, Restricted Access

In this case, the condition $K < M$ requires explicit consideration of both noise and interference. The M codes in use at any given time are taken to be a subset of a total population of L user codes. For example, there may be $L = 1000$ potential users, of which only $M = 500$ are allowed simultaneous access.

The receiver correlates the incoming signal against $c_i \cos(\omega_c t)$ (where ϕ_1 has been taken to be zero), and uses the sign of the result as an estimate of b_i. The test statistic against which the bit decisions are made is proportional to

$$y_1 = b_1 + \sum_{i=2}^{M} b_i \rho_{1i} \cos\phi_i + n_i \quad (4)$$

where ρ_{ij} is the normalized cross-correlation

$$\rho_{ij} = \frac{1}{E} \int_0^T c_i(t) c_j(t) \, dt$$

and

$$n_1 = \int_0^T n(t) c_1(t) \cos\omega_c t \, dt$$

is the noise component at the correlator output. The predetection SNR (power ratio) is

$$\frac{2E}{N_0'} = \frac{2}{\sum_{i=2}^{M} \rho_{1i}^2 + N_0/E} \quad (5)$$

Because of the presence of ρ_{1i} in equations (4) and (5), the performance depends on the particular set of codes in use, as well as the number of links in use. However, for any intelligently-selected large set of codes, there are bounds on ρ_{ij}^2. As a lower bound, the Welch bound on inner products (A19) requires that

$$\rho_{max}^2 > \frac{1}{L-1}\left(\frac{L}{K} - 1\right)$$

where ρ_{max}^2 is the maximum ρ_{ij}^2 for a set of L codes, each consisting of K binary elements.

The Welch bound does not provide a direct constraint on the predetection SNR, except for the very weak bound

$$\frac{2E}{N_0'} \le \frac{2}{\rho_{max}^2 + N_0/E}$$

and even this holds only for the code subsets containing ρ_{max}. It is therefore useful to consider the ensemble of all subsets of M codes out of the population L. The ensemble average of ρ_{ij}^2 will then be close to the arithmetic average of ρ_{ij}^2 for a given set of codes as long as M is sufficiently large. The "typical" predetection SNR is then

$$\frac{2E}{N_0'} = \frac{2}{(M-1)\overline{\rho^2} + N_0/E} \quad (6)$$

where $\overline{\rho^2}$ is the ensemble average of ρ_{ij}^2.

The spectrum efficiency is then

$$\eta = \frac{M}{2K} = \frac{1}{2K}\left(\frac{\beta}{\overline{\rho^2}} + 1\right)$$

where the notation

$$\beta = N_0'/E - N_0/E$$

has been introduced for brevity.

It happens that the Welch bound on inner products (A19) was derived by placing a bound on

$$\Sigma \rho_{ij}^2$$

where the summation is taken over all i,j. The result is that $\overline{\rho^2}$ is subject to the same bound as ρ_{max}^2. The impact of this on spectrum efficiency is that

$$\eta \le \frac{1}{2}\left[\frac{\beta(L-1)}{L-K} + \frac{1}{K}\right]$$

It will now be shown that this bound can be achieved, at least for integer values of L/K. Consider the following strategy for assigning M codes out of a population L. Let the first K codes be orthogonal ($\rho_{ij} = 0$). These same codes are then uniformly re-assigned for the remaining L-K codewords. Because of the re-use of codewords, $\rho_{max}^2 = 1$. Also,

$$\overline{\rho^2} = 0 \cdot \text{Pr (2 randomly selected codewords are different)}$$
$$+ 1 \cdot \text{Pr (2 randomly selected codewords are identical)}$$

If we consider only values of L that are integer multiples of K, then the number of times that each of the K codewords is assigned is L/K. Then it is easily shown that

$$\overline{\rho^2} = K\left[\frac{\frac{L}{K}\left(\frac{L}{K} - 1\right)}{L(L - 1)}\right] = \frac{1}{L - 1}\left(\frac{L}{K} - 1\right) \quad K < L$$

which is precisely the lower bound on $\overline{\rho^2}$ that was derived above. Since the bound is clearly attainable, at least in principle, the achievable spectrum efficiency is

$$\eta = \frac{1}{2}\left[\frac{\beta(L - 1)}{L - K} + \frac{1}{K}\right] \approx \frac{\beta}{2}\left(\frac{L/K}{L/K - 1}\right) \quad 1 \ll K \ll L$$

This expression seems to suggest that the spectrum efficiency can be increased to any desired value by operating at low signal-to-noise ratios (which correspond to high values of β).

The problem is that in defining η, a low error rate has been assumed. At sufficiently low SNRs, η is high, but the amount of information being conveyed is small. This deficiency can be corrected by defining

$$\eta' = \eta(1 - H_L) \tag{7}$$

where H_L is the average loss of information per transmitted bit. That is, for binary signaling,

$$\begin{aligned} H_L &= -P_e \log_2 P_e - P_c \log_2 P_c \\ &= -P_e \log_2 P_e - (1 - P_e)\log_2(1 - P_e) \end{aligned} \tag{8}$$

This modification does not account for the impact of error correction coding (if used) or repeated transmissions in an ARQ system, but it does allow a more accurate information-theoretic view of spectrum utilization.

The bit error probability, P_e, can be easily computed if the interference term $\sum b_i \rho_{li} \cos\phi_i$ in equation (4) is taken to be a normally distributed random variable, as suggested by the central limit theorem and the condition of equal received power levels. Then

$$P_e = Q\left(\sqrt{\frac{2E}{N_o'}}\right)$$

where $Q(x) = \frac{1}{\sqrt{2\pi}} \int_x^\infty e^{-t^2/2} dt$

The normal approximation to the bit error rate can be used to plot spectrum efficiency (η') in terms of error probability. The curves $\eta'(P_e)$ are shown in Figure A1 for representative values of L/K and E/N_o. It can be seen that the spectrum efficiency rises rapidly as the error rate is allowed to rise above 10^{-3}. The curves extend only to bit error rates of 0.10, since this approaches the useful limit for practical systems. However, the curves continue to rise to a maximum theoretical efficiency at error rates of about 0.4. At such high error rates, the upper family of curves (L/K = 2) peaks at about 0.88, while the lower family of curves (L/K = ∞) peaks at half that value. It is reasonable to question whether such high efficiencies could ever be attained in a practical system because the overhead associated with error detection and correction is not included in η'.

In using Figure A1, it is important to keep in mind that P_e is computed using the mean squared cross correlation for the code population L. It is not a guaranteed error rate for any particular subset of M codes, but rather an indication of performance for an "average" set of M codes.

A.2.1.3 Interference-Limited Case, Unrestricted Access

In some cases, it may not be possible to limit access to the network. For this reason, it is interesting to examine the results when all assigned user codes are simultaneously active, i.e., when L = M. Using the arguments put forth in the previous section, the mean squared cross correlation between codes is subject to the bound

$$\overline{\rho^2} \geq \frac{1}{M - 1}\left(\frac{M}{K} - 1\right)$$

where, as shown earlier, the equality is attainable (for integer values of M/K, at least) using simple (though not necessarily practical) code assignment strategies. Using equation (6),

$$\frac{2E}{N_o'} \approx \frac{2}{\left(\frac{M}{K} - 1\right) + N_o/E}$$

From which

$$\eta = \frac{M}{2K} = (\beta + 1)/2$$

where, again,

$$\beta = N_o'/E - N_o/E$$

As before, $\eta' = \eta(1 - H_L)$ can be computed based on a normal approximation to the interference-plus-noise. This is shown in Figure A2 as a function of the bit error rate for $E/N_o = 10$ and $E/N_o = \infty$.

It is interesting to note that η approaches 0.5 from above as β approaches zero. The condition $\beta = 0$ corresponds to an absence of interference, and this is a condition that can occur only for M $\leq$ K. As noted under the "noise-limited case," the spectrum efficiency for M = K is precisely 0.5. For the "restricted access case," M < L. Under these conditions, a lack of interference *cannot* be guaranteed for K = M, since there is a good chance that the M codes selected at random out of the population L will not be orthogonal. Clearly, increasing the number of potential users, L, over the maximum number of simultaneous users, M, is not always beneficial.

A second interesting feature of Figure A2 is that, unlike Figure 1, the maxima occur within a potentially practical range of error rates. Specifically, the maxima occur for error rates of about 0.02.

A.2.2 Asynchronous Direct Sequence

In most cases, it is not practical to establish precise timing between all terminals in the network. Indeed, one of the advantages of SSMA with respect to time division multiple access (TDMA) is that such network synchronization is not required for SSMA.

As before, consider the case where M users are active out of a potential user population L. However, in this case, any interfering code is displaced from the desired code by a random time delay. In general, neither bit transitions nor the

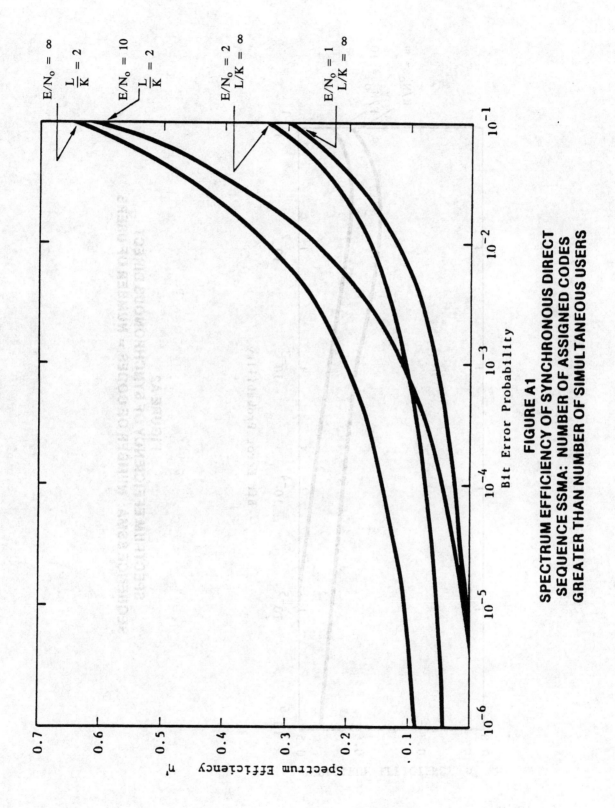

**FIGURE A1
SPECTRUM EFFICIENCY OF SYNCHRONOUS DIRECT
SEQUENCE SSMA: NUMBER OF ASSIGNED CODES
GREATER THAN NUMBER OF SIMULTANEOUS USERS**

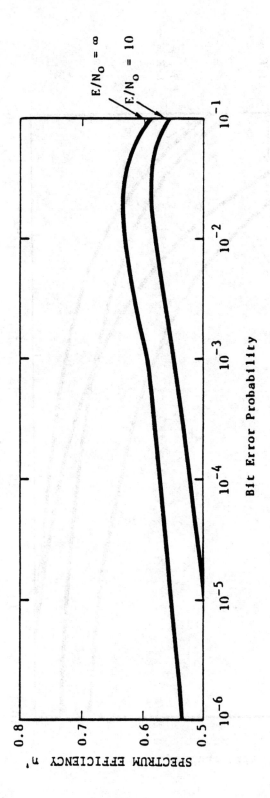

**FIGURE A2
SPECTRUM EFFICIENCY OF SYNCHRONOUS DIRECT
SEQUENCE SSMA: NUMBER OF CODES = NUMBER OF USERS**

chip transitions of the various codes are aligned. The impact of the non-aligned bit transitions will be examined first.

There are $K - 1$ ways in which the bit transition of a particular interfering signal can fall within the interval $0 \leq t \leq T$, assuming the chip transitions are aligned. Furthermore, there are 8 possible values of correlator output for each of these positions, depending on the signs of the two interfering bit segments and the "desired" bit. Thus, for two codes, the number of possible cross correlations is $(K - 1)8$ when the bit transitions are not simultaneous. If the bit transitions are simultaneous, then there are just 4 possible cross correlations, since only one interfering bit occurs during the "desired" bit. The total number of possible cross correlations is then $(K - 1)8 + 4 = 8K - 4$. Each of these cross correlations is an inner product between two binary vectors, so the total set of $(L9L - 1)(8K - 4)$ such correlations is subject to the Welch inner product bound.

$$\rho_{max}^2 \geq \frac{1}{L(L-1)(8K-4)} \left[\frac{L(L-1)(8K-4)}{K} - 1 \right] \approx \frac{1}{K}$$

As noted earlier, this bound also applies to $\overline{\rho^2}$. Also $1/K$ is just the mean squared cross correlation between random bit streams when the chip transitions between the two sequences correspond. Thus, the bound on $\overline{\rho^2}$ should be approachable for $1 << K$ and $L << 2^K$.

As noted in (A9), the impact of nonaligned bit transitions is to reduce the mean squared cross correlation of the random sequences from $1/K$ to $2K/3$. Thus, the mean predetection SNR is

$$\frac{2E}{N_o'} \approx \frac{2}{(M-1)\overline{\rho^2} + N_o/E} = \left[\frac{M-1}{3K} + \frac{N_o}{2E} \right]^{-1}$$

from which

$$\eta = \frac{M}{2K} = \frac{3}{4}\left(\frac{N_o'}{E} - \frac{N_o}{E} \right) + \frac{1}{2K} \approx \frac{3\beta}{4}$$

As before, the efficiency, modified to include the loss of information in the channel, is $\eta' = \eta(1 - H_L)$. As shown in Figure A3, the efficiency rises with increasing error rate within the range of "practical" error rates. The maxima of η, for the two values of E/N_o plotted in Figure 3, occur at error rates of about 0.4, yielding a theoretical maximum efficiency of about 0.66.

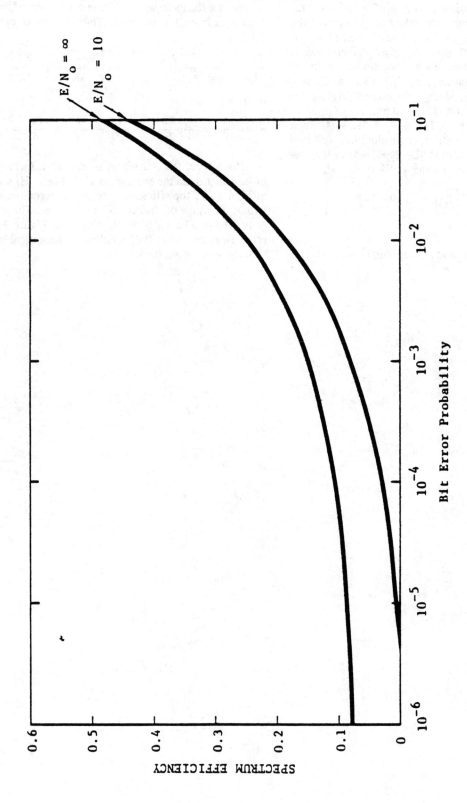

**FIGURE A3
SPECTRUM EFFICIENCY OF ASYNCHRONOUS DIRECT SEQUENCE SSMA**

A.3 FREQUENCY HOPPING

Frequency hopping (FH) systems operate by simultaneously changing the transmit and receive carrier frequencies at intervals throughout each message. This process can be visualized with the aid of Figure A4, which is a frequency versus-time plot for a single frequency-hopping link. Ideally, multiple links can operate with no mutual interference as long as their frequency-time plots never overlap. This is obviously a simplification, since some adjacent-channel interference will inevitably occur in practical systems.

Because of the difficulty of maintaining the phase coherence of the carrier from one hop to another, frequency shift keying (FSK) is often used in FH systems. For the purpose of this analysis, each FH transmitter sends one bit per hop using binary FSK. The receiver uses noncoherent detection to recover the transmitted signal. All signal levels are assumed to be equal at the receiver input. The effect of spectral "splatter" is not considered; for the case of equal signal levels, this simplification can be expected to have a minor impact. Also, for simplicity, the number of assigned hopping patterns is taken to be equal to the maximum number of simultaneous users.

The spectrum efficiency of a FH system, using the definition of equation (7), is

$$\eta' = \eta(1 - H_L) = \frac{M}{2N}(1 - H_L) \quad (9)$$

where $N = B/2R$ is taken to be the number of sub-channels, each sub-channel containing a mark frequency and a space frequency. In systems where the sub-channel spacing is greater (or less) than $2R$, equation (9) must be adjusted accordingly.

A.3.1 Synchronous FH

If the hopping of the various links can be synchronized to a master timing source, then interference between links can be averted for $M \leq N$ by assigning nonoverlapping time-frequency sequences to the M users. Again, this condition is of primarily theoretical interest. The limiting value of η for this case is $\eta(M = N) = 0.5$. Then the network throughput per unit bandwidth is

$$\eta'(M = N) = \eta(M = N)(1 - H_L)$$
$$= 0.5[1 + P_e \log_2 P_e + (1 - P_e)\log_2(1 - P_e)]$$

where $P_e = 0.5 \exp(-E/2N_o)$.

Note that of the N^N possible time-frequency trajectories, only N have been used for the first $M = N$ users. For $N < M \leq 2N$ users, interference-free operation can no longer be guaranteed, but it is still possible to assign a unique hopping pattern to each user such that $2(M - N)$ signals suffer interference on each hop, with all signals suffering interference for an equal fraction of any long message.

This can be accomplished, in principle, by constructing the M codes such that on each hop, the $2(M - N)$ "unlucky" signals are selected at random out of the population M. The remaining $2N - M$ signals are assigned the "preferred" "noncolliding" sub-channels. Although this may not be a practical design procedure, it demonstrates that an interference probability (per signal) of $2(M - N)/M$ is attainable in principle. Furthermore, by this procedure no "collision" involves more than two signals. The average error probability is then

$$P_e = (2N - M)P_o/M + 2(M - N)P_1/M \quad (10)$$
$$\approx \left(\frac{2N - M}{M}\right)0.5\exp(-E/2N_o) + \left[\frac{2(M - N)}{M}\right]0.25$$

where P_i is the bit error probability with i interferers.

Given M, N, and E/N_o, the values of P_e and η' can be determined from equations (8) through (10). The upper set of curves in Figure A5 show η' as a function of P_e for $N = 100$. It can be seen that the curves for $E/N_o = 10$ and $E/N_o = \infty$ are essentially the same for $0.015 < P_e$. Note that meaningful values of the curves exist only for integral values of M, as shown by the staircase function under the $E/N_o = 10$ curve. The smooth curves actually represent the envelopes of discrete functions.

A.3.2 Asynchronous FH

As noted earlier, it is usually not practical to establish mutual synchronization between all active links in the system. For asynchronous FH systems, the bit error probability has been shown to be (A12).

$$P_e = \sum_{j=0}^{J}\left(1 - \frac{b}{N}\right)^{J-j}\left(\frac{b}{N}\right)^j \binom{J}{j} P_j \quad (11)$$

where $J = M - 1$ = number of potential interferers
b = transmit duty factor and
P_j = probability of error with j interferers.

Equations (8), (9), and (11) thus provide for the evaluation of P_e and η' given M, N, b, and E/N_o. The computation is simplified by taking P_e to be 0.5 for $2 \leq j$. This is reasonable approximation since binary FSK alone is not a particularly interference-resistant technique. The error rate for two or more interferers (each having the same power as the desired signal) can be expected to rapidly approach 0.5.

The lower curves in Figure A5 show the spectrum efficiency (η) of asynchronous FH for E/N_o of 100 (i.e., 20 dB) and 10. Again, the variation in η' due to changes in E/N_o is negligible except at low error rates. The maximum value of η' is about 0.115 for error rates in the vicinity of 0.18.

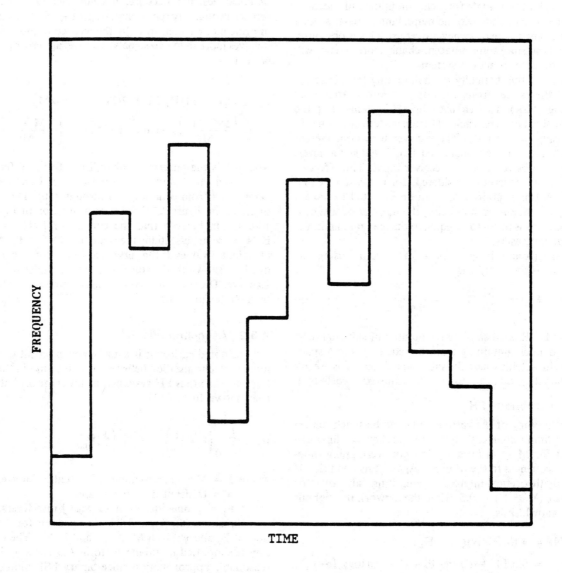

**FIGURE A4
TYPICAL FREQUENCY HOPPING SEQUENCE**

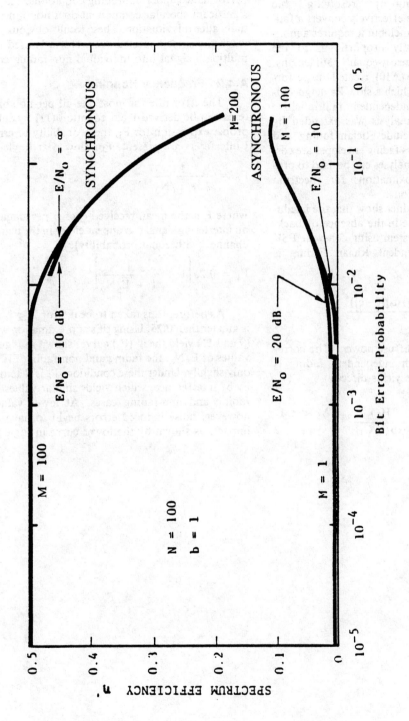

**FIGURE A5
SPECTRUM EFFICIENCY OF FREQUENCY HOPPING SSMA
SPECTRUM EFFICIENCY**

A.4 PERFORMANCE OF ASYNCHRONOUS DS AND FH IN FADING

A.4.1 Direct Sequence

Fading can be caused by a wide variety of mechanisms. In some cases, direct sequence systems provide protection against multipath-induced fading by "resolving" the individual paths. This strategy can effectively convert a fading channel into a non-fading channel, but it requires a minimum bandwidth that is inversely proportional to the difference between the two "most nearly equal" path lengths.

Gardner and Orr, in reference (A10), treated the performance of a DS SSMA system in which a slow Rayleigh fading envelope is applied as an independent multiplicitive constant to each signal. Their analysis was extended by Hanlon and Gardner (10) to include Rician fading and Rayleigh fading in which the various fading envelopes are correlated. These two performances analyses can be used to provide simple, closed form approximations for spectrum efficiency under restricted conditions.

Specifically, Hanlon and Gardner show that the irreducible error rate (i.e., the error rate in the absence of background noise) for a DS SSMA system using coherent PSK through a channel with independent Rician Fading is approximately

$$P_e(N_o = 0) \approx \frac{(M-1)(1+C_1)\exp(-c_1)}{6K}$$

where C_1 is the ratio specular to diffuse power. (The corresponding expression for Rayleigh independent fading is obtained by setting $C_1 = 0$.) This yields directly

$$\eta' = \frac{M}{2K}(1 - H_L) \approx \frac{3P_e(1 - H_L)}{(1 + C_1)\exp(-C_1)} \quad \begin{array}{l} M \ll L \\ C_1 < 2 \end{array}$$

Since this expression is based on the error rate in the absence of background noise, it represents an upper bound on η' for the stated conditions. Figure A6 shows $\eta'(P_e)$ for $C_1 = 0$ (Rayleigh fading) and for $C_1 = 1$ (specular component and diffuse component having equal power). As might be expected, the spectrum efficiency in Rayleigh fading channels is substantially reduced from the corresponding performance under nonfading conditions. The addition of a significant specular component does not appear to substantially alter this situation. These results obviously do not apply when the spread spectrum waveform is used to resolve the multipath signal into individual non-fading components.

A.4.2 Frequency Hopping

The error rate for most one-bit-per-hop binary FH systems can be derived from equation (11) by substituting the proper expression for P_j, the probability of error given that j interferers are present. For slow, flat Rayleigh fading,

$$P_o = \frac{1}{2 + \overline{E}/N_o}$$

where $\overline{E}$ is the mean received energy per signal. Also, since an interfering signal can appear either in the mark or the space channel (with equal probability),

$$P_1 = 0.25\left[1 + \frac{1}{1 + \overline{E}/N_o}\right]$$

As before, P_j is taken to be 0.5 for $2 \leq j$, since even P_1 is greater than 0.25. Using this approximation with equations (9 and 11) yield the $\eta'(P_e)$ curves shown in Figure 7. At high values of E/N_o, the fading and non-fading FH curves differ only slightly. Under these conditions P_e is determined primarily by interference, which yields similar values of P_1 for the fading and non-fading cases. At lower values of E/N_o, however, noise induced errors begin to have a substantial impact, as shown by the lower curve in Figure A7.

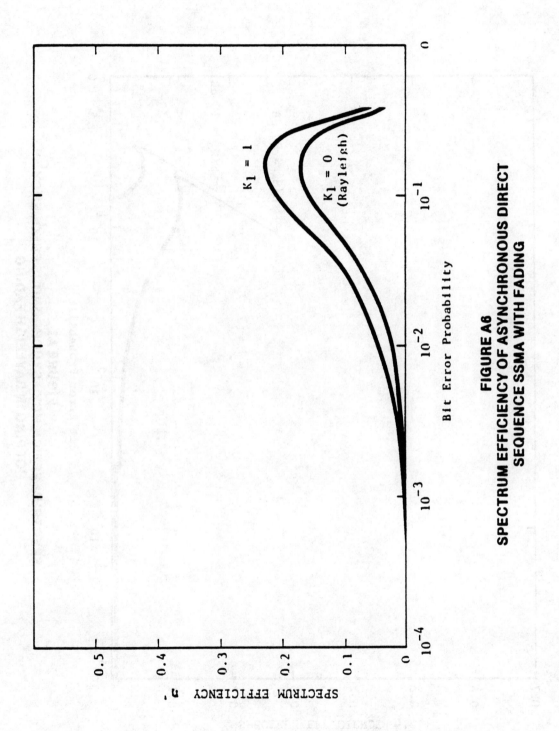

**FIGURE A6
SPECTRUM EFFICIENCY OF ASYNCHRONOUS DIRECT SEQUENCE SSMA WITH FADING**

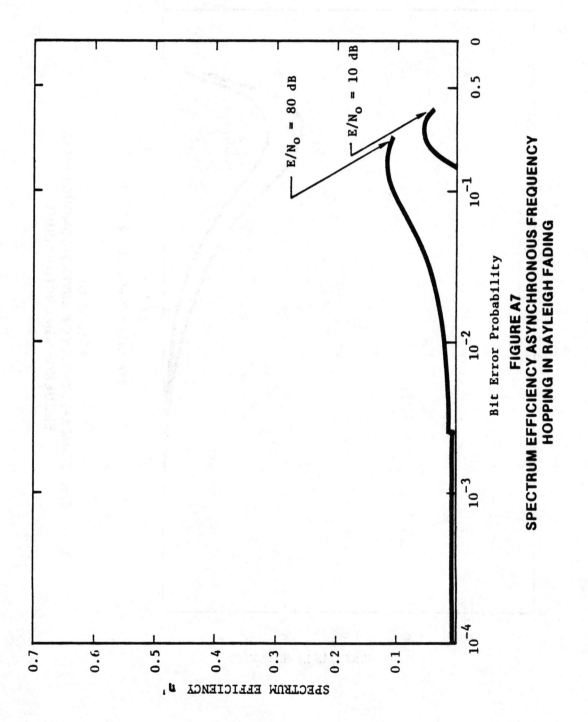

**FIGURE A7
SPECTRUM EFFICIENCY ASYNCHRONOUS FREQUENCY HOPPING IN RAYLEIGH FADING**

A.5 ERROR-CORRECTION CODING: FUNDAMENTAL LIMITATIONS

It can be argued that the results presented above are unduly restrictive because they are limited to binary signaling without error correction. It is clear that error correction coding can increase the spectrum efficiency of SSMA networks relative to the networks analyzed in the preceding sections. However, if error correction coding is allowed on SSMA networks it must also be allowed on the FDMA or TDMA networks which serve as a basis of comparison.

The Shannon capacity for channels with additive white Gaussain noise is

$$C = B_c \log_2\left(1 + \frac{P}{N_0 B_c}\right)$$

where B_c is the RF channel bandwidth allowed to each user and P is the received signal power. For FDMA, $B_c = B/M$ and P is the average received power per user. For TDMA, $B_c = B$ and P is M times the average power for one user. For SSMA, $B_c = B$, P is the average power, and N_0 becomes N_0', as defined earlier.

Thus the total network channel capacities for FDMA, TDMA, and SSMA are

FDMA: $C_T = M(B/M)\log_2(1 + MP/N_0 B)$
TDMA: $C_T = B\log_2(1 = MP_{AV}/N_0 B)$
SSMA: $C_T = MB\log_2\left[1 + \frac{P}{N_0 B + (M-1)P}\right]$
$= MB\log_2\left[1 + \left(\frac{N_0 B}{P} + M - 1\right)^{-1}\right]$

Obviously, the first two expressions are the same. The best achievable spectrum efficiencies at arbitrarily low error rates are

FDMA: $\eta = MC/B = C_T/B = \log_2(1 + MA)$
TDMA: $\eta = MC/B = C_T/B = \log_2(1 + MA)$
SSMA: $\eta = MC/B = C_T/B$
$= M\log_2[1 + (A^{-1} + M - 1)^{-1}]$
$\approx M/(M-1)$ for $1 \ll A$
≈ 1 for $1 \leq A$ and $1 \ll M$

where $A = P/N_0 B$ has been used for brevity. The major result here is that the maximum spectrum efficiencies for FDMA and TDMA increase monotonically with M, while the maximum spectrum efficiency for SSMA decreases asymptotically to 1.0 for increasing M. This is basically due to the assumed lack of self-interference in FDMa and TDMA systems. The channel capacities for such systems can theoretically be raised to any desired level by increasing the transmitter power. Increasing the transmitter power in SSMA systems quickly results in a point of diminishing returns; the system becomes interference-limited. Moreover, increasing the number of SSMA users decreases the effective SNR for each user. For FDMA or TDMA systems, the SNR per user increase as the number of users increases (assuming that the average power per user and noise power density remain fixed), since adding users either decreases the bandwidth per user (in FDMA) or increases the ratio of peak to average power (in TDMA).

When the number of active users, M, is not constant, the average network channel capacity can be computed as

$$C_T = \sum_{M=0}^{\infty} C_T(M) P(M)$$

where P(M) is the probability that M users are active. Under these conditions it can be shown that SSMA can yield a higher average network capacity (for a fixed bandwidth, B) when the average number of active users is low (A17, A18). This is due to the assumption that even when few users are active, each FDMA user can occupy only 1/M of the system bandwidth; each TDMA user can occupy the total system bandwidth only 1/M of the time. If this constraint is removed (by trunking the available TDMA or FDMA channels, for example), then it is easily shown that the average SSMA channel capacity is always below the average TDMA or FDMA capacity.

A.6 DISCUSSION

The preceding derivations are designed to provide some basic constraints on spectrum efficiency under the stated conditions. In Sections A.1 through A.4, the assumption of a normal predetection distribution of noise-plus-interference was invoked in order to relate spectrum efficiency to error rate.

The spectrum efficiency of synchronous networks is primarily of theoretical interest. Most practical systems provide synchronization only between pairs of terminals that are communicating with one another. The results given above for synchronous networks indicate that direct sequence SSMA provides better spectrum efficiency than frequency hopping SSMA for the stated conditions. However, part of this apparent difference in performance is simply due to the difference in communication performance between coherently demodulated PSK and noncoherently demodulated FSK. Furthermore, it is easy to find examples in which frequency hopping with noncoherent binary FSK outperforms direct sequence with coherent binary PSK.

Consider, for example, the case in which the received power levels are no longer equal. Specifically, let

$$E_1 = E_2 = E_3 = \ldots E_{M-1} = E_M/\alpha, \quad 1 << \alpha$$

It is easily shown, starting from equation (4) that the spectrum efficiency for asynchronous DS is reduced from $\eta = 3\beta/4$ to

$$\eta \approx \left[\frac{M}{M - 2 + \alpha}\right]\left(\frac{3\beta}{4}\right)$$

$$\approx \frac{M}{\alpha}\left(\frac{3\beta}{4}\right)$$

In many situations, it is not unusual to find 40 dB differences between the weakest and the strongest signal. Taking the hypothetical case of $M = 10$, $\alpha = 10^4$, and $N_o = 0$, and requiring $\beta = 1$ (a generously high value) we find

$$\eta \approx 7.5 \times 10^{-4}$$

For the slow frequency hopping approach treated above, the impact of the single strong interferer is restricted to a single sub-channel (although in extreme cases, splatter from adjacent sub-channels would also have to be considered). If $N = 100$, then a receiver not synchronized to the strong interferer "sees" it only one percent of the time. The error rate due to the strong interference alone is therefore on the order of 5×10^{-3}, which has only a second order effect when the total error rate is on the order of .02, for example, as in the case of 8 interferers of equal power in a 100 channel FH system. The spectrum efficiency for FH is then approximately as shown in Figure A9 for $P_e = 0.02$, or $\eta' \approx 0.05$. Although this efficiency may seem quite low, it is almost two orders of magnitude higher than the efficiency of asynchronous DS under the same conditions.

Regardless of spread spectrum modulation technique, it is clear that the spectrum efficiency of any given SSMA system will improve with decreasing signal-to-interference ratio until the point is reached at which so many errors are being made that the loss of information throughput overwhelms any gain in the number of links operating in a given bandwidth. For binary signals without forward error correction, this point corresponds to a relatively high error rate. The fundamental requirement for practical spectrum-efficient SSMA systems is the ability to operate at low signal-to-interference ratios. However, expanding the system bandwidth in order to achieve this goal by conventional spread spectrum techniques is not typically effective in terms of spectrum efficiency, since the permissible number of users is roughly proportional to the bandwidth, all other things being equal. What is needed, then, is an improvement in basic communication efficiency relative to uncoded binary signaling and the usual spread spectrum techniques, by themselves, do not provide this, since the required E_b/N_o for a given error rate in white Gaussain noise is not changed by the addition of spread spectrum modulation.

Any discussion of theoretical spectrum efficiency needs to be tempered by caveats on practical limitations. For spread spectrum systems, the attainable performance can be limited by the requirement for fast acquisition and reliable synchronization with economically feasible implementations. TDMA and FDMA systems also have practiced limitations. TDMA requires timing gaps to allow for finite propagation delays and can be expensive to implement. FDMA systems are generally easy to implement but, in practice, are usually not free of interference. In addition, their spectrum efficiencies are limited by the guard bands required between channels. Examples have been produced in which hypothetical spread spectrum systems provide better spectrum utilization than conventional frequency-channelized approaches (see Reference A7 and Appendix B). The outcomes of such comparisons inevitably depend on the particular characteristics and parameters of the systems being compared. These characteristics and parameters, in turn, reflect the potential cost and communication performance of the two systems. In practical situations, it may not be meaningful to compare SSMA with other techniques without specifying particular practical designs as part of the comparison process. A loss of generality in such specific comparisons is inevitable, but this is the price of improving the reliability of the result.

A.7 REFERENCES

A1. G. R. Cooper and R. W. Nettleton, "A Spread Spectrum Technique for High Capacity Mobile Communications," *Conference Record of the 27th Vehicular Technology Conference*, Orlando, Florida, March 16-18, 1977, pp. 98-103.

A2. R. W. Nettleton and G. R. Cooper, "Mutual Interference in Cellular LMR Systems Narrowband and Broadband Techniques Compared," *Conference Record of MIDCON/77*, Chicago, Illinois, November 8-10, 1977.

A3. R. W. Nettleton and G. R. Cooper, "Error Performance of a Spread-Spectrum Mobile Communications System in a Rapidly-Fading Environment," *Conference Record of National Telecommunications Conference*, Los Angeles, California, December 5-7, 1977.

A4. G. R. Cooper, "Multi-Service Aspects of Spread-Spectrum Mobile Communications System," *Conference Record of MIDCON/77*, Dallas, Texas, December 12-14, 1978.

A5. G. R. Cooper and R. W. Nettleton, "A Spread-Spectrum Technique for High-Capacity Mobile Communications," *IEEE Transactions on Vehicular Technology*, Vol. VT-27, No. 4, November 1978.

A6. P. S. Henry, "Spectrum Efficiency of a Frequency-Hopped-DPSK Mobile Radio System," *Conference Record of the 30th IEEE Vehicular Technology Conference*, 1977, pp. 7-12.

A7. D. J. Goodman, et al., "Frequency Hopped Multilevel FSK for Mobile Radio," 1980 International Zurich Seminar on Digital Communications.

A8. J. E. Mazo, "Some Theoretical Observations on Spread Spectrum Communications," *Bell System Technical Journal*, Vol. 58, No. 9, November 1979, pp. 2013-2023.

A9. M. B. Pursley, "Performance Evaluation for Phase-Coded Spread Spectrum Multiple-Access Communication—Part 1: System Analysis," *IEEE Trans. Comm.*, Vol. COM-25, No. 8, August 1977, pp. 795-799.

A10. C. S. Gardner and J. A. Orr, "Fading Effects on the Performance of a Spread Spectrum Multiple Access Communication System," *IEEE Trans. Comm.*, Vol. COPM-27, No. 1, January 1979, pp. 143-149.

A11. R. C. Hanlon and C. S. Gardner, "Error Performance of Direct Sequence Spread Spectrum on Nonselective Fading Channels," *IEEE Trans. Comm.*, Vol. COM-27, No. 11, November 1979, pp. 1696-1700.

A12. S. A. Musa and W. Wasylkiwskyj, "Co-Channel Interference of Spread Spectrum Systems in a Multiple User Environment," *IEEE Trans. Comm.*, Vol. COM-26, No. 10, October 1978, pp. 1405-1413.

A13. D. J. Toerrieri, *Simultaneous Mutual Interference and Jamming in a Frequency Hopping Network*, DARCOM Countermeasures/Counter-Countermeasures Office, DARCOM CM/CCM 79-6, October 1979.

A14. J. P. Costas, "Poisson, Shannon, and the Radio Amateur," *Proc. IEEE*, December 1959.

A15. S. H. Cameron, "The Efficiency of Spread Spectrum Signaling with Respect to Spectrum Utilization," 19th IEEE Electromagnetic Compatibility Symposium, August 1977.

A16. L. A. Berry, "Spectrum Metrics and Spectrum Efficiency: Proposed Definitions," *IEEE Trans. EMC*, Vol. EMC-19, No. 3, August 1977, pp. 254-260.

A17. R. C. Sommer, "Asynchronously Multiplexed Channel Capacity," *Proc. IEEE*, January 1966.

A18. R. C. Sommer, "The Noisy Asynchronously Multiplexed Channel," *IEEE Trans. Information Theory*, January 1967.

A19. L. R. Welch, "Lower Bounds on the Minimum Cross Correlation of Signals," *IEEE Trans. Information Theory*, May 1974.

APPENDIX B

SPECTRUM EFFICIENCY OF FDMA AND SSMA NETWORKS: THE IMPACT OF FREQUENCY TOLERANCES AT LOW INFORMATION RATES

B.1 INTRODUCTION

In Frequency Division Multiple Access, guard bands are used to separate adjacent frequency assignments. If all transmitters and receivers are synchronized to a single carrier source, the guard bands need only compensate for the finite roll-off characteristics of the channel filters. In the more usual situation, however, each transmitter/receiver has its own frequency source. In this case, each guard band must be expanded to provide for the frequency tolerances.

With Spread Spectrum Multiple Access (SSMA), users are separated by codes, so only a single pair of guard bands at the edges of the SSMA channel is required. When frequency tolerances are considered, SSMA can surpass FDMA in spectral efficiency, i.e., in the number of users per unit bandwidth for a fixed set of system parameters. This statement can be translated to quantitative terms as follows.

B.2 FDMA NOTATION

Let f_i = assigned frequency of the ith carrier
f_c = center frequency of the FDMA band
d = relative frequency tolerance of the transmitters
R = information bandwidth
K_f = bandwidth expansion factor = RF bandwidth/R
b_f = required bandwidth per user
b_g = guard bandwidth per channel in the absence of frequency errors

Spectrum Utilization

It is assumed that the FDMA band is contiguous and that its bandwidth, B, is much smaller than its center frequency, f_c. It is also assumed that the frequency errors of the various user terminals are independent, and there is exactly one user per channel. The frequency separation between channels is then:

$$f_{i+1} - f_i = K_f R + f_i d + f_{i+1} d + b_g \approx K_f R + 2f_c d + b_g \quad (1)$$

That is, the bandwidth per user is just the occupied bandwidth plus the two tolerances plus the minimal guard band, b_g.

B.3 ASYNCHRONOUS DIRECT SEQUENCE SSMA (CDMA)

Notation

Let M = number of active users
R = information bandwidth
K_s = bandwidth expansion factor = RF bandwidth/R
d = relative frequency tolerance
N_o = noise power density
A_i = average output SNR with i active users
p = 1 for noncoherent detection, 2 for coherent detection
B = total bandwidth of channel
b_s = bandwidth per user
B_e = guard bandwidth at the channel edges for d = 0
n = number of code chips per code period

Spectrum Utilization

In this case, each user has a unique code. Again, the frequency errors of the various user terminals are taken to be independent. All users are assumed to be active simultaneously in the worst case. All signals are assumed to produce equal power levels at the receiver input. Here, the total width of the assigned channel is taken to be:

$$B = K_s R + 2f_c d + B_e$$

i.e., the occupied bandwidth plus twice the tolerance plus B_e. The bandwidth per user is then:

$$b_s + \frac{K_s R + 2f_c d + B_e}{M} \quad (2)$$

The required bandwidth expansion factor, K_s, is determined by the processing gain required to achieve a given value of mean signal-to-noise ratio. The exact relationship between processing gain and output SNR in a direct sequence SSMA system depends on the specific set of codes and modulation technique being used. However, for a system using binary PSK, in which the codes can be approximated as random binary sequences, Pursley (1) has shown that the mean output SNR (power ratio) is approximately

$$\left[\frac{i-1}{3n} + \frac{N_o}{2E} \right]^{-1}$$

where i is the number of users, n is number of code chips per information bit and coherent detection has been assumed. In general,

$$A_i = \frac{p}{2} \left[\frac{i-1}{3n} + \frac{N_o}{2E} \right]^{-1}$$

solving for the code length, n, yields

$$n = \frac{2(i-1)}{3p} \left[\frac{1}{A_i} - \frac{1}{A_1} \right]^{-1}$$

Because the basic phase-shift keying process has a bandwidth expansion factor of 2, the overall bandwidth expansion factor, K_s, is 2n.

Thus, given the available output SNR for a single user (A_1), the required output SNR for full system loading (A_m), and the system capacity (M), the required bandwidth expansion factor (K_s) can be estimated. This determines b_s as shown in equation (2).

Break-Even Point

In examining equations (1) and (2), it is clear that both are linear in f_c, d, and R. However, the coefficients of these variables in b_s are smaller than their counterpart in b_f by the factor M. Allowing d and R to remain fixed, the lines $b_s(f_c)$ and $b_f(f_c)$ always intersect at one and only one point for $1 < M$.

Furthermore, since both slopes are positive, $b_s < b_f$ above this point and $b_f < b_s$ below. Points above the break-even point require less bandwidth per user with SSMA; points below the break-even point require less bandwidth per user with FDMA. Analogous results hold if d or R is allowed to vary with all other parameters constant.

The break-even point is readily found by setting $b_f = b_s$ and solving for f_c, d, or R. For example,

$$f_o = \frac{R(K_s - K_f M) + B_e - M b_g}{2d(M - 1)} \qquad (3)$$

Note that positive solutions exist only for $M(K_f R + b_g) < K_s R + B_e$, since the opposite condition would imply that the total required bandwidth for FDMA would be larger than the total bandwidth for SSMA even with perfect frequency control.

B.4 SSMA WITH SLOW FREQUENCY HOPPING

For this case, equations (1) through (3) are still valid. However, a modified approach to determining the required K_s may be needed. For slow-hopping systems, the signal-to-interference ratio can change drastically from one moment to the next, as the time-frequency slots used by the desired signal and the interferer briefly "collide." Instead of describing the system in terms of average SNR, it may be more meaningful to describe the probability that interference will occur in a given slot. This is especially significant for the "equal-signal-levels" model, since any interference can momentarily render the channel useless. For a single interferer, the probability of this for any particular "state" of the system is $1/L$, where L is the number of channels (shared by the desired signal and the interferer). For an active user population M, the probability of interference is:

$$P = 1 - \left(1 - \frac{1}{L}\right)^{M-1}$$

The required number of channels is then

$$L = \left[1 - (1-P)^{(M-1)^{-1}}\right]^{-1} \simeq \frac{M-1}{P} \quad \text{for small P} \qquad (4)$$

The total bandwidth expansion factor, K_s, is then L times the "un-hopped" bandwidth per channel, assuming that the channels are spaced as closely as possible.

It is interesting to apply these results in two hypothetical examples where the information rate is low.

Example 1

Let us use the following parameters to compare FDMA with direct sequence SSMA.

d = 2×10^{-6} ($\pm 0.0002\%$)
R = 100 Hz
M = 100 Users
A_1 = 30 dB
A_m = 10 dB
K_f = 2
b_g = 100 Hz
p = 2 (coherent detection)

Then,

$$n = \frac{2(M-1)}{3P}\left[\frac{1}{A_M} - \frac{1}{A_1}\right]^{-1} = 333$$

The spread spectrum signal bandwidth is then $K_s R = 2nR = 66.6$ kHz. For convenience, we take the combined guard bandwidth at the edges of the spread spectrum band to be equal to the spread spectrum signal bandwidth. Thus, by equation (3) f_o is 261 MHz. The model indicates that at higher frequencies, SSMA is more spectrum-efficient. The total required bandwidth at 261 MHz, by either equation (1) or equation (2), is just over 134 kHz, or about 1 kHz per user.

Example 2

Now consider an asynchronous SSMA system with slow frequency hopping. Let

P = 0.05
M = 100 users
R = 100 Hz
d = 2×10^{-6}
K_f = 2
b_g = 100 Hz
B_e = 4R

Then, by equation (4), L = 1930 channels. The "un-hopped" bandwidth is $K_f R = 200$ Hz, so the total spread spectrum signal bandwidth is 386 kHz. That is,
K_s = 386 kHz/100 Hz = 3860.

By equation (3), the break-even frequency is
f_o = 900 MHz. The total required bandwidth at this frequency is 390 kHz, or 3.9 kHz per user.

Discussion

It can be seen that the factors favoring SSMA over FDMA are:

(1) Loose frequency tolerances.
(2) Low information rates.
(3) High operating frequencies.
(4) Low marginal cost of transmitter power (relative to the cost of precision oscillators or TDMA), since SSMA typically requires either increased transmitter power or coding gain, or both, to combat the combined effects of background noise and self-interference. This is reflected in the requirement for direct sequence SSMA that $A_m < A_1$. If A_m is an acceptable SNR, then, in a permanently interference-free system, A_1 could be reduced to this level. In an SSMA system, however, the "spread" between A_1 and A_m fixes the relationship between the required processing gain and the number of users.

In comparing SSMA with FDMA, the effect of frequency tolerances can be neglected only if the absolute tolerances are small with respect to the information rate, or if all network terminals share a common frequency standard. Otherwise, ignoring frequency tolerances can bias the trade-off in favor of FDMA.

B.5 REFERENCES

1. Michael B. Pursley, "Performance Evaluation for Phase-Coded Spread Spectrum Multiple-Access Communication—Part I: System Analysis," IEEE Trans. Comm., Vol. COM-25, No. 8, August 1977.

APPENDIX C

MONTE-CARLO SIMULATION OF LAND MOBILE RADIO WITH SLOW FREQUENCY HOPPING

C.1 OBJECTIVES

The purpose of this simulation is to provide a preliminary assessment of the feasibility of using slow FH for land mobile services. The sensitivity of the model to change in certain parameters is investigated.

C.2 SCOPE

In the development of the simulation models described here, it soon became evident that the specific parameters of the models would depend on the type of service being provided. Because of the potential expense of studying a number of potential types of services in depth, a hypothetical Citizens Band service was selected for simulation. However, the basic models described here are easily modified to embody the characteristics of other services.

C.3 DESCRIPTION OF MODEL

Every meaningful model requires the use of approximations, simplifications, or other constraints. The following description documents these and provides some comments on their impact.

C.3.1 Frequency Hopping

One of the basic features of the model presented here is that all active transmitters are assumed to hop in mutual synchronism. That is, they are assumed to change channels simultaneously. In actual operation, the various transmitters will hop asynchronously. The impact of this simplification is that it results in a predicted average frequency of interruption (by interference) that is too low. However, it also results in a prediction of the average duration of each interruption that is too high by the same factor. So prediction of the average signal-to-interference ratio is unbiased in this respect. However, the other performance parameter predicted in this study — articulation — is based on previously-published empirical evidence (see Section C3.4) so it is hard to say to what extent, if any, it may be biased by this simplification.

No attempt is made to model the hopping codes of the various transmitters. Instead, each transmitter is assumed to select a channel at random on each hop. All channels have a equal probability of being selected by any given transmitter on each hop. The hops of the various transmitters are statistically independent. This is expected to be an accurate representation for the "average" selection of M active user codes out of a larger population of K well-designed user codes. For the worst-case selection of M out of K codes, the performance predicted by the simulation may be slightly optimistic.

However, most of the simulation runs are based on the use of 100 channels, and this results in an extremely large potential set of user codes. (Approximately $100! = 10^{158}$ codes, if all sequences using each channel once and only once per code cycle are allowed). Thus, the probability of randomly selecting two codes that are highly correlated is small, unless this code set is poorly designed.

Hopping rates of 1 hop per second and 20 hops per second are investigated. Hopping rates lower than 1 hop per second will provide reduced voice privacy and will engender long synchronization lags, which are unacceptable in real-time voice communication systems. Hopping rates much in excess of 20 hops per second may be difficult to implement with inexpensive synthesizers.

C.3.2 Sources of Performance Degradation

The major source of performance degradation is assumed to be interference from frequency-hopping transmitters. Sources of degradation that are neglected include:

Background noise
Fading
Interference from other services
Random FM
Harmonic distortion
"Click" noise resulting from the phase discontinuities between hops
Receiver synchronization errors

In addition, the model neglects the possibility of intermodulation products generated by the passage of multiple FH signals through nonlinear circuits. However, the model does compute the effect of "splatter" from adjacent-channel interferers and from interferers that are two channels removed from the intended signal. Splatter resulting from interference that is more than two channels removed from the intended signal is assumed to have a negligible impact.

The overall impact of these simplifications obviously bias the predicted performance in a positive direction. Thus, when a particular simulation run predicts poor performance, it can be concluded that the performance of an operational system would be poor. When the run predicts good results, the only conclusion is that good performance *might* be achievable in an operational system.

C.3.3 Propagation Model

The path losses between transmitters and receiver are computed by the Longley-Rice propagation model (Reference A-1). This model predicts the long-term median transmission loss between points, given parameters such as antenna heights, terrain irregularity, frequency, and distance. The specific parameters used in computing transmission loss are presented in Table C-1.

TABLE C-1
PROPAGATION PARAMETERS

PARAMETER	VALUE
Surface Refractivity	301.
Ground Permittivity	15. esu
Ground Conductivity	0.005 mho/meter
Surface Irregularity	90 meters
Frequency	915 MHz
Antenna Siting	Random (Mobile & Base)
Mobile Antenna Height	2. meters
Base Antenna Height	7. meters

The values of refractivity, ground permittivity, and ground conductivity shown in Table C-1 are recognized as "typical" values; transmission loss at 915 MHz is relatively insensitive to changes in these parameters, except under conditions of unusual refractivity. Random antenna siting (with respect to the surrounding terrain) is assumed for both mobiles and base stations, on the hypothesis that most base stations would be located in the user's home or office. Antenna heights (above ground) of 2 and 7 meters are assumed for the mobile and base, respectively. The surface irregularity (90m) is typical of hilly terrain.

In order to minimize the cost of running the simulation, a lookup table of transmission loss versus distance was constructed. The table covers distances from 1.0 km to 100. km in logarithmically spaced increments of about two percent. The largest difference in transmission loss between adjacent entries in the table is 1.0 dB. However, the quantization interval for the most frequently encountered distances is less than 0.5 dB.

The geographic variability in path loss for a given transmitter-receiver distance is estimated by Reference A-2 as:

$$\sigma_L = 6 + 0.55\sqrt{d_H/L} - 0.004(d_H/L)\,\text{dB}$$

where d_H is the terrain irregularity and L is the wavelength. This value is used to produce a normally-distributed, zero mean adjustment to the median transmission loss.

For transmitter-receiver separations less than 1.0 km, the Longley-Rice model is not applicable. Free-space loss is assumed for this case, which occurs infrequently in most of the simulation runs.

The propagation models described above are based largely on data taken in areas free of urban development. In a recent paper (Reference C-3), Longley suggested modifications for use in urban areas. The suggested correction to the median transmission loss (in dB) contains a constant term, a frequency-dependent term, and a distance-dependent term. In computing signal-to-interference ratios, only the distance-dependent term is significant, since the other two terms are the same for the signal and the interference. The distance-dependent term has a negative sign, which implies that despite higher overall transmission losses in urban areas, the transmission loss increases with distance at a somewhat lower rate than in undeveloped areas. However, the overall change in the rate of fall-off is small for the range of distances used here.

C.3.4 Antennas and Power Levels

All stations are assumed to have antennas that are omnidirectional in azimuth and of equal gain. Equal transmitter power levels are assumed. Since the system is assumed to be interference-limited, no other assumptions about antenna gains or transmitter powers are required.

C.3.5 Basic Simulation Models

Two basic simulation models have been developed. The simplest computes the statistics of the RF signal-to-interference ratio for given statistical distributions of transmitter/receiver distances over a specified coverage area. The second model goes one step further and estimates the intelligibility of the resulting speech based on the pattern of interruptions, using empirical results by Miller and Licklider (C-4)

Both models assume that mobile-to-base transmissions and base-to-mobile transmissions occupy separate, disjoint frequency bands. Primary emphasis is on evaluating the performance of the mobile-to-base links, since, in the absence of repeaters, mobile-to-mobile operations would probably also be conducted in this band, thus making the traffic intensity higher than in the base-to-mobile band. A few simulation runs were also performed for mobile-to-mobile operation, although it is expected that this mode of operation would be of less importance than it is in the 27-MHz citizens band because of the discrete-address nature of the system.

C.3.5.1 Signal-to-Interference Model

A simplified flow chart of the signal-to-interference simulation model is shown in Figure C-1. After the various parameters have been read in, a random trial is conducted to select the distance between the receiver and the transmitter generating the intended signal. Another set of random trials is conducted to select the distances between the receiver and all other active transmitters. Next, a random trial is conducted to select a channel for each active transmitter. The absolute value of the frequency difference (measured in number of channels) between the intended signal and each potential interference is computed and stored. For co-channel interference or interferences within two channels of the intended signal, the path loss is computed. Finally, the signal-to-interference ratio is computed, ignoring those active transmitters that are more than two channels removed from the intended signal.

This process is iterated 200 times, and the distribution of the resulting values of RF signal-to-interference ratio is computed and plotted.

The second simulation model estimates a measure of speech intelligibility by a procedure that will be described in the next section. In order to perform these computations it is necessary to first estimate the total fraction of the voice message that is interrupted by interference and the average rate of interruption.

Figure C-2 is a simplified flow chart of the second simulation model. It should be viewed as a functional description of the computer program, because some operations deviate slightly from the format of Figure C-2 for reasons of computational efficiency.

After the parameters are read in, all temporary storage arrays are cleared. A random trial is then conducted to assign a channel to each active transmitter. On the first pass through this inner loop, random trials are also performed to select transmitter/receiver distances. Path losses are computed and stored so that they will not have to be recomputed on future hopping trials. The r.f. signal-to-interference ratio (C/I) is computed and a counter is incremented if the C/I exceeds 6 dB. This is a simplification based on the threshold/capture characteristics of narrowband FM. It would be a serious oversimplification if most of the observed C/I values were in the region of 6 dB, but because the observed trial values of C/I are spread over a wide range, the approximation should be acceptable.

One hundred such passes are made through the inner loop, with the transmitter/receiver distances and path losses fixed. The result is the number of passes, k, for which C/I < 6 dB. The fraction of hops for which significant interference occurs is then k/100. A count is also kept of the number of interference bursts, i.e., the number of runs of interference-contaminated hops. If this count is L, then the average rate at which interference bursts occur is L/100 times the hopping rate. These two measures — the fraction of the

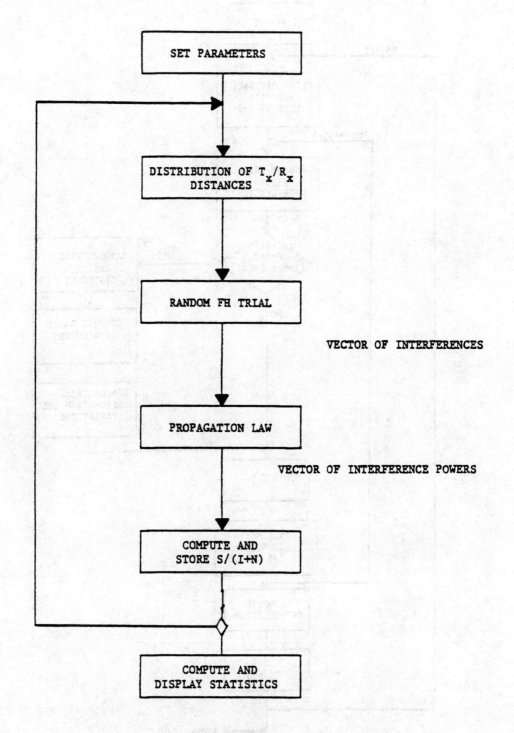

**FIGURE C1
SIGNAL-TO-INTERFERENCE SIMULATION:
SIMPLIFIED FLOW CHART**

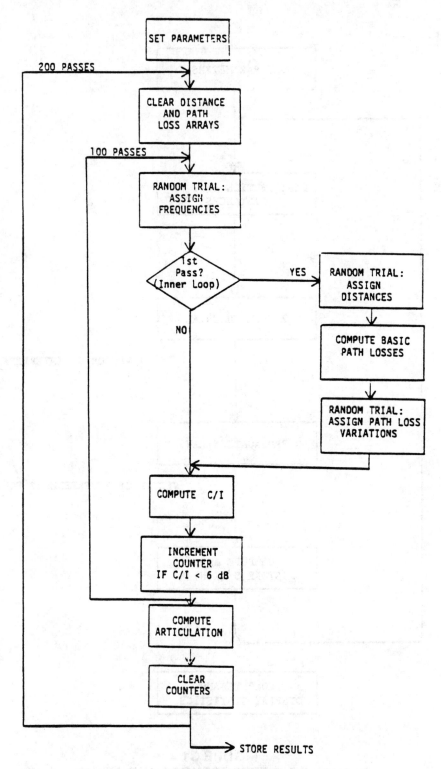

**FIGURE C2
SIMPLIFIED FLOW CHART OF
INTELLIGIBILITY SIMULATION**

hops on which interference (C/I < 6 dB) occurs and the average number of interference bursts per second are used to compute the articulation, based on procedures to be described in the next section.

This entire process is repeated up to 200 times, with a different randomly-selected set of transmitter/receiver distances on each pass. The distribution of the resulting articulation scores is computed and plotted.

C.3.5.2 Intelligibility Model

The intelligibility of interrupted speech has been studied empirically by Miller and Licklider (Reference C-4). The basic measure of intelligibility used in their study is the Harvard phonetically balanced (PB) articulation test (Reference C-5), which scores the ability of listeners to correctly interpret monosyllable words.

Figure C-3 illustrates one result of the Miller and Licklider study. It shows word articulation scores for speech interrupted by noise, as a function of "noise-time fraction," i.e., the duty factor of the noise. The results of interrupting the speech 1, 10, 100 and 1000 times per second are shown. Although these data are based on periodic interruptions, interruptions with irregularly spaced bursts of noise yields essentially the same results.

In order to use the results to predict intelligibility, the average rate of interruption and the average "noise-time fraction," or noise duty factor, must be known. Since the Miller and Licklider results are empirical, some form of interpolation is required when the rate of interruption and "noise-time fraction" produced by the simulation do not coincide with the corresponding values used in the empirical intelligibility tests. In order to satisfy this requirement, a table of articulation versus "noise-time fraction" was computed for each of the four rates of interruption shown in Figure C-3. Since each series of 100 hopping trials has only 101 possible outcomes in terms of "noise-time fraction," the size of this lookup table is $4 \times 101 = 404$ elements. Linear interpolation is used between the empirical data points at each rate of interruption. For a given noise-time fraction and rate of interruption, the articulation is estimated by linearly interpolating between points in the table corresponding to the rates of interruption that bracket the observation. Rates of interruption below one per second are assumed to produce the same articulation scores that result for one interruption per second.

The PB word articulation test used by Miller and Licklider is significantly more demanding than several other scoring methods that have been used (Reference C-6). A PB word articulation score of 75 percent is deemed to be acceptable for "typical" communication purposes.

C.3.6 Distribution of Transmitter-to-Receiver Distances

In all of the simulation runs, the receiver is assumed to be located at the center of a circular region having a radius of 30 km. Interfering transmitters are assumed to have a uniform probability of being within any differential area within this region. In order to account for the fact that mobiles cannot, in most circumstances, approach within a few meters of the base station antenna, an interference-free ring having a radius of 100 meters is created around the receiver. Although these assumptions are obviously simplifications, they are similar to the ones that were used in a recent analysis by Torrieri (Reference C-7).

The uniform annular geographic distribution of interferers leads to the distribution of transmitter-to-receiver distances shown in Figure C-4 for a receiver in the center of the region.

It is reasonable to expect that the distribution of distances between the receiver and the transmitter with which communication is being attempted will not be in agreement with Figure C-4. In particular, users will soon learn that the range of the system is limited (even in the absence of interference), so it is unreasonable to assume that the longest distances between the receiver and the intended transmitter are the most likely to be attempted. In fact, one would suppose that, after a certain point, the probability that of user attempting communication over a given distance will decrease monotonically with distance. On the other hand, it is also unlikely that users will often attempt communication over very short ranges because a moving vehicle will quickly leave such a short-range region. These arguments suggest a unimodel distribution communication distances like the one shown in Figure C-5. The functional form of this distribution is:

$$p(x) = (x/x_0^2) e^{-x/x_0}$$

where x is the distance over which communication is being attempted and $2x_0$ is the mean of the distribution.

In an operational scenario, there is likely to be a significant amount of "coupling" between the number of interferers and the mean distance over which communication is attempted. No attempt has been made to model this effect, and further investigation in this area is needed.

C.3.7 Output Format

The output produced by both simulation models is in the form of a cumulative distribution function (CDF) for each simulation run. One model estimates the CDF of the signal-to-interference ratio; the other estimates the CDF of the word articulation. Each type of CDF represents the probability (taken over the ensembles of transmitter locations and frequency hops) that the computed variable falls below any particular value.

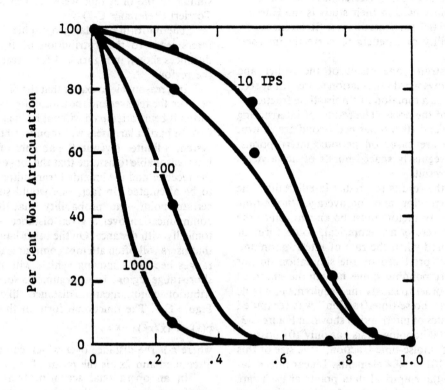

**FIGURE C3
EMPIRICAL BASIS OF INTELLIGIBILITY MODEL**

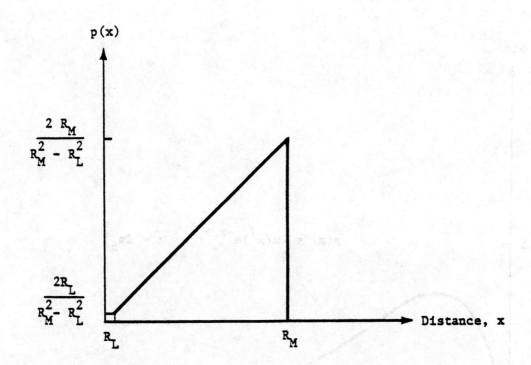

**FIGURE C4
RADIAL DENSITY FOR UNIFORM DEPLOYMENT OF INTERFERERS**

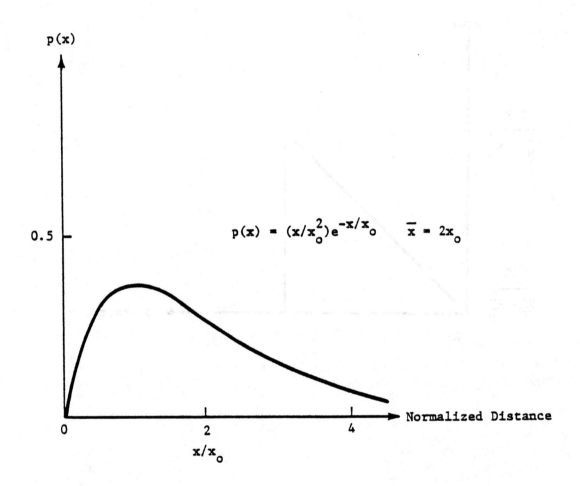

**FIGURE C5
DENSITY OF DISTANCE FROM RECEIVER TO INTENDED TRANSMITTER**

C.4 SIMULATION RESULTS

C.4.1 Signal-to-Interference Simulation Model

All simulation runs on this model are mobile-to-base runs, with a mean communication distance of 8 km (5 miles). A hopping rate of one per second is used. "Splatter" from adjacent-channel interference is neglected in this preliminary model.

In order to provide a baseline against which to compare the performance of a frequency hopping system, several runs were made for a single channel system. The first run, shown in Figure C-6, provided the CDF of the RF signal-to-interference ratio when one interferer is present continuously on the single channel. In order to facilitate interpretation of the CDF, the locations median and the lower decile are marked on the abscissa by a triangle (∇) and a vertical arrow ($\uparrow$), respectively. This convention will be followed on all of the graphs.

It can be argued, heuristically, that an RF signal-to-interference ratio of about 10 dB is required for adequate communications. This rather arbitrary criterion is based on the threshold of a narrowband FM receiver, which is typically a few dB below the 10 dB level. Note that the signal-to-interference ratio varies not only with the various transmitter/receiver distances, but also with time (i.e., with hopping) for any given configuration of transmitter-receiver distances.

It can be seen from Figure C-6 that for a single channel and a single interferer, the 10 dB S/I criterion was met on only about 62 percent of the trials. For two interferers on a single channel (see Figure C-7) this fraction reduces to about 37 percent. For four interferers (see Figure C-8) it is further reduced to about 26 percent. The implication of this is that the degree of frequency reuse (by geographic separation) possible in a conventional narrowband 900-MHz system with the assumed distributions of distance will be limited. However, this issue needs to be addressed in more detail, particularly if the dominant mode of operation in such a system turns out to be mobile-to-mobile (without repeater).

The results of the signal-to-interference simulation for a 100-channel frequency-hopping are illustrated in Figures C-9 through C-15, for 5, 10, 20, 40, 80, 160 and 200 interferers, respectively. In computing the signal-to-interference ratio on each trial, provision must be made for the case in which no interference occurs, due to a particularly favorable selection of channels on that hop. In this case, an RF signal-to-interference ratio of 100 dB has been arbitrarily assigned. In practice, the difference between a 100 dB RF signal-to-interference ratio and an infinite RF signal-to-interference ratio will not be discernible to the listener. The system will become limited by noise and distortion long before this point. Note also that the program used to plot the CDFs automatically scales the abscissa according to the range of the data points. Thus, the abscissa scales on the various figures are generally different.

One result of the runs made on this model is that a single-channel system with two interferers (Figure C-7) and a 100-channel system with 200 frequency-hopping interferers (Figure C-15) yield roughly the same signal-to-interference statistics (in the absence of splatter), although the FH statistics suggest slightly inferior performance. On the basis of such a comparison, some degradation should be expected for the FH system, since the instantaneous channel loadings for FH are not uniform, as with N equally loaded conventional channels.

From Figures C-9 through C-15, it is possible to plot the probability of attaining a 10 dB S/I ratio as a function of the number of active interferers. Such a plot is shown in Figure C-16. Also shown on this figure are the corresponding points for the single-channel simulation results. In both cases, it is possible to interpolate between the data points using piecewise linear segments. The largest residual for either case is about .04. Of course, the continuous line segments are meaningful only for integer values of M (the number of interferers). Extrapolations of these segments beyond the data points would seem to be a conservative procedure, since the probability of achieving a 10 dB C/I ratio must approach zero asymptotically as M increases. The extrapolations, on the other hand, show a zero probability of achieving the desired performance at M = 35 (for one channel) and M = 640 (for 100 channels). The cost of replacing the extrapolations with simulation results was not judged to be justified, since the cost of the simulation rises rapidly for large numbers of interferers if accurate estimates of low-probability events are required.

A simple traffic model (Poisson arrivals, exponential message durations) can now be used to predict the probability of achieving a 10 dB C/I ratio as a function of traffic intensity. Recognizing that the statement "$10 \leq C/I$" has no meaning unless there is at least one call in progress or about to be made:

$$P[10 < C/I] = \sum_{M=0}^{\infty} P[10 < C/I | M] P[M]$$

$$= \sum_{M=0}^{\infty} P[10 < C/I | M] \left[\frac{\rho M_e^{-\rho}}{M!} \right]$$

where ρ = traffic intensity, Erlangs

The results of inserting the simulation-produced estimates of p(10 < C/I) into this traffic model are shown in Figures C-17 and C-18 for single-channel and 100-channel operation, respectively. It can be seen that, for a 90 percent probability of achieving a 10 dB C/I ratio, the maximum traffic intensity is about 0.25 Erlangs for a single channel and 12.5 Erlangs for a 100-channel frequency hopping system having the characteristics described above. Thus, the channel-by-channel version (i.e., nonfrequency hopping) would appear to be about twice as efficient in its use of the spectrum as the FH version, based on the simple models used here. This comparison embodies the implicit assumption that the channels in the conventional system are loaded equally, on the average. Constraints that cause the channels to be loaded unequally (due to regulatory factors, for example) may shift the balance in favor of the FH approach. In addition, it will be shown in the next section that the spectrum efficiency of the FH system improves significantly when the mean attempted communication distance is reduced.

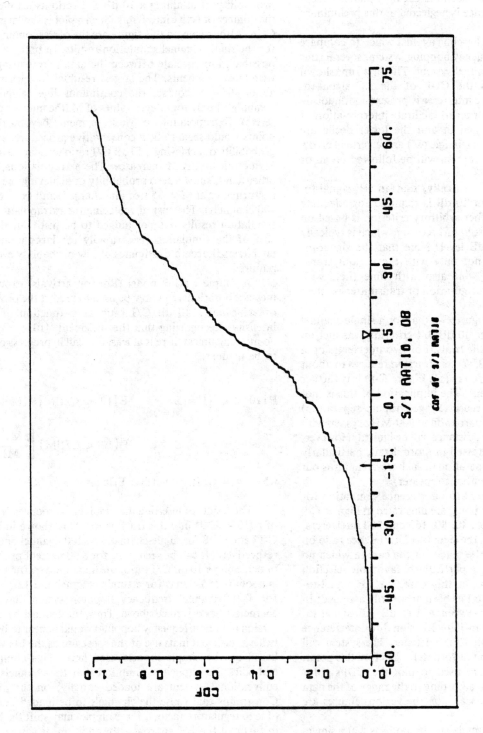

FIGURE C6
CUMULATIVE DISTRIBUTION OF R.F.
SIGNAL-TO-INTERFERENCE RATIO: 1 CHANNEL,
1 INTERFERER

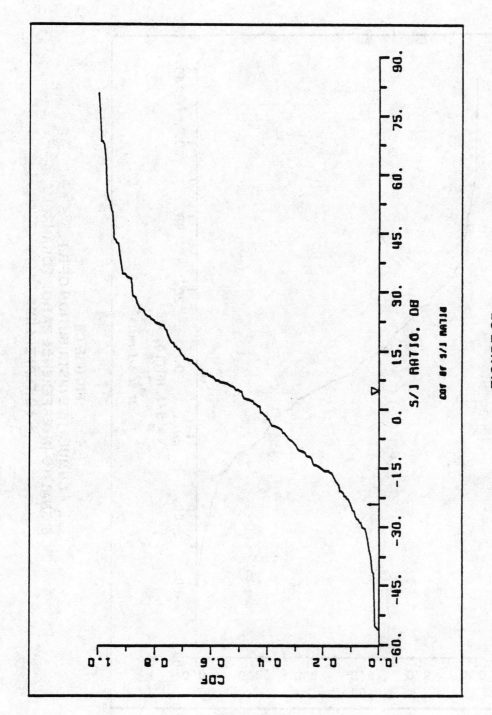

**FIGURE C7
CUMULATIVE DISTRIBUTION OF R.F.
SIGNAL-TO-INTERFERENCE RATIO: 1 CHANNEL,
2 INTERFERERS**

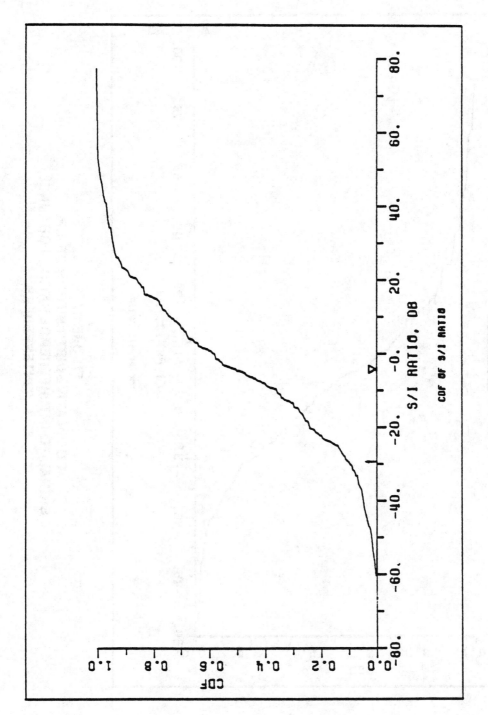

**FIGURE C8
CUMULATIVE DISTRIBUTION OF R.F.
SIGNAL-TO-INTERFERENCE RATIO: 1 CHANNEL,
4 INTERFERERS**

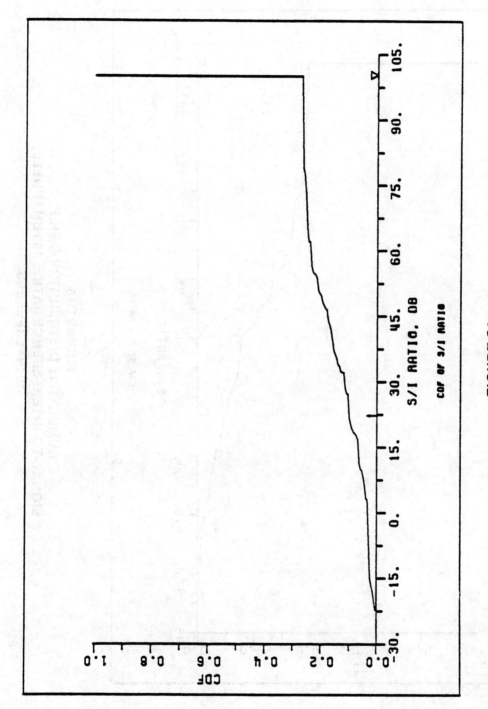

FIGURE C9
CUMULATIVE DISTRIBUTION OF R.F.
SIGNAL-TO-INTERFERENCE RATIO: 100 CHANNELS,
5 INTERFERERS

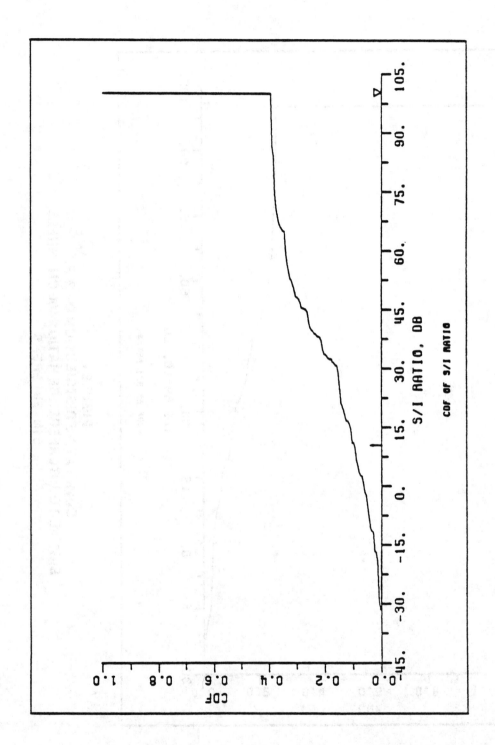

**FIGURE C10
CUMULATIVE DISTRIBUTION OF R.F.
SIGNAL-TO-INTERFERENCE RATIO: 100 CHANNELS,
10 INTERFERERS**

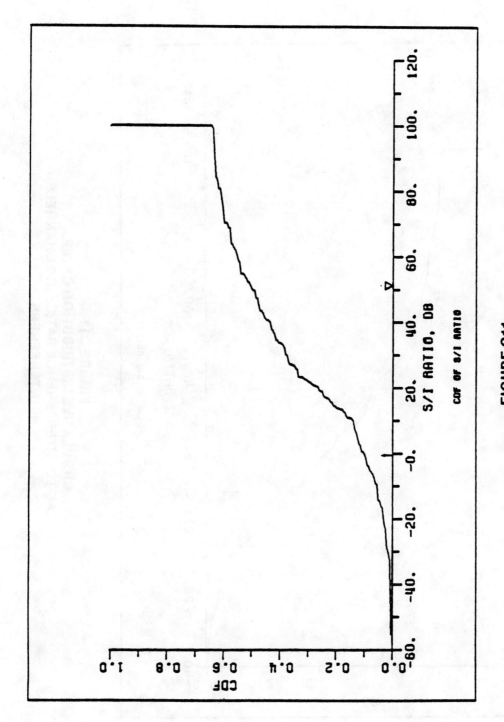

**FIGURE C11
CUMULATIVE DISTRIBUTION OF R.F.
SIGNAL-TO-INTERFERENCE RATIO: 100 CHANNELS,
20 INTERFERERS**

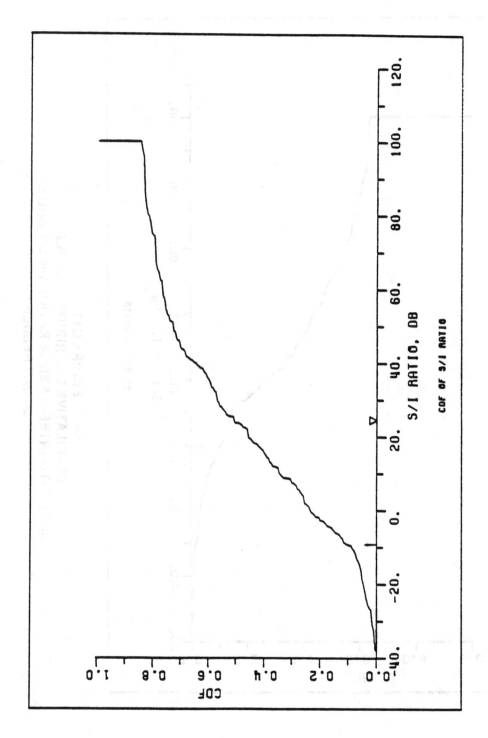

**FIGURE C12
CUMULATIVE DISTRIBUTION OF R.F.
SIGNAL-TO-INTERFERENCE RATIO: 100 CHANNELS,
40 INTERFERERS**

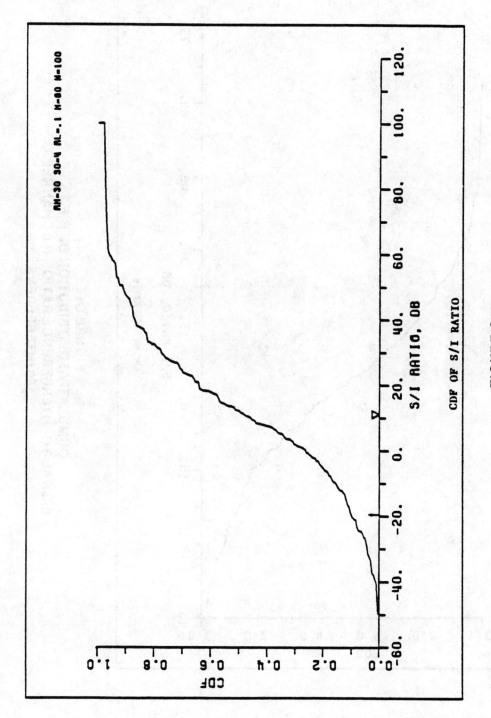

**FIGURE C13
CUMULATIVE DISTRIBUTION OF R.F.
SIGNAL-TO-INTERFERENCE RATIO: 100 CHANNELS,
80 INTERFERERS**

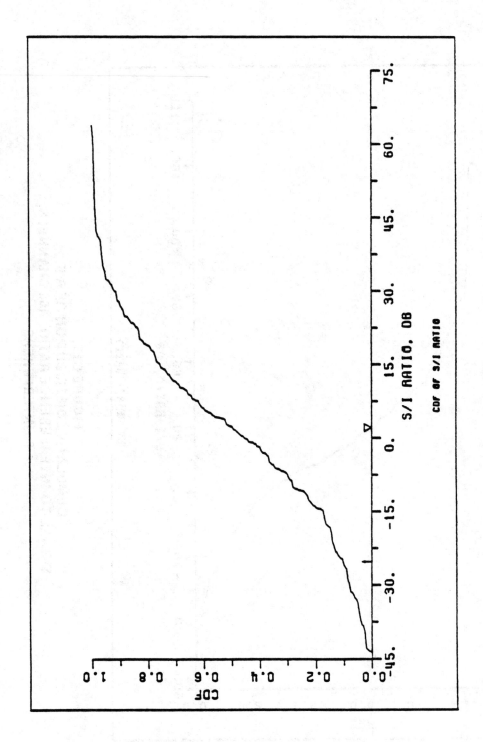

**FIGURE C14
CUMULATIVE DISTRIBUTION OF R.F.
SIGNAL-TO-INTERFERENCE RATIO: 100 CHANNELS,
160 INTERFERERS**

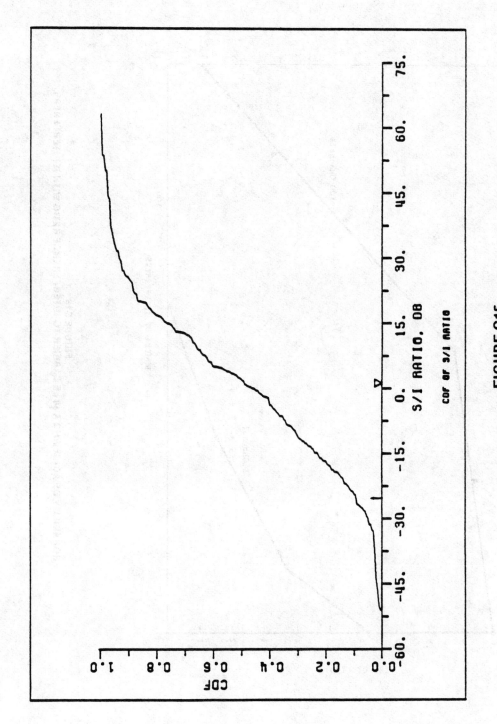

**FIGURE C15
CUMULATIVE DISTRIBUTION OF R.F.
SIGNAL-TO-INTERFERENCE RATIOS:
100 CHANNELS, 200 INTERFERERS**

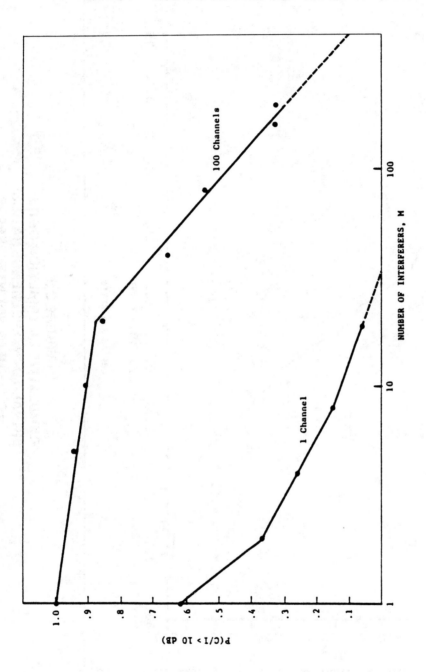

**FIGURE C16
PROBABILITY OF ACHIEVING A 10 dB CARRIER-TO-INTERFERENCE RATIO WITH M INTERFERERS**

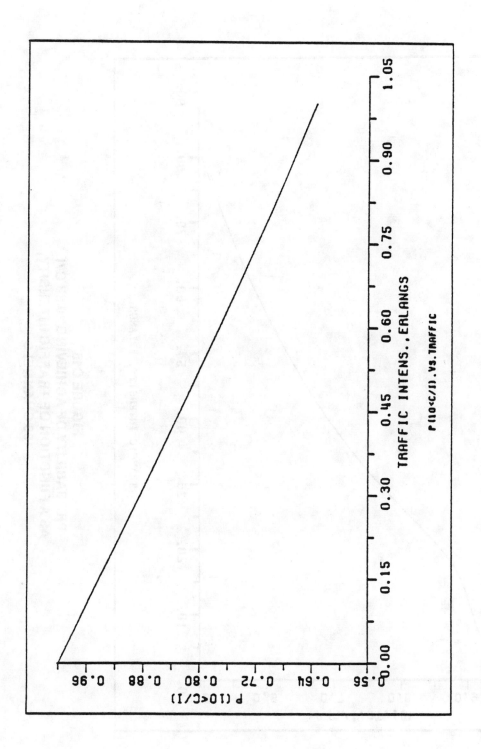

**FIGURE C17
PROBABILITY OF ACHIEVING 10 dB C/I
AS A FUNCTION OF TRAFFIC INTENSITY,
1 CHANNEL**

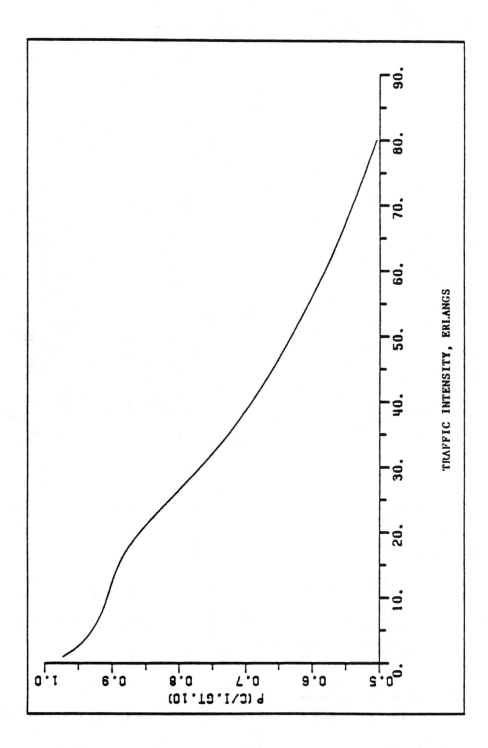

**FIGURE C18
PROBABILITY OF ACHIEVING 10 dB C/I
AS A FUNCTION OF TRAFFIC INTENSITY,
100 CHANNELS**

C.4.2 Intelligibility Model

The intelligibility model incorporates a number of refinements that allow the impact of parameter changes to be assessed. Specifically, it provides for the hopping rate, the amount of "splatter," the mean communication distance, and the link type (mobile-to-base or mobile-to-mobile) to be set for each simulation run. All runs are based on a 100-channel system.

C.4.2.1 Mobile-to-Base Operation

A tabulation of the principal mobile-to-base simulation runs and their parameter settings is shown in Table C-2. The table is self-explanatory except for the column marked "splatter." The figures in this column represent the assumed attenuation on the adjacent and next-adjacent channels, respectively. For example, 60/80 means that an adjacent-channel signal is attenuated by 60 dB relative to a co-channel signal and that a signal two channels removed is attenuated 80 dB relative to a co-channel signal. Instead of trying to group the runs into sets having all but one parameter in common, the runs will simply be presented in chronological order.

Figures C-19 and C-20 show the impact of raising the hopping rate from one per second (Figure C-19; this rate was also used in the signal-to-interference model) and twenty per second (Figure C-20; this rate is used for all of the simulation runs that follow). For both runs, the mean distance at which communication is attempted is 8.0 km (5 miles). The "splatter" attenuation is assumed to be 10 dB and 30 dB for adjacent and next-adjacent channels, respectively. This is representative of what can be achieved using channel spacings of 15 to 20 kHz with the least expensive two-pole monolithic crystal filters (see, for example, reference C-8).

It can be seen from Figures C-19 and C-20 that the impact of increasing the hopping rate from 1/s to 20/s is not great under these conditions. This is primarily because the rate of interruption is usually low with only 5 interferers, even at 20 hops/s. For low rates of interruption, the dependence of articulation on interruption rate is small.

Figure C-21 shows the result of increasing the number of interferers to 20 with all other parameters unchanged. Although the degradation of performance is clearly visible, there is still a 91 percent chance of achieving an articulation score of 0.75 or better.

If the number of interferers is increased to 40 (without changing any other parameters), the CDF of the word articulation score is as shown in Figure C-22. If the criterion for acceptable performance is a 90 percent chance of achieving a 0.75 word articulation score, then this run fails to yield acceptable performance, since the probability of achieving a word articulation score of 0.75 is only 0.79.

In an effort to determine the effect of the assumed 8 km mean attempted communication distance, this parameter was reduced to 4 km (2.5 miles) and the simulation was re-run with 20 and 40 interferers (other parameters, except for the number of trials, were unchanged from the conditions that produced Figures C-21 and C-22). The results are shown in Figures C-23 and C-24, respectively. Although some improvement is evident, the amount of improvement is not great. The probability of attaining an acceptable articulation score (0.75) is still below 90 percent for 40 interferers.

In Figure C-25, the "splatter" parameters have been changed to show the effect of substantially removing splatter. This accomplished by setting adjacent-channel and next-adjacent channel attenuation to 200 dB. Since the maximum possible interference-to-signal ratio for a single interferer (interferer 0.1 km from the receiver and intended signal source 100 km from the receiver) is on the order of 115 dB, there is no way that a single interferer (or any reasonably likely combination of multiple interferers) can significantly contribute to pushing the net signal-to-interference ratio across the 6 dB threshold unless it is already extremely close to that threshold with co-channel interference alone.

The results of this change are significant, indicating that the 10 dB/30 dB attenuation figures used in the initial simulation runs had contributed to a measurable performance degradation. With 200 dB/200 dB splatter, the lower decile of the distribution of articulation scores is about 0.84, compared with 0.73 for the equivalent run with 10/30 splatter. (Both runs were for 40 interferers.)

TABLE C-2
LIST OF SIMULATION RUNS (INTELLIGIBILITY MODEL)

FIGURE	NO. OF CHANNELS	NO. OF INTERFERERS	SPLATTER, DB ATTENUATION	MEAN COMM. DIST, KM	HOPPING RATE, /s	NO. OF TRIALS	LINK DIRECTION
C-19	100	5	10/30	8	1	200	MB
C-20	100	5	10/30	8	20	200	MB
C-21	100	20	10/30	8	20	200	MB
C-22	100	40	10/30	8	20	200	MB
C-23	100	20	10/30	4	20	100	MB
C-24	100	40	10/30	4	20	50	MB
C-25	100	40	200/200	4	20	50	MB
C-26	100	80	200/200	4	20	50	MB
C-27	100	120	200/200	4	20	50	MB
C-28	100	80	60/80	8	20	50	MB
C-29	100	80	60/80	4	20	50	MB
C-30	100	100	60/80	2	20	50	MB
C-31	100	130	60/80	2	20	50	MB
C-32	100	160	60/80	2	20	50	MB
C-33	100	20	60/80	16	20	50	MB
C-34	100	40	60/80	16	20	50	MB
C-35	100	60	60/80	16	20	50	MB
C-36	100	60	60/80	8	20	50	MB

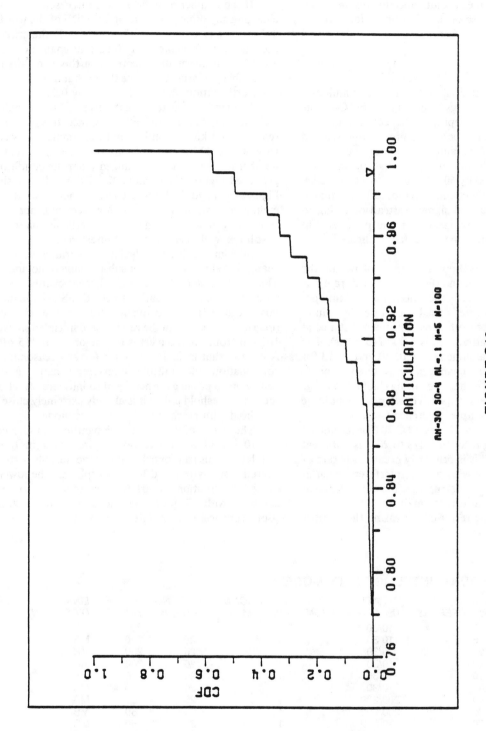

**FIGURE C19
CUMULATIVE DISTRIBUTION OF PB WORD
ARTICULATION: 5 INTERFERERS, 10/30
SPLATTER, 8 KM MEAN COMM. DIST., 1 HOP/S,
200 TRIALS, MOBILE-TO-BASE**

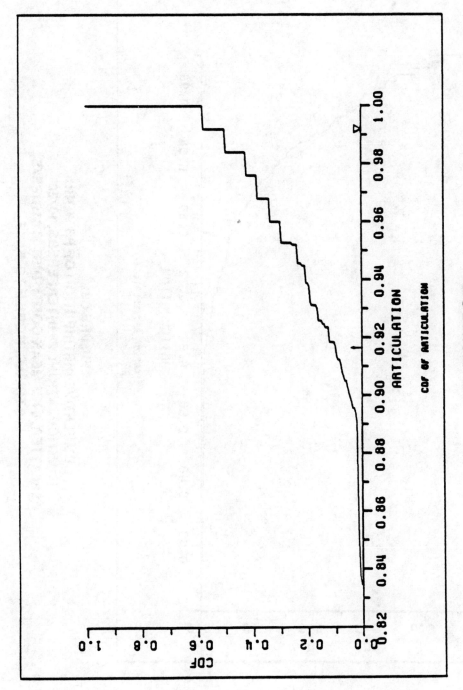

FIGURE C20
CUMULATIVE DISTRIBUTION OF PB WORD
ARTICULATION: 5 INTERFERERS,
10/30 SPLATTER, 8 KM MEAN COMM. DIST.,
20 HOPS/S, 200 TRIALS, MOBILE-TO-BASE

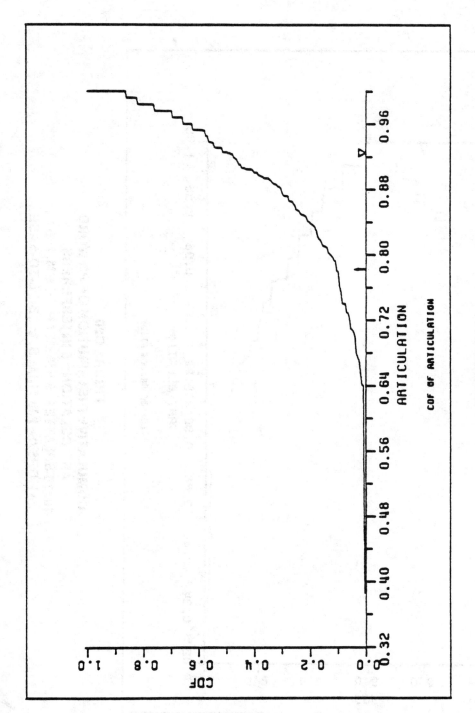

**FIGURE C21
CUMULATIVE DISTRIBUTION OF PB WORD
ARTICULATION: 20 INTERFERERS, 10/30
SPLATTER, 8 KM MEAN COMM. DIST., 20 HOPS/S,
200 TRIALS, MOBILE-TO-BASE**

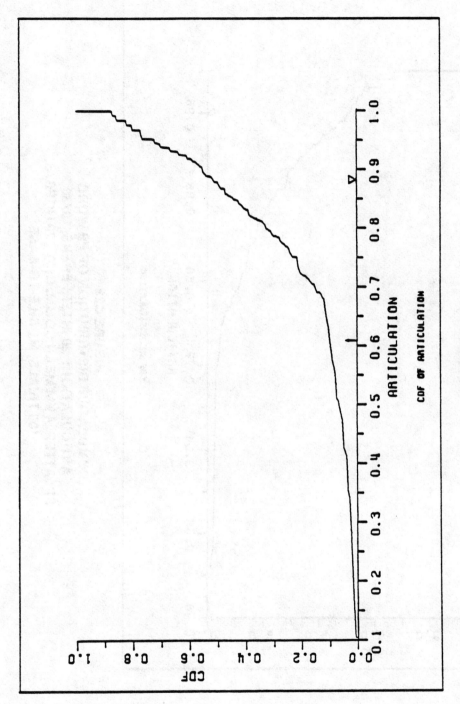

**FIGURE C22
CUMULATIVE DISTRIBUTION OF PB WORD
ARTICULATION: 40 INTERFERERS, 10/30
SPLATTER, 8 KM MEAN COMM. DIST., 20 HOPS/S,
200 TRIALS, MOBILE-TO-BASE**

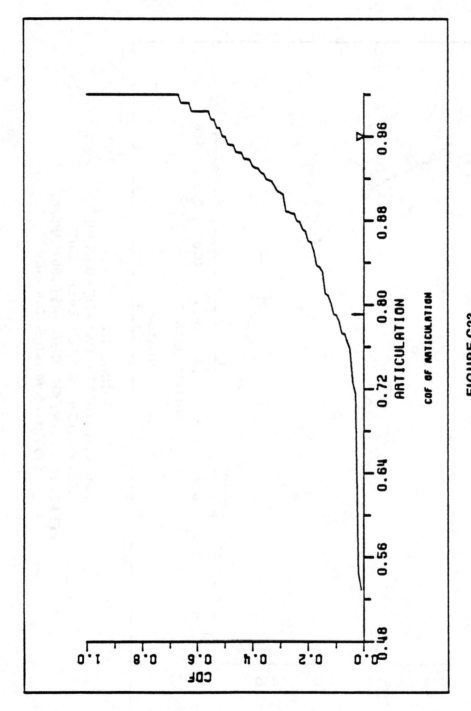

**FIGURE C23
CUMULATIVE DISTRIBUTION OF PB WORD
ARTICULATION: 20 INTERFERERS, 10/30
SPLATTER, 4 KM MEAN COMM. DIST., 20 HOPS/S,
100 TRIALS, MOBILE-TO-BASE**

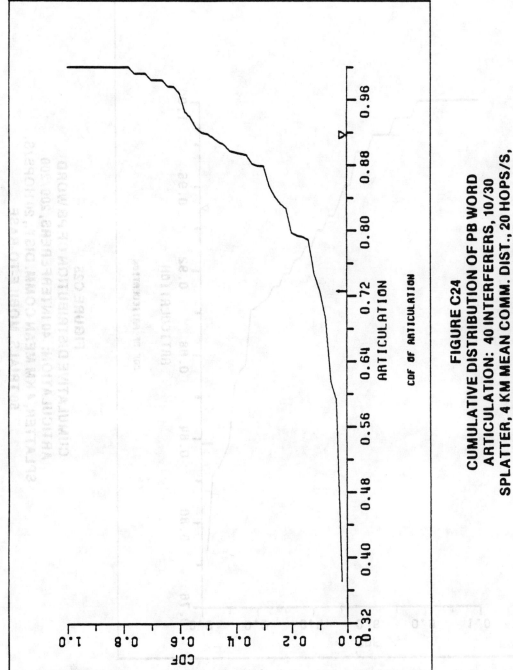

FIGURE C24
CUMULATIVE DISTRIBUTION OF PB WORD
ARTICULATION: 40 INTERFERERS, 10/30
SPLATTER, 4 KM MEAN COMM. DIST., 20 HOPS/S,
50 TRIALS, MOBILE-TO-BASE

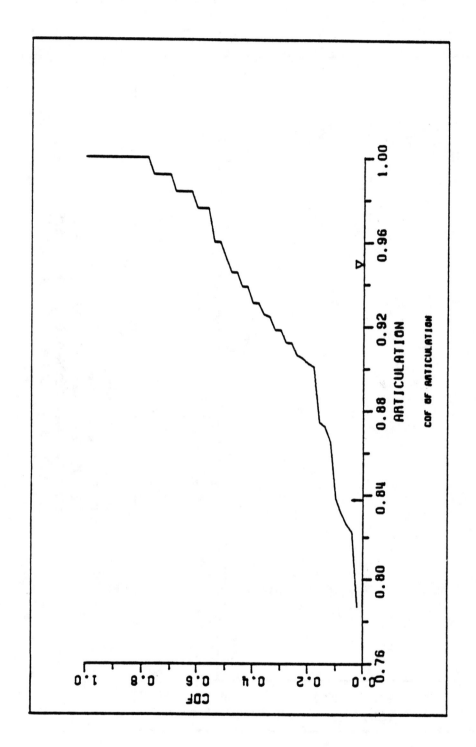

**FIGURE C25
CUMULATIVE DISTRIBUTION OF PB WORD
ARTICULATION: 40 INTERFERERS, 200/200
SPLATTER, 4 KM MEAN COMM. DIST., 20 HOPS/S,
50 TRIALS, MOBILE-TO-BASE**

The impact of increasing the number of interferers from 40 to 80 is shown in Figure C-26. All other parameters are unchanged from the previous run. The performance at this level still appears to be acceptable; the lower decile of the distribution is 0.77. When the number of interferers is increased to 120 (Figure C-27), the lower decile decreases to about 0.63; there is only an 82 percent chance of achieving an articulation score of 0.75. Thus, the maximum number of users for this case (given the performance constraint stated above) is between 80 and 120, or about one per channel.

Although this result is encouraging, it is also unrealistic. The potential cost of developing and producing (1) receivers with 200 dB selectivity and (2) transmitters with 200 dB suppression of sideband noise and other spurious emissions, is likely to dissuade any entrepreneur from undertaking such an effort, particularly for a consumer-oriented application.

In order to see whether similar results could be attained with more realistic "splatter" parameters, the adjacent channel and next-adjacent channel attenuations were changed to 60 dB and 80 dB, respectively. Simultaneously, the mean distance over which communication is attempted was restored to 8.0 km (5 miles) and the number of interferers was set to 80. The results are shown in Figure C-28. It can be seen that the combination of changes has put the performance back into the unacceptable region, since the probability of achieving a 75 percent PB word articulation score is only about 0.76.

The effect of maintaining the 60/80 splatter configuration while reducing the mean attempted communication distance 4 km (2.5 miles) is shown in Figure C-29. This run is comparable to the run displayed in Figure C-26 except for the increased splatter.

It can be seen that the performance is now back in the acceptable region; the chance of attaining a 75 percent word articulation score is about 0.93. Thus, the maximum number of users per channel for this case is again close to unity.

Receiver selectivities of 60 dB (adjacent channel) and 80 dB (next-adjacent channel) are attainable using multipole crystal filters in conjunction with 25-kHz channel spacing. However, reducing transmitter sideband splatter to these levels may require careful design efforts or unusually wide channel spacings. The impact of receiver-generated intermodulation products also needs to be investigated.

Additional simulation runs with 60/80 dB splatter are shown in Figures C-30 through C-36 for mean communication distress of 2.0 km through 16.0 km. These results were used to provide a coarse plot of maximum number of simultaneous users versus mean communication distance (see Section C.4.4).

C.4.2.2 Mobile-to-Mobile Operation (Without Repeaters)

In order to simulate direct mobile-to-mobile operation, it is only necessary to generate a new path loss table based on a lower antenna height at the receiver. This assumes that all interferers are mobiles. Antenna heights of 2.0 meters were assumed at both ends of the link. By the Longley-Rice model, this had the net effect of increasing the median path loss by about 5 dB for distance greater than 1.0 km. For shorter distances, free space path loss is used, so the received power is independent of antenna height in this range.

The results are shown in Figures C-37 and C-38 for mean attempted communication distances of 4 km (2.5 miles) and 2 km (1.25 miles), respectively. Both runs are based on 80 interferers and 60/80 dB splatter. At 4 km, the performance is not significantly different from the corresponding mobile-to-base performance (Figure C-29). At 2 km, performance is substantially improved.

C.4.2.3 Base-to-Mobile Operation

The mobile-to-base and base-to-mobile links in land mobile radio systems are normally not equivalent, because the antenna gains and noise levels can differ. However, all of the simulation runs described above are based on the assumptions of interference-limited operation, so neither background noise nor antenna gains are reflected in the results. Also, the (Longley-Rice) propagation model is reciprocal with respect to transmitter and receiver antenna heights, so the uplink and downlink are equivalent in terms of signal-to-interference ratio for a given set of transmitter-to-receiver distances.

Thus, the results presented above for mobile-to-base operation are also applicable, in principle, to the base-to-mobile link. In order for this equivalence to hold, however, it is necessary to imagine the mobile receiver at the center of the coverage area, with the base station transmitters distributed around it according to the density functions describe for mobile transmitters in Section C.3.7.

C.4.3 The Impact of Background Noise and Fading

The assumption of interference-limited operation is a useful mechanism that has been applied in other analyses and simulations of FH systems. Nevertheless, real systems are limited in the attached communication range in the absence of interference. Table C-3 illustrates a simplified 915-MHz mobile-to-base link budget for a quiet receiver location. Based on the assumed parameters, the maximum permissible path loss is 155 dB, including a 20 dB fade margin. In order for this loss to be achieved at 90 percent of the mobile locations, a maximum range of about 8.3 km (5.2 miles) is predicted by the Longley-Rice model, based on the parameters shown in Table C-1. This 155 dB path loss will be achieved 50 percent of the time at a range of about 19.3 km (12.1 miles). Noisy receiver locations, lower transmitter powers, line losses, and urban shadowing will all contribute to reduced ranges. Such range limitations obviously apply to conventional narrowband systems as well as frequency-hopping systems.

TABLE C-3

SIMPLIFIED MOBILE-TO-BASE LINK BUDGET FOR QUIET RECEIVER LOCATION

Receiver Sensitivity:	−146 dBW (.35 NV)
Receiver Antenna Gain:	9 dB
Transmit Antenna Gain:	5 dB
Transmit Power: 15 dBW	
Maximum Nonfading Path Loss:	175 dB
Fade Margin 20 dB	
Maximum Path Loss: 155 dB	
8.3 km, 90%	
19.3 km, 50%	

C.4.4 Summary of Simulation Results

The simulations described here indicate that it may be feasible to provide acceptable communication performance with slow frequency hopping FM. The spectrum efficiency of such an approach, in terms of number of active users per channel over a given geographic area depends on the degree

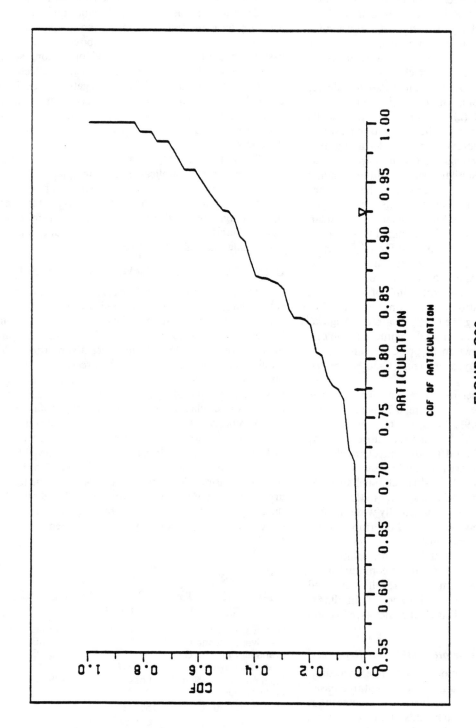

**FIGURE C26
CUMULATIVE DISTRIBUTION OF PB WORD
ARTICULATION: 80 INTERFERERS, 200/200
SPLATTER, 4 KM MEAN COMM. DIST., 20 HOPS/S,
50 TRIALS, MOBILE-TO-BASE**

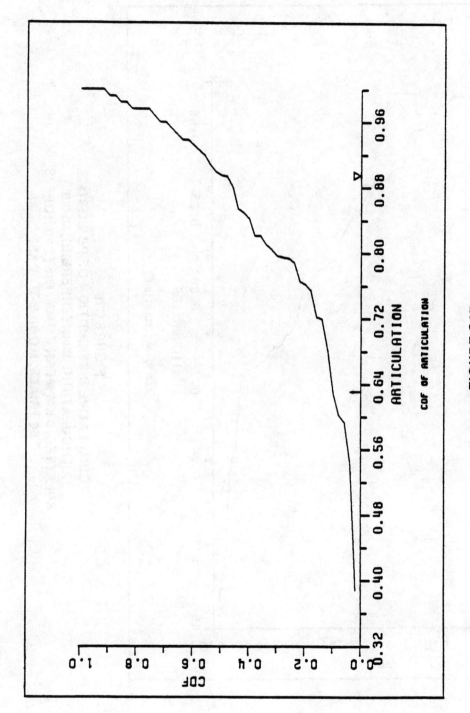

**FIGURE C27
CUMULATIVE DISTRIBUTION OF PB WORD
ARTICULATION: 120 INTERFERERS, 200/200
SPLATTER, 4 KM MEAN COMM. DIST., 20 HOPS/S,
50 TRIALS MOBILE-TO-BASE**

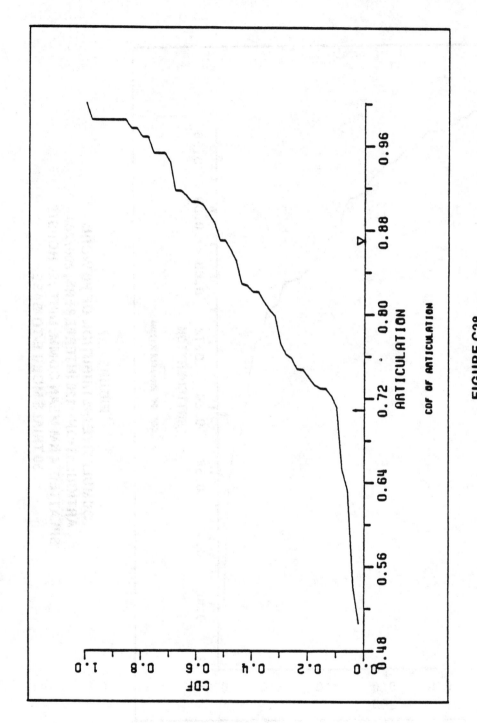

FIGURE C28
CUMULATIVE DISTRIBUTION OF PB WORD
ARTICULATION: 80 INTERFERERS, 60/80
SPLATTER, 6 KM MEAN COMM. DIST., 20 HOPS/S,
50 TRIALS, MOBILE-TO-BASE

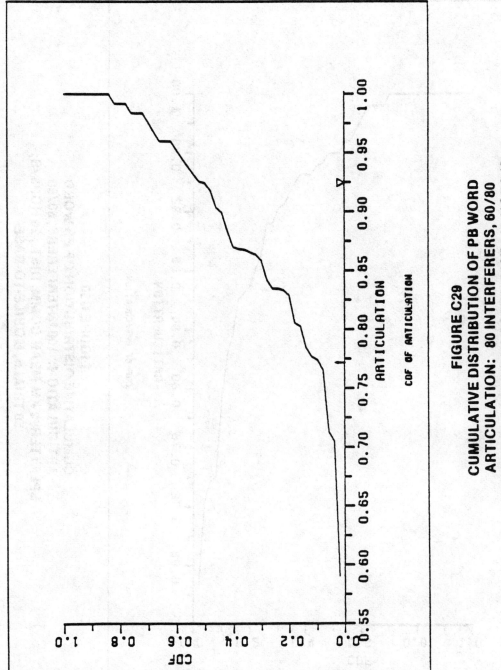

FIGURE C29
CUMULATIVE DISTRIBUTION OF PB WORD
ARTICULATION: 80 INTERFERERS, 60/80
SPLATTER, 4 KM MEAN COMM. DIST., 20 HOPS/S,
50 TRIALS, MOBILE-TO-BASE

9-113

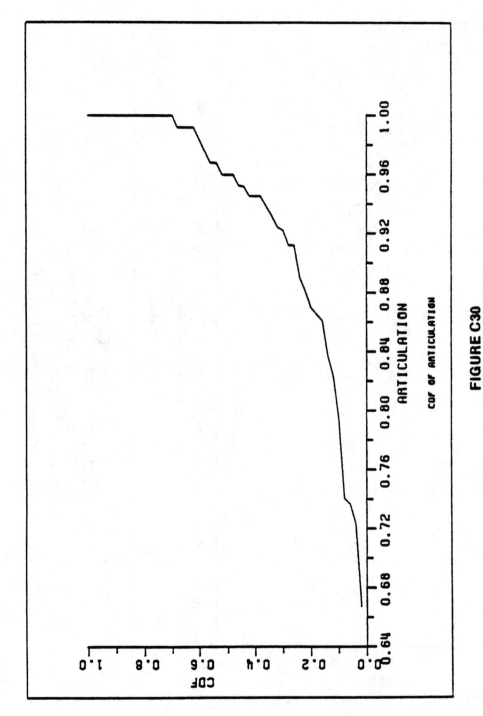

FIGURE C30
CUMULATIVE DISTRIBUTION OF PB WORD
ARTICULATION: 100 INTERFERERS, 60/80
SPLATTER, 2 KM MEAN COMM. DIST., 20 HOPS/S,
50 TRIALS, MOBILE-TO-BASE

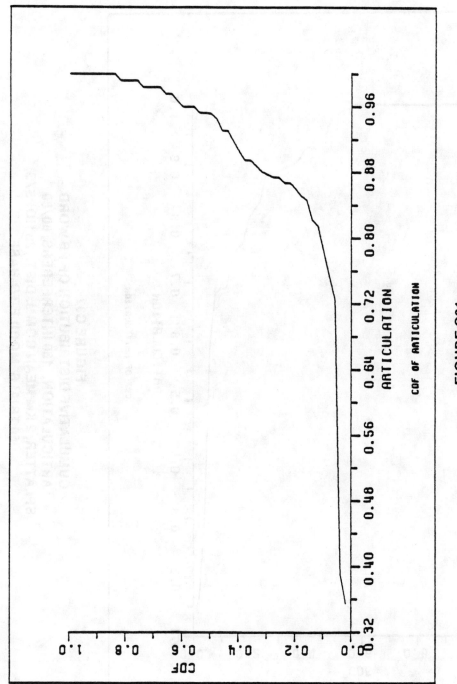

**FIGURE C31
CUMULATIVE DISTRIBUTION OF PB WORD
ARTICULATION: 130 INTERFERERS, 60/80
SPLATTER, 2 KM MEAN COMM. DIST., 20 HOPS/S,
50 TRIALS, MOBILE-TO-BASE**

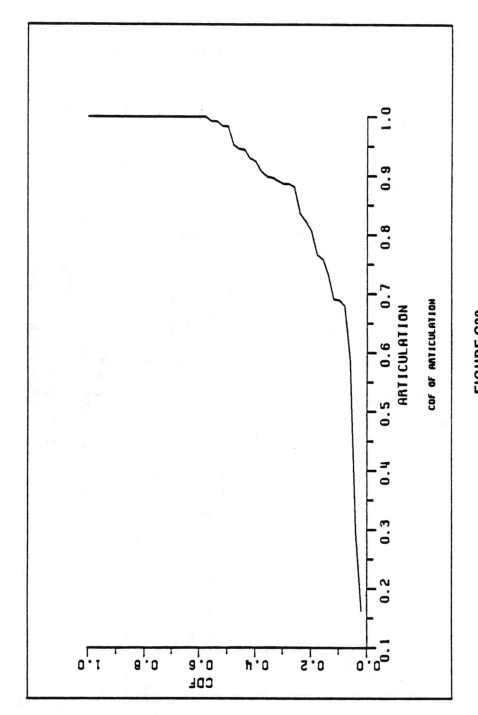

**FIGURE C32
CUMULATIVE DISTRIBUTION OF PB WORD
ARTICULATION: 160 INTERFERERS, 60/80
SPLATTER, 2 KM MEAN COMM. DIST., 20 HOPS/S,
50 TRIALS, MOBILE-TO-BASE**

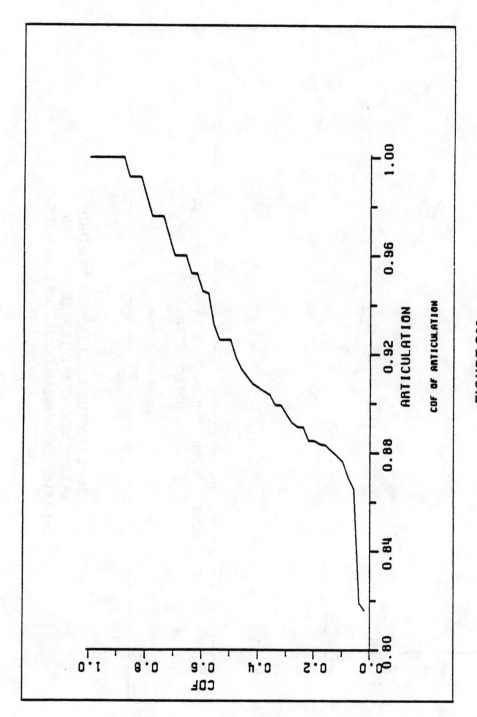

FIGURE C33
CUMULATIVE DISTRIBUTION OF PB WORD
ARTICULATION: 20 INTERFERERS, 60/80
SPLATTER, 16 KM MEAN COMM. DIST., 20 HOPS/S,
50 TRIALS, MOBILE-TO-BASE

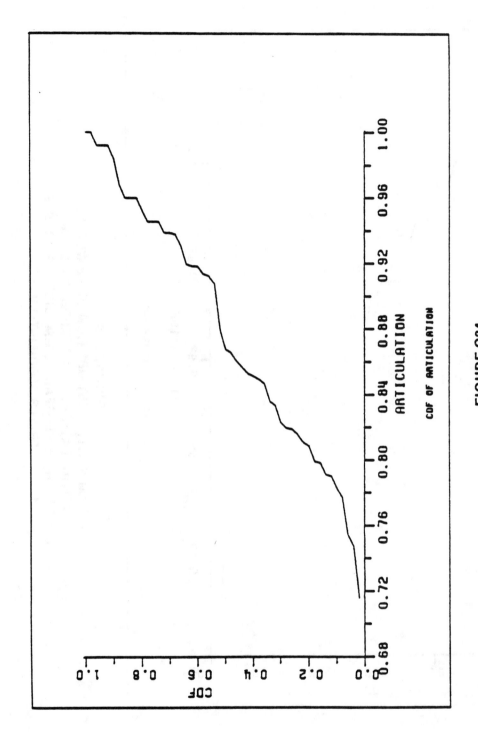

**FIGURE C34
CUMULATIVE DISTRIBUTION OF PB WORD
ARTICULATION: 40 INTERFERERS, 60/80
SPLATTER, 16 KM MEAN COMM. DIST., 20 HOPS/S,
50 TRIALS, MOBILE-TO-BASE**

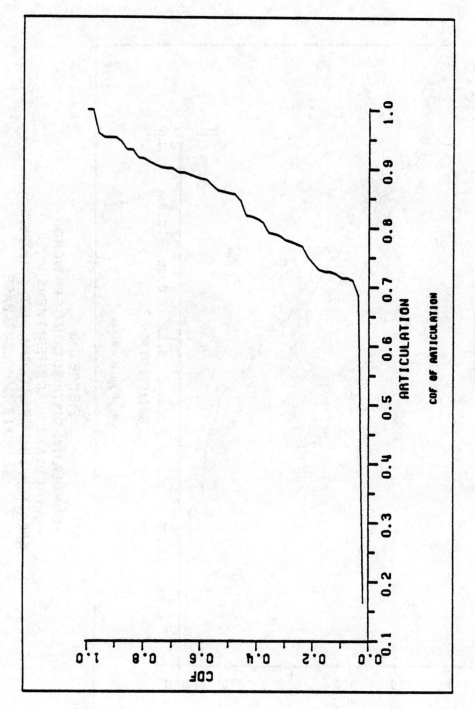

**FIGURE C35
CUMULATIVE DISTRIBUTION OF PB WORD
ARTICULATION: 60 INTERFERERS, 60/80
SPLATTER, 16 KM MEAN COMM. DIST., 20 HOPS/S,
50 TRIALS, MOBILE-TO-BASE**

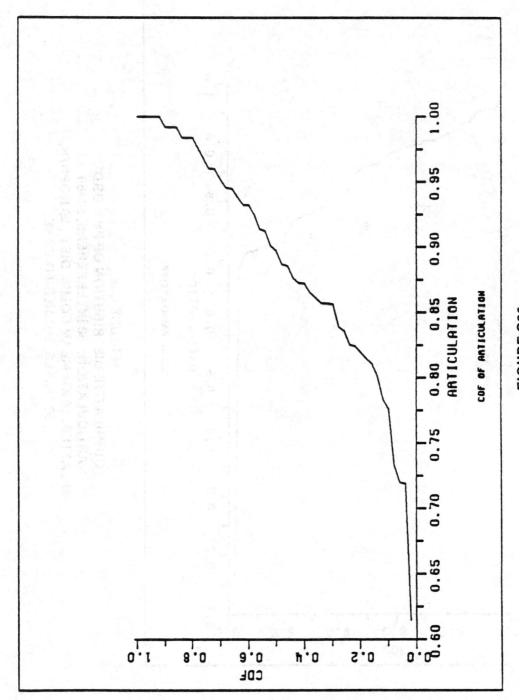

FIGURE C36
CUMULATIVE DISTRIBUTION OF PB WORD
ARTICULATION: 60 INTERFERERS, 60/80
SPLATTER, 8 KM MEAN COMM. DIST., 20 HOPS/S,
50 TRIALS, MOBILE-TO-BASE

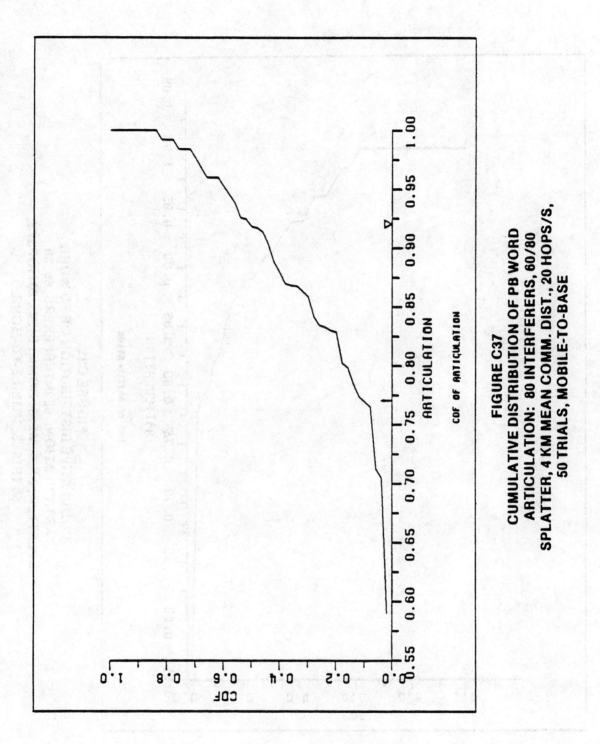

**FIGURE C37
CUMULATIVE DISTRIBUTION OF PB WORD
ARTICULATION: 80 INTERFERERS, 60/80
SPLATTER, 4 KM MEAN COMM. DIST., 20 HOPS/S,
50 TRIALS, MOBILE-TO-BASE**

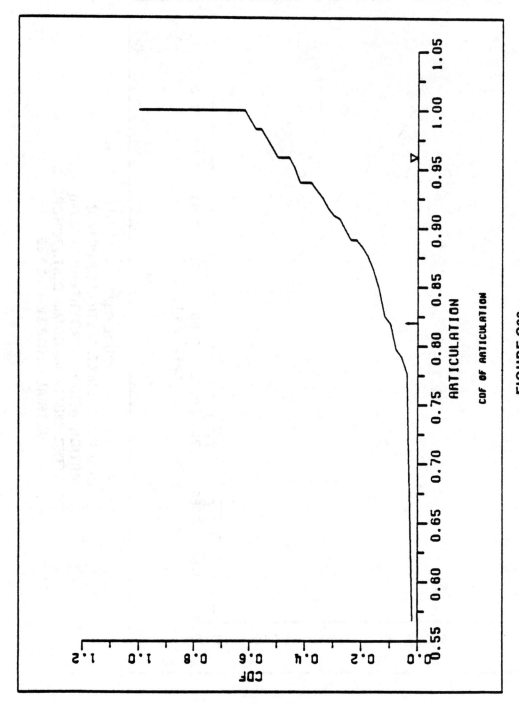

FIGURE C38
CUMULATIVE DISTRIBUTION OF PB WORD
ARTICULATION: 80 INTERFERERS, 60/80
SPLATTER, 2 KM, MEAN COMM. DIST., 20 HOPS/S,
50 TRIALS, MOBILE-TO-MOBILE

9-122

TABLE C-4

ESTIMATED MAXIMUM NUMBER OF USERS PER CHANNEL FOR INTERFERENCE-LIMITED OPERATION

Splatter, dB Attenuation	Mean Communication Distance, km	Type of Link	Estimated Maximum Number of Users Per Channel
10/30	8.0	MB	0.3
10/30	4.0	MB	0.3
200/200	4.0	MB	1.0
60/80	16.0	MB	0.5
60/80	8.0	MB	0.6
60/80	4.0	MB	0.9
60/80	2.0	MB	1.4
60/80	4.0	MM	0.9
60/80	2.0	MM	1.4

of adjacent and next-adjacent channel interferences as well as the mean communication distance. The slow FH approach may be able to accommodate roughly one active user per channel under favorable conditions. The degree of adjacent-channel and next-adjacent channel "splatter" can have a significant impact on communication performance and spectrum efficiency. The estimated maximum number of users per channel is summarized in Table C-4 for various operating conditions. for 60/80 dB splatter, the estimated maximum number of users per channel is plotted versus the mean communication distance in Figure C-39.

C.5 REFERENCES

1. A. G. Longley and P. L. Rice, "Prediction of Tropospheric Radio Transmission Loss Over Irregular Terrain, A Computer Method-1968," ESA Technical Report ERL-79-ITS67, AD-676874, July 1968.
2. Anita G. Longley, "Location Variability of Transmission Loss — Land Mobile and Broadcast Systems," OT Report 76-87, May 1976.
3. Anita G. Longley, "Radio Propagation in Urban Areas," 2nd IEEE Vehicular Technology Conference, Denver, March 22-24, 1978.
4. George A. Miller and J.C.R. Licklider, "The Intelligibility of Interrupted Speech," J. Acoustical Soc. of America, Vol. 22, No. 2 March 1950.
5. J. P. Egan, "Articulation Testing Methods," Laryngoscope 58, 1948.
6. Carl E. Williams, et al., "Intelligibility Test Methods and Procedures for the Evaluation of Speech Communication Systems," Report No. ESD-TR-66-677, by Bolt, Beranek, and Newman for Decision Sciences Laboratories, Electronic Systems Division, U.S. Air Force, December 1966.
7. Don J. Torrieri, "Simultaneous Mutual Interference and Jamming in a Frequency Hopping Network," DARCOM Countermeasures/Counter-Countermeasures Office, DARCOM CM/CCM 79-6, October 1979.
8. Publication R29-3-15, Motorola, Inc., Communications Systems Division, Component Products, Franklin Park, Illinois.

Department Approval: _R. Braff_

Author Approval: _W. Scales_

Technical Report Documentation Page

1. Report No.	2. Government Accession No.	3. Recipient's Catalog No.

4. Title and Subtitle	5. Report Date
Potential Use of Spread Spectrum Techniques in Non-Government Applications	December 1980
	6. Performing Organization Code

7. Author(s)	8. Performing Organization Report No.
Walter C. Scales	MTR-80W335

9. Performing Organization Name and Address	10. Work Unit No. (TRAIS)
The MITRE Corporation Metrek Division 1820 Dolley Madison Blvd. McLean, VA 22102	
	11. Contract or Grant No.
	FCC-0320

12. Sponsoring Agency Name and Address	13. Type of Report and Period Covered
Federal Communications Commission 2025 M Street, N.W. Washington, DC 20554	Final Report December 1979–December 1980
	14. Sponsoring Agency Code
	FCC/OST

15. Supplementary Notes

16. Abstract

Spread spectrum technologies have evolved primarily in military and aerospace environments, where the usual objectives are to provide resistance to jamming, multipath and interception. Spread spectrum techniques may also have applications that fall under the jurisdiction of the FCC. In some civil applications, spread spectrum can provide communication privacy and multiple access on a transmit-at-will basis. In certain cases, spread spectrum may be a pratical, spectrum-efficient approach, although spread spectrum hardware is not currently available on a commercial basis. An example is given of a land mobile communication system using slow frequency hopping. Simulation results indicate that this technique might provide short-range service in applications when random channel access is an important feature.

17. Key Words	18. Distribution Statement
Spread Spectrum, Spectrum Efficiency, Frequency Hopping	Document is available to the public through the National Technical Information Service, Springfield, VA 22161

19. Security Classif. (of this report)	20. Security Classif. (of this page)	21. No. of Pages	22. Price
U	U		

Form DOT F 1700.7 (8-72)   Reproduction of completed page authorized

Chapter 10 NTIA Report 82-111

Proposed Direct Sequence Spread Spectrum Voice Techniques for the Amateur Radio Service

J. E. Hershey

U.S. DEPARTMENT OF COMMERCE
Malcolm Baldrige, Secretary

Bernard J. Wunder, Jr., Assistant Secretary
for Communications and Information

November 1982

TABLE OF CONTENTS

	Page
LIST OF TABLES	10-3
LIST OF FIGURES	10-3
ABSTRACT	10-4
1. INTRODUCTION	10-4
2. CHARACTERIZATION OF THE ISSUE	10-4
2.1 The Present Status of the Three ARS Bands in Question	
2.2 Brief Comments on the Proposed Changes to the FCC's Rules and Regulations	
2.3 Concerns	
2.4 The General Issues Surrounding DS Spread Spectrum and the Purpose of This Report	
3. A DS SPREAD SPECTRUM SYSTEM PROPOSED FOR CONSIDERATION FOR ARS USE	10-7
3.1 Introduction	
3.2 The DS Spread Spectrum Technique	
3.3 Proposed DS Design	
4. AN EXAMPLE AND A CURSORY ANALYSIS	10-17
4.1 Example Parameters	
4.2 Operational Parameters	
4.3 A Cursory Analysis of the Above Example	
5. BELIEFS AND RECOMMENDATIONS	10-18
6. ACKNOWLEDGMENTS	10-22
7. REFERENCES	10-22
APPENDIX A: m-SEQUENCES: WHAT THEY ARE AND HOW THEY CAN BE IMPLEMENTED	10-24
1. INTRODUCTION	10-24
2. SEQUENCE THEORY	10-24
3. M-SEQUENCE ARCHITECTURE	10-26
3.1 Introductory Remarks	
3.2 An Example and Its Analysis Via Matrices	
3.3 Special Purpose Architectures	
3.4 A Curious Architectural Property	
4. M-SEQUENCE MANIPULATIONS	10-35
4.1 The "Shift and Add" Property	
4.2 Phase Shifts and the Delay Operator Calculus	
4.3 Large Phase Shifts and the Art of Exponentiation	
4.4 Decimation	
4.5 Decimation by a Power of Two	
4.6 General Decimation	
4.7 Inverse Decimation	
5. GENERATION OF HIGH-SPEED M-SEQUENCES	10-45
6. REFERENCES: APPENDIX A	10-53
APPENDIX B: A CURSORY LOOK AT SYNCHRONIZATION FOLLOWING CLOCK RECOVERY	10-55
1. INTRODUCTION	10-55
1.1 Epoch Synchronization	
1.2 Phase Synchronization	
2. REFERENCES: APPENDIX B	10-84
APPENDIX C: SPECTRAL SHAPING	10-85
1. INTRODUCTION	10-85
2. MARKOV FILTERING	10-85
3. CONCLUSION	10-90
4. REFERENCES: APPENDIX C	10-90
APPENDIX D: PROPOSED CHANGES TO THE FCC'S RULES AND REGULATIONS	10-92

LIST OF TABLES

		Page
Table 1.	Abbreviated Description of Emission Designators	10-5
Table A-1.	Architectural "Richness"	10-30
Table A-2.	All Phase Shifts of $x^3 + x^2 + 1$	10-37
Table B-1.	Examples of Unique Words When Bit Sense is Known	10-61
Table B-2.	Examples of Unique Words When Bit Sense is Unknown	10-64

LIST OF FIGURES

Figure 1.	Generic DS biphase spread spectrum transmitter.	10-8
Figure 2.	Spectrum shape of a DS spread spectrum signal.	10-9
Figure 3.	The most common DS spread spectrum receivers.	10-10
Figure 4.	Costas loop demodulator.	10-12
Figure 5.	Proposed transmitter structure.	10-14
Figure 6.	Proposed receiver structure.	10-15
Figure 7.	Gold code generator.	10-17
Figure 8.	Gold code power spectral density (upper curve), weighted by sinc-squared (lower curve).	10-19
Figure 9.	The near/far problem.	10-20
Figure 10.	The power spectral density showing interstices.	10-21
Figure A-1.	The Companion matrix structure.	10-28
Figure A-2.	The eight field representations.	10-29
Figure A-3.	Implementation of $x^8 + x^6 + x^5 + x^3 + 1$.	10-32
Figure A-4.	Parallel clocked structure.	10-33
Figure A-5.	Realization of $x^8 + x^6 + x^5 + x^3 + 1$ by parallel clocked structure.	10-34
Figure A-6.	Realization of $x^3 + x^2 + 1$.	10-36
Figure A-7.	Phase switchable m-sequence generator.	10-38
Figure A-8.	A phase shift of $x^4 + x^3 + 1$.	10-39
Figure A-9.	The "binary algorithm".	10-41
Figure A-10.	VanLuyn's algorithm (modified).	10-42
Figure A-11.	Phase switchable generator for $x^5 + x^2 + 1$.	10-43
Figure A-12a.	Companion matrix realization of $x^4 + x^3 + 1$.	10-47
Figure A-12b.	Delay line realization of $x^4 + x^3 + 1$.	10-47
Figure A-13.	"Analog" (delay line) shift register.	10-48
Figure A-14.	High speed m-sequence generation by decimation and modulo-two addition.	10-49
Figure A-15.	High speed m-sequence generation by decimation and multiplexing.	10-50
Figure A-16.	The vanestream structure.	10-51
Figure A-17.	The WINDMILL high-speed m-sequence generator.	10-52
Figure B-1.	Autocorrelation.	10-56
Figure B-2.	A matched filter.	10-57
Figure B-3.	Matched filter action.	10-58
Figure B-4.	Matched filter action.	10-59
Figure B-5.	Matched filter action.	10-60
Figure B-6.	Matched filter action on random input bitstream.	10-62
Figure B-7.	Autocorrelation.	10-63
Figure B-8.	Autocorrelation.	10-66
Figure B-9.	Type I/Type II errors.	10-67
Figure B-10.	Phase synchronization, generic diagram.	10-68
Figure B-11.	m-sequence generator.	10-69
Figure B-12.	Sliding correlator.	10-70
Figure B-13.	Crosscorrelation of first Rademacher sequence.	10-72
Figure B-14.	Crosscorrelation of second Rademacher sequence.	10-73
Figure B-15.	Crosscorrelation of third Rademacher sequence.	10-74
Figure B-16.	Crosscorrelation of fourth Rademacher sequence.	10-75
Figure B-17.	Crosscorrelation of fifth Rademacher sequence.	10-76
Figure B-18.	Coalescence of crosscorrelations identifying the characteristic sequence.	10-77
Figure B-19.	Rapid Acquisition Sequence (RAS) generator.	10-79
Figure B-20.	Normalized crosscorrelation of the 2^n – long RAS.	10-80
Figure B-21.	The Thue-Morse (TM) sequence generator.	10-82
Figure B-22.	A-B counts to determine: first bit of TMS counter (top block); second bit of TMS counter (middle block); and third bit of TMS counter (bottom block).	10-83
Figure C-1.	The basic DS spread spectrum system.	10-86
Figure C-2.	The Markov filter window.	10-87
Figure C-3.	The DeBruijn diagram.	10-88
Figure C-4.	The Markov process resulting from an excision of states in the DeBruijn diagram.	10-89
Figure C-5.	Power spectral density.	10-91

PROPOSED DIRECT SEQUENCE SPREAD SPECTRUM VOICE TECHNIQUES FOR THE AMATEUR RADIO SERVICE

J. E. Hershey[1]

General Docket 81-414, Notice of Inquiry and Proposed Rulemaking, proposes allowing the Amateur Radio Service to use spread spectrum techniques in three bands. This report reviews the Docket's proposals and the public's reaction, reviews direct sequence spread spectrum techniques, and proposes (for purposes of further discussion) a direct sequence spread system suitable for voice communications.

Key words:
Amateur Radio Service; direct sequence spread spectrum; General Docket 81-414; spread spectrum

1. INTRODUCTION

Title 47, Section 303 of the US Code sets forth some of the powers and responsibilities of the Federal Communications Commission (FCC). Among numerous responsibilities is ensconced the directive:

...the Commission from time to time, as public convenience, interest, or necessity requires, shall-
...
(g) Study new uses for radio, provide for experimental uses of frequencies. and generally encourage the larger and more effective use of radio in the public interest...

On September 18, 1981, the FCC released a Notice of Inquiry and Proposed Rule Making under General Docket No. 81-414 which addressed the concept of Amateur Radio Service (ARS) use of spread spectrum techniques within certain of their frequency bands. The question has sparked a healthy public comment both from neighboring band users who profess to be legitimately concerned over potential interference problems and members of the ARS who are themselves divided on the issue or key parts thereof.

The issue is an extremely interesting one as no radio technique is surrounded by more mythology than is spread spectrum. This class of modulation techniques was swaddled in secrecy at its inception over three decades ago and still maintains its aura despite numerous open literature examinations and primers. Essentially, spread spectrum is any technique that demands a substantially greater spectral domain than that required by its information baseband. The two most common techniques are frequency hopping (FH) and direct sequence (DS). In an FH system members of a wide range of discrete frequencies are used for short "dwell" times and thus the signal appears to hop about through an extended frequency domain. In a DS system, the relatively narrow spectral baseband is added to a wide spectral pseudorandom (and therefore deterministic or predictable) digital process. The baseband energy is thus "spread" over a wide frequency range.

We have prepared this report in an effort to coalesce the proposed actions and the public's reactions. We have included

[1]The author is with the Institute for Telecommunication Sciences, National Telecommunications and Information Administration, U.S. Department of Commerce, Boulder, CO 80303.

a number of appendices, written in the style of primers, that explore some of the basic tools of spread spectrum systems. Finally we have also attempted to review one class of spread spectrum systems, the DS biphase technique, and we have suggested an architecture for the ARS. We hope our proposed architecture will prompt interest, thought, and comment to help lead to a viable vehicle for ARS experimentation. We believe that the ARS can make significant contributions to the study of spread spectrum techniques particularly in the Code Division Multiple Access (CDMA) and local networking arenas. The ARS has long been recognized as a responsible and contributing member of the body of radio spectrum users and has been a consistent source of innovation—a valuable national asset for furthering technological development in the radio engineering disciplines.

For the interested reader, there are a number of excellent and easily understandable pieces of literature on spread spectrum. In particular, the following are recommended: Scholtz (1982), Dixon (1976a), Viterbi (1979), AGARD (1973) and Utlaut (1978). There is even a paper on the subject at hand: Rinaldo (1980). For further technical indoctrination see: Holmes (1982), Dixon (1976b), Scholtz (1977), Pursley (1977), and Pickholtz el al. (1972).

2. CHARACTERIZATION OF THE ISSUE

The issue is the Notice of Inquiry and Proposed Rule Making General Docket 81-414. The issue proposes to affect ARS operations in the three frequency bands 50-54 MHz, 144-148 MHz, and 220-225 MHz. This chapter attempts to collate the facts without comment.

2.1 The Present Status of the Three ARS Bands in Question

The presently allowed types of modulation in the three bands (and their sub-bands, where appropriate) are presented below per Section 97.61, Authorized frequencies and emissions, of Part 97 of the FCC's Rules and Regulations.

Frequency Band	Allowed Emissions
50.0 - 54.0 MHz	A1
50.1 - 54.0 MHz	A2, A3, A4, A5, F1, F2, F3, F5
51.0 - 54.0 MHz	A0
144 - 148 MHz	A1
144.1 - 148.0 MHz	A0, A2, A3, A4, A5, F0, F1, F2, F3, F5
220 - 225 MHz	A0, A1, A2, A3, A4, A5, F0, F1, F2, F3, F4, F5

Type A0 emission where not designated may be used for short durations if required for remote control purposes or experimental work. Other exceptions are presented in Section 97.65, Emission limitations, of Part 97. The maximum authorized power input is not to exceed one kilowatt to the plate circuit of the final amplifier stage of an amplifier oscillator transmitter or to the plate circuit of an oscillator transmitter as per Section 97.67, Maximum authorized power, of Part 97.

Table 1. Abbreviated Description of Emission Designators

Emission Designator	Main Carrier Modulation	Transmission Type	Other Details
A0	Amplitude	Unmodulated	
A1	Amplitude	On-off keyed telegraphy (no audio freq. modulation)	
A2	Amplitude	On-off keyed telegraphy (amplitude modulated audio freq. or freqs.)	
A3	Amplitude	Telephony	Double sideband, full carrier
A4	Amplitude	Facsimile (modulation of main carrier directly or by freq. modulated subcarrier)	
A5(C)	Amplitude	Television	Vestigial sideband
F0	Frequency*	Unmodulated	
F1	Frequency*	Telegraphy (FSK, i.e., one of two freqs. at any instant)	
F2	Frequency*	Telegraphy (on-off keying of a freq. modulating audio freq.)	
F3	Frequency*	Telephony	
F4	Frequency*	Facsimile by direct freq. modulation of the carrier	
F5	Frequency*	Television	

*Can also be phase modulation.

The above emission designators are described in Table 1 (from the Reference Data for Radio Engineers, 1968, pp. 1-16 and 1-17).

2.2 Brief Comments on the Proposed Changes to the FCC's Rules and Regulations

The proposed changes are recorded in full detail in Appendix D. This present section attempts to review, with comment, the significant aspects of the proposed changes.

The main aspect, indeed the crux of the issue, is that the FCC has proposed to allow the ARS the use of spread spectrum techniques in three frequency bands: (1) 50-54 MHz, (2) 144-148 MHz, and (3) 220-225 MHz. What is not addressed, however, is guidance relative to spread spectrum genre. The FCC's proposed rules will admit frequency hopping, direct sequence, time hopping or hybrid spread spectrum techniques. Further, a method of synchronization is not specified nor suggested. Rather, the method of synchronization is merely to be detailed in the station's log along with the other signal parameters of concern such as center frequency, code rate, chip rate, and others. Indeed, the only technical requirements directly affecting the pseudorandom portion of the spreading mechanism concern those "pseudorandom sequences [that] may be used to generate the transmitted signal." The FCC specifies that only "binary linear feedback shift register[s]" of particular lengths and modulo-two added feedback connections be used.

The second, and the only other major aspect, concerns station identification. The FCC proposes that "identification in telegraphy shall be given on the center frequency of the transmission." Identification of a transmitter is a classical concern of all public radio usage. The FCC's intent is clear but the specifics are somewhat murky. Does the spread spectrum station identify itself using spread spectrum modulation or by a conventional, more nearly universal, narrowband transmission? If the former, can we expect someone who is not spread spectrum equipped to identify an interfering transmitter? If the latter, would it not be better to agree on a common "check-in" frequency vice the center frequency of the spread spectrum transmission?

All things considered, it seems, in sum, that the FCC's proposals are novel and in the best interest of the radio arts; however, they appear to lack specificity in some critical areas. The issue is a most complex one as the equities involved touch, at least tangentially, the national security and, more mundanely, the rights of other radio amateurs of the "narrowband persuasion." For these reasons, it is perhaps advisable that any new allowed mode of ARS operation be carefully posited, evaluated, and rigorously specified at least in the early phases of its use.

2.3 Concerns

The following report includes most of the concerns expressed in the public commentaries submitted to the FCC regarding the General Docket.

2.3.1 FCC Monitoring Capability

Interference: A number of persons have expressed worry that the FCC will have difficulty locating spread spectrum transmitters if they interfere with other users. One commentator stated that because of the "difficulty of detection (of spread spectrum signals)" there would be a concomitant difficulty of discerning the source of interference.

Monitoring Message Content: The concerns here are perceptions that clandestine radios inimical to the national security might spring up and also that other, less exotic interlopers, such as business concerns, might arrogate ARS spectrum by using spread spectrum modulations that are difficult to monitor.

Cost: One respondent stated his belief that proper implementation of the proposals contained in the Docket might lead to outlandish expenditures of funds by the Government.

2.3.2 Interference

The 50-54 MHz Band and Channel 2 (54-60 MHz): A number of respondents profess concern that spread spectrum activities within the 50-54 MHz band might adversely impact the quality of Channel 2 reception.

Locating Interferers: Concern was expressed that the ARS as well as the FCC would have significant difficulties in determining the location and identities of spread spectrum interferers.

Repeaters: There is some concern that spread spectrum activity would adversely affect repeater service especially in the 144-148 MHz band.

Marginal Terrestrial Links: Concern was expressed that low power VHF line-of-sight and diffraction paths should be protected from undue interference related to spread spectrum activity.

Alert Frequency: It was suggested that the spot frequency of 145.695 MHz be kept clear of spread spectrum energy. This frequency is used by the Radio Amateur Civil Emergency Service (RACES) as an alert frequency.

Moonbounce: Much concern was expressed by moonbounce experimenters in the 144-148 MHz band.

2.3.3 Overlay

The concerns here are the questions of practical coexistence of spread spectrum transmissions overlaid on narrowband emissions, i.e., sharing the same spectrum space. One commentator was concerned that the FM capture effect might cause the suppression of either the channelized FM station or the spread spectrum station. The general question of spectrum efficiency, its meaning, and its potential for realization for overlaid systems was raised.

2.3.4 Miscellaneous ARS Sensitivities

Some ARS commentators expressed:

a) the view that the proposed shift register codes were not sufficiently flexible.

b) the view that the spread spectrum privileges should also be extended to Technician Class licensees.

c) the hope that the FCC would permit international spread spectrum experimentation in some cases. This point addresses Article 32 of the Radio Regulations of the International Telecommunication Union (ITU). Paragraph 2732, Section 2, part (1) of the 1982 Edition reads as follows:

> When transmissions between amateur stations of different countries are permitted, they shall be made in plain language and shall be limited to messages of a technical nature relating to tests and to remarks of a personal character for which, by reason of their unimportance, recourse to the public telecommunications service is not justified.

As pointed out in the public commentary, Paragraph 2734 (3) does provide as follows:

> The preceding provisions may be modified by special arrangements between the administrations of the countries concerned.

2.4 The General Issues Surrounding DS Spread Spectrum and the Purpose of This Report

There are five large problem areas that need attention when considering DS spread spectrum usage:

a) Spectral efficiency
b) the overlay problem
c) the near/far problem
d) synchronization
e) user identification.

We will briefly consider these five items and then outline what this report will attempt to cover and contribute towards the resolution of some of these areas.

Spectral efficiency is a measure of the maximum number of users that can simultaneously use a portion of spectrum. (It is often defined relative to a geographical area.) The DS spread spectrum method we propose in Chapter 3 is designed to exhibit some degree of spectral efficiency. This is proposed to be attained through the use of Gold codes as the spreading sequences. A Gold code family is a set of sequences of bits that possesses bounded crosscorrelations between the family members. The crosscorrelations are bounded in the absolute value sense and thus are independent of bit sense. The Gold code family used has sequences whose periods are half as long as a data baud of the DATA stream; the narrowband information which is to be spread.

The overlay problem addresses the compatibility issues involved with using spread spectrum and narrowband, channelized communications within the same spectrum space in the same, or in a closely neighboring, geographical area. Juroshek (1979) and others have studied this problem, although the DS spread spectrum system that they posited used a pseudorandom spreading sequence vice a Gold, or other, short periodic code. This will lead to very distinct and important differences that must be accounted for in a similar analysis for the proposed system of Chapter 3. This analysis has not yet been made for our proposed system, but it is, in our judgment, worth recounting what Juroshek determined. Juroshek's analysis indicates that DS spread spectrum overlay with channelized FM communications is not practical as the spread spectrum signals will cause unacceptable interference to the narrowband FM links. It is not worthwhile to be any more than qualitative in this brief recount because of the peculiar spectral differences between Juroshek's DS power spectral density model and that which would result from the system proposed in Chapter 3. We do believe, however, that even the system proposed in Chapter 3 will not overlay well with narrowband communications, i.e., there will probably be interference to some of the narrowband links. However, as we will later recommend, the potential of this interference should not summarily rule out the use of DS spread spectrum techniques by the ARS as the experimental results and experience gained is expected to be of especial benefit to the ARS and to the radio engineering arts. There is a clear need to conduct a thorough study of the overlay problem for the proposed system and also to aggressively pursue other avenues such as the spectrum shaping technique advanced in Appendix C.

The near/far problem is perhaps the most difficult problem facing ARS usage of DS spread spectrum techniques. Simultaneous usage of spectrum in the same geographical area by two or more DS spread spectrum links is possible only because the spreading codes can be made to exhibit low crosscorrelations. This "processing gain" is easily offset, however, by the power disadvantage that obtains when a transmitter is much closer to a victim's receiver than the victim receiver's transmitter. In many situations the near/far problem is simply not present. The case of multiple earthbound users communicating via DS spread spectrum through a satellite is an example of such a case. In this instance, the users all keep their power levels equal. The fact that the users are all essentially equidistant *to the satellite* results in equal transmitter powers incident at the satellite. For the ARS, unfortunately, the situation is not so simple. Receivers and transmitters are distributed in helter-skelter relationship and the near/far problem is a cruel reality. The system proposed in Chapter 3 has a provision that should help overcome, at least to a limited degree, the near/far problem. By choosing Gold codes vice a pseudorandom spreading sequence, and by further assigning two (or perhaps more) sequence periods per narrowband data baud, we will leave equally spaced interstices in the power spectrum. Other users can, by a minimal adjustment of their center frequencies, fit their power spectra into the interstices of the interfering transmitter's power spectrum.

The synchronization problem is, in general, a very difficult problem for DS spread spectrum systems. The difficulty is further heightened for those systems that must operate in a jamming environment. The system proposed for the ARS need not, of course, be so rigorous and we have therefore opted to use a classical epoch determination scheme (see Appendix B) and train the receiver's clock so that the receiver can operate open-loop in the post-synchronization phase. This will provide a further degree of system immunity to interference.

The identification problem, i.e., determining the identity of a transmitter, is of critical importance to the FCC and the ARS. We propose an automatic identification mechanism which is resident in the synchronization process. The method proposed is robust, with regard to interference, and simple in deference to engineering complexity and cost.

3. A DS SPREAD SPECTRUM SYSTEM PROPOSED FOR CONSIDERATION FOR ARS USE

3.1 Introduction

In this chapter we present a short review of DS spread spectrum techniques. We follow with a candidate DS spread spectrum that we developed in order that the following advantages accrue.

a) Station identification and monitoring of transmission content can be done with a minimum of effort by the enforcement authority and by the ARS.

b) The near/far problem, although not eliminated, can be possibly overcome at least for a small number of users.[2]

c) The spectrum can be shared more easily among a small number of users, i.e., there will be a modicum of spectral efficiency.

d) There will be ample latitude for the ARS experimenters to be creative and have the potential to contribute towards the solution of important, contemporary problems and add to the "state-of-the-art."

e) A workable system should be able to be constructed with a reasonable budget albeit with much dedicated labor.

f) The system can be improved to higher and higher levels of quality through investment in the receiver system alone.

Some disadvantages are the following:

a) There is no latitude available for varying the spreading codes. They are limited for two reasons. The first is to provide a degree of spectral efficiency. The second is to ensure that content monitoring can be easily effected.

b) The chip rate of the transmissions must be approximately uniform for all the users.

c) The synchronization and automatic identification architecture is fixed for the same two reasons as per (a).

d) Some special hardware items may have to be commercially developed.

---

[2]The near/far problem refers to the problems encountered by a receiver attempting to demodulate a weak spread spectrum signal in the presence of a stronger one.

3.2 The DS Spread Spectrum Technique

The simplest block diagram of a DS biphase spread spectrum transmitter is shown in Figure 1. (It is a slight modification to that used in AGARD, 1973, p. 5-7.) What Figure 1 depicts is a relatively slow rate DATA stream; let us assume the data are bauds, zeros and ones, with duration T_b. These data are modulo-two added (exclusive-ored) with a relatively high speed binary spreading code—a stream of zeros and ones, sometimes called "chips," with bit durations T_c. The period of the spreading code is denoted by T_p. The sum of the data and spreading code bit streams is presented to a balanced modulator with its other input set to the carrier frequency F_c. The output of the balanced modulator is amplified and radiated.

Let us now examine the above parts in a bit more detail. First, it is desirable to have an integral number of chips per data baud and, further, the chip clock should be phased with the data clock so that the data baud transition times coincide with a chip transition time. To do otherwise would lead to pulses "skinnier" than T_c and thus would cause a wider bandwidth process than is desirable. Second, let us assume that the balanced modulator allows the carrier to pass without a phase change if the digital stream bit into the balanced modulator is a zero and changes the phase by π radians if the digital stream bit is a one. The carrier should be phased with the chip clock in the sense that the carrier will not change phase until near a zero crossing when a chip transition occurs. To do otherwise will also result in needless and wasteful high frequency components. (Notice, incidentally, as Pasupathy (1979) and others point out, this scheme is equivalent to AM modulation of a carrier by a stream of plus and minus ones.) If the spreading code bit stream is independent of the data stream and resembles a random and balanced bit stream (a stream of bits such that the probability is one-half that any particular bit is zero) such as a long m-sequence (see Appendix A), then the normalized power spectral density at the antenna is

$$S(f) = T_c \left[\frac{\sin \pi (f - f_c) T_c}{\pi (f - f_c) T_c} \right]^2 \qquad (eq\ 1)$$

This power spectral density is depicted in Figure 2. Note that the main lobe is $2/T_c$ Hz wide. Dixon (1976a) relates that 90 percent of the signal power is contained within the main lobe. The main lobe is all we need to demodulate the spread spectrum signal and thus we may filter out the higher lobes before transmitting to prevent their spilling into unauthorized spectrum space.

The receiver for a biphase spread spectrum signal may take a number of forms. Figure 3 depicts two of the most common approaches (Unkauf, 1977). The two schematisms of Figure 3 seem straightforward enough but they contain one component that is usually "easier said than done" and this is the SYNCHRONIZATION MODULE. Listen to what Dixon (1976a, p. 177) has to say about synchronization:

"Now we must talk about the hardest part. Throughout this book we have assumed good synchronization (on the part of the code) between transmitters and receivers...we have assumed that the codes in the systems...were already synchronized and would remain so.

"What an assumption! More time, effort, and money has been spent developing and improving synchronizing techniques than in any other area of spread spectrum systems. There is not reason to suspect that this will not continue to be true in the future."

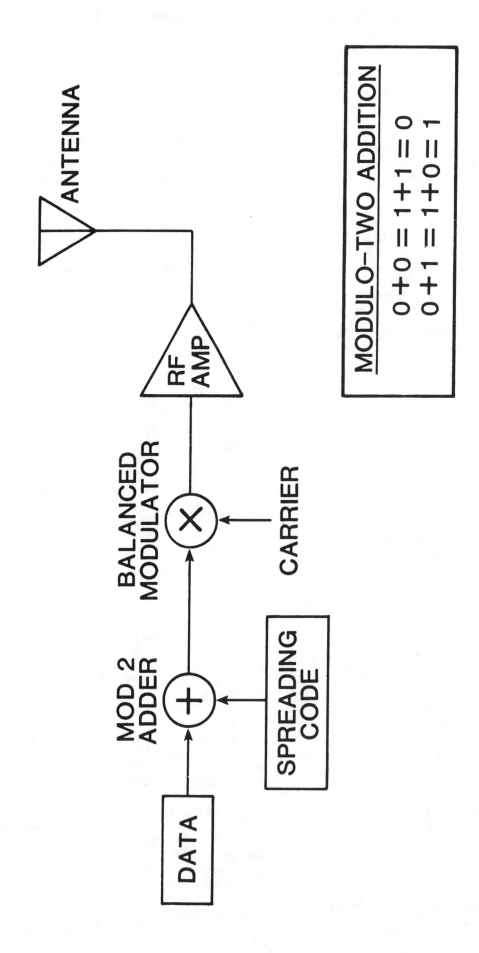

Figure 1. Generic DS biphase spread spectrum transmitter.

10-8

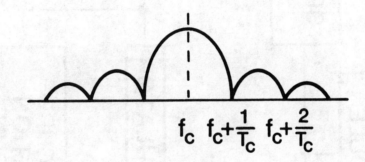

Figure 2. Spectrum shape of a DS spread spectrum signal.

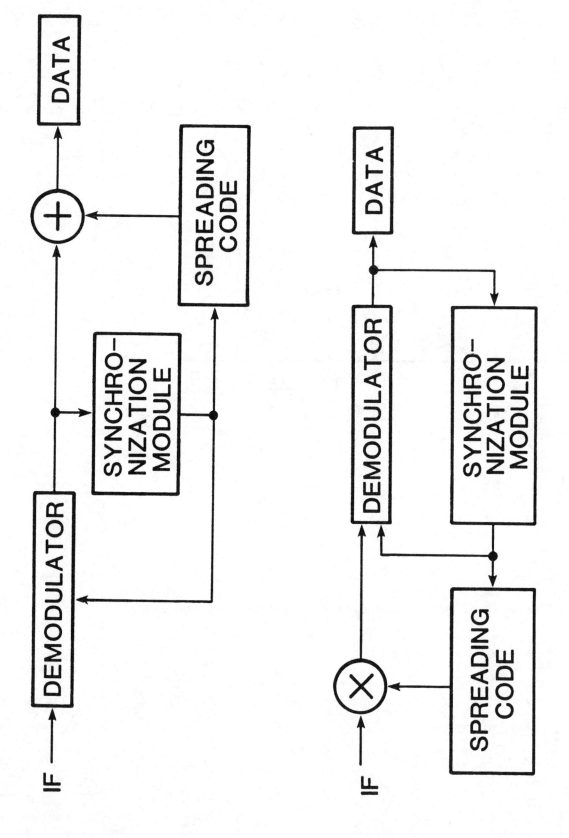

Figure 3. The most common DS spread spectrum receivers. Top receiver performs postdemodulation despreading. Bottom receiver performs predemodulation despreading.

Synchronization has two components, the coarse and the fine. Coarse synchronization gets the receiver's spreading code synchronized with the transmitter's within one chip time, T_c. Fine synchronization reduces the error to within a small fraction of a chip time and dynamically works to keep it there throughout the reception. This dynamic refinement is accomplished by a feedback loop structure which is easily discerned in the diagrams of Figure 3.

But even before we work on synchronizing the spreading codes, however, we must first be sure we know what the center frequency of our spread spectrum signal is. Recall that a balanced modulator removes (actually it suppresses) the carrier frequency and therefore we cannot use a phase locked loop (PLL) to track the carrier as a PLL requires at least a residual carrier. (For an excellent discussion of this and what is to come see Chapters 4 and 5 of Holmes, 1982.)

What we can use, however, is either a squaring loop or a Costas loop; the loops are equivalent and optimum at low signal-to-noise ratios (Riter, 1969). Figure 4 (from Dixon, 1976a) shows a Costas loop operating on a baseband modulated carrier. As shown, the OUTPUT is

$$\frac{A}{2} \cos \phi.$$

When the loop has achieved lock, $\phi \simeq 0$ and $\cos \phi \simeq 1$. What is important to note is that we have recovered the sign of the input process without explicit knowledge of ω and therefore the Costas loop is functioning as a demodulator. Dixon (1976a) notes that there is an ambiguity between zeros and ones. The loop will allow us to extract the difference between the bit at time t and the bit at time $t + 1$ but not the absolute values of the bits and therefore differential schemes such as differential phase shift keying (DPSK) are used in transmitting the data. By observing the shortest period between the transitions we can recover the clock. Now, having recovered carrier and clock, we are ready to work on synchronizing the receiver's spreading code sequence to the transmitter's. We examine various techniques for doing this in Appendix B.

3.3 Proposed DS Design

3.3.1 Introduction

There are fundamental decisions that must, or at least should, be made before a DS spread spectrum system is designed; fundamental parameters that must be decided. We are not talking about clock rate, center frequency, or any single, simple parameter but rather the philosophy of the system—what its purpose is to be and how it should be designed to fulfill that purpose. This is a most basic question and it is not well answered, or even asked, in the present case concerning the ARS. The American Radio Relay League, in their comments to the FCC (received 1 March 1982) concerning General Docket No. 81-414, crystallized this thought:

"At this time, the interest of the amateur community in spread spectrum techniques is primarily experimental. While those techniques are of considerable interest to inquiring amateurs now that advances in technology have brought them within practical reach of individual experimenters, their major advantages do not particularly promote the *communications* objectives of the Amateur Service. Looking at the matter from the point of view of the amateur *qua* communicator, message privacy is not a desired feature of amateur communication. To the contrary, as the commission has noted, techniques which provide privacy raise difficulties in monitoring and enforcement. Selective addressing and multiple access are desirable in certain situations, but may be achieved more easily by other means and with presently authorized modulation techniques. Thus, the primary motivation for amateur utilization of spread spectrum techniques is simply the desire to better understand and develop the concepts which make spread spectrum a useful communications medium. While it is unlikely that these techniques will be widely used in the Amateur Service in the near future, even a modest level of experimentation in this service will significantly expand the body of knowledge about spread spectrum techniques in non-government applications."

The answer to the question of system purpose will deeply affect three areas: synchronizing procedure, spreading code and the ratio T_b/T_p. Let us comment on these items in order.

Synchronizing Procedure: The easiest synchronization method to implement in hardware is epoch determination via a unique word as discussed in the first part of Appendix B. If, however, we are concerned with communications subject to jamming or spoofing, then a unique word system may be very vulnerable. Dixon (1976a, p. 185) puts it very well:

"With the exception of the vulnerability problem, however, preamble synchronization is by far the least critical, easiest to implement, least complex, and best for all around use."

Spreading Code: Many authors have stated that if the spreading code is an m-sequence or Gold Code or similar linearly generated sequence, then any other suitably equipped party could, after just a modicum of analysis of a small stretch of the code, predict the future spreading code sequence without error. Such knowledge would enable the "outsider" to read the communications and intelligently, and therefore effectively, jam the communications whenever desired.

The Ratio T_b/T_p: It is in consideration of this ratio that the greatest differences in philosophy are reflected. There are in reality two cases. The first displays $T_b/T_p \ll 1$; the second $T_b/T_p \simeq 1$.

One use of spread spectrum techniques is to allow communications to be conducted in secrecy, not just denial of message comprehension by an interceptor, but also denial of the knowledge that communications are indeed taking place. This is sometimes referred to as LPI for *Low Probability of Intercept*. The acronym is poorly used since in most cases, the communications are intercepted, i.e., gathered in by the antenna and receiver systems. They are just sufficiently buried in the noise that the interceptor does not recognize their presence. Some folks have argued that to more accurately reflect this fine difference one should use the acronym LPR for *Low Probability of Recognition*. The author believes that it is this type of spread spectrum that most people consociate with DS spread spectrum and for which $T_b/T_p \ll 1$. This is the type of usage that Scholtz (1977) had in mind when he advised:

"(a) Make sure that the data modulation bandwidth is much larger than...the reciprocal of the SS [spread spectrum] code modulation's period...

[or]

(b) If it is impossible to guarantee data modulation of sufficient bandwidth, then make sure that [the code modulation period] is very large."

What Scholtz is warning us about is a fallout of Fourier analysis. Because our spreading code is periodic (because it is deterministic) it will, if unmodulated by the DATA stream (e.g.,

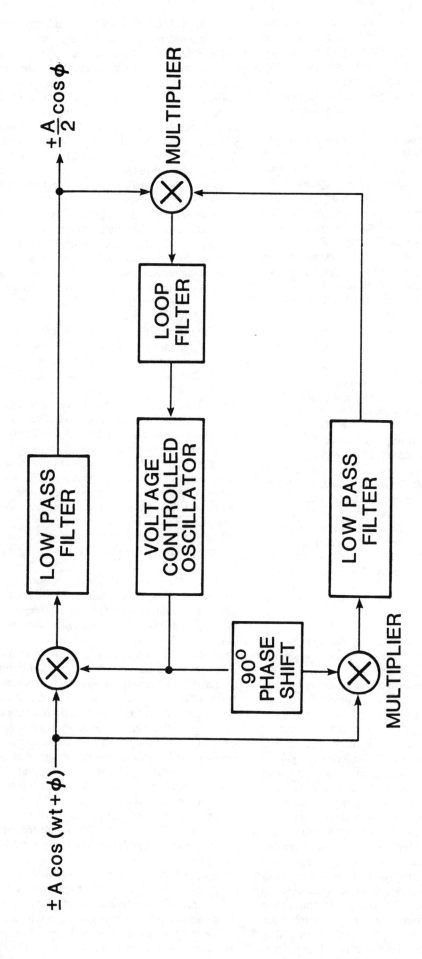

Figure 4. Costas loop demodulator.

consider that the DATA stream is an unchanging stream of all zeros or ones), result in a process exhibiting a line spectrum rather than a continuous spectrum. The lines in the power spectrum will be separated by a frequency of $1/T_p$. The larger T_p, the closer the lines will be together. Also, as the DATA stream begins to vary and modulate the spreading code, the lines will broaden and thereby move closer together until, in the limit, the continuous power spectral density of (1) obtains.

If, on the other hand, we are not interested in coverture for our communications but rather spectral efficiency, then we may elect to set $T_b/T_p = 1$, or a small integer, and employ spreading codes that exhibit excellent crosscorrelation properties. Such procedures are outlined by Pursley (1977) among others.

3.3.2 System Architecture

The system architecture that we are proposing will at first blush appear to be an unusual hybridization of various techniques. As we stated in the Introduction to this chapter, we wanted to design a system that could be realized within a reasonable budget, would allow at least some bandsharing under the near/far problem, allow much latitude for creativity and novel designs (particularly in the receiver system), and allow for easy monitoring and identification of transmitters.

The system we designed is characterized as follows:
- Bi-Phase Shift Keying/Differential Encoding
- Costas Loop Training Sequence
- Epoch Synchronization
- Open Loop Timing
- A Gold Code Family of Spreading Codes with Randomized PPM Coded Selection of Family Members[3]

We chose biphase shift keying essentially because it is easy to implement. We will have a few more words to say about this later in the section entitled Implementation. Figure 5 shows the proposed transmitter structure. At the figure's bottom is a sequency flow depicting the transmission preamble (which begins at t = 0) and the eventual start of traffic. The preamble starts with a stretch of alternating ones and zeros denoted by A. This is continued until $t = t_1$ at which time a Unique Word (UW) (see Appendix B) is sent. This is followed by τ_1 random bits and another UW. This second occurrence of the UW is followed by another stretch of random bits and another UW and so on until the UW following τ_6 has been sent after which τ_c random bits are sent followed by a final UW which marks the start of traffic. The function of this "kludgey" looking procedure, which is in a sense Pulse (UW) Position Encoding is twofold. It identifies, with relatively strong, i.e., noise-resistant, coding, the transmitter's identity and his spreading code. It does these things by the following conventions:

---

[3]Gold Codes are codes of length $2^n - 1$ which can be easily generated by adding together, term-by-term two m-sequences generated by "preferred" polynomials. The Gold Codes have the following properties: (a) a Gold Code family has $2^n + 1$ members (sequences of length $2^n - 1$) and (b) the (full-period) crosscorrelation R(k) of two Gold code family members satisfies the following

$$|R(k)| \leq \begin{cases} 2^{(n+1)/2} + 1, & n \text{ odd} \\ 2^{(n+2)/2} + 1, & n \text{ even and not divisible by 4.} \end{cases}$$

Suitable references are contained in Gold (1967), Gold (1968) and a very helpful step-by-step ("cookbook") approach given in Holmes (1982).

a) $\tau_1, \tau_2, \tau_3, \tau_4, \tau_5, \tau_6$ are each 0 to 36 chip times in length and decode according to the following.

τ (chip times)	Meaning	τ (chip times)	Meaning
0	0	18	I
1	1	19	J
2	2	20	K
3	3	21	L
4	4	22	M
5	5	23	N
6	6	24	O
7	7	25	P
8	8	26	Q
9	9	27	R
10	A	28	S
11	B	29	T
12	C	30	U
13	D	31	V
14	E	32	W
15	F	33	X
16	G	34	Y
17	H	35	Z
		36	(NULL)

The quantities τ_1-τ_6 serve to identify the transmitter. As an example, assume the transmitter's call sign is K2QRM. For this case $\tau_1 = 20$, $\tau_2 = 2$, $\tau_3 = 26$, $\tau_4 = 27$, $\tau_5 = 22$, $\tau_6 = 36$.

b) The final pulse position encoded variable, τ_c, ranges from 0 to $2^n - 1$ where n is the number of stages of one of the two (equal length) Gold Code Generator (GCG) linear feedback shift registers. The way τ_c selects the code is as follows. At t = 0, both the upper and the lower feedback (m-sequence) shift registers in the GCG modules are set to all ones. After the UW following τ_6 has been detected (this UW determines τ_6), the top feedback shift register in the GCG starts clocking at the chip rate clock. When the next, and final UW is detected, the bottom register of the GCG also starts clocking at the chip rate clock. This point in time also determines the start of a baud time for the DATA stream.[4]

c) We chose to use two periods of the GCG per data baud, for reasons to be discussed later, and therefore the chip rate is equal to two times the DATA stream rate times the number of bits in a GCG period which is of the form $2^n - 1$.

d) The alternations generator is part of the synchronization preamble and is used to train the receiver's Costas loop.

e) The random bit generator provides randomly derived bits where needed, i.e., to specify the length of τ_c and also the chip necessary to fill in the periods $\tau_1 - \tau_6$.

Figure 6 depicts the proposed receiver's structure. We chose a Costas loop to recover center frequency and chip timing and boundaries. We train the Costas loop, however, by sending a preamble to the traffic and other preamble parts of our transmission. The training sequence is simply a string of alternations 01010101.... The output of the modulator will be a very narrowband (a spectral pair of lines in the main lobe in the limit) energy signal centered at

$$f_c \pm \frac{0.5}{T_c}.$$

---

[4]This range gives a very slight preference to one of the Gold Code Family members. It is easy to implement in hardware, however.

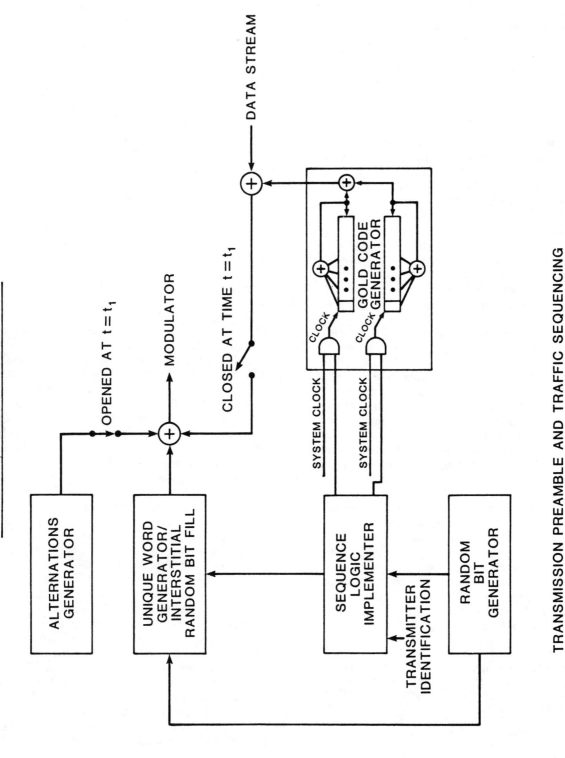

Figure 5. Proposed transmitter structure.

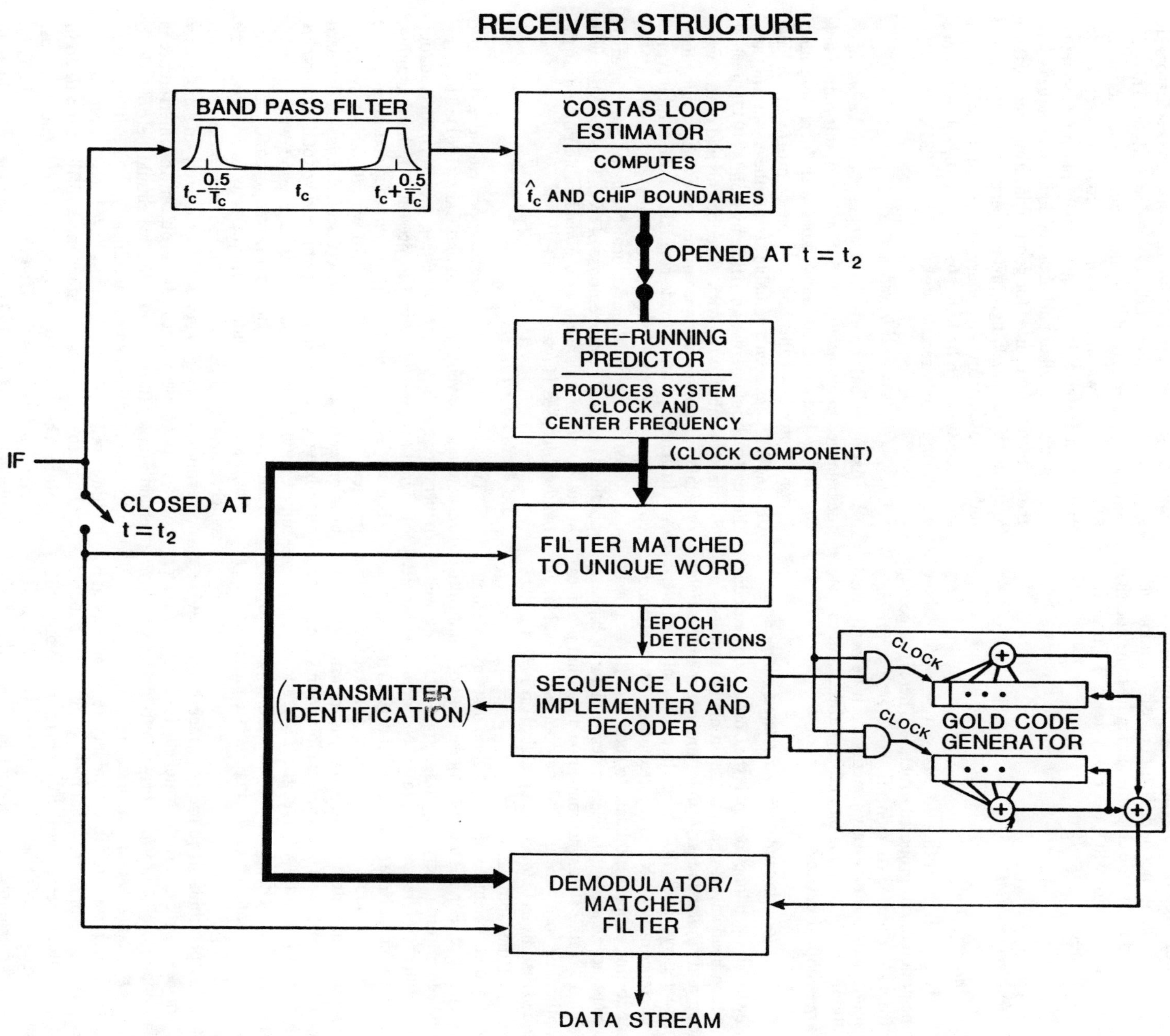

Figure 6. Proposed receiver structure.

To help the loop estimate f_c and T_c as accurately as possible, we preface the input to the loop with a band-pass-filter (BPF). The receiver opens its loop before the training segment (A) comes to an end. The receiver then uses an estimate of f_c and the chip boundaries to process the remainder of the synchronization preamble and the traffic. We chose open loop operation to allow the receiver to obtain and retain synchronization even in a high noise environment. An open loop operation should not be difficult or costly to implement when we consider that the chip rates are quite tame and that the longest an operator will speak ("key down") is probably a minute at most.

3.3.3 Implementation

3.3.3.1 Introduction

We believe it is fitting to say a few words about implementation of our proposed structure. There is no getting away from some analog components, however, a surprisingly large number of the necessary functions can be performed digitally or with hybrid analog and digital techniques such as charge coupled devices.

3.3.3.2 Open loop architecture and related components

Key to our open loop concept is the Costas loop demodulator. Much headway has been made on digital synthesis of this loop and similar loops. A milestone paper in the field was presented by Natali (1972) at the International Telemetering Conference in 1972. Natali reported on analytical methods and supporting experimental data to show that it was possible to construct an all digital coherent demodulator of biphase PSK signals. Natali's bit synchronization algorithm provides near optimal performance at low signal-to-noise ratios. Natali concluded that his techniques were practical (in 1972) for data rates up to 1 megabit-Hz.

A key component to any open loop system is a stable frequency source. For our modest requirements of bandwidth and open loop timing, crystals should perform quite well. Three excellent references are: (a) a National Bureau of Standards Technical Note (Walls and Stein, 1976) that reviews analog servo techniques for oscillators, (b) a study of an electromechanical system which, after synchronization could run open loop for twelve hours and exhibit an rms time error of only 70 nanoseconds (Allan, et al., 1968), and (c) a paper by Walls and Stein (1978) that reviews and improves upon the traditional technique of slaving and frequency locking crystal oscillators.

Another possibility is to use a Surface Acoustic Wave (SAW) device. A recent article (Mitchell, 1982) reports on a SAW-resonator oscillator with short-term frequency stability for 2 to 3 seconds of about one part in 10^{-11}.

Even though our frequency stability requirements are modest, we must bear in mind that many people have alluded to and that is the extreme sensitivity in performance of BPSK systems to frequency offset. (See Bhargava, et al., 1981, Chapter 5, for example.) One way of looking at this is via the ambiguity function of radar theory. In radar, one of the variables is motion and hence a Doppler frequency. In the AGARD (1973) publication we find that the peak amplitude of a matched filter for a BPSK receiver is

$$\left| \frac{\sin \pi f_\phi T}{\pi f_\phi T} \right| \quad \text{(eq 2)}$$

where f_ϕ is the frequency offset and T is the duration of integration.

For performing the digital clocking functions, the author subscribes to the belief that standard, off-the-shelf, inexpensive and widely available TTL logic should be sufficient. It is probably best (Hayward, 1982) to use frequency division techniques (divide down counters) rather than frequency multiplication. Noise in oscillators is, unfortunately, frequently disregarded in analysis. This noise may be dichotomized into amplitude and phase components. Phase noise grows on the order of $20 \log_{10} N$ where N is the multiplication factor in a frequency multiplication technique.

Finally, the full or partial use of digital techniques brings extra attention to the need for good engineering practices such as isolation and shielding. This is especially true for circuits such as those performing phase-locked synthesis. An excellent discussion is to be found in Chapter 10, entitled "Spectral Purity," of a recent book by Egan (1981).

3.3.3.3 The balanced mixer

Dixon (1976a) reports that Carson obtained a patent on the balanced modulator in 1915. Since then, this remarkable circuit has been refined many times with increasing innovation. Apropos of this and as a bit of proffered evidence that the ARS has been a valuable asset to the engineering arts, the reader should peruse an article by Rohde (1977) in which Rohde, a member of the ARS, described his and others' work on achieving double-balanced mixers of high dynamic range.

In general, simple divide-ring mixers will exhibit typical conversion losses of only 6 to 7 dB. A bandwidth of 0.5-500 MHz for the RF and local oscillator ports is easily obtained as is an IF port response of from 0 to 500 MHz with about 40 dB of balance over a considerable portion of the bandwidth. (See Hayward, 1982.)

3.3.3.4 Matched filters

Two very important papers appeared more than two decades ago: Davenport (1953) and Cahn (1961). These two papers dealt with the very interesting question of degradation of signal-to-noise ratio in the presence of hard limiting. Abstruse as this sounds it has a very decided input to the problem at hand. In Appendix B we talk about synchronization words and we compute crosscorrelations by first hard quantizing each bit and then computing total agreements minus total disagreements. This is a good way to *study* synchronization words and what Davenport and Cahn tell us is that, although it is not an optimal way to *implement* the synchronization process, it is not too bad for low signal-to-noise ratios in *Gaussian* noise. Unfortunately, in a multi-user CDMA system with a near/far problem, the interference is certainly not gaussian and what we should be doing is to perform our crosscorrelations linearly, i.e., without hard limiting until total disagreements have been subtracted from total agreements. This is a much more difficult job to do in hardware. There are two very promising techniques to aid us, however, surface acoustic wave (SAW) devices and charge coupled devices (CCDs). These two families of solid state devices are quite different and each possesses its own advantages. For example, CCDs are somewhat easier to interface with ancillary logic circuits but harder to fabricate. SAWs can operate at RF; CCDs only at baseband. The following literature is recommended as introductory. Milstein and Das (1979) give an excellent, easily read, view of theory, hardware, and snapshots of the state-of-the-technology for

SAWs. Bell et al. (1973) and Milstein and Das (1977) show how SAWs are of direct benefit to spread spectrum problems. Morgan et al. (1976) discuss a spread spectrum synchronization problem using a SAW that processes 730 microseconds of a 9 megabit-Hz signal and which realizes a 40 dB processing gain. Another good reference is Unkauf (1977). The CCD literature is also abundant. Two worthwhile pieces are Grant (1981) who reviews SAWs and CCDs as to their performance when used for fixed and programmable analog matched filters for spread spectrum work, and Collins et al. (1972), who also present empirical results of using CCDs to perform the analog matched filter function.

4. AN EXAMPLE AND A CURSORY ANALYSIS

4.1 Example Parameters

For our example we:

— choose a unique word (UW) of length 1023 bits (perhaps an m-sequence phased for best results—see Appendix B)

— assume that the DATA stream is 8000 bits/second. This rate would support fair quality voice that was digitized by an adaptive delta modulation technique such as a continuously variable slope delta modulator (CVSD). Such devices are commercially available in LSI (chip) form at nominal cost. (For a good tutorial and review of delta modulation techniques see Rabiner and Schafer (1978) and Flanagan et al. (1979) respectively.)

— choose a Gold code of length 127 derived from the preferred primitive polynomials

$$x^7 + x^3 + x^2 + x + 1$$
and
$$x^7 + x^5 + x^4 + x^3 + x^2 + x + 1 \quad \text{(eq 3)}$$

as shown in Figure 7.

— choose the center frequency of the DS spread spectrum transmissions to be adjustable over the range 222.492 MHz $\leq f_c \leq$ 222.508 MHz.

— require that all power outside the main lobe be filtered out.

4.2 Operational Parameters

— The length of the Alternations (A) segment of the transmitter's preamble is yet to be determined by the channel conditions. Theoretically and empirically supported guidelines must be developed.

— The operator shall, in addition to the auto-identification performed by the transmitter, identify himself in Morse code, using narrowband A1, A2, F1 or F2 modulation, by his call sign, his center frequency and other pertinent information on a standard "check-in" frequency within the 220-225 MHz band before beginning spread spectrum transmissions, at the end of a spread spectrum session and at intervals not to exceed 10 minutes during any single transmission or exchange of transmissions of more than 10 minutes duration to keep within the spirit of Section 97.84, Station identification, of Part 97 of the FCC's Rules and Regulations.

4.3 A Cursory Analysis of the Above Example

For our proposed system, the ability to easily monitor the spread spectrum transmissions, which is of paramount

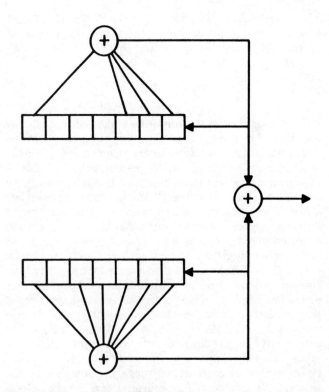

Figure 7. Gold code generator.

importance, is totally resident in the synchronization process. The quantities τ_1-τ_6 identify the transmitter and τ_c identifies the transmitter's code. It is important, then, that the Pulse Position Encoding delimiter, i.e., the UW, survive much channel degradation. We chose a length for the UW that is long compared to the system baud period for this reason. As we stated previously, we chose an open loop system because most one way transmissions would probably not last for more than a minute. This is a very modest requirement for open loop estimation. The advantage of an open loop system is that interference of any type will not affect clock generation after the preamble or synchronization phase. Thus, in improving the receiver, the operator need not be concerned about loop pull-out or other false-lock or capture problems.

Our mainlobe will be 4.064 MHz wide as we are using two 127 bit code periods per 8 kilobit-Hz data bauds. The form of power spectral density, is, of course, what is of interest. Following Scholtz (1977), let the spreading code sequence be denoted by $c_0, c_1, c_2, \ldots c_{126}$. We define the normalized (full period) autocorrelation of the spreading code sequence by

$$R(k) = \frac{1}{127} \sum_{i=0}^{126} c_i c_{i+k} \quad \text{(eq 4)}$$

where k is the "lag" or degree of slip. (We have mapped 0 into a 1 and a 1 into a minus 1 for these computations; also note that the values of $\{R(k)\}$ are all real and we have therefore dropped the conjugation operation.) The power spectral (line) density, S(k), for the spreading code is found by taking the Fourier transform of (eq 4) and we obtain

$$S(k) = \begin{cases} \dfrac{1}{127} \sum_{i=0}^{126} R(i), & k = 0 \\ \dfrac{2}{127} \sum_{i=0}^{126} R(i) \cos \dfrac{2\pi i k}{127}, & k \neq 0 \end{cases} \quad \text{(eq 5)}$$

The power spectral (line) density (eq 5) is displayed in Figure 8. The top set of points are the $\{S(k)\}$ and the bottom set are the $\{S(k)\}$ weighted by the "sinc-squared" envelope, i.e.,

$$S(k)\left[\frac{\sin(\pi k/127)}{\pi k/127}\right]^2 \quad (eq\ 6)$$

The maximum cross-correlation of the Gold code family members[5] for length 127 is 17. A fair question would be: "Why not just use pseudorandomly generated strings of 127 bits (or even 254 bits, if we forget about the $T_b/T_p = 2$ requirement) such as those generated by taking successive stretches of a long m-sequence?" This is a good and legitimate question. Consider that we were to do this. After all, we have the process synchronized so using a long m-sequence, or even an exotic nonlinear generator will not, in any way, impair our ability to monitor the transmissions and identify the transmitter. Consider now that we had indeed done so. The cross-correlation of one segment with that from another transmitter would be described, in first order analysis, by a variable, x, that possesses the statistics of a Binomial process with range $x = -127, -125, \ldots -3, -1, +1, +3, \ldots +125, +127$; agreements minus disagreements. The probability that $|x| > 17$ is only about 13%. This is not a highly significant percentage and we conclude that using short Gold codes, while we do accrue some correlation gain advantage, we do not have a compelling edge over the pseudorandom technique. Why then bother with Gold codes and, especially, why set $T_b/T_p = 2$? The answer lies in the "near-far" problem. Consider the situation of Figure 9. Assume that A_1 and A_2 are engaged in half-duplex communications and similarly for B_1 and B_2. The code isolation or "processing gain" provided by 127 or even 254 pseudorandomly chosen bits is at *best* $10 \log_{10} 254 \cong 24$ dB. If $d_2 > 16d_1$,[6] communications will be impossible or severely degraded. The nearness of A_1 to B_1 or A_2 to B_2 increases the effective interfering power received. Cahn (undated) puts it very well:

"Interest in such techniques [Code Division Multiple Access; CDMA] typically is relative to random access communication systems where the number of potential users (subscribers) is much larger than the maximum number of users simultaneously active in the channel.

•
•
•

A pure CDMA approach is not capable of accommodating signals with large power differentials, which we term the "near-far" problem since it typically arises from the large variation in ranges between users in a geographically dispersed network. In other words, the processing gain of the receiver determines the tolerable total interference, and it does not matter whether there are 1000 signals at the same power level, 100 signals each at ten times the power level, or 10 signals each at 100 times the power level as the desired signal."

But. By using two periods of a Gold code per DATA stream baud we have an interesting fallback position. We have already examined, in Figure 8, the power spectral (line) density of one period of one phase of the Gold code. The chip rate at which the 127-bit Gold code is run is 16 kilobits-Hz. This yields a line spacing of 16 kHz. Because we use two periods per DATA stream baud, we will cause the Gold code power spectral density to broaden and resemble that shown in Figure 10. Note that we have interstices between our bands of energy. It is within these interstices that another, similar power spectral density, can be inserted with a relatively high degree of interference isolation. All the second half-duplex user pair have to do is to offset their center frequency by 8 kHz from the first half-duplex user pair.

It is even conceivable that an operator could construct a second receiver system to subtract out an interfering signal. All that would be needed would be a DATA stream baud estimator that ran in real time. If the interfering signal were strong, such an estimation should be practicable.

Finally, although it is doubtful that collisions, i.e., two users attempting to synchronize during the same time, would ever be a serious problem, the mathematics applicable to its analysis can probably be borrowed from analyses of the ALOHA system. (See Abramson and Kuo (1973), Chapter 14.)

5. BELIEFS AND RECOMMENDATIONS

The author believes that:

a) DS spread spectrum is not profitably overlaid with channelized communications (see Juroshek, 1979, and deHaas and Watterson 1981) and is not a good choice for ARS communications qua communications but is a valid vehicle for research by the ARS.

b) the Technician Class licensees should be allowed any spread spectrum privileges allocated to higher class licensees as the license class is specifically for competent ARS experimenters.

c) monitoring and identification can be taken care of through a well chosen synchronization mechanism that includes an auto-identification technique.

The author recommends that:

a) if the Amateur Extra, Advanced, and Technician Class licensees are allowed to experiment with the DS spread spectrum techniques outlined in Chapter 3, the bands, or portions thereof, must be carefully chosen. The following are cursorily outlined key considerations. The 50-54 MHz band is adjacent to a TV channel allocation and while the operators of spread spectrum units may conscientiously keep their emissions within their assigned limits, TV receiver selectivity is not a uniform parameter. (See Allnatt et al. (1963).) The extent and nature of potential interference is unknown and should be studied. The 144-148 MHz band is partly used for communications that operate under very low signal-to-noise ratios. The experimentation associated with moonbounce, path losses of up to 254 dB (see Tynan, 1981), and other herculean communication tasks that have low S/N deserve protection. While FH spread spectrum techniques might be reasonably tailored for this band, and the author is impressed with the partial but encouraging results reported by AMRAD to the public file, a serious question remains concerning the suitability of DS systems for this band. Finally, we note that the 220-225 MHz band is being jointly planned by the NTIA and FCC for Amateur, Fixed and Mobile Services. The joint planning will consider technical standards and sharing requirements between the authorized services for maximum effective use of the band. This planning will also consider the use of spread spectrum modulation by the ARS.

b) other spreading code schemes be investigated in a

---

[5]Note, incidentally, that we are using only 127 out of the 129 possible Gold code family members.

[6]This assumes a very simple line-of-sight case.

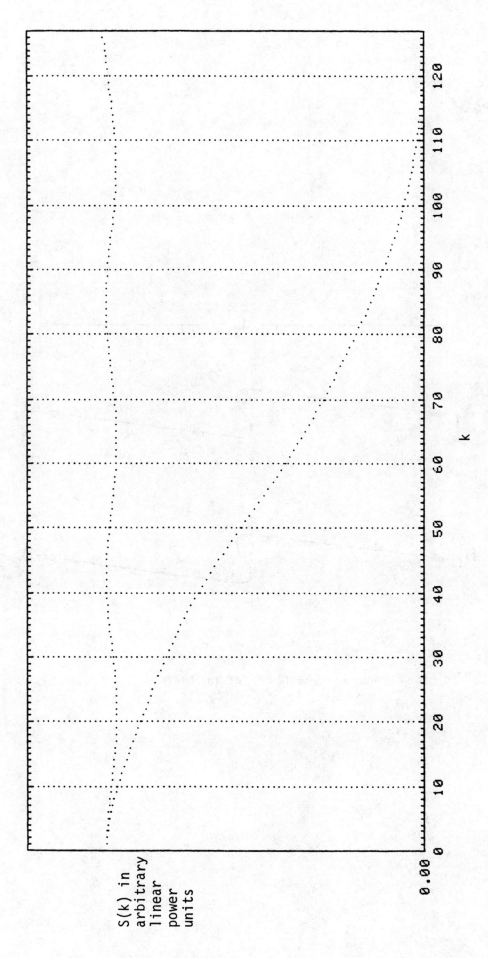

Figure 8. Gold Code power spectral density (upper curve), weighted by sinc-squared (lower curve).

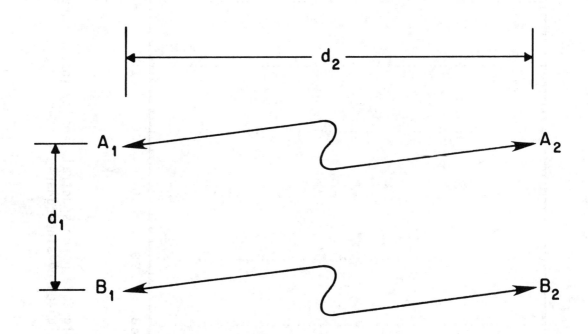

Figure 9. The near/far problem.

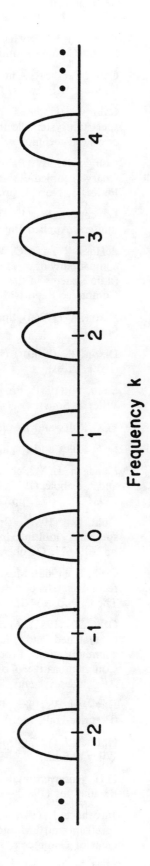

Figure 10. The power spectral density showing interstices.

search for higher spectral efficiency. An interesting, and largely unknown concept, has been proposed by Olsen (1977). It is also suggested that spectrum shaping techniques, such as Markov filtering proposed by the author in Appendix C, be researched.

c) section 97.67, Maximum authorized power, of Part 97 of the FCC's Rules and Regulations, which presently states, in part,

...

> (b) Notwithstanding the provisions of paragraph (a) of this section, amateur stations shall use the minimum amount of transmitter power necessary to carry out the desired communications.

...

be rewritten to stress the extra need for ARS compliance when using spread spectrum communications.

d) detailed study be undertaken to determine whether DS spread spectrum users should be subject to lower power limits than those imposed on narrowband communications. The study should probably be sensitive to local geography and local traffic types and densities.

e) DS spread spectrum be operated from fixed sites only, i.e., that mobile operation not be permitted.

f) a scheme similar to that suggested in Chapter 3 be studied for use with narrowband (nonvoice) DATA streams such as TTY. Such a system could probably overcome the near/far problem for a many user community if T_b/T_p were set to k where k was 10, 20 or perhaps even greater.

g) a prototype system, making extensive use of digital techniques, be constructed and tested as an aid to further studies supporting this new frontier.

6. ACKNOWLEDGMENTS

The author especially thanks Dr. Peter McManamon for technical direction on this project, Dr. William Utlaut for his encouragement, and Mr. Douglass Crombie (NTIA) for his interest and guidance. The author also expresses his gratitude to the following, randomly ordered, people for giving their time and thoughts in conversation: Martin Nesenbergs, Mike Kennedy (FCC), Don Spaulding, Gene Ax (NTIA), Les Berry, Bill Hartman, Gene Adams, Bill Pomper, Val Pietrasiewicz, Rao Yarlagadda (Oklahoma State University), Dave Wortendyke, Ted deHass, John Juroshek, Clark Watterson, Dwight Melcher, Carole Ax, and Charlene Cunningham.

7. REFERENCES

Abramson, N., and F. Kuo (1973), Computer-Communication Networks (Prentice-Hall, Inc., Englewood Cliffs, NJ).

AGARD (1973), NATA Advisory Group for Aerospace Research and Development, Lecture Series No. 58, Spread Spectrum Communications.

Allan, D., L. Fey, H. Machlan, and J. Barnes (1968), An ultra-precise time synchronization system designed by computer simulation, Frequency, January.

Allnatt, J., D. Mills, and E. Loveless (1963), The subjective effect of co-channel and adjacent-channel interference in television reception, The Institution of Electrical Engineers, Paper No. 3941E, originally presented at the International Television Conference, 4 June 1962, pp. 109-117.

Bell, D., J. Holmes, and R. Ridings (1973), Application of acoustic surface-wave technology to spread spectrum communications, IEEE Trans. Microwave Theory Tech. *MTT-21*, No. 4, pp. 263-271, April.

Bhargava, V., D. Haccoun, R. Matyas and P. Nuspl (1981), Digital Communications by Satellite (John Wiley & Sons).

Cahn, C. (1961), A note on signal-to-noise ratio in band-pass limiters, IRE Trans. Inform. Theory, pp. 39-43, January.

Cahn, C. (Undated), Notes from a short course on spread spectrum systems lecture entitled: Theoretical concepts and advanced techniques, Section 4, Multiple Access.

Collins, D., W. Bailey, W. Gosney, and D. Buss (1972), Charge-coupled-device analogue matched filters, Electron. Letters *8*, No. 13, pp. 328-329, June.

Davenport, W. (1953), Signal-to-noise ratios in band-pass limiters, Applied Physics *24*, No. 6, pp. 720-727, June.

deHass, T., and C. Watterson (1981), An analysis of the compatibility of spread-spectrum and narrowband FM mobile radio systems in the 156 to 162 MHz band, U.S. Dept. of Commerce Report MA-RD-940-81011, April.

Dixon, R. (1976a), Spread Spectrum Systems (John Wiley & Sons).

Dixon, R. (Editor) (1976b), Spread Spectrum Techniques (IEEE Press).

Egan, W. (1981), Frequency Synthesis by Phase Lock (John Wiley & Sons).

FCC Rules and Regulations (1981), Part 2, July.

FCC Rules and Regulations (1981), Part 97, 1 October.

Flanagan, J., M. Schroeder, B. Atal, R. Crochiere, N. Jayant, and J. Tribolet (1977), Speech coding, IEEE Trans. Commun. *COM-27*. no. 4, pp. 710-737, April.

Gold, R. (1967), Optimal binary sequences for spread spectrum multiplexing, IEEE Trans. Inform. Theory, pp. 619-621, October.

Gold, R. (1968), Maximal recursive sequences with 3-valued recursive cross-correlation functions, IEEE Trans. Inform. Theory, pp. 154-156, January.

Grant, P. (1981), Application of analogue signal processors to matched and adaptive filtering for spread spectrum communications, Proceedings (No. 50) of the Clerk Maxwell Commemorative Conference on Radio Receivers and Associated Systems, 7-9 July.

Hayward, W. (1982), Introduction to Radio Frequency Design (Prentice-Hall, Inc.).

Holmes, J. (1982), Coherent Spread Spectrum Systems (John Wiley & Sons).

ITU (International Telecommunication Union) Radio Regulations (1982).

Juroshek, J. (1979), A compatibility analysis of spread-spectrum and FM and mobile radio systems, U.S. Department of Commerce, NTIA Report 79-23, August.

Milstein, L., and P. Das (1977), Spread spectrum receiver using surface acoustic wave technology, IEEE Trans. Commun. *COM-25*, No. 8, pp. 841-847, August.

Milstein, L., and P. Das (1979), Surface acoustic wave devices, IEEE Communications Magazine, September, pp. 25-33.

Mitchell, B. (1982), SAW oscillators: an alternative to quartz-crystal sources, Microwaves, February, pp. 14, 18.

Morgan, D., J. Hannah, and J. Collins (1976), Spread-spectrum synchronizer using a SAW convolver and recirculation loop, Proc. IEEE, pp. 751-759, May.

Natali, F. (1972), All-digital coherent demodulator techniques, Proceedings of the 1972 International Telemetering Conference, Vol. *VIII*, 10-12 October, pp. 89-107.

Olsen, J. (1977), Nonlinear Binary Sequences with Asymptotically Optimum Cross-correlation, PhD. Dissertation, USC, December.

Pasupathy, S. (1979), Minimum shift keying: a spectrally efficient modulation, IEEE Communications Magazine, pp. 14-22, July.

Pickholtz, R., D. Schilling, and L. Milstein (1982), Theory of spread-spectrum communication—a tutorial, IEEE Trans. Commun. *COM-30*, No. 5, pp. 855-884, May.

Pursley, M. (1977), Performance evaluation for phase-coded spread-spectrum multiple-access communication—Part I: system analysis, IEEE Trans. Commun. *COM-25*, No. 8, pp. 795-799, August.

Rabiner, L. and R. Schafer (1978), Digital Processing of Speech signals (Prentice-Hall).

Reference Data for Radio Engineers (1968), (Howard W. Sams & Co. Inc.).

Rinaldo, P. (1980), Spread spectrum and the radio amateur, *QST, LXIV*, No. 11, pp. 15-17, November.

Riter, S. (1969), An optimum phase reference detector for fully modulated phase shift keyed signals, IEEE Trans. Aerospace and Electron. Systems *AES-5*, No. 4, pp. 627-631, July.

Rohde, U. (1977), High-dynamic range active double-balanced mixer, Ham Radio *10*, No. 11, pp. 90-91, November.

Scholtz, R. (1977), The spread spectrum concept, IEEE Trans. Commun. *COM-25*, No. 8, pp. 748-755, August.

Scholtz, R. (1982), The origins of spread-spectrum communications, IEEE Trans. Commun. *COM-30*, No. 5, pp. 822-854, May.

Tynan, W. (1981), A VHF/UHF primer—EME, *QST, LXV*, No. 11, p. 84, November.

Unkauf, M. (1977), Surface Wave Devices in Spread Spectrum Systems, Chapter 11 of Surface Wave Filters, H. Matthews (Editor), (John Wiley & Sons), pp. 477-509.

Utlaut, W. (1978), Spread-spectrum principles and possible application to spectrum utilization and allocation, ITU Telecommunication J., January.

Viterbi, A. (1979), Spread spectrum communications—myths and realities, IEEE Communications Magazine, pp. 11-18, May.

Walls, F., and S. Stein (1976), Servo techniques in oscillators and measurement systems, NBS Technical Note 692, December.

Walls, F., and S. Stein (1978), A frequency-lock system for improved quartz crystal oscillator performance, IEEE Trans. Instr. Measr. *IM-27*, No. 3, pp. 249-252, September.

APPENDIX A: M-SEQUENCES: WHAT THEY ARE AND HOW THEY CAN BE IMPLEMENTED[7]

1. INTRODUCTION

This appendix is both a primer on maximal length recursive sequences (m-sequences) and a discourse on their implementation architectures and properties they exhibit under certain manipulations. The style is that of an annotated bibliography that attempts to cover most of the significant contributions of the last two decades. The appendix also contains some new material, however, and new ways of looking at old material. The authors have deliberately chosen this style as it has been their experience that innovation most likely proceeds by viewing a problem or situation from many and diverse perspectives.

The appendix, then, attempts to bring together most of the developments of m-sequence architectures and properties following the publication of Golomb's milestone book (1967), This appendix is concerned with binary m-sequences only, and therefore the mathematics is to be understood as almost exclusively modulo-two computations. Further, we are concerned with single m-sequences and not interactions of different m-sequences. The subject of interactions is rich in its own right and has been recently well examined by Sarwate and Pursley (1980a, 1980b), In addition to attempting to collocate results which have been scattered through scores of different sources, we have attempted to present some new items and, perhaps more important, to recast some of the extant theory in a different form. Specifically, the authors believe that m-sequence theory can often be profitably viewed in matrix terms vice strictly polynomial algebra. The trend has been away from matrix representation and this has been motivated largely by the error correction coding theorists who maintain, and rightly so, that most useful properties can be handled exclusively by polynomial algebra and to employ matrices is both cumbersome and inefficient. Yet, circuit engineers, system engineers, and sequential machine theorists find matrix formulation a far more intuitive vehicle than abstract algebra. We have therefore attempted to balance our approach in an attempt to motivate as well as to coalesce the important aspects of the subject for a broad audience.

The appendix has four main parts: (a) a propaedeutic on recursive sequence theory: (b) sequential machine architectures for implementing m-sequences; (c) manipulation of m-sequences; and (d) implementation of high-speed m-sequences.

2. SEQUENCE THEORY

The theory of maximum length (linearly generated) binary sequences, or m-sequences is one of the most mathematically aesthetic disciplines of finite field theory. The theory offers far more than aesthetics, however, as m-sequences are extensively used by electrical engineers, particularly in the communications, radar, navigation, and computer disciplines. The theory behind m-sequences is sufficiently well developed and, for lack of a better word, "modular," so that even one who is not a mathematician can manipulate and apply powerful results in order to create new useful architectures and uncover new truths.

The sequences are an important subclass of recursively generated binary sequences which are defined by

$$s(t) = f(s(t-1), s(t-2), \ldots s(t-n)), \quad s(i)\epsilon \{0,1\} \quad (A-1)$$

which states that the bit at time t is precisely dependent on the n-bits preceding it. The sequence so produced is sometimes said to be of "span-n" (Golomb, 1980). We see from (A-1) that the sequence is deterministically generated and we immediately deduce that the sequence will eventually give rise to a cycle, or recurrent (in the Markov sense) set of states where a state is defined as the n-tuple

$$(s(\tau-1), s(\tau-2), \ldots s(\tau-n)) \quad (A-2)$$

We also deduce that the maximum possible cycle length must be bounded by the number of possible tuples of the form (A-2) which is 2^n. We also observe that because f in (A-1) is a function of n binary terms, f may be viewed as a boolean function of n variables. There are 2^{2^n} possible functions and thus 2^{2^n} possible recurrences. (See p. 12 of Golomb, 1967.)

If and only if f is expressible as a modulo-two sum of terms, i.e.,

$$f = \sum_{i=1}^{n} \alpha_i s(t-i), \quad \alpha_i \epsilon \{0,1\} \quad (A-3)$$

then f is said to be a linear function and the recursively generated sequence is said to be linearly generated. We can, incidentally, view a linearly generated recursive sequence as a nonlinear difference equation using regular (nonmodular) mathematics. For example, the modulo-two linear recursive sequence

$$s(t) = s(t-3) + s(t-5) \quad (A-4)$$

where, according to our convention, the plus sign is modulo-two addition can be written as

$$s(t) = s(t-3) + s(t-5) - 2s(t-3)s(t-5) \quad (A-5)$$

where the plus and minus signs in (A-5) imply regular addition and subtraction. When, and only when, all the s(i) are zeros and ones will (A-5) be the same as (A-4). In this one special case, a periodic solution obtains to the nonlinear difference equation (A-5).

One further general remark is in order. If and only if f can be written as

$$f = g(s(t-1) s(t-2), \ldots s(t-n+1)) + s(t-n) \quad (A-6)$$

where g may be any boolean function of n − 1 variables, will the sequence of tuples (A-2) be such that every tuple has a unique predecessor. This is an obvious, yet very powerful truth and is well presented by Golomb (1967, p. 116). Note that all span-n linear functions are of the form (A-6).

The study of sequences is in and of itself a tremendous undertaking. We do not pretend to even try. Why then do

---

[7]This Appendix coauthored with Professor Rao Yarlagadda, School of Electrical Engineering, Oklahoma State University, Stillwater, Oklahoma 74078.

we select one particular class of sequences, viz the m-sequences? The answer is twofold. First, the theory behind m-sequences is seasoned, tractable, and rich. Second, and most important to an engineer, m-sequences are useful, primarily because of their random-like qualities. To motivate further we again return to Solomon Golomb, who, in his famous book (1967, pp. 25-26), sets three "randomness postulates" or three properties or characteristics one would expect or demand from sequences purported to be random. Before recounting these properties we must comment that there is a subtle "doublethink" involved. Because our sequences will be deterministically generated, they will exhibit a period. They are thus anything but random. What we are addressing is a study of their "short-term" behavior which is taken to be their statistical analysis over a single period only. Thus we must (silently) preface our use of the word random with the prefix "pseudo." Golomb's postulates for sequence randomness are, then,

a) $p - 2 \sum_{i=1}^{p} s(i) \leq 1$ where p is the sequence period.

b) To the extent that the period can be subdivided, the number of runs of zeros and ones exhibited must fall in inverse geometric proportion to their lengths, i.e., half the runs should be 1-long, one quarter 2-long, etc.

c) The autocorrelation,

$$R_{ss}(\tau) = \sum_{i=1}^{p} s(i)s(i + \tau),$$

must be two valued, i.e.,

$$R_{ss}(0) = p \text{ and } R_{ss}(\tau) = \rho \neq p, \tau \neq 0.$$

If a sequence comports to the above requirements, it is termed a pseudonoise or PN sequence. The following 31 bit period sequence (this sequence and all other sequences should be read from left to right)

1111100011011101010000100101100   (A-7)

meets all three postulates, i.e.,

  a) There are sixteen ones and 15 zeros.
  b) There are sixteen runs distributed as follows:
    1) four runs of zeros and ones each of length 1
    2) two runs each of length 2
    3) one run each of length 3
    4) one run of zeros of length 4
    5) one run of ones of length 5.
  c) The autocorrelation is 16 for $\tau = 0$ and 8 otherwise.

The PN sequence (A-7) was generated by the recursion

$$s(t) = s(t - 3) + s(t - 5). \quad (A-8)$$

Because f is of the form (A-3), the recursion is linearly generated.

We thus have a hint that linearly generated sequences might be useful as PN sequence generators. Let us try another (arbitrarily chosen) linear generator, say, the recursion

$$s(t) = s(t - 4) + s(t - 5). \quad (A-9)$$

We easily find that (A-9), if started with the all ones tuple, gives rise to the sequence

111110000100011001010.   (A-10)

The sequence (A-10) is of period 21. Checking, we find that it satisfies randomness postulate (a). The sequence (A-10) exhibits a total of 10 runs. Of the 10 runs, half are of length one but it is required that the number of 1-long runs of ones must equal the number of 1-long runs of zeros which is impossible as 5 is an odd number. Hence, the sequence (A-10) fails to meet the second randomness postulate. The sequence (A-10) also fails to meet the third postulate as it exhibits autocorrelation values of {10, 5, 4 and 3}.

Why has one linear generator of span equal to 5 produced a PN sequence and another linear span-5 generator failed? With further experimentation, we would come to the hypothesis that only and all linear generators of span-n whose sequences exhibit periods equal to $2^n - 1$ produce PN sequences. Golomb (1967, pp. 43-45) establishes the sufficiency of the hypothesis, i.e., a linear generator of span-n that exhibits a period of length $2^n - 1$ must indeed produce a PN sequence. Not established, but *conjectured* by Golomb (1980) is the (even strengthened) necessity, viz, the PN sequences are solely composed of maximum length linearly generated sequences. Consider for example, the nonlinearly generated sequence produced by the recursion

$$s(t) = s(t - 1) + s(t - 5) + \bar{s}(t - 1)s(t - 2)s(t - 3) \quad (A-11)$$

(where the super-bar on the s denotes complementation). Starting (A-11) with the all ones tuple we obtain the sequence

1111100101110101001101100010000.   (A-12)

The sequence (A-12) meets randomness postulates (a) and (b) but exhibits autocorrelation values of {16, 9, 8 and 7} and hence fails postulate (c).

The longest, or maximum length, sequence that can be produced by a recursion of the form

$$s(t) = \sum_{i=1}^{n} \alpha_i s(t - i) \quad (A-13)$$

is clearly $2^n - 1$ (as the all zero n-tuple will immediately perpetuate itself), hence the term m-sequence is given to such a sequence. (The authors believe that the term "m-sequence" was coined by Zierler (1959, p. 39).)

The period of a linearly generated recursive sequence can be straight-forwardly analyzed by using generating functions (Golomb, 1967, pp. 30-33) or by z-transform theory (Charney and Mengani, 1961). The essence of the theory is that the polynomial

$$1 + \sum_{i=1}^{n} \alpha_i x^i \quad (A-14)$$

where the α_i's in (A-14) are the same as in (A-13), is either reducible, divisible without remainder by a polynomial of degree d, $1 < d < n$, or irreducible. The recursion corresponding to a reducible polynomial, such as $x^4 + x^2 + 1 = (x^2 + x + 1)^2$, cannot produce an m-sequence. The recursion corresponding to an irreducible polynomial will not necessarily produce an m-sequence, however. Irreducible polynomials are further dichotomized into those that are termed primitive, such as $x^4 + x + 1$, and those that are not, such as $x^4 + x^3 + x^2 + x + 1$. An irreducible degree n polynomial is primitive if and only if the smallest m for which it divides $x^m + 1$ is $m = 2^n - 1$. It is therefore redundant to say "irreducible and primitive" as primitivity implies irreducibility. Irreducibility implies primitivity only for Mersenne primes, i.e., when $2^n - 1$ is prime.

The number of primitive polynomials of degree n is well known and equal to

$$\frac{\phi(2^n - 1)}{n} \qquad (A-15)$$

where ϕ is Euler's "totient" or phi function, an extremely important number theoretic function. This function when applied to a positive integer m, $\phi(m)$, gives the count of the number of integers relatively prime (sharing no factors save unity) to m starting with 1 (which is relatively prime to all positive integers) and incrementing by 1 up to m. Thus $\phi(6) = 2$ and $\phi(8) = 4$, for examples. For a prime, p, $\phi(p) = p - 1$. The function ϕ is said to be "weakly multiplicative." This means that $\phi(mn) = \phi(m)\phi(n)$ if m and n are relatively prime. It is easy to show that $\phi(p^n) = p^{n-1}(p - 1)$ for p a prime. Knowing this, we can easily calculate $\phi(q)$ for any positive integer q. All we need do is:

a) Canonically decompose q into its (unique) product of powers of primes, i.e.,

$$q = p_1^{\alpha_1} p_2^{\alpha_2} \ldots p_r^{\alpha_r}$$

b) Use the fact that ϕ is a weakly multiplicative function and write

$$\phi(q) = \phi(p_1^{\alpha_1}) \phi(p_2^{\alpha_2}) \ldots \phi(p_r^{\alpha_r})$$

c) Sequentially evaluate all right-hand terms using the result

$$\phi(p_k^{\alpha_k}) = p_k^{\alpha_k - 1}(p_k - 1).$$

For example, $\phi(108) = \phi(2^2 \cdot 3^3) = \phi(2^2)\phi(3^3)$
$= 2 \cdot (2 - 1) \cdot 3^2 \cdot (3 - 1) = 36$.

The count of primitive polynomials (A-15) can be easily derived from an algebraic argument (Golomb, 1967, pp. 48-49) or by a less formal and intuitive argument such as that given by Birdsall and Ristenbatt (1958, pp. 34-35).

Finding primitive polynomials, then, is of great importance. Attempting to factor polynomials is, of course, the first step in proving primitivity as a primitive polynomial must be irreducible. Factoring has become an exciting field in and of itself. An early important method is due to Berlekamp (1967, 1970). A good review of the early methods along with some excellent exercises is given in Knuth (1969). For more recent work, the reader is invited to review Moenck's method (1977) which incorporates a time-saving refinement of Berlekamp's method. Another recent contribution is a new factoring algorithm by Cantor and Zassenhaus (1981) that is of the "probabilistic genre." The probabilistic algorithms are presently in vogue in all sorts of fields and portend, in the authors' opinion, to be a powerful and revolutionary approach to some classically difficult computations.

Tables of primitive polynomials are readily available. The earliest, and most famous compendium, is Marsh's set of tables (1957) which contains an exhaustive listing of all primitive polynomials through degree 19. Watson's list (1962) provides a primitive polynomial for each $n \leq 100$ and $n = 107$ and $n = 127$. Watson's list was followed by Stahnke's list (1973) which presents a primitive polynomial for all $n \leq 168$. Unlike Watson's table, Stahnke's uses trinomials when a primitive trinomial exists for a particular n and a pentanomial otherwise. (All tetranomials are, of course, reducible.) The trinomial listed is of the form $x^n + x^a + 1$ and the "a" listed is as small as possible. When Stahnke was forced to choose a pentanomial, he chose it to be of the form $x^n + x^{b+a} + x^b + x^a + 1$ with $0 < a < b < n - a$ and "a" as small as possible with "b" also as small as possible following the selection of "a". Stahnke chose this pentanomial form in order to comport with Scholefield's architecture (1960) which we shall examine in the next part of the report. If we examine Stahnke's valuable list we are struck by the lack of regularity or patterns in the polynomials. What, for instance, determines whether or not there exists a primitive trinomial for a given n? This and other similar questions broach the frontiers of knowledge of this corner of abstract algebra. There have been rents in the curtain of ignorance, however. Swan (1962) uncovered a number of criteria under which certain polynomial forms must be reducible and hence not primitive. Perhaps Swan's most general, at least most easily remembered, and, to the authors, most exciting, rule is that the trinomial

$$x^{8k} + x^m + 1, \quad m < 8k \qquad (A-16)$$

is always reducible. Thus we at least understand the absence of primitive trinomials for $n = 8, 16, 24 \ldots$

Zierler and Brillhart (1968, 1969) have catalogued all of the irreducible trinomials for $n \leq 1000$ and, where the factorization of $2^n - 1$ was known, have indicated those irreducible trinomials which were found to be primitive. Zierler (1969) adding mainly to work done by Rodemich and Rumsey (1968), has catalogued all primitive trinomials for the first 23 Mersenne primes. Finally, Zierler (1970) has shown that the trinomial $x^n + x + 1$ is primitive for $n = 2, 3, 4, 6, 7, 15, 22, 60, 63, 127, 153,$ and 532.

3. M-SEQUENCE ARCHITECTURE

3.1 Introductory Remarks

Once we have a primitive polynomial we can construct a sequential machine out of memory elements and exclusive-or gates which will exhibit a cycle of $2^n - 1$ distinct states. Our first construction is a "natural" construction and to do it we use elementary matrix theory. As an introduction, consider the following statement by Garrett Birkhoff (1964, p. 299):

"Each square matrix A has a determinant; though the determinant can be used in the elementary study of the rank of a matrix and in the solution of simultaneous linear equations, its most essential application in matrix theory is to the definition of the characteristic polynomial of a matrix."

The characteristic polynomial of a matrix, M, is, of course, obtained by evaluating the determinant

$$|M + \lambda I|. \qquad (A-17)$$

One magnificent result from matrix theory is the Cayley-Hamilton theorem. Simply stated by Perlis (1952, p. 136), "Every (square) matrix satisfies its characteristic equation." The characteristic equation is created by simply setting the characteristic polynomial (A-17) to zero:

$$|M + \lambda I_n| = \lambda^n + \lambda^{n-1} c_{n-1} + \ldots + \lambda c_1 + c_0 = 0.$$
$$(A-18)$$

By the Cayley-Hamilton theorem, then, we have

$$M^n + c_{n-1}M^{n-1} + \ldots + c_1 M + c_0 I_n = 0. \quad \text{(A-19)}$$

We observe that (A-19) guarantees (constructively) that powers of M greater than or equal to n can be expressed by a linear combination of the set of matrices

$$\{I_n, M, M^2, \ldots, M^{n-2}, M^{n-1}\} \quad \text{(A-20)}$$

If the characteristic polynomial is primitive then each member of the set

$$\{I_n, M, M^2, \ldots, M^{2^n-2}\} \quad \text{(A-21)}$$

will be distinct and $M^{2^n-1} = I_n$.

Given a primitive polynomial

$$f(\lambda) = \lambda^n + \lambda^{n-1}c_{n-1} + \ldots + \lambda c_1 + 1 \quad \text{(A-22)}$$

we can immediately devise a matrix whose characteristic polynomial will be $f(\lambda)$. This matrix is termed a companion matrix and is defined as

$$M_c = \begin{pmatrix} 0 & \ldots & 0 & 1 \\ & & & c_1 \\ & & & c_2 \\ I_{n-1} & & & \cdot \\ & & & \cdot \\ & & & c_{n-1} \end{pmatrix} \quad \text{(A-23)}$$

Observe that M_c is merely the state transition matrix for the most common realization of an m-sequence generator shown in Figure A-1. If we denote the content of the n-stage shift register's stage i at time t as

b_i^t and let $\underline{b}^t = (b_{n-1}^t, b_{n-2}^t, \ldots b_0^t)$, then

$$\underline{b}^{t+1} = \underline{b}^t M_c. \quad \text{(A-24)}$$

We must provide a note of caution. It has become traditional to list primitive polynomials as polynomials in x, i.e., $f(x)$. To create a matrix of the form (A-23) which will implement the sequence corresponding to these polynomials we must make the transformation

$$f(x) \rightarrow \lambda^n f\left(\frac{1}{\lambda}\right)$$

to convert $f(x)$ to the polynomial of form (A-22), see (Golomb, 1967, p. 35). Thus, if we wish to construct the companion matrix for the primitive polynomial $x^3 + x^2 + 1$ we would use

$$\lambda^3 \left(\frac{1}{\lambda^3} + \frac{1}{\lambda^2} + 1\right) = \lambda^3 + \lambda + 1$$

as the characteristic polynomial.

It is at this juncture that the pure mathematician loses interest for, as he correctly asserts, there is only one finite field of 2^n elements and the set of elements

$$\{I_n, M_c, M_c^2, \ldots M_c^{2^n-2}\} \quad \text{(A-25)}$$

is "as good as" any other choice as all finite fields of the same order are isomorphic. See, for example (MacDuffie, 1940, p. 180). But the isomorphisms of the Galois field can be of significant practical importance and should not be dismissed as mere mathematical curiosities. The following from Berlekamp (1968, p. 104) captures the essence of this thought:

"From an engineering standpoint, it is misleading to overstress the uniqueness of $GF(p^k)$, for this field may have many different representations... The design and cost of circuitry to perform calculations in $GF(p^k)$ depend critically on the representation. For this reason, some engineers prefer to think of different representations as different fields. This viewpoint is particularly justified in solutions where the cost of transforming from one representation to another is large."

Well, just how many field representations are there and how can they help us? Consider the general 3×3 matrix:

$$E = \begin{pmatrix} e_{11} & e_{12} & e_{13} \\ e_{21} & e_{22} & e_{23} \\ e_{31} & e_{32} & e_{33} \end{pmatrix} \quad \text{(A-26)}$$

If we take the determinant of $E + \lambda I$ we obtain:

$$\lambda^3 + \lambda^2(e_{11} + e_{22} + e_{33}) + \lambda(e_{11}e_{22} + e_{11}e_{33} + e_{22}e_{33} + e_{23}e_{32} + e_{12}e_{21} + e_{13}e_{31}) + (e_{11}e_{22}e_{33} + e_{11}e_{23}e_{32} + e_{12}e_{21}e_{33} + e_{12}e_{23}e_{31} + e_{13}e_{21}e_{32} + e_{13}e_{22}e_{31}). \quad \text{(A-27)}$$

There are two primitive polynomials of degree 3, viz

$$\lambda^3 + \lambda + 1, \lambda^3 + \lambda^2 + 1. \quad \text{(A-28)}$$

Direct solution of the $\{e_{ij}\}$ for these cases yielding the polynomials in (A-28) uncovers no fewer than 48 distinct matrices which are arrayed in eight fields each with 2^3 members (each field contains I_3 and 0_3, the multiplicative and additive identities, respectively, which do not, of course, exhibit a primitive characteristic polynomial). The eight fields are shown in Figure A-2.

The counting problem has been solved for general n. Reiner (1961) has derived an expression for the number of matrices that exhibit a particular characteristic polynomial. Following a little manipulation of Reiner's result, we find that the number of nxn matrices that possess a specific primitive characteristic polynomial is

$$\frac{R(n)}{2^n - 1} \quad \text{(A-29)}$$

where $R(n)$ is the number of regular or non-singular nxn matrices, i.e., the number of matrices whose determinant is unity. Explicitly,

$$R(n) = 2^{n^2}\left(\frac{1}{2}\right)\left(\frac{3}{4}\right)\left(\frac{7}{8}\right)\ldots\frac{2^n-1}{2^n}. \quad \text{(A-30)}$$

Recalling that there are

$$\frac{\phi(2^n - 1)}{n}$$

primitive polynomials of degree n, we find that the number of matrices that can serve as finite field generators, or equivalently, "wiring schematics" for m-sequence generators, is

$$c(n) = \frac{\phi(2^n - 1)}{n} \cdot \frac{R(n)}{2^n - 1}. \quad \text{(A-31)}$$

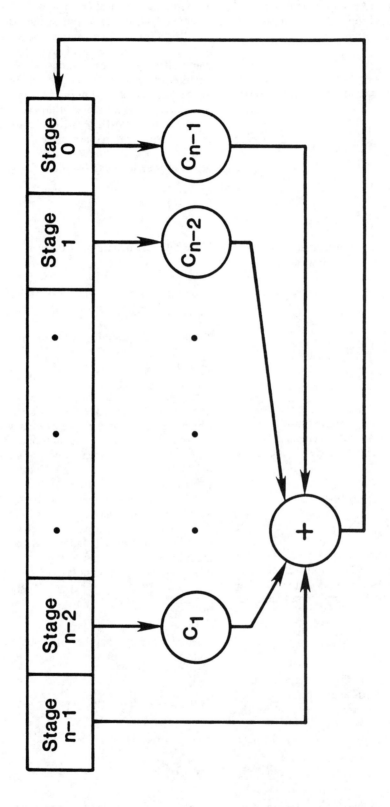

Figure A-1. The companion matrix structure.

	FIELD A	FIELD B	FIELD C	FIELD D	FIELD E	FIELD F	FIELD G	FIELD H
0_3	000 000 000	000 000 000	000 000 000	000 000 000	000 000 000	000 000 000	000 000 000	000 000 000
M	111 110 100	111 110 011	111 101 100	111 101 011	111 100 110	111 100 101	111 011 010	111 011 101
M^2	101 001 111	010 001 101	110 011 111	001 100 110	101 111 011	110 111 010	010 101 100	001 110 010
M^3	011 100 101	110 011 100	010 001 110	011 111 010	001 101 011	011 110 010	011 001 111	101 100 011
M^4	010 111 011	001 101 111	011 111 101	110 001 101	110 001 100	001 011 111	101 110 010	010 111 110
M^5	110 101 010	011 100 010	010 001 100	010 011 100	011 010 111	101 001 110	001 100 011	011 001 100
M^6	001 011 110	101 111 110	001 110 011	101 110 111	010 011 101	010 101 011	001 100 111	110 101 111
$M^7=I_3$	100 010 001	100 010 001	100 010 001	100 010 001	100 010 001	100 010 001	100 010 001	100 010 001

Figure A-2. The eight field representations.

Table A-1. Architectural "Richness"

n	$\frac{\phi(2^n-1)}{n}$	$\frac{R(n)}{2^n-1}$	c(n)
2	1	2	2
3	2	24	48
4	2	1,344	2,688
5	6	322,560	1,935,360

Note that for Mersenne primes (p and $2^p - 1$ both prime),

$$c(p) \to \frac{R(p)}{p}.$$

Table A-1 demonstrates just how very rich the potential architectural schematisms are.

3.2 An Example and Its Analysis Via Matrices

As an example of a specific architecture, different from the companion matrix structure of Figure A-1, let us consider and analyze the following machine:

a) There are n flip-flop memory elements. Their states at time t are denoted by

$b^t_{n-1} \ldots b^t_0$

b) Their states at time $t + 1$ are derived as follows. We add

b^t_0 to b^t_{n-1}

The result is

b^{t+1}_{n-1}

We then add

b^{t+1}_{n-1} to b^t_{n-2}

The result is

b^{t+1}_{n-2}

We add

b^{t+1}_{n-2} to b^t_{n-3}

The result is

b^{t+1}_{n-3}

We proceed in this fashion until we have attained

b^{t+1}_0

by adding

b^{t+1}_1 to b^t_0

A little thought will convince that

$$\underline{b}^{t+1} = (b^t_{n-1}, b^t_{n-2}, \ldots, b^t_0) = \underline{b}^t M \quad \text{(A-32)}$$

where

$$M = \begin{pmatrix} 1 & 1 & 1 & \cdots & 1 & 1 & 1 \\ 0 & 1 & 1 & \cdots & 1 & 1 & 1 \\ 0 & 0 & 1 & \cdots & 1 & 1 & 1 \\ & & & \vdots & & & \\ 0 & 0 & 0 & \cdots & 0 & 1 & 1 \\ 1 & 1 & 1 & \cdots & 1 & 1 & 0 \end{pmatrix} \quad \text{(A-33)}$$

To find that characteristic polynomial of M given in (A-33) we write the determinant

$$f_n(\lambda) = \begin{vmatrix} \lambda+1 & 1 & 1 & \cdots & 1 & 1 & 1 \\ 0 & \lambda+1 & 1 & \cdots & 1 & 1 & 1 \\ 0 & 0 & \lambda+1 & \cdots & 1 & 1 & 1 \\ & & & \vdots & & & \\ 0 & 0 & 0 & \cdots & 0 & \lambda+1 & 1 \\ 1 & 1 & 1 & \cdots & 1 & 1 & \lambda \end{vmatrix}$$
(A-34)

Expanding (A-34) by evaluating the minors specified by the two non-zero entries of the first column and using simple matrix algebra we obtain the following recursion:

$$f_n(\lambda) = (\lambda + 1)f_{n-1}(\lambda) + \lambda^{n-2} \quad \text{(A-35)}$$

By direct computation we find that

$$f_2(\lambda) = \lambda^2 + \lambda + 1 \quad \text{(A-36)}$$

and then by recursive computation we find

$$f_3(\lambda) = \lambda^3 + \lambda + 1$$
$$f_4(\lambda) = \lambda^4 + \lambda^3 + 1. \quad \text{(A-37)}$$

Thus we know that the finite state linearly sequential machine defined by (A-32) and (A-33) will exhibit a maximum length cycle of states if and only if (A-35) is primitive.

Let us examine the succession of states for one of the primitive polynomials, $f_4(\lambda) = \lambda^4 + \lambda^3 + 1$. We start the machine in the all ones state and observe the following state progression:

1111
0101
1001
0001*
1110
1011
0010*
0011
1101
0110
0100*
0111
1010
1100
1000*

The unity density states are starred and we note that their positions seem to be at approximately equidistant spacings throughout the cycle. Will this be true in general for those cases in which $f_n(\lambda)$ is primitive, or is it merely fortuitous for this case?

First, let us define the density one or unit weight vectors as

$$u_i = (00\ldots010\ldots0) \quad (A\text{-}38)$$

where the single 1 is in the position i and $1 \leq i \leq n$. Second, consider the immediate successor states of the unit weight vectors. On multiplying u_n by M we obtain:

$$u_n M = u_1 + u_2 + \ldots + u_{n-1} \quad (A\text{-}39)$$

where the sums in (A-39) are vector (modulo two sums, component by component). Similarly, multiplying M by u_1 yields:

$$u_1 M = u_1 + u_2 + \ldots + u_{n-1} + u_n \quad (A\text{-}40)$$

Summing (A-39) and (A-40) we get:

$$u_n M + u_1 M = u_n \quad (A\text{-}41)$$

In a manner similar to the above we also derive the equation:

$$u_1 M + u_2 M = u_1. \quad (A\text{-}42)$$

Now suppose that u_n and u_1 are on the same cycle, which, of course, they will be if $f_n(\lambda)$ is primitive. There then exists an integer d such that

$$u_1 = u_n M^d \quad (A\text{-}43)$$

Now, if we add (A-41) to (A-42) we obtain:

$$u_n M + u_2 M = u_n + u_1 \quad (A\text{-}44)$$

Because the inverse of M exists we can convert (A-44) to

$$u_2 + u_n + u_n M^{-1} = U_1 M^{-1} \quad (A\text{-}45)$$

On substituting (A-43) into (A-45) we get

$$u_2 + u_n + u_n M^{-1} = u_n M^{d-1} \quad (A\text{-}46)$$

Post-multiplying (A-41) by M^{-1} we find that

$$u_n + u_n M^{-1} = u_1 \quad (A\text{-}47)$$

Substituting (A-47) into (A-46) we get:

$$u_2 + u_1 = u_n M^{d-1} \quad (A\text{-}48)$$

Substituting (A-43) for u_1 in (A-48) we get:

$$u_2 = u_n M^d + u_n M^{d-1} = (u_n + u_n M) M^{d-1} \quad (A\text{-}49)$$

Using (A-41) we can rewrite (A-49) as:

$$u_2 = (u_1 M) M^{d-1} = u_1 M^d \quad (A\text{-}50)$$

Using (A-43) we immediately rewrite (A-50) as:

$$u_2 = u_n M^{2d} \quad (A\text{-}51)$$

Equation (A-51) demonstrates that u_2 is the same distance from u_1 as u_1 is from u_n. For our example, we note that this distance is 11 steps. We can easily extend the above argument for all u_i up to $i = n-1$, i.e., if

$$u_1 = u_n M^d,$$

then

$$u_2 = u_n M^d,$$

$$u_3 = u_2 M^d, \ldots, u_{n-1} = u_{n-2} M^d.$$

The final unit weight vector, u_n, however, is found to satisfy

$$u_n = u_{n-1} M^{d+1} \quad (A\text{-}52)$$

Thus, we have found that for the particular sequential machine defined by (A-32) and (A-33), the unit weight vectors are distributed at approximately equal distances around the cycle. This fact can be deduced without the matrix oriented argument presented but is intended as an example of how matrix arguments can be simply and efficaciously used.

3.3 Special Purpose Architectures

As we have seen in Figure A-1, the companion matrix shift register realization is the simplest realization of an m-sequence generator. It may not be the "best," however; that all depends on what constitutes value to the designer or implementer. As an example, let us assume that we wish to realize an m-sequence of period 255. As we have noted previously, Swan's criterion states that there are no primitive trinomials of degree n where n is divisible by 8. Thus an implementation of the form shown in Figure A-1 will require more than one two-input exclusive-or logic gate. As an example, let us choose the primitive pentanomial $x^8 + x^6 + x^5 + x^3 + 1$. It may be implemented as shown in Figure A-3. Note that this implementation requires three modulo-two adders. More important, the adders are layered to a depth of two. The exclusive-or boolean function is not threshold realizable and, consequently, is often the time-limiting basic element in a logic family. How then can we obviate this annoying layering of relatively slow logic?

The answer can often be found through special architectures. Scholefield (1960) considers a variety of interconnected, parallel-clocked structures. For example, he shows that each stage of the structure shown in Figure A-4 exhibits the recursion specified by the polynomial $x^{p+q} + x^{p+v} + x^{q+u} + x^{u+v} + 1$. By properly selecting the tetrad (p,q,u,v) it is possible to synthesize some polynomials in many ways with this particular structure. For our example, we can effect the realization of the pentanomial $x^8 + x^6 + x^5 + x^3 + 1$ with $(p,q,u,v) = (5,3,2,1)$ as shown in Figure A-5. The realization presented in Figure A-5 has two advantages over the realization presented in Figure A-3. First, there is one less exclusive-or gate required. Second, and perhaps more important, there is no layering of exclusive-or gates.

3.4 A Curious Architectural Property

Recall that the Cayley-Hamilton theorem (A-19) requires that each element of M, the finite field generator matrix, exhibit the same recursion as the matrix as a whole as its successive powers are computed. In later sections, it will become clear that the n sequences that describe any row or

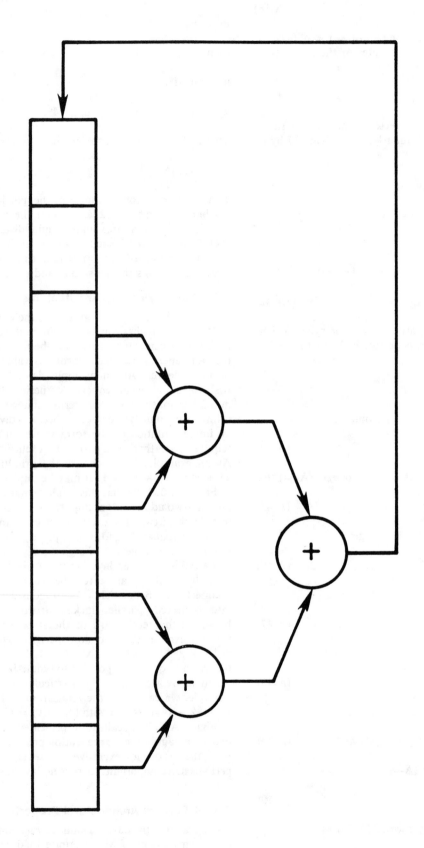

Figure A-3. Implementation of $x^8+x^6+x^5+x^3+1$.

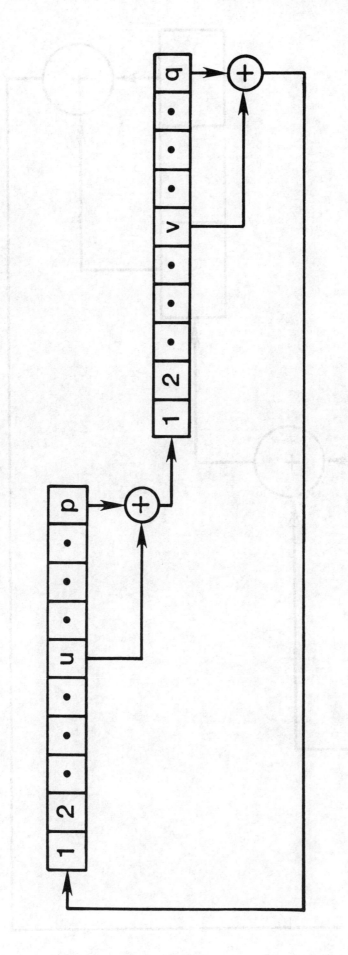

Figure A-4. Parallel clocked structure.

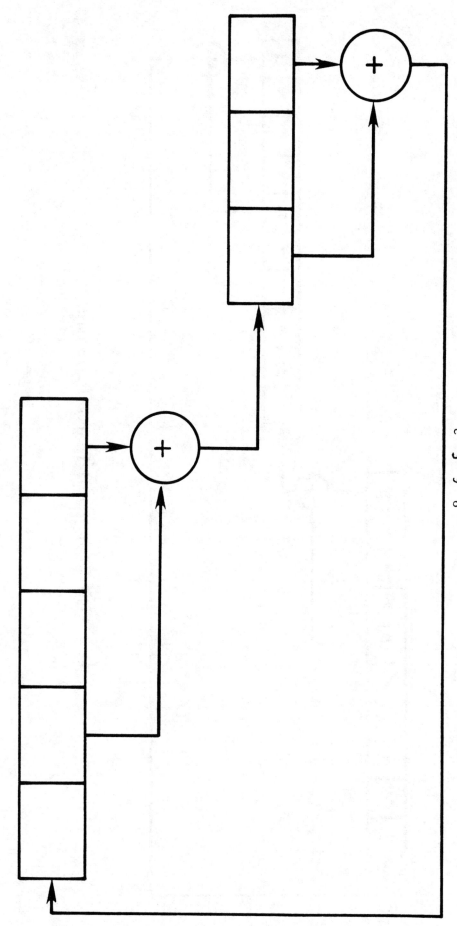

Figure A-5. Realization of $x^8+x^6+x^5+x^3+1$ by parallel clocked structure.

10-34

column of successive powers of M will be independent of each other, i.e., no term-by-term sums of up to n − 1 or any of the row or column sequences will yield the nth remaining row or column sequence. These facts immediately yield the following theorem: Every power of M is expressible as the following matrix, i.e., the following form is invariant over exponentiation:

$$\begin{pmatrix} m_1 & m_2 & \ldots & m_n \\ L_{2,1}(\underline{m}) & L_{2,2}(\underline{m}) & \ldots & L_{2,n}(\underline{m}) \\ \cdot & \cdot & & \cdot \\ \cdot & \cdot & & \cdot \\ \cdot & \cdot & & \cdot \\ L_{n-1,1}(\underline{m}) & L_{n-1,2}(\underline{m}) & \ldots & L_{n-1,n}(\underline{m}) \end{pmatrix} \quad (A-53)$$

The $m_i \epsilon \{0,1\}$. The $L_{i,j}(\underline{m})$ are linear combinations of $m_1, m_2, \ldots m_n$.

As an example, consider FIELD B of Figure A-2. By solving some elementary linear equations, the general form (A-53) of the matrices in the field is easily shown to be

$$\begin{pmatrix} m_1 & m_2 & m_3 \\ m_3 & m_1 & m_2 + m_3 \\ m_2 + m_3 & m_3 & m_1 + m_2 + m_3 \end{pmatrix} \quad (A-54)$$

Notice that all of the FIELD B matrices can be derived by letting (m_1, m_2, m_3) assume all possible $2^3 = 8$ binary triples.

4. m-SEQUENCE MANIPULATIONS

4.1 The "Shift and Add" Property

One of the most celebrated properties of an m-sequence is the so-called "shift and add property." It deserves study not only because it is of theoretical interest but also because it lies at the heart of special architectural techniques. Essentially the shift and add property states that if an m-sequence is added, term-by-term, to a shift or phase of itself, the resulting sequence will be the same m-sequence but at yet another shift of itself. For example, consider the sequence generated by the primitive polynomial $x^3 + x^2 + 1$: {1001011}. Let us delay the sequence by one clocktime: {1100101}. Adding these two sequences, we obtain: {0101110} which is the first sequence delayed by five clocktimes.

The proof of the shift and add property is simple and it is instructive to show it by two different methods, via algebra and via matrices. First, consider that the linear recurrence generating the m-sequence is the same as used in (A-13), viz,

$$s(t) = \sum_{i=1}^{n} \alpha_i s(t - i) \quad (A-55)$$

Let us create a set of 2^n elements whose members are the set of sequences:

$$\begin{aligned} &s(1), s(2), \ldots s(2^n - 1) \\ &s(2), s(3), \ldots s(1) \\ &s(3), s(4), \ldots s(2) \\ &\quad \vdots \\ &s(2^n - 1), s(1), \ldots s(2^n - 2) \\ &0, 0, \ldots 0 \end{aligned} \quad (A-56)$$

Each of the 2^n sequences in (A-56) satisfies (A-55), and because (A-55) generates an m-sequence, we know that all n-tuples excepting the all zero n-tuple exist somewhere in the sequence $s(1), s(2), \ldots s(2^n - 1)$. We realize, then, that all possible n-tuples, including the all zero n-tuple, exist as the first n bits of one of the sequences in (A-56). Thus we claim that all possible sequences or solutions to (A-55) are present in the set (A-56). These sequences or solutions are rotations or phases of each other. Now consider the term-by-term sum of any two of the sequences in (A-56). Let these sequences be denoted by

$$\{a_1, a_2, \ldots a_{2^n - 1}\} \text{ and } \{b_1, b_2, \ldots b_{2^n - 1}\}.$$

Because

$$a(t) = \sum_{i=1}^{n} \alpha_i a(t - i)$$

and

$$b(t) = \sum_{i=1}^{n} \alpha_i b(t - i)$$

linearity assures that the term-by-term sum sequence

$$\{a_1 + b_1, a_2 + b_2, \ldots a_{2^n - 1} + b_{2^n - 1}\}$$

also satisfies the recursion and is therefore contained in (A-56) thus proving the shift and add property. Furthermore, it is clear that the set of sequences in (A-56) forms an abelian group under the operation of term-by-term addition. This proof and observation is given by Golomb (1967, pp. 44-45).

Weathers (1972, pp. 13, 15) has given a matrix proof. Assume that two m-sequence generators have the same state transition matrix, M. Assume that one machine is started with the initial state $\underline{b}^0$ and that the other machine started at d clock times away from $\underline{b}^0$, i.e., at state $\underline{b}^0 M^d$. If an observer were to sum the contents of identical stages at each clock time, he would observe the sequence

$$\underline{b}^0 + \underline{b}^0 M^d, \underline{b}^0 M + \underline{b}^0 M^{d+1}, \underline{b}^0 M^2 + \underline{b}^0 M^{d+2}, \ldots \quad (A-57)$$

which we can rewrite as

$$\underline{b}^0(I + M^d), \underline{b}^0(I + M^d)M, \underline{b}^0(I + M^d)M^2, \ldots \quad (A-58)$$

But, as Weathers points out, (A-58) is equivalent to starting the machine at the state $\underline{\beta}^0 = \underline{b}^0(I + M^d)$ and observing the sequence $\underline{\beta}^0, \underline{\beta}^0 M, \underline{\beta}^0 M^2, \ldots$ which is either the all zero sequence or a rotation or phase shift of the m-sequence.

4.2 Phase Shifts and the Delay Operator Calculus

Consider again an m-sequence realized by the companion matrix of Figure A-1. For specifics, let us consider the shift register arrangement of Figure A-6 corresponding to the primitive trinomial $x^3 + x^2 + 1$. We have labeled the stages 0, 1, and 2. Consider that the observed m-sequence is taken from stage 0. Let D be the unit delay operator applied to a sequence. If the register of Figure A-6 is started with all ones, the following m-sequence is observed

$$S_0 = \{1001011\}. \quad (A-59)$$

As stage 1, the following sequence is observed

$$S_1 = \{1100101\} \quad (A-60)$$

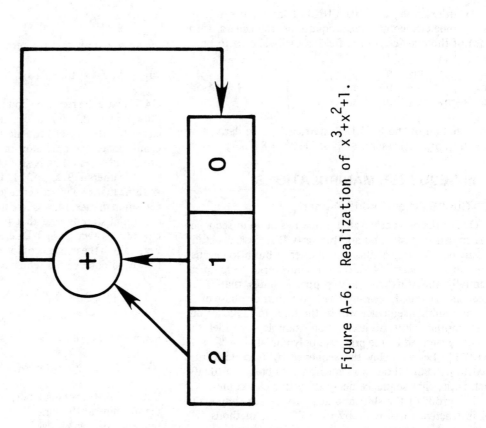

Figure A-6. Realization of x^3+x^2+1.

Table A-2. All Phase Shifts of $x^3 + x^2 + 1$

S_0	S_1	S_2	Σ
0	0	0	0000000
0	0	1	1110010
0	1	0	1100101
0	1	1	0010111
1	0	0	1001011
1	0	1	0111001
1	1	0	0101110
1	1	1	1011100

which is D operating on (A-59). At stage 2 we observe

$$S_2 = \{1110010\} \tag{A-61}$$

which is D^2 operating on (A-59). Consider, now, that we have a set of three switches $\{s_0, s_1, s_2\}$ as shown in Figure A-7. Table A-2 shows the sequences which are produced at point Σ according to the eight possible switch configurations. Note that the set of eight sequences of Table A-2 contain all seven phase shifts of the m-sequence produced by the primitive polynomial $x^3 + x^2 + 1$ and the all-zero sequence. This set is of the form (A-56). The noteworthy point here is that all phase shifts are present and can be generated by summing, modulo two, various combinations of the three stages of the machine shown in Figure A-6. That this is true in general, that any phase of a $2^n - 1$ bit m-sequence can be formed by a linear combination of stages of the n-stage companion matrix shift register realization, is an amazing and useful property. [Tsao (1964) gives an excellent, elementary argument proving this.] Also implied is uniqueness, i.e., no two linear combinations of stages will produce the same phase. This follows immediately from the "pigeon-hole principle" (Schubeinfachprinzip) as there are

$$\sum_{i=0}^{n} \binom{n}{i} = 2^n$$

possible linear combinations and 2^n possible phases including, of course, the all-zero sequence.

We can now move naturally into the delay operator calculus. Observe that the machine in Figure A-8 obeys the primitive polynomial $x^4 + x^3 + 1$. Observe that it can be viewed in terms of the delay operator notation by observing that

$$D^3\sigma + D^2\sigma = D^{-1}\sigma \tag{A-62}$$

where σ is the m-sequence observed at stage 0. Rewriting (A-62) as

$$(D^4 + D^3 + 1)\sigma = 0 \tag{A-63}$$

we see that the polynomial in x converts directly to a polynomial in D and this is true in general.

Davies (1965) presents the following algorithm to derive the numbers of the stages which must be added together to achieve a phase delay of d steps:

a) Divide the degree n primitive polynomial, written left-to-right with terms of *decreasing* powers of D, into D^d.

b) Continue division until the remainder consists of powers of D *all of which* are less than n.

c) The powers of D in the remainder correspond to those stages that are to be summed modulo two.

For example, let us say that we wish to generate the m-sequence according to the primitive polynomial $x^4 + x^3 + 1$ and that we simultaneously wish to generate the sequence delayed by six steps. Following Davies' algorithm we proceed as follows:

$$
\begin{array}{r}
D^2 + D + 1 \\
D^4 + D^3 + 1 \overline{\smash{\big)} D^6 } \\
\underline{D^6 + D^5 + D^2} \\
D^5 + D^2 \\
\underline{D^5 + D^4 + D} \\
D^4 + D^2 + D \\
\underline{D^4 + D^3 + 1} \\
D^3 + D^2 + D + 1
\end{array}
$$

The machine depicted in Figure A-8 will produce an m-sequence at point Σ that is delayed by six steps from the sequence observed at stage 0.

Douce (1968) and Davies (1968) worked the inverse problem, i.e., given the m-sequence generator polynomial and the stages of the companion matrix that are summed, determine the phase delay of the sum. This algorithm is also very simple:

a) Divide the degree n primitive polynomial, written left-to-right with terms of *increasing* powers of D, into the stage specifying powers of D, also written as a polynomial is ascending powers (this polynomial will be of degree <n).

b) Continue division until a one term remainder is obtained.

c) The exponent of D of the remainder above is the phase delay. Using our previous example, we perform the following division:

$$
\begin{array}{r}
1 + D + D^2 \\
1 + D^3 + D^4 \overline{\smash{\big)} 1 + D + D^2 + D^3 } \\
\underline{1 + D^3 + D^4} \\
D + D^2 + D^4 \\
\underline{D + D^4 + D^5} \\
D^2 + D^5 \\
\underline{D^2 + D^5 + D^6} \\
D^6
\end{array}
$$

The first single-term remainder encountered is D^6 and our delay is therefore 6.

The above algorithms depend on the simple relation as expressed by Davies (1965):

$$D^d = f(D) \cdot q(D) + r(D) \tag{A-64}$$

where f(D) is the primitive polynomial generating the m-sequence and q(D) and r(D) are the quotient and remainder polynomials, respectively. The 2^n possible residues, or equivalence classes that obtain upon the division correspond to either the all-zero sequence or the $2^n - 1$ possible phase delays.

Gardiner (1965) has devised, and Davis (1966) has further generalized, a sequential circuit that derives the numbers of the stages which must be added together to achieve a given phase shift delay.

4.3 Large Phase Shifts and the Art of Exponentiation

The preceding material is valuable in that it constructively demonstrates the calculability of the relation and inverse relation between phase shift and stage connections needed to be summed to achieve the phase shift. For large phase shifts, however, Davies' method (1965) is not recommended as the computation time required to determine the stages and be summed grows linearly with the magnitude of the phase shift.

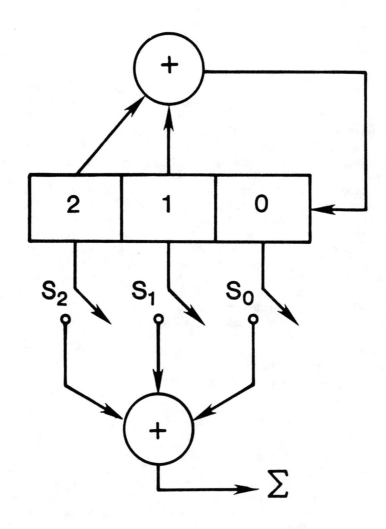

Figure A-7. Phase switchable m-sequence generator.

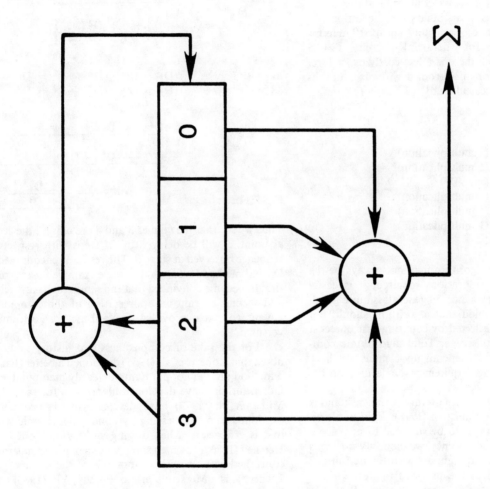

Figure A-8. A phase shift of x^4+x^3+1.

10-39

Other preferred methods, such as (Ireland and Marshall, 1968a, 1968b, 1976) and (Latawiec, 1974, 1975, 1976) are complex and are not presented by the authors as efficient for handling very large phase delays. Only recently has a series of papers dealt with the problem of the computational complexity of calculating the stages required to achieve very large phase delays.

To review, what we are trying to determine is D^d modulo the primitive polynomial generating the m-sequence. Given d, how many multiplications are required to obtain D^d? The answer in general, as far as the authors are aware, is unknown. Knuth (1969, pp. 401+) considers the problem at length and presents the algorithm, slightly modified by the authors shown in Figure A-9 which he terms the "binary algorithm," for accomplishing the exponentiation $Y = D^d$. The binary algorithm requires $[\log_2 d] + s(d) - 1$ multiplications where $s(d)$ is the number of ones in d's binary representation. As stated, the algorithm is not necessarily the "cheapest" in terms of multiplications required for general d. Knuth cites $d = 15$ as the smallest d for which there is a less costly procedure. The binary algorithm forms d^{15} from d with six multiplications. Let us, however, calculate d^{15} with only five multiplications as follows:

START: $\delta \leftarrow D$
         $\delta \leftarrow \delta^2$ (1 multiplication)
         $\delta \leftarrow \delta \cdot D$ (1 multiplication)
         $\gamma \leftarrow \delta$ (save D^3)
         $\delta \leftarrow \delta^2$ (1 multiplication)
         $\delta \leftarrow \delta^2$ (1 multiplication)
         $\delta \leftarrow \delta \cdot \gamma$ (1 multiplication)
END: $\delta = D^{15}$

Although the binary algorithm may not always be the cheapest in terms of multiplications required, it is easily programmed and its performance, in general, is quite good. Also, one must bear in mind that algorithms should not always be evaluated and selected by counting just one cost item, multiplications in this case. Total algorithmic complexity, and "convenience," depends upon many ancillary considerations such as storage requirements, indexing, sorting, and other housekeeping tasks.

The binary algorithm is quite useful for our task and is at the heart of a paper by VanLuyn (1978). VanLuyn's algorithm determines the stages to be summed to produce a phase shift of d steps. For consistency, we modify VanLuyn's algorithm slightly in order to comport with the definitions and architectures implied by the previous figures. The algorithm is motivated by the following: Assume that d is expressed in binary form as

$$d = a_0 2^0 + a_1 2^1 + a_2 2^2 + \ldots + a_s 2^s;$$

then we have

$$D^d = D^{a_0 + a_1 2^1 + a_2 2^2 + \ldots + a_s 2^s}$$
$$= D^{a_0}(D^{a_1}(\ldots(D^{a_s})^2\ldots)^2)^2. \quad \text{(A-65)}$$

But (A-65) is the recursive form of the binary algorithm. The embellishment needed is to reduce powers of D that equal or exceed n following each squaring and this can be done using Davies' (1965) long division method. VanLuyn's algorithm is then as shown in Figure A-10.

As an example of VanLuyn's algorithm, consider that we wish to determine the setting of the switches s_0, s_1, s_2, s_3, s_4 of the machine depicted in Figure A-11 so that the sequence at point Σ is delayed by 21 steps with respect to the sequence observed at stage 0. The machine of Figure A-11 obeys the recursion $D^5 + D^2 + 1$ and the delay $s = 2^0 + 2^2 + 2^4$, hence $a_0 = 1$, $a_1 = 0$, $a_2 = 1$, $a_3 = 0$, $a_4 = 1$. The algorithm proceeds as follows:

START: $r(d) = 1$
$a_4 = 1$: $r(D) = D^{a_4} \cdot (1)^2 + D$
$a_3 = 0$: $r(D) = D^{a_3} \cdot (D)^2 = D^2$
$a_2 = 0$: $r(D) = D^{a_2} \cdot (D^2)^2 = D^5$

$$D^5 + D^2 + 1 \overline{\smash{\big)} D^5}$$
$$\underline{D^5 + D^2 + 1}$$
$$D^2 + 1 = r(D)$$

$a_1 = 0$: $r(D) = D^{a_1} \cdot (D^2 + 1)^2 = D^4 + 1$
$a_0 = 1$: $r(D) = D^{a_0} \cdot (D^4 + 1)^2 = D^9 + D$

$$D^5 + D^2 + 1 \overline{\smash{\big)} \begin{array}{l} D^4 + D \\ D^9 + D \end{array}}$$
$$\underline{D^9 + D^6 + D^4 }$$
$$D^6 + D^4 + D$$
$$\underline{D^6 + D^3 + D}$$

FINISHED: $D^4 + D^3 = r(D)$

Thus we see that if switches 3 and 4 are closed, the sequence at point Σ will be delayed by 21 steps with respect to the sequence observed at stage 0. The reader is encouraged to try Davies' method (1965) on the above example for two reasons: first to become convinced that the same results are achieved, and second to gain an appreciation of the computational advantage that is provided by VanLuyn's algorithm, log d versus d.

For the sake of completeness and history it should be noted that Roberts et al. (1965) developed a better than linear d method but did not put it into an easily manipulable form. Other authors have developed algorithms for special cases. Miller et al. (1977) have devised what is probably best captioned as a "coalescing pyramid" algorithmic structure that is efficacious when dealing with primitive trinomials. Hershey (1980) presented some timesaving shortcuts for those concerned with primitive trinomials of the form $x^n + x + 1$ where n is a Mersenne prime. Finally, Yiu (1980) has also developed a fast algorithm that is similar to VanLuyn's.

4.4 Decimation

Decimation, the creation of a new sequence by the periodic sampling of an old, is a process of both great theoretical and practical import. Consider the m-sequence generated by the trinomial $x^3 + x + 1$:

1001110...   (A-66)

Taking every second bit from (A-66) and starting from the zeroth, we get the sequence:

1010011...   (A-67)

But (A-67) is merely a phase shift of (A-66) and therefore dis-

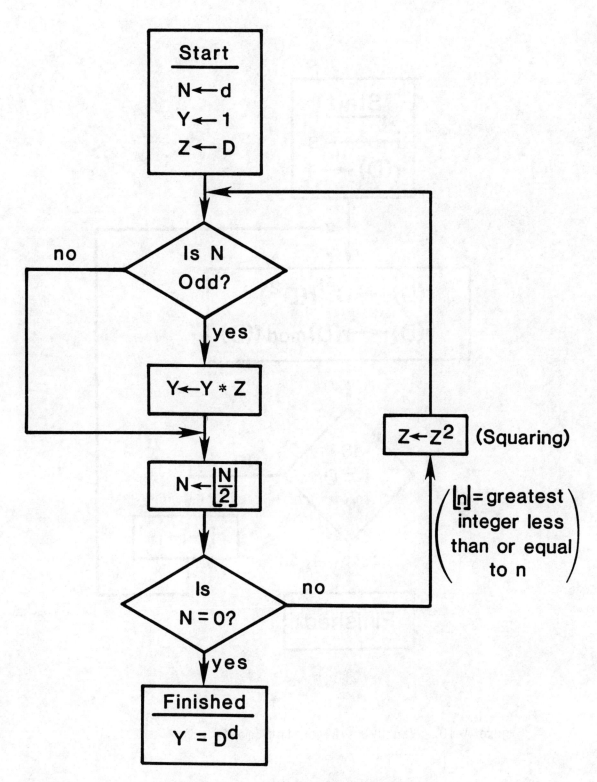

Figure A-9. The 'binary algorithm'.

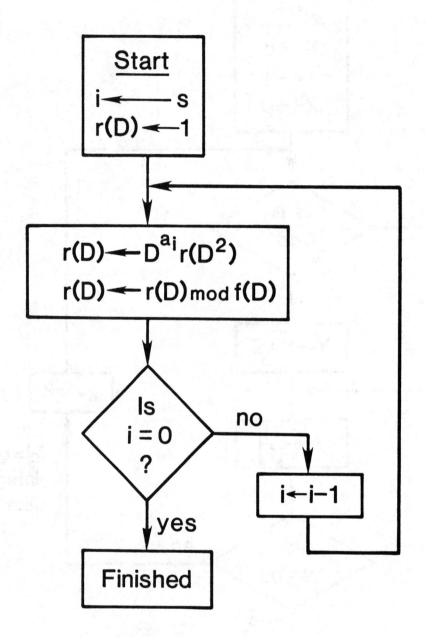

Figure A-10. VanLuyn's algorithm (modified).

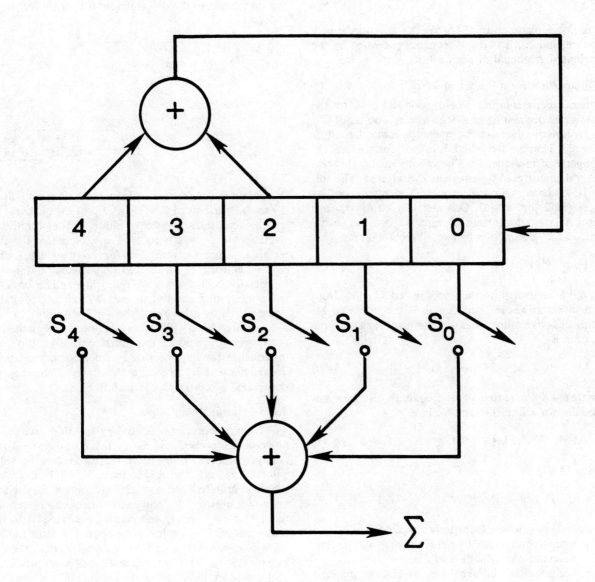

Figure A-11. Phase switchable generator for x^5+x^2+1.

plays the same primitive trinomial. If, however, we take every third bit from (A-66), again starting from the zeroth, we obtain:

1100101..., (A-68)

which is not a phase shift of (A-66) but an entirely new m-sequence. Indeed, it is the m-sequence generated by the other primitive trinomial of degree 3, $x^3 + x^2 + 1$.

4.5 Decimation by a Power of Two

Other experimentation would soon lead us to the hypothesis that decimating an m-sequence by a power of two always gives rise to the same m-sequence at some phase shift from the old. That this hypothesis is indeed true is a remarkable property of m-sequences. The validity of this theorem is quickly demonstrated by using abstract algebra. The following is, however, an excellent matrix-oriented proof by Weathers (1972, pp. 12-13). Consider that an m-sequence generator is started at state $\underline{b}^0$ and we consider every other state, i.e., decimation by two. We observe

$$\underline{b}^0, \underline{b}^0 M^2, \underline{b}^0 M^4, \underline{b}^0 M^6, \ldots \quad \text{(A-69)}$$

But (A-69) is equivalent to the undecimated stepping from $\underline{b}^0$ of a machine whose transition matrix is M^2 vice M. Following this observation, consider now that the characteristic equation is

$$M^n + M^{n-\alpha_1} + M^{n-\alpha_2} + \ldots + I = 0 \quad \text{(A-70)}$$

We square (A-70), note that all the crossterms disappear under modulo-two arithmetic, and we have

$$M^{2n} + M^{2(n-\alpha_1)} + M^{2(n-\alpha_2)} + \ldots + I = 0 \quad \text{(A-71)}$$

Rewriting (A-71) as

$$(M^2)^n + (M^2)^{n-\alpha_1} + (M^2)^{n-\alpha_2} + \ldots + I = 0 \quad \text{(A-72)}$$

we see that M^2 satisfies the characteristic equation (A-70) and thus the decimated sequence (A-69) is the original, undecimated m-sequence merely shifted in phase.

An m-sequence is a bit like the proverbial and material ring. It has no beginning, no end. A natural benchmark does exist however. It is called the "characteristic sequence." Recall that an m-sequence is changed in phase only, when decimated by two. It turns out that there is a phase of the m-sequence that is left invariant upon decimation by two. This phase is the characteristic sequence and was discovered by Gold (1966). Gold's straightforward rules that construct the characteristic sequence for the nth degree primitive polynomial $f = f(x)$ is to perform the divisions:

$$\frac{\frac{d}{dx}(xf)}{f} \quad \text{if n is odd} \quad \text{(A-73)}$$

$$\frac{\frac{d}{dx}(xf)}{f} + 1 \quad \text{if n is even} \quad \text{(A-74)}$$

For example, consider the trinomial $x^3 + x^2 + 1$ which gives the sequence presented in (A-68).

As n is odd, we apply the formula (A-73) and we see that

$$\frac{\frac{d}{dx}(xf)}{f} = \frac{\frac{d}{dx}(x + x^3 + x^4)}{1 + x^2 + x^3} = \frac{1 + x^2}{1 + x^2 + x^3} \quad \text{(A-75)}$$

We now perform the division specified in (A-75)

$$
\begin{array}{r}
1 + x^3 + x^5 + x^6 + \ldots \\
1 + x^2 + x^3 \overline{\smash{)}\, 1 + x^2 } \\
\underline{1 + x^2 + x^3} \\
x^3 \\
\underline{x^3 + x^5 + x^6} \\
x^5 + x^6 \\
\underline{x^5 + x^7 + x^8} \\
x^6 + x^7 + x^8 \\
\underline{x^6 + x^8 + x^9} \\
x^7 + x^9
\end{array}
\quad \text{(A-76)}
$$

We obtain the characteristic sequence by sequentially reading the coefficients of $1, x, x^2, x^3, \ldots$ of the quotient. Thus, the characteristic sequence, or phase, of the primitive polynomial $x^3 + x^2 + 1$ is $1001011\ldots$ Note that the characteristic sequence starts with zero for n even and one for n odd.

So enticing is this beautiful benchmark it has been independently discovered by Weinrichter and Surbock (1976) in an interesting and instructive way and will probably be rediscovered by other researchers in the future.

Finally, Arazi (1977) has developed the mathematical machinery to compute the initial setting or tuple of an m-sequence for its power of two decimation to achieve a desired phase shift. Arazi cites the characteristic sequence as the special case of zero phase shift.

4.6 General Decimation

If an m-sequence produced by an nth degree primitive polynomial is properly decimated, that is, if the period of decimation is relatively prime to $2^n - 1$, then another m-sequence will be produced. This new m-sequence will be described by the same primitive polynomial if and only if the decimation is a power of two. For decimation other than a power of two, there is no easy "paper-and-pencil" method of determining the polynomial with a few exceptions. The most famous of these exceptions is the "cubic transformation" introduced by Marsh (1957). This transformation is that produced by decimating by three, or, as Golomb (1967, p. 79) calls it, "tertiation." Golomb (1969, pp. 363-366) has prepared a lucid algorithm for implementing Marsh's transformation. Golomb's algorithm proceeds as follows:

a) Create three "bins" A, B and C. We will place, but not reduce modulo-three, all numbers that we encounter that are 0 modulo-three into bin A. Into bin B we will place all numbers that are 1 modulo-three. Finally, we will place all numbers that are 2 modulo-three into bin C.

b) The first set of numbers to be sorted and placed into the bins is the exponents of x in f(x), the primitive polynomial.

c) For all distinct pairs (α_1, α_2) of exponents in f(x) that are in the same bin (which is in reality a residue class) we form $(2\alpha_1 + \alpha_2)/3$ and $(2\alpha_2 + \alpha_1)/3$ and place the results into the appropriate bin.

d) For all distinct tuples $(\alpha_A, \alpha_B, \alpha_C)$ where α_A is an exponent of x of f(x) which has been placed into bin A, α_B is an exponent of x of f(x) which has been placed into bin B, etc., we compute $(\alpha_A + \alpha_B + \alpha_C)/3$ and place it into the appropriate bin.

e) The occupants of the bins are now examined. If a

particular occupant is present an odd number of times it is copied onto a list, L, otherwise it is discarded.

f) The new polynomial, the polynomial that describes the decimated-by-three recursion, is formed by summing together the terms consisting of an x raised to each of the powers in the list, L.

An example is in order. Consider the primitive pentanomial

$$F(x) = x^5 + x^4 + x^2 + x + 1 \quad (A\text{-}77)$$

It produces the sequence

$$11101100111000011010100010001011 \quad (A\text{-}78)$$

Decimating the sequence (A-78) by three, we obtain the sequence

$$10010110011111000110111010000 \quad (A\text{-}79)$$

Let us use Golomb's method to discover the primitive polynomial which (A-79) obeys:

	bin A	bin B	bin C
STEP (b)	0	1,4	2,5
STEP (c)	3,3	4	2
STEP (d)	3	1	2,2

$$L = \{0, 3, 5\}$$

Thus the polynomial that generates (A-79) is $g(x) = x^5 + x^3 + 1$. The reader should note that in order for decimation-by-three to yield a primitive polynomial, n, the degree of the polynomial generating the original sequence, must be odd as $2^n - 1$, where n is even and greater than 0, is divisible by three and the decimation would therefore be improper.

Golomb (1969, pp. 366-369) generalizes the cubic transformation to a general kth power transformation but deriving the generator polynomial for the decimated sequence quickly becomes overly involved with increasing k. An easy way to derive the generating polynomial of a decimated sequence is to solve a set of simultaneous equations derived as follows. We know that if an m-sequence, that is generated by an nth degree primitive polynomial, is properly decimated, the resulting sequence will also be an m-sequence generated by an nth degree polynomial. Thus, all we need to do in order to uncover the polynomial is to produce a sequence of bits of the decimated recursion

$$\{d(i)\}, \, i = 0, 1, 2, \ldots \quad (A\text{-}80)$$

and then solve for the α_i's from n − 1 independent equations of the form

$$d(n + i) = \alpha_1 d(n + i - 1) + \ldots \\ + \alpha_{n-1} d(i + 1) + d(i) \quad (A\text{-}81)$$

This method, for a different application, was suggested by Meyer and Tuchman (1972). As an example, let us apply the method to the sequence given in (A-79). The first four 6-tuples are

$$\begin{array}{l} 100101 \\ 001011 \\ 010110 \\ 101100 \end{array} \quad (A\text{-}82)$$

From the four 6-tuples of (A-82) we immediately derive the four equations

$$\begin{array}{rcl} 1 \quad\;\; + \alpha_2 \quad\quad\quad\quad\;\; & = & 1 \\ \alpha_1 \quad\;\; + \alpha_3 \quad\quad\; & = & 1 \\ \alpha_1 + \alpha_2 \quad\;\; + \alpha_4 & = & 0 \\ 1 \quad\;\; + \alpha_2 + \alpha_3 \quad\;\; & = & 0 \end{array} \quad (A\text{-}83)$$

The solution of the equations in (A-83) yields $\alpha_1 = \alpha_2 = \alpha_4 = 0$ and $\alpha_3 = 1$. Thus, the primitive polynomial which generates the sequence in (A-79) is $x^5 + x^3 + 1$ as previously determined by the cubic transformation.

4.7 Inverse Decimation

The inverse problem of determining the undecimated sequence given just a short segment of the decimated sequence is considered by Arazi (1977). The following problem development and solution is different from Arazi's but more in consonance with our overall development. We assume that every kth bit is taken from the zeroth stage of a shift register which is implementing an nth degree primitive polynomial by the companion matrix architecture. The contents of the shift register, $\underline{b}^t$, change according to (t ≤ 1)

$$\underline{b}^{t+1} = \underline{b}^t M \quad (A\text{-}84)$$

$$a^t = \underline{b}^t \underline{C} \quad (A\text{-}85)$$

where

$$\underline{C} = \text{col}[1, 0, 0 \ldots, 0] \quad (A\text{-}86)$$

and a^t corresponds to the output bit at time t. From the hypothesis, we assume that we know

$$a^t, a^{t+k}, a^{t+2k}, \ldots, a^{t+(n-1)k} \quad (A\text{-}87)$$

and we need to compute

$$\underline{b}^1 = [a^1, a^2, \ldots, a^n] \quad (A\text{-}88)$$

from (A-87). In control theory, this is usually referred to as the observability problem (Luenberger, pp. 285-287).

From (A-84) and (A-85), we have

$$a^t = \underline{b}^1 M^t \underline{C} \quad (A\text{-}89)$$

and

$$\left[a^t, a^{t+k}, \ldots, a^{t+(n-1)k}\right] = \underline{b}^1 M^{t-1} \left[\underline{C} M^k \underline{C} \ldots M^{(n-1)k} \underline{C}\right] \quad (A\text{-}90)$$

since it is known that the solution exists and is unique, it is clear that

$$\underline{b}^1 = \left[a^t, a^{t+k}, \ldots, a^{t+(n-1)k}\right] \left[M^{t-1}\underline{C} M^{t+k-1}\underline{C} \ldots M^{t+(n-1)k-1}\underline{C}\right]^{-1} \quad (A\text{-}91)$$

5. GENERATION OF HIGH-SPEED M-SEQUENCES

The ability to generate high-speed m-sequences is important to a number of disciplines, especially direct sequence spread spectrum systems and precise ranging systems. The maximum rate at which the companion matrix realization of an m-sequence generator, Figure A-1, can be run is determined by two parameters, the propagation delay of a shift register stage and the computation time required by the modulo-two adder.

An "electronic" approach to speed increase was proferred by Harvey (1974) for certain trinomial feedback functions in which an analog delay line was substituted for the zeroth stage, which must be untapped for feedback to the modulo-two adder. The delay of the analog line is chosen to be the difference of the clocking period and the computation time of the modulo-two adder. The "before-and-after" diagrams illustrating Harvey's method for implementing the m-sequence specified by $x^4 + x^3 + 1$ are shown in Figures A-12a and A-12b. Clearly, Harvey's method also applies to the substitution of more than one shift register stage by analog elements.

Ball et al. (1975) extended the above method and showed that for trinomials of the form $x^n + x^{n-1} + 1$, no shift register stages are needed at all as is shown in Figure A-13. The delays indicated in the figure are determined by

$$D_1 = T$$
$$D_2 = (n - 1)T - D_g \quad \text{(A-92)}$$

where T is the bit period, n is the degree of the trinomial ($x^n + x^{n-1} + 1$) and D_g is the computation time of the modulo-two adder. Ball, et al., noted that the configuration shown in Figure A-13 will run free and is, in their words, "a special case of the delay-line oscillator."

Another approach to high-speed generation was offered by Lempel and Eastman (1971). Unlike the methods just reviewed, Lempel and Eastman's method depends on a mathematical, vice electronic, basis to achieve high-speed generation. The method depends upon a curious aspect of decimated m-sequences. Essentially, what Lempel and Eastman proposed was the creation of k shift registers each of which would be driven by the feedback polynomial corresponding to the kth decimation of the sequence $\{s(i)\}$, i = 0, 1, 2, ... which is generated by the nth degree polynomial that it is desired to implement at high speed. (The decimations need not be proper, that is, k and $2^n - 1$ need not be relatively prime.) What is done is to step, in succession, each of the k registers and modulo-two add together the last stage from each. The registers are started with initial conditions so that register j exhibits the decimated sequence $\{s(ik + j)\}$, i = 0, 1, 2, ... The sequence produced by this addition exhibits the desired polynomial and runs at a rate k times the rate of the individual shift registers. As an example, consider that we wish to synthesize the sequence produced by the primitive trinomial $x^3 + x + 1$ at a rate two times faster than we shift our registers. For this example, k = 2, the decimation is a proper one and, additionally, the polynomial driving our two registers must also be $x^3 + x + 1$ as the decimation is a power of two. (Recall that this was shown in the previous section.) Thus, the diagram shown in Figure A-14 will implement the recursion at twice the clock rate. If the top register is started with 010 and the bottom with 111, the machine will progress as follows:

CLOCK: 10101010101010
TOP REGISTER STAGE 2: 01100001111110
BOTTOM REGISTER STAGE 2: 11111100110000
MOD-TWO ADDER OUTPUT: 10011101001110

The three implementations so far reviewed assume that the limiting parameter is the propagation delay of the shift register stage. Often this is not the case but it is rather the computation time required by the modulo-two adder that caps the operating speed. One solution against this inherently different problem was given by Quan (1974) who proposed a modification of Lempel and Eastman's method. He suggested time division multiplexing the decimated sequences instead of adding them modulo-two. Figure A-15 depicts the scheme shown in Figure A-14 recast under Quan's modification. The initial conditions of the registers are the same as for Eastman and Lempel's version.

One final method for generating high-speed m-sequences is that proposed by Warlick and Hershey (1980). These authors first developed a special architecture for m-sequence generators based on primitive trinomials

$$x^n + x^a + 1 \quad \text{(A-93)}$$

They found that the general architecture depicted in Figure A-16, which they termed the "vanestream structure," (The repeated register blocks with feedback are termed the "vanes.") will progress through a maximum length cycle of $2^{v\ell + 1} - 1$ states for a surprisingly rich set of triples (v, ℓ, t) which they catalog, along with the primitive trinomials (A-93) upon which the structures are based, for $v\ell \leq 99$. The architecture of Figure A-16 is then coupled with a parallel in (broadside load), serial out multiplexer as shown in Figure A-17 to form what they term the "WINDMILL m-sequence generator." (The name "WINDMILL" derives from a predecessor sequential machine of theoretical interest only.) The slower logic vanestream generator is stepped at a rate R. After each step, the contents of the v stages are copied into the v high-speed shift register stages indicated and the v high-speed shift register is shifted v times in the direction indicated. The high-speed stream exhibits the m-sequence specified by the trinomial

$$x^n + x^{n-a} + 1 \quad \text{(A-94)}$$

(not $x^n + x^a + 1$ as erroneously reported in the paper).

The WINDMILL thus achieves a speed ratio increase of v to attain a high-speed rate Rv. Warlick and Hershey briefly considered the electronic architectural implications of realizing a WINDMILL and concluded that the WINDMILL offers power advantages for certain hybrid MOS layouts and its periodic substructures are especially amenable to LSI blacksmithery.

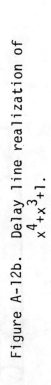
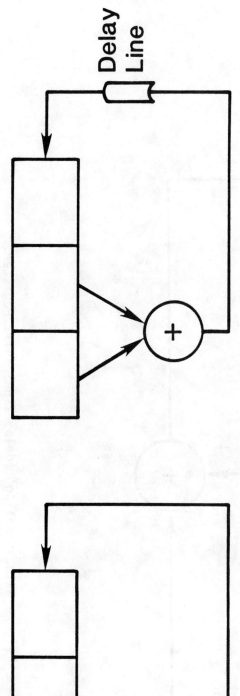
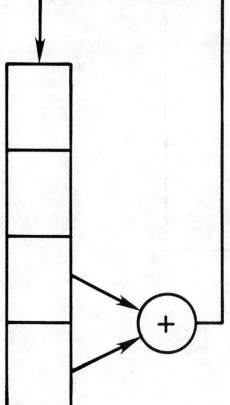

Figure A-12a. Companion matrix realization of x^4+x^3+1.

Figure A-12b. Delay line realization of x^4+x^3+1.

10-47

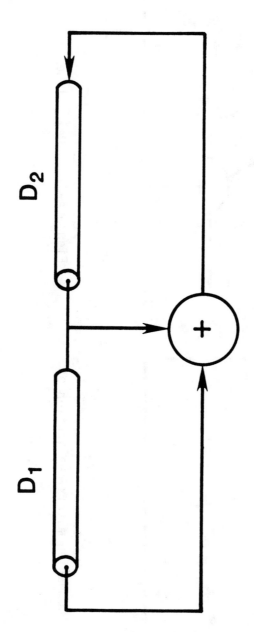

Figure A-13. 'Analog' (delay line) shift register.

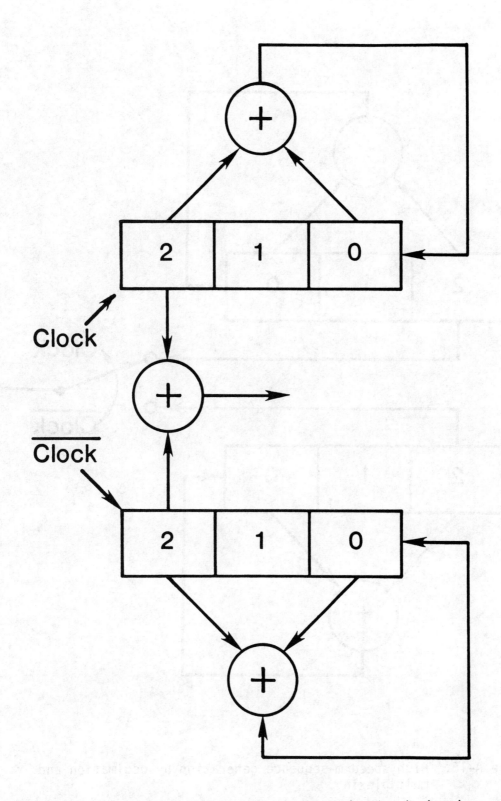

Figure A-14. High speed m-sequence generation by decimation and modulo-two addition.

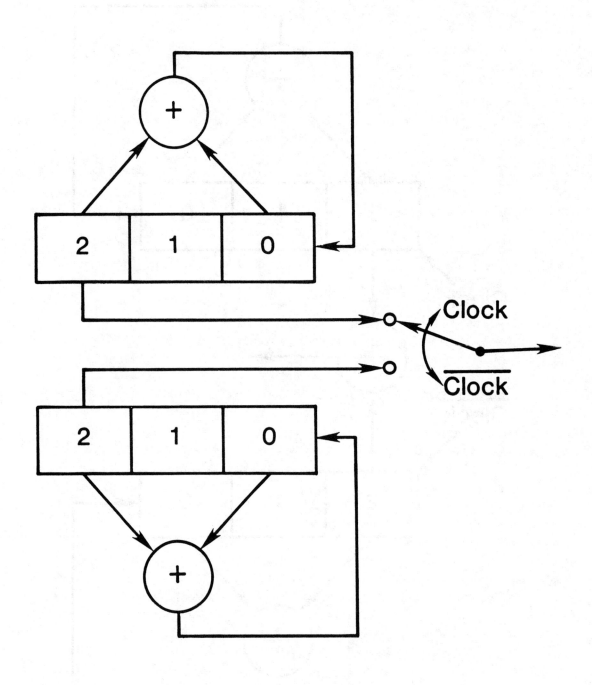

Figure A-15. High speed m-sequence generation by decimation and multiplexing.

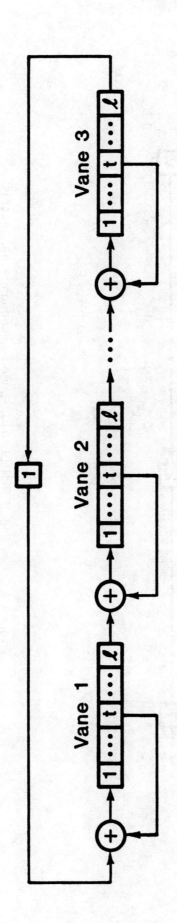

Figure A-16. The vanestream structure.

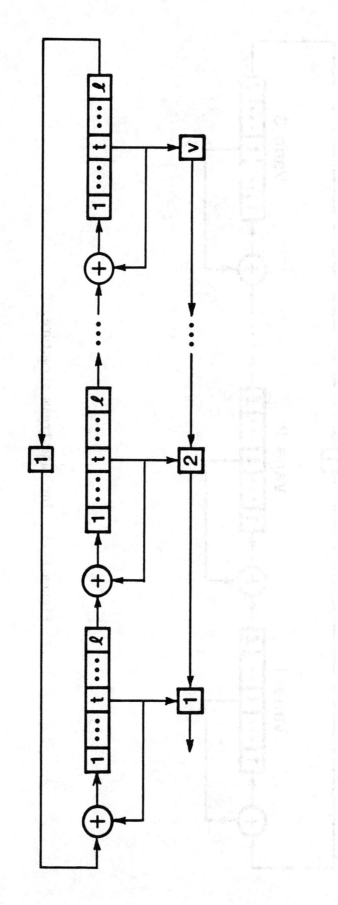

Figure A-17. The WINDMILL high-speed m-sequence generator.

6. REFERENCES: APPENDIX A

Arazi, B. (1977), Decimation of m-sequence leading to any desired phase shift, Electron. Letters *13*, No. 7, pp. 213-215, March.

Ball, J., A. Spittle, and H. Liu (1975), High-speed m-sequence generation: A further note, Electron. Letters *11*, No. 5, pp. 107-108, March.

Berlekamp, E. (1967), Factoring polynomials over finite fields, BSTJ, pp. 1853-1859, October.

Berlekamp, E. (1968), Algebraic Coding Theory (McGraw-Hill Book Co., New York, NY).

Berlekamp, E. (1970), Factoring polynomials over large finite fields, Mathematics of Computation *24*, No. 111, pp. 713-735, July.

Birdsall, T., and M. Ristenbatt (1958), Introduction to linear shift-register generated sequences, Technical Report No. 90, Electronic Defense Group, Dept. of Electrical Engineering, The University of Michigan Research Institute, Ann Arbor, MI. (Work performed for the Signal Corps, Dept. of the Army, Control No. DA-36-039 sc-63203, Task Order No. EDG-3, Project 2262.)

Birkhoff, G., and S. MacLane (1964), A survey of Modern Algebra, Revised Edition (MacMillan Company).

Cantor, D., and H. Zassenhaus (1981), A new algorithm for factoring polynomials over finite fields, Mathematics of Computation *36*, No. 154, pp. 587-592, April.

Charney, H., and C. Mengani (1961), Generation of linear binary sequences, RCA Rev., pp. 420-430, September.

Davies, A. (1965), Delayed versions of maximal-length linear binary sequences, Electron. Letters *1*, No. 3, pp. 61-62, May.

Davies, A. (1968), Calculations relating to delayed m-sequences, Electron. Letters *4*, No. 14, pp. 291-292, July.

Davis, W. (1966), Automatic delay changing facility for delayed m-sequences. Proc. IEEE *54*, pp. 913-914, June.

deVisme, G. (1971), Binary Sequences (The English Universities Press Ltd.).

Douce, J. (1968), Delayed versions of m-sequences, Electron. Letters *4*, No. 12, p. 254, June.

Gardiner, A. (1965), Logic P.R.B.S. delay calculator and delayed-version generator with automatic delay-changing facility, Electron. Letters *1*, No. 5, pp. 123-125, July.

Gold, R. (1966), Characteristic linear sequences and their coset functions, Journal SIAM *14*, No. 5, pp. 980-985, September.

Golomb, S. (1967), Shift Register Sequences (Holden-Day, Inc.).

Golomb, S. (1969), Irreducible polynomials, synchronization codes, primitive necklaces, and the cyclotomic algebra, Chapter 21, Combinatorial Mathematics and Its Applications (The University of North Carolina Press), pp. 358-370. (Proceedings of the conference held at the University of North Carolina, Chapel Hill, NC, April.)

Golomb, S. (1980), On the classification of balanced binary sequences of period $2^n - 1$, IEEE Trans. Inform. Theory *26*, No. 6, pp. 730-732, November.

Harvey, J. (1974), High-speed m-sequence generation, Electron. Letters *10*, No. 23, pp. 480-481, November.

Hershey, J. (1980), Implementation of MITRE public key cryptographic system, Electron. Letters *16*, No. 24, pp. 930-931, November.

Ireland, B., and J. Marshall (1968a), Matrix method to determine shift-register connections for delayed pseudorandom binary sequences, Electron. Letters *4*, No. 15, pp. 309-310, July.

Ireland, B., and J. Marshall (1968b), Matrix method to determine shift-register connections for delayed pseudorandom binary sequences, Electron. Letters *4*, No. 21, pp. 467-468, October.

Ireland, B., and J. Marshall (1976), New method of generating shifted linear pseudorandom binary sequences, Proc. Inst. Elec. Engrs. (London) *123*, p. 182, February.

Knuth, D. (1969), The art of computer programming, Seminumerical Algorithms, (Addison-Wesley), Vol. 2, pp. 381-396.

Latawiec, K. (1974), New method of generation of shifted linear pseudorandom binary sequences, Proc. Inst. Elec. Engrs. (London) *12*, No. 8, pp. 905-906, May.

Latawiec, K. (1975), New method of generation of shifted linear pseudorandom binary sequences, Proc. Inst. Elec. Engrs. (London) *122*, No. 4, p. 448, April.

Latawiec, K. (1976), New method of generation of shifted linear pseudorandom binary sequences, Proc. Inst. Elec. Engrs. (London) *123*, No. 2, p. 182.

Laxton, R., and J. Anderson (1972), Linear recurrences and maximal length sequences, The Mathematical Gazette *LVI*, No. 398, pp. 299-309, December.

Lempel, A., and W. Eastman (1971), High speed generation of maximal length sequences, IEEE Trans. Computers, pp. 227-229, February.

Luenberger, D. (1979), Introduction to Dynamic Theory—Theory, Models and Applications (Wiley), pp. 285-289.

MacDuffie, C. (1940), An Introduction to Abstract Algebra (John Wiley & Sons, Inc.).

Marsh, R. (1957), Table of irreducible polynomials over GF(2) through degree 19, U.S. Department of Commerce, Office of Technical Services, Washington, DC.

Meyer, C., and W. Tuchman (1972), Pseudorandom codes can be cracked. Electronic Design *20*, No. 23, pp. 74-76, November.

Meyer, F. (1976), Gigabit/s m-sequence generation, Electron. Letters *12*, No. 14, p. 353, July.

Miller, A., A. Brown, and P. Mars (1977), A simple technique for the determination of delayed maximal length linear binary sequences, IEEE Trans. Computers *C-26*, No. 8, pp. 808-811, August.

Moenck, R. (1977), On the efficiency of algorithms for polynomial factoring, Mathematics of Computation *31*, No. 137, pp. 235-250, January.

Perlis, S. (1952), Theory of Matrices (Addison-Wesley).

Quan, A. (1974), A note on high-speed generation of maximal length sequences, IEEE Trans. Computers, pp. 201-203, February.

Reiner, I. (1961), On the number of matrices with given characteristic polynomial, Illinois Journal of Mathematics *5*, pp. 324-329.

Roberts, P., A. Davies, and Tsao (1965), Discussion on generation of delayed replicas of maximal-length binary sequences, Proc. Inst. Elec. Engrs. (London) *112*, No. 4, pp. 702-704, April.

Rodemich, E., and H. Rumsey (1968), Primitive trinomials of high degree, Mathematics of Computation *22*, pp. 863-865.

Sarwate, D., and M. Pursley (1980a), Crosscorrelation properties of pseudorandom and related sequences, Proc. IEEE *68*, No. 5, pp. 593-619, May.

Sarwate, D., and M. Pursley (1980b), Correction of "Crosscorrelation properties of pseudorandom and related sequences," Proc. IEEE *68*, No. 12, p. 1554, December.

Scholefield, P. (1960), Shift registers generating maximum-length sequences, Electronic Technol., pp. 389-394, October.

Stahnke, W. (1973), Primitive binary polynomials, Mathematics of Computation *27*, No. 124, pp. 977-980, October.

Swan, R. (1962), Factorization of polynomials over finite fields, Pacific Journal of Mathematics *12*, pp. 1099-1106.

Tsao, S. (1964), Generation of delayed replicas of maximal-length linear binary sequences, Proc. Inst. Elec. Engrs. (London) *111*, No. 11, pp. 1803-1806, November.

VanLuyn, A. (1978), Shift-register connections for delayed versions of m-sequences, Electron. Letters *14*, No. 22, pp. 713-715, October.

Warlick, W., and J. Hershey (1980), High-speed M-sequence generators, IEEE Trans. on Computers *C-29*, No. 5. pp. 398-400, May.

Watson, E. (1962), Primitive polynomials (Mod 2), Mathematics of Computation *16*, No. 79, pp. 368-369, July.

Weathers, G. (1972), Statistical properties of filtered pseudorandom digital sequences, Sperry Rand Corporation, Technical Report No. SP-275-0599, January. (Prepared for NASA George C. Marshall Space Flight Center, Huntsville, AL.)

Weinrichter, H., and F. Surbock (1976), Phase normalized m-sequences with the inphase decimation property $\{m(k)\} = \{m(2k)\}$, Electron. Letters *12*, No. 22, pp. 590-591, October.

Yiu, K. (1980), A simple method for the determination of feedback, shift register connections for delayed maximal-length sequences, Proc. IEEE *68*, No. 4, pp. 537-538, April.

Zierler, N. (1959), Linear recurring sequences, Journal SIAM *7*, No. 1, pp. 31-48, March.

Zierler, N., and Brillhart (1968), On primitive trinomials (Mod 2), Information and Control *13*, pp. 541-554.

Zierler, N., and J. Brillhart (1969), On primitive trinomials (Mod 2), II, Information and Control *14*, pp. 566-569.

Zierler, N. (1969), Primitive trinomials whose degree is a Mersenne exponent, Information and Control *15*, pp. 67-69.

Zierler, N. (1970), On $x^n + x + 1$ over GF(2), Information and Control *16*, pp. 502-505.

APPENDIX B: A CURSORY LOOK AT SYNCHRONIZATION FOLLOWING CLOCK RECOVERY

1. INTRODUCTION

Synchronization is usually performed to establish either epoch or phase. Epoch synchronization refers to those techniques that seek to establish agreement on the epoch or occurrence of a particular instant of time. Phase synchronization comprises those processes that endeavor to determine the phase of a cyclic (steady state deterministic) digital process. We will consider a variety of epoch and phase synchronization methods, starting with epoch synchronization followed by the phase synchronization genres.

1.1 Epoch Synchronization

Epoch synchronization is a process by which a transmitter communicates a reference time mark to a receiver. This is usually done in the time domain by sending a carefully constructed sequence. The sequence is such that it possesses an autocorrelation that has low sidelobes thus allowing a large peak to sidelobe ratio when passed through a matched filter. There is some confusion in the literature that centers on sequence design according to the presence or absence of bit sense or "bit ambiguity." This difficulty is explored.

1.1.1 Autocorrelation

The autocorrelation function most commonly used for sequence design is as follows. Let the sequence be denoted by $s(1), s(2), \ldots s(n)$ where $s_i \epsilon \{0, 1\}$. We compute the autocorrelation, $r(k)$, by counting the agreements, A, between $s(i)$ and $s(i + k)$ over the range $0 \leq i \leq n - k$ and subtracting the disagreements over the same range. From this definition it is clear that $r(k) = r(-k)$. As an example, let us determine the autocorrelation of the sequence

$$0000011010111001 \qquad \text{(B-1)}$$

We calculate $r(1)$ by lining up the slipped sequence with the unslipped sequence as follows

```
0000011010111001
 0000011010111001
----------------
 AAAADADDDDAADAD     A indicates an
                     AGREEMENT:
                     D a DISAGREEMENT
```

from the above we see that $r(1) = 8 - 7 = 1$. Figure B-1 depicts $r(k)$ for $k = 0, \pm 1, \pm 2, \ldots \pm 15$. Note that max $(A - D)$ for $k \neq 0$ is 1. Note further that at $k = 8$, $A - D = -6$. This is a consequence of designing the sequence (B-1) in order to achieve

$$\min_{k \neq 0} (\max (A - D)) \qquad \text{(B-2)}$$

What (B-2) implies is that bit "sense" is known, i.e., the receiver knows the received bits exactly and not within the

```
0   0
1   1
```
(B-3)

ambiguity of some differentially coded systems. This is a reasonable and practical assumption for many communications circuits and signaling architectures.

1.1.2 Bit Sense Known

Sequences, or as they are often called, "unique words," which are designed under (B-2) may be used in various telemetry schemes, see Maury et al. (1964), and for Time Division Multiple Access (TDMA) satellite communications. See, for example Sekimoto (1968), Gabard (1968), Schrempp and Sekimoto (1968), and Nuspl et al. (1977). A common implementation that can be used for synchronization using sequences wherein bit sense is known, is a three step process (Schrempp and Sekimoto, 1968):

I. Allow carrier recovery by sending the receiver a stream of all zeros

II. Provide bit timing recovery by transmitting a stream of zero/one alternations

III. Mark epoch by sending a unique word.

As an example, let us assume we have carrier and clock and let us examine a few cases where we identify the epoch with the unique word specified in (B-1). We assume, for this case, that we are provided with a zero/one quantized bit stream and that we are passing this bit stream through a matched filter or replica of the unique word (B-1). (This is not the optimum method of detection, incidentally, but it is easy to realize in hardware.) Our experiment is depicted in Figure B-2. The bitstream that is passed into the Bitstream Analysis Window is a stream of ones and zeros taken from a balanced Bernoulli source which is a source of zeros and ones such that the probability of a one at time t equals the probability of a zero at time t equals one-half; further, the bit at time t is independent of all bits preceding it. At $t = 30$, the unique word (B-1) is sent and thus at $t = 45$, the unique word is lined up with its replica in the filter. At this point, if the unique word has not been corrupted by the channel, it will completely agree with its stored replica and cause a large pulse. Figure B-3 depicts the action showing $A - D$ as a function of time. Time is assumed to flow left to right.

As Figure B-3 depicts, the epoch is clearly defined by the A-D pulse of height 16 at $t = 45$. (The correlation is not observed for the first 15 clock times as the window is being initially filled during this time.) To detect the epoch, then, we would set a threshold on the $A - D$ waveform and make the rule that the epoch is declared when the $A - D$ waveform meets or exceeds the threshold. This is obviously doable and effective if the unique word is received, detected and quantized without errors. Such is not the case in general, however. Figure B-4 depicts the action wherein the unique word is passed through a binary symmetric channel (BSC) with $p = 1/16$. (A BSC with parameter p is a simple channel model that dictates that the bit transmitted at the t is received correctly with probability $1 - p$ and is independent of previous channel errors.) In this case, the unique word suffered a single bit error causing A to decrease by one at epoch and D to increase by one causing $A - D$ to drop from 16 to 14. For this case we would have correctly detected the epoch only if we had set our threshold in the range $8 <$ threshold ≤ 14.

Figure B-5 repeats the above experiment with a BSC with

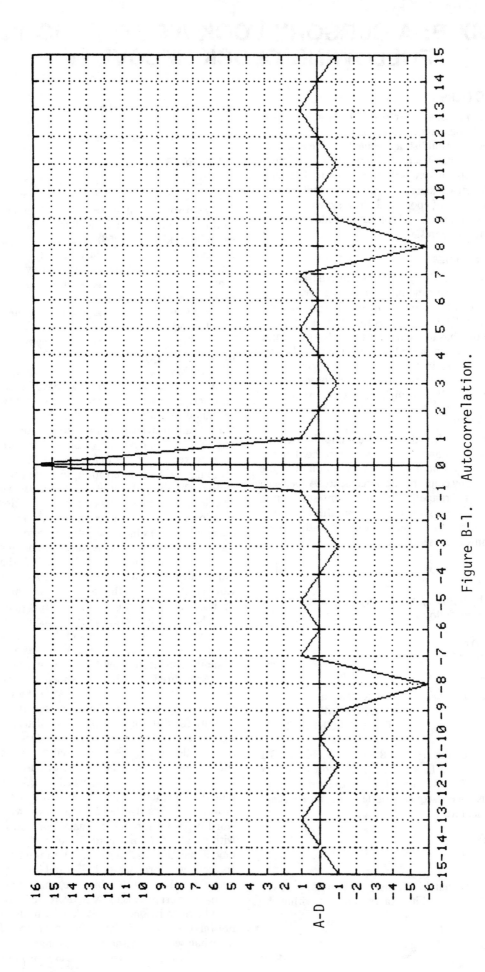

Figure B-1. Autocorrelation.

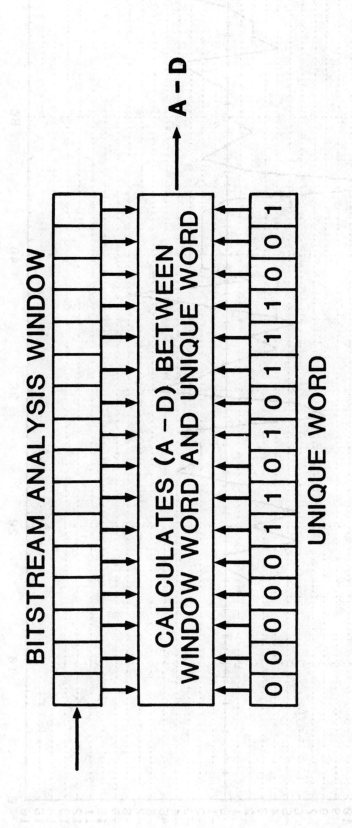

Figure B-2. A matched filter.

10-57

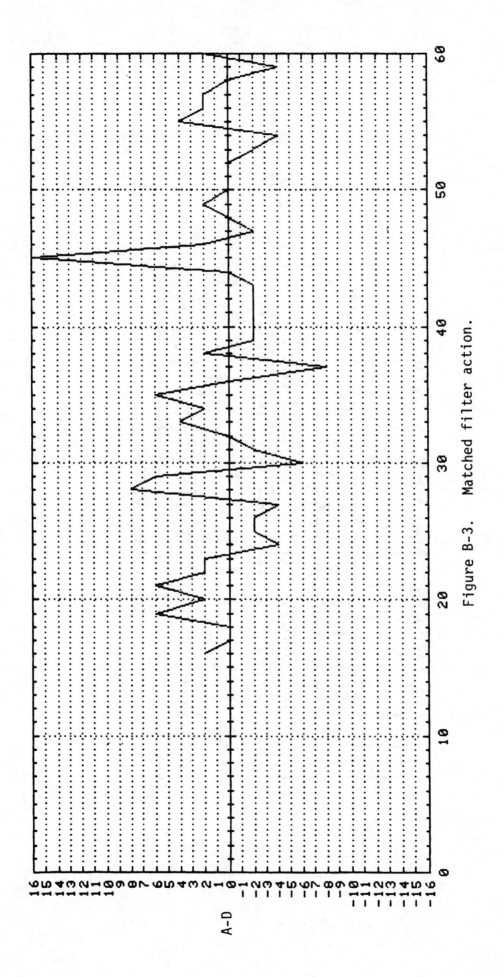

Figure B-3. Matched filter action.

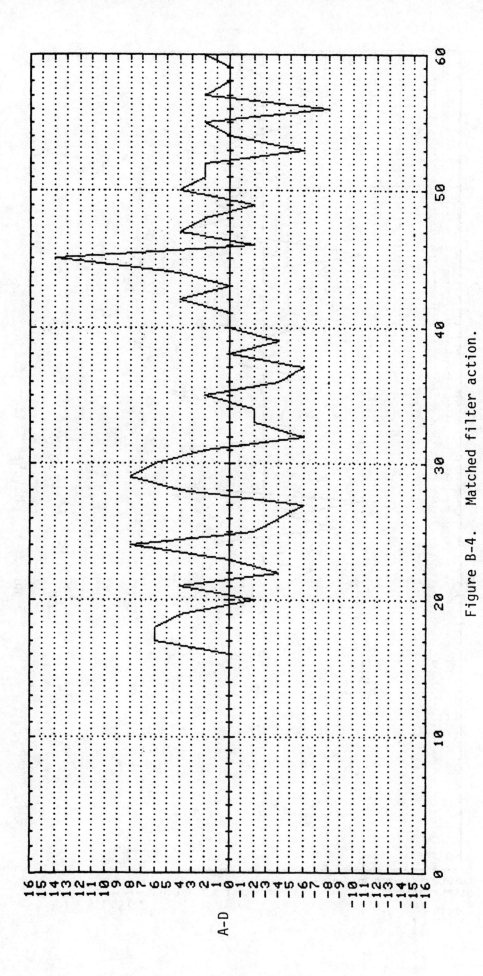

Figure B-4. Matched filter action.

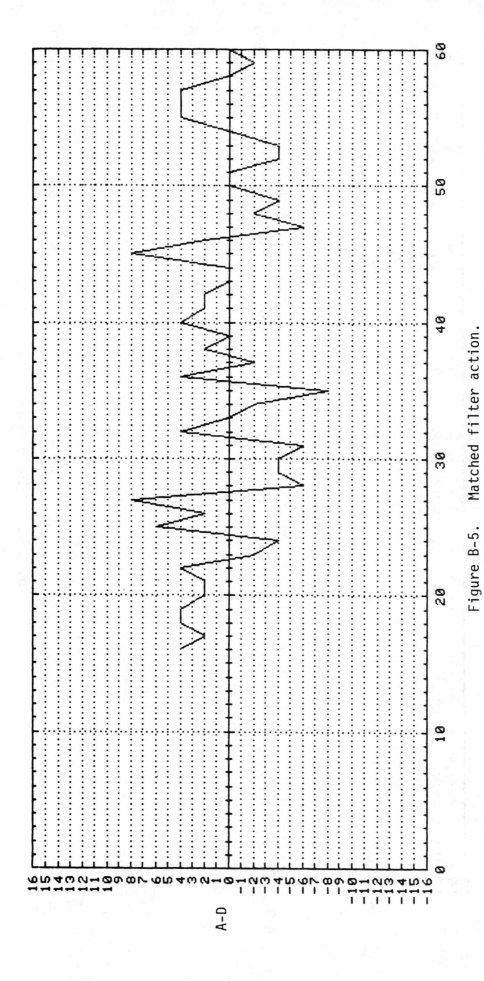

Figure B-5. Matched filter action.

Table B-1. Examples of Unique Words When Bit Sense is Known

Length	Max ACF	Sequence
3	0	001*
4	1	0001*
5	1	00010*
6	1	000110
7	0	0001101
8	1	00001101
9	1	000011010
10	1	0001110010
11	0	00011101101*
12	1	000010110011
13	1	0000011001010*
14	1	00011011110010
15	2	000000110010100
16	1	0000011010111001
17	1	00000101100111010
18	1	000010101101100111
19	2	0000000101100111 0010
20	1	00000101101001 11001
21	1	000000111001101101010
22	1	0001000111110011011010
23	2	00000011001011011001
24	1	000001100111010101 10110
25	1	0000000110110111 0001101010
26	2	00000001101010110011110010
27	2	000000011011001111001010100
28	2	0000000110110110001110101011
29	1	00000010110010011100111101010

*Indicates Barker code

$p = \frac{1}{4}$. In this example the unique word suffered 4 errors and the $A - D$ waveform exhibited a peak value of only 8 at the correct epoch, $t = 45$. Note however that a randomly occurring peak of height 8 also occurred at $t = 27$. For this case it would have been impossible to select a threshold which would have correctly identified the epoch.

So far we have learned that we must be careful in choosing the structure of our unique word and we must be careful to correctly set the epoch detection threshold. There is another consideration that is worthwhile and this refers to "look time." The look time is defined as the amount of time a matched filter will be allowed to search for an epoch. The random local maxima that occur may cause a false epoch determination. The probability of a false epoch determination varies in a direct relationship to the look time. If a long look time is required, it may be necessary to select a high threshold. Doing so may require the user to pick a longer unique word because a higher threshold may allow for fewer channel errors. As an example, Figure B-6 depicts the action of the matched filter of Figure B-2 for an input bitstream produced by a balanced Bernoulli source of 1000 bits. Note the many relatively large maxima that occur randomly.

Table B-1 lists examples of unique words meeting condition (B-2) for lengths 3-29 bits.

1.1.3 Bit Sense Unknown

In many cases bit sense is unknown and unique word sequences are then designed by modifying (B-2) to the following

$$\min (\max (|A - D|)) \quad k \neq 0 \quad \text{(B-4)}$$

By way of introduction, let us examine a 16 bit sequence designed according to (B-4).

$$0000011001101011 \quad \text{(B-5)}$$

Figure B-7 depicts $r(k)$ for this word for $k = 0, \pm 1, \pm 2, \pm 15$. Note that for this word, the best that can be achieved under condition (B-4) is $|r(k)| \leq 2$ in contrast to $r(k) \leq 1$ for the word (B-1) which was designed under condition (B-2).

The great majority of research on unique words has addressed the case where bit sense is unknown. Probably this was a result of the great influence that the radar field has had on special sequence development. A PSK modulated radar return is a good example of a case in which one would instinctively choose a pulse compression code or unique word designed under (B-4).

The most famous of the sequences designed under (B-4) are the Barker codes. These sequences are those designed under (B-4) for which

$$\max (|A - D|) = 1 \quad k \neq 0 \quad \text{(B-6)}$$

Some of the sequences that meet (B-6) were introduced by R. H. Barker (1953) and bear his name. No Barker sequences beyond length 13 are known but it is known that if any do indeed exist, then they must have a length that is an even square. All possibilities up to $78^2 = 6084$ have been eliminated. (See Petit, 1967.)

Lindner (1975a) has compiled an exhaustive listing of all binary sequences or unique words that meet (B-4) for lengths 3-40 bits. Table B-2 lists one of these sequences for each word length for $n = 3$ to $n = 40$ bits along with the

$$\max (|A - D|) \quad k \neq 0$$

value achievable for n (Lindner, 1957b).

Many efforts have been made to create longer sequences from the Barker sequences that exhibit good values for the autocorrelation. Klyuyev and Silkov (1976), for example, proposed the construction

$$\alpha \alpha \alpha \bar{\alpha} \quad \text{(B-7)}$$

where α is a Barker sequence and the superbar indicates complementation. Constructions such as (B-7) can be viewed as examples of Kronecker constructions which are produced as follows (see Stiffler, 1971, or Turyn, 1968):

a) Let $s(1), s(2), \ldots s(\ell)$ and $t(1), t(2), \ldots t(m)$ be two sequences with $r(k)$ and $r'(k)$ the autocorrelations of $s(\ )$ and $t(\ )$ respectively.

b) Form the ℓm long sequence $s(1) + t(1), s(1) + t(2), \ldots s(1) + t(m), s(2) + t(1), \ldots s(\ell) + t(m)$. (B-8)

Unfortunately, the Kronecker constructed sequence (B-8) possesses an autocorrelation function $r''(k)$ such that

$$\max r''(k) \geq (\max r(k))(\max r'(k)) \quad \text{(B-9)}$$

and thus can never lead to a better normalized autocorrelation function than either of its component sequences.

A very interesting result due to Moser and Moon and cited on p. 198 of Turyn (1968) is that if a sequence is generated at random (from a balanced Bernoulli source) then one can expect $\max (|A - D|)$ to be on the order of the square root of the length of the sequence. Keeping this thought in

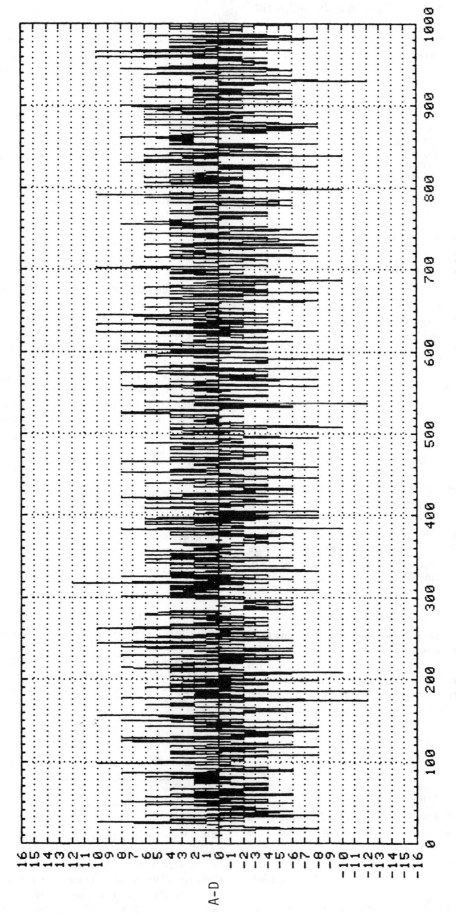

Figure B-6. Matched filter action on random input bitstream.

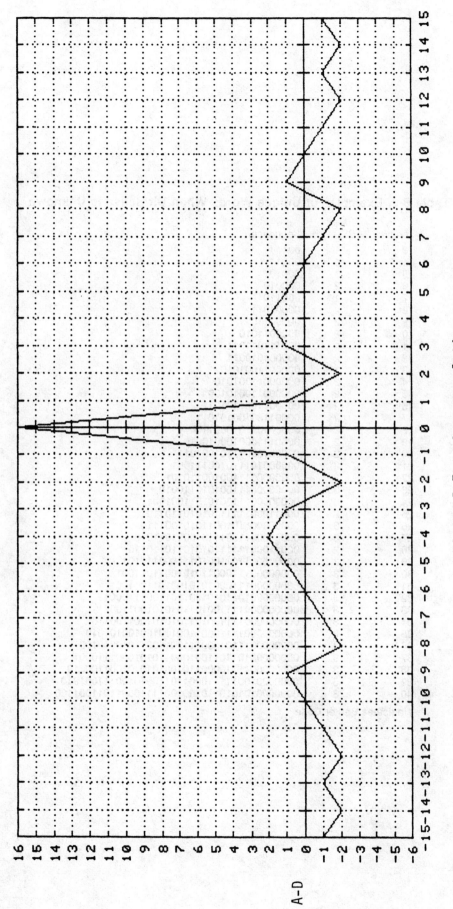

Figure B-7. Autocorrelation.

10-63

Table B-2. Examples of Unique Words When Bit Sense is Unknown

Length	Max ABS ACF	Sequence
3	1	001*
4	1	0001*
5	1	00010*
6	2	000010
7	1	0001101*
8	2	00001011
9	2	000001101
10	2	0000011010
11	1	00011101101*
12	2	000001010011
13	1	0000011001010*
14	2	00000011001010
15	2	000000110010100
16	2	0000011001101011
17	2	00001100100101011
18	2	000001011010001100
19	2	0000111000100010010
20	2	00000100011101001011
21	2	000000101110100111001
22	3	0000000011100100110101
23	3	00000000111001010110010
24	3	000000001110010101100100
25	2	0001100011111101010110110
26	3	00000000110100110101001110
27	3	000000001100111010001011010
28	2	0001100011111010101101101110
29	3	00000000110110010111000110101
30	3	000000001110001101010011011001
31	3	0000000011100010101001011011001
32	3	00000000101101010011011001110001
33	3	000000001111001011010101001100011
34	3	0000000011110010110101011001100110
35	3	00000000111000110011010010010101010
36	3	000000001110100011100110011010010100
37	3	0000000010110110010001101010001110001
38	3	00000000111100001101001010101001100110
39	3	000000010011111000110110010100111001010
40	3	0000000010101011010010011110001101100110 0

*Indicates Barker Code

mind, let us examine a technique that has often been suggested to finding a long sequence with good autocorrelation behavior. This technique is to generate an m-sequence and then pick the best phase to minimize the maximum value of $|A - D|$. The author has done this with the 63-bit m-sequence generated according to $x^6 + x^5 + 1$. The best phase (one of several, actually) is

101101110110011010101111100000100001100010100111101000111001000 (B-10)

Figure B-8 depicts the autocorrelation of this sequence. Note that the maximum absolute value is 6 which is a bit better than the square root of 63.

1.1.4 Concluding Remarks

The reader will note that we have not fully examined the behavior of epoch synchronization under channel noise nor have we even delved into the simple statistics relevant to choosing an appropriate threshold. The reasons for these omissions are twofold:

a) Epoch synchronization can be implemented by first hard quantizing the incoming data stream into zeros and ones. This hard quantization method may be practical but it is certainly not optimal.

b) The channel noise processes will heavily influence the choice of parameters for the unique word once the matched filter or unique word detection system has been agreed upon. It is only then that the necessary statistics for describing the behavior of the unique word detection system should be derived.

One final observation is that unique word detection is a classical TYPE I/TYPE II decision process as diagrammed in Figure B-9. (See Kreyszig, 1967, p. 802.) In picking the unique word, the designer must assess appropriate costs for the TYPE I/TYPE II errors. For example, for an Amateur Radio transmission to be occasionally lost (TYPE II errors) may be far less annoying (costly) than the occurrence of frequent false synchronizations (TYPE I errors).

1.2 Phase Synchronization

Figure B-10 depicts the generic diagram for the phase synchronization process. The top box represents a cyclic digital process, in effect a repetitive sequence of binary n-tuples. The middle box is a time invariant (fixed) mapping from the binary n-tuples to a single binary unit. The sequence of bits thus produced constitutes the sequence. The period of the sequence is, of course, upperbounded by the cycle length of the cyclic digital process.

1.2.1 m-Sequence Synchronization

This phase synchronization process uses a shift register of length n with a primitive polynomial for feedback as the cyclic digital process of Figure B-10, i.e., an m-sequence generator. The combinatorial logic is simply a single tapped stage of the shift register. The sequence is an m-sequence. This is depicted in Figure B-11 to show comportment to the canonical structure of Figure B-10.

As we have seen in the previous appendix, an m-sequence generated by a primitive polynomial of degree n has $2^n - 1$ distinct phases. To determine the phase and thereby establish synchronization it is necessary to know without error n consecutive bits and the polynomial that generates the m-sequence. If the signal-to-noise ratio is low, this may not be possible by direct demodulation. Consequently we may have to rely on methods that work with more than n bits in making a decision.

The following sections deal with the three possibilities denoted by X's:

	DEMODULATED BITS ERRORLESS	SOME DEMODULATED BITS IN ERROR
PRIMITIVE POLYNOMIAL OF DEGREE n KNOWN	X	X
PRIMITIVE POLYNOMIAL OF DEGREE n UNKNOWN	X	

Polynomial known: errorless reception

This is the simplest of the four possible cases and also the least likely as the result of a spread spectrum system is to spread the signal energy so that the individual chip possesses a low signal-to-noise investment thus making errorless acquisition of n consecutive bits very unlikely.

Assuming, however, that n consecutive bits are obtained without error, then the future behavior of the m-sequence is determined and synchronization is established.

Polynomial known: errors in reception

This is the most common case and actually consists of two subcases. First, if $2^n - 1$ is sufficiently small that one can expect to see all phases of the m-sequence during synchronization, then one need only create a matched filter to look for m consecutive bits of the m-sequence where $m > n$ for moderate signal-to-noise ratios to $m >> n$ for very low signal-to-noise ratios. When the m-sequence generates the m-bits, the matched filter will detect this epoch and phase synchronization will be accomplished.

If $2^n - 1$ is not sufficiently small that one can expect to see all of the m-sequence phases during the synchronization period, then the above method is not guaranteed and we must have a better than random estimate of the phase of the m-sequence before we attempt synchronization. Our job, then, is to search through a limited number of phase candidates and winnow the correct one.

The way this is usually done is by means of a serial search procedure implemented via what has become known as a "sliding correlator." The sliding presumed to be the correct phase of the m-sequence. If the phase is correct, the integral will show a large departure from the mean value. As an example, consider that we are trying to synchronize with the m-sequence generated by $x^6 + x^5 + 1$ and let us assume that our estimate of the phase is 4 clock times ahead of the true phase. For this experiment we correlate 6 bits at a time. Our rule for synchronization will be that we have achieved synchronization only if all 6 bits agree. Our procedure will be to test 6 bits. If all 6 agree, we declare that we are in synchrony; if all 6 do not agree, we retard our reference m-sequence by one clock time and compare (integrate) for another 6 bits. This process is depicted in Figure B-12.

The number of bits to be crosscorrelated against the reference m-sequence (6 in this example) will depend upon the type

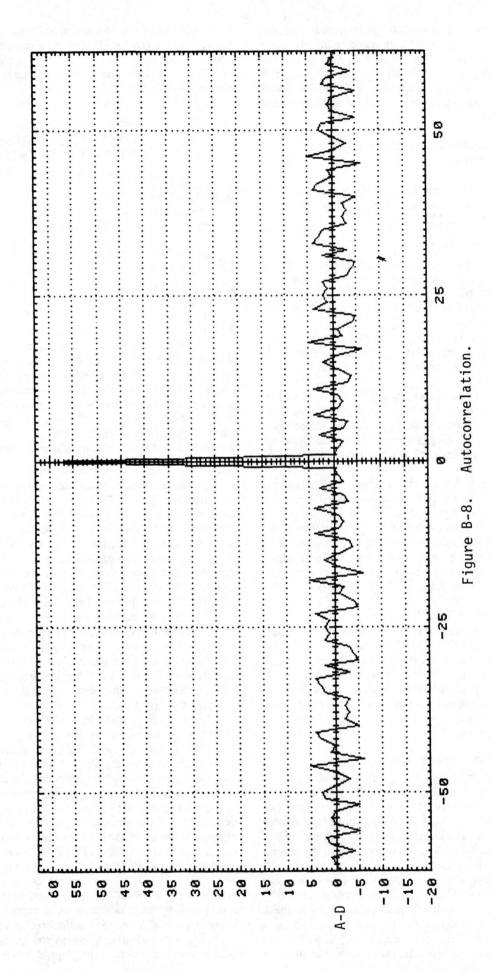

Figure B-8. Autocorrelation.

		UNIQUE WORD NOT PRESENT IN NOISY TRANSMISSION	UNIQUE WORD PRESENT IN NOISY TRANSMISSION
UNIQUE WORD NOT PRESENT	DETECTION THRESHOLD NOT MET OR EXCEEDED	CORRECT DECISION	TYPE II ERROR
UNIQUE WORD PRESENT	DETECTION THRESHOLD MET OR EXCEEDED	TYPE I ERROR	CORRECT DECISION

Figure B-9. Type I/Type II errors.

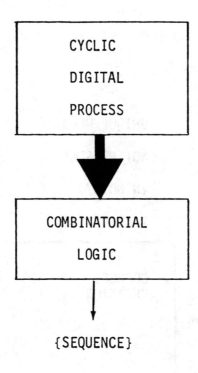

Figure B-10. Phase synchronization, generic diagram.

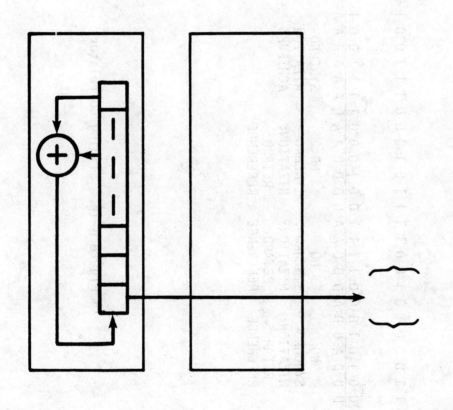

Figure B-11. m-sequence generator.

RECEIVED m-SEQUENCE: 1 0 0 0 0 1 1 0 0 0 1 0 1 0 0 1 1 1 1 0 1 0 0 0 1 1 1 0 0 1 0 0 1 0 1 1 . . .

REFERENCE m-SEQUENCE: 0 1 1 0 0 0 0 1 0 1 0 0 0 1 1 1 1 0 0 1 0 0 0 1 1 1 1 0 0 1

D D D A A D D D A D D A D D D A A D D D D A A D A A A A A A

NO SYNCHRO-NIZATION: RETARD REFERENCE	NO SYNCHRO-NIZATION: RETARD REFERENCE	NO SYNCHRO-NIZATION: RETARD REFERENCE	NO SYNCHRO-NIZATION: RETARD REFERENCE	SYNCHRO-NIZATION ACHIEVED

THE ARROW INDICATES RETARDATION OF THE REFERENCE BY ONE CLOCK TIME

Figure B-12. Sliding correlator.

of noise, the appropriate signal-to-noise ratio, the decision threshold and the look time (in this case, the time allowed to achieve synchronization. Clearly, the sliding correlator can require an enormous amount of search time if the number of phase candidates is very large or if the signal-to-noise is very low. Dixon (1976, pp. 181-183) discusses implementation of the sliding correlator by sliding the reference m-sequence in a continuous fashion (i.e., not discrete retardation per our example). Braun (1982) provides an excellent relevant theoretical analysis. He also points out that schemes such as that used in our example are "single dwell time" procedures and he considers a multiple dwell time approach that dynamically changes the width of the correlation window, i.e., the number of bits correlated for each decision. This concept of sequential hypothesis testing can bring far greater efficiency to our search. For background on sequential testing the reader is referred to Posner and Rumsey (1966).

Polynomial unknown: errorless reception

Yarlagadda and Hershey (1982) have proposed that Gold's characteristic sequence (see Appendix A) be used as a benchmark for m-sequence synchronization. They found that there is a curious crosscorrelation between the truncated Rademacher sequences and an m-sequence at its characteristic sequence phase. The Rademacher sequences are those sequences that describe the bit sequences of a normal binary counter started from zero. For example, consider the 3-bit binary counter below:

000
001
010
011
100
101
110
111

The first Rademacher sequence is the sequence exhibited by the counter's first bit, viz:

01010101   (B-11)

the second Rademacher sequence is

00110011   (B-12)

and the third is

00001111   (B-13)

We drop the last bits of the Rademacher sequences to form sequences of length $2^n - 1$ where n is the number of bits or stages of the counter. These sequences are termed the truncated Rademacher sequences. We form a matrix R which is composed or partitioned of the n truncated Rademacher sequences of length $2^n - 1$. For our example:

$$R = \begin{bmatrix} 0101010 \\ 0011001 \\ 0000111 \end{bmatrix} \quad (B\text{-}14)$$

Now consider any m-sequence of period 7. Let us arbitrarily select the m-sequence generated by the primitive polynomial $x^3 + x + 1$:

1001110   (B-15)

We now construct the $(2^n - 1) \times (2^n - 1)$ matrix, S, of all phases of (B-15)

$$S = \begin{bmatrix} 1001110 \\ 0011101 \\ 0111010 \\ 1110100 \\ 1101001 \\ 1010011 \\ 0100111 \end{bmatrix} \quad (B\text{-}16)$$

We now multiply R and S using conventional matrix multiplication, i.e., we do not modularly reduce the row-column dot products:

$$RS = \begin{bmatrix} 2131212 \\ 1321221 \\ 2211123 \end{bmatrix} \quad (B\text{-}17)$$

Notice that the fourth column's entries are all equal. This will be so only of the characteristic sequence of the m-sequence which for our polynomial is:

1110100   (B-18)

A bank of n-correlators will thus be able to determine the epoch at which an m-sequence, *regardless of the generating polynomial*, passes through its characteristic sequence phase. Figures B-13 through B-17 show the output of the crosscorrelation

$$\sum_{i=0}^{30} r_k(i) s(i+j) \quad (B\text{-}19)$$

where $r_k(i)$ is the ith bit of the kth Rademacher sequence and $\{s()\}$ is the m-sequence generated by $x^5 + x^3 + 1$. The m-sequence is

```
1111100011011101010000100101100
         1111111111222222222233    (B-20)
1234567890123456789012345678901
```

The phases of the m-sequence are arbitrarily assigned by the smaller subscript numbers. The abscissa of B-13 through B-18 are the phase numbers. Figure B-18 is an overlay of Figures B-13 through B-17. This figures allows the reader to spot the point at which the crosscorrelations are all equal and thereby identifies the phase (23) at which the characteristic sequence begins.

1.2.2 Rapid Acquisition Sequences

This synchronization process was introduced by Stiffler (1968). It uses a normal binary counter of n stages as the cyclic digital process. The counter is started at zero and incremented by one every clock time. When the count has reached $2^n - 1$, the next incrementation causes the counter to be reduced modulo 2^n and have all zeros in its stages. The contents of the counter stages $x_1, x_2, \ldots x_n$ (x_1 is the least significant stage) are input to the following combinatorial logic.

$$f(x_1, x_2, \ldots x_n) = \begin{cases} 0 \text{ if } \sum_{i=1}^{n} x_i \leq \left[\frac{n}{2}\right] \\ 1 \text{ otherwise} \end{cases} \quad (B\text{-}21)$$

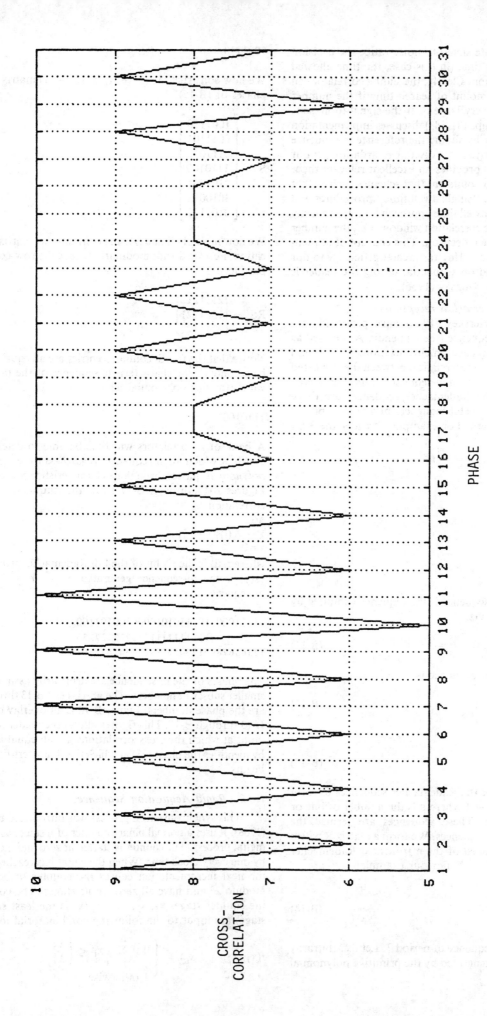

Figure B-13. Crosscorrelation of first Rademacher sequence.

10-72

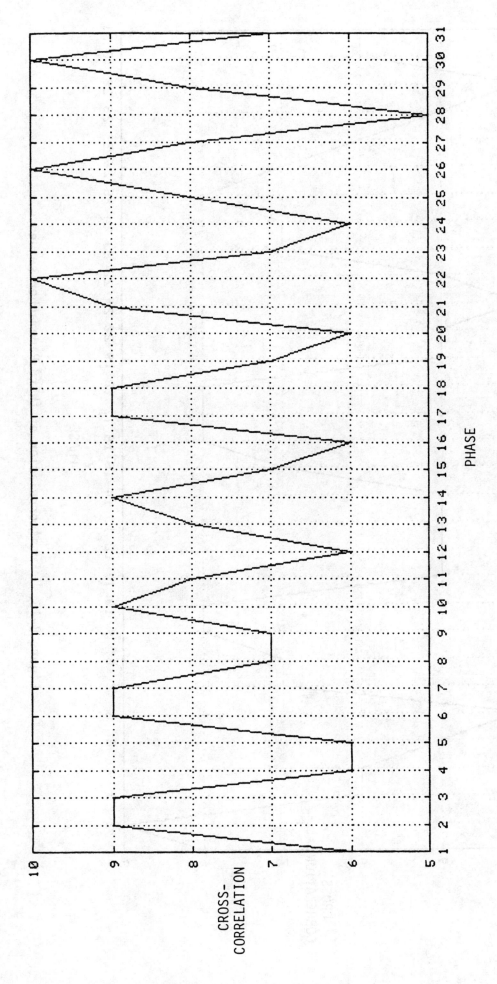

Figure B-14. Crosscorrelation of second Rademacher sequence.

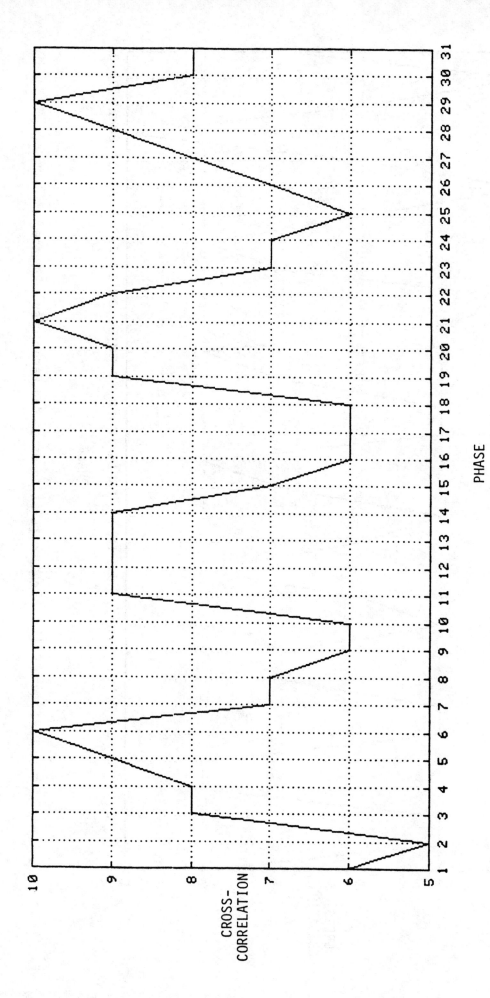

Figure B-15. Crosscorrelation of third Rademacher sequence.

10-74

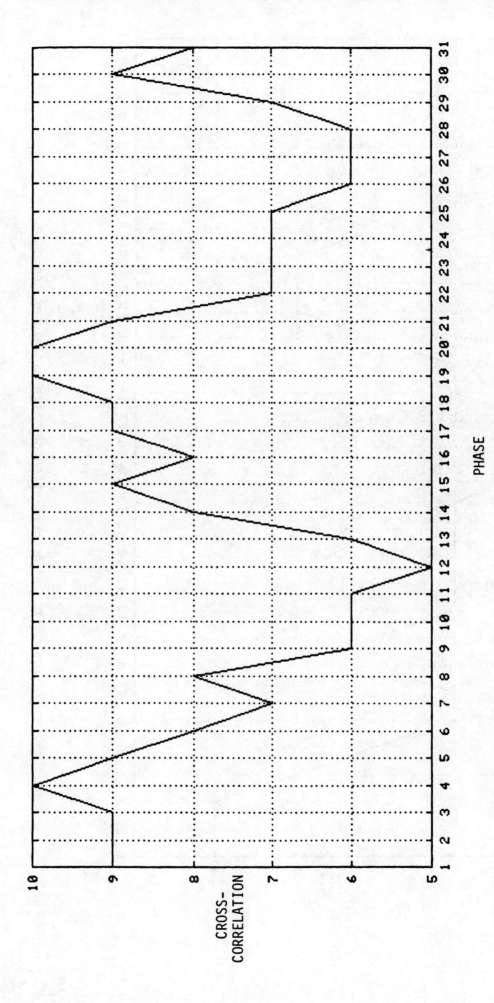

Figure B-16. Crosscorrelation of fourth Rademacher sequence.

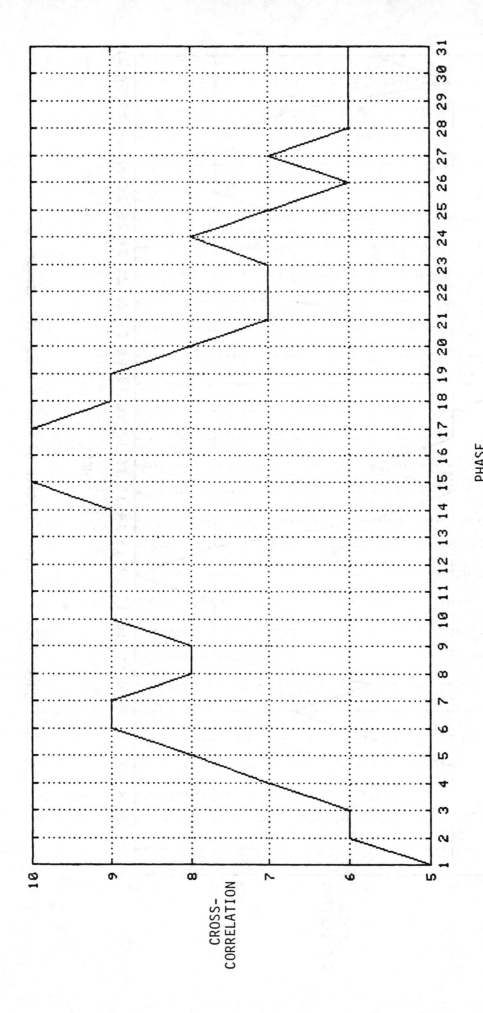

Figure B-17. Croscorrelation of fifth Rademacher sequence.

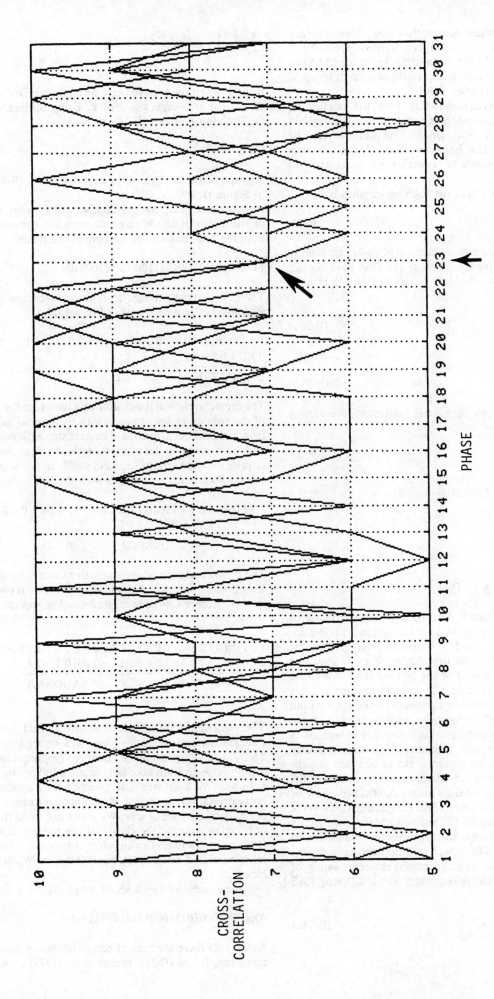

Figure B-18. Coalescence of crosscorrelations identifying the characteristic sequence.

10-77

where [ ] is the greatest integer function. The sequence produced, the output of the boolean function specified in (B-21), is termed the Rapid Acquisition Sequence (RAS) of length 2^n. Figure B-19 depicts this configuration in terms of our canonical model (Figure B-10).

Two items are worth mentioning. First, the function f() of (B-21) is a non-linear boolean function that is threshold realizable. Because it is threshold realizable, it can be implemented in very fast hardware. Second, the sequence described by the successive contents of x_i is the ith Rademacher sequence.

It is easy to verify that the RAS of length 8 is

$$00010111 \tag{B-22}$$

Consider the full period crosscorrelation of (B-22) against the first Rademacher sequence. Because the first Rademacher sequence has only two distinct phases, i.e., the "normal" phase

$$01010101 \tag{B-23a}$$

and the other phase

$$10101010 \tag{B-23b}$$

it is clear that the crosscorrelation will exhibit only two values. By direct computation

```
0 0 0 1 0 1 1 1
0 1 0 1 0 1 0 1
```
ADAAAADA $A - D = 6 - 2 = 4$

and

```
0 0 0 1 0 1 1 1
1 0 1 0 1 0 1 0
```
DADDDDAD $A - D = -4$

Thus the crosscorrelation is 4 for the Rademacher phase of (B-23a), the "normal" phase and -4 for the phase of (B-23b). The synchronization procedure begins to emerge. The first step is to crosscorrelate the first Rademacher sequence of length 2^n over a full period of the 2^n long RAS. This first step will give us one of two equally probable answers and this "bit" of information resolves the phase of the RAS within modulo 2. Once the phase has been resolved modulo 2, the second Rademacher sequence is crosscorrelated in both of its possible phases. (There are, of course, four distinct phases of the second Rademacher sequence, but we consider only the two phases that "survive" the first test, i.e., only two of the four phases will be consonant with the determined phase of the first Rademacher sequence.) This computation tells us the phase of the RAS modulo 4. Thus at each stage we gain a bit of information and sequentially recover the counter's stage sequences. Stiffler (1968) showed that the (normalized, i.e., $A - D$ divided by the sequence length) crosscorrelation of the kth 2^n long Rademacher sequences with the 2^n long RAS is

$$\frac{1}{2^{n-1}} \binom{n-1}{\frac{n-1}{2}} \quad \text{n odd} \tag{B-24a}$$

and

$$\frac{1}{2^n} \binom{n}{\frac{n}{2}} \quad \text{n even} \tag{B-24b}$$

if the $k - 1$ phases have been correctly resolved. Note that (B-24a, b) are *independent* of k. Using Stirling's formula, Stiffler approximates (B-24a,b) by

$$(2/\pi)^{\frac{1}{2}} n^{-\frac{1}{2}} \tag{B-25}$$

A graph of (B-24a,b) and the approximation (B-25) is given in Figure (B-20).

As an example, consider that we wish to recover the phase of the 8-long RAS. We assume errorless reception for the example and that we are looking at the phase

$$11100010111000101110001011100010...$$

The first step is to compute the crosscorrelation of the first Rademacher sequence over a full period:

```
1 1 1 0 0 0 1 0 1 1 1 0 0 0 1 0 1 1 1 0 0 0 1 0 1 1 1 0 0 0 1 0 ...
0 1 0 1 0 1 0 1
```
DADDADDD $A - D = -4$

The crosscorrelation is negative and therefore the first stage of the counter is out of phase with the normal phase of the first Rademacher sequence. The next step is to delay one bit and thus bring us into phase with the first Rademacher sequence component of the RAS and then crosscorrelate the second Rademacher sequence with the RAS:

```
1 1 1 0 0 0 1 0 1 1 1 0 0 0 1 0 1 1 1 0 0 0 1 0 1 1 1 0 0 0 1 0 ...
0 1 0 1 0 1 0 1   0 0 1 1 0 0 1 1
```
DDDDADDA $A - D = -4$

The crosscorrelation is again negative so we bring our second Rademacher sequence into phase by delaying it two bits and then crosscorrelate the third Rademacher sequence with the RAS:

```
1 1 1 0 0 0 1 0 1 1 1 0 0 0 1 0 1 1 1 0 0 0 1 0 1 1 1 0 0 0 1 0
0 1 0 1 0 1 0 1   0 0 1 1 0 0 1 1   0 0 0 0 1 1 1 1
```
AAADDAAA $A - D = +4$

The crosscorrelation is positive so here we find that our third Rademacher sequence is in phase and we are synchronized. Thus, a 2^n-long sequence can be synchronized with n decisions. A decision need not require a full period as per the example. We also need not "waste" time by delaying the crosscorrelations as shown, we could instead have advanced the Rademacher sequences. We chose not to do these things in order to enhance the clarity of the presentation. Stiffler (1968) also examines the question of how many bits need be crosscorrelated when operating in Additive White Gaussian Noise.

Now consider the RAS of length 32:

$$00000001000101110001011101111111 \tag{B-26}$$

Notice that there are runs of bits (stretches of identical bits) up to length 7 in (B-26). Ipatov et al. (1975) noted that the

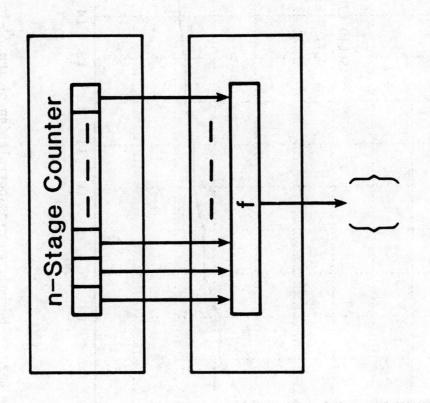

Figure B-19. Rapid Acquisition Sequence (RAS) generator.

10-79

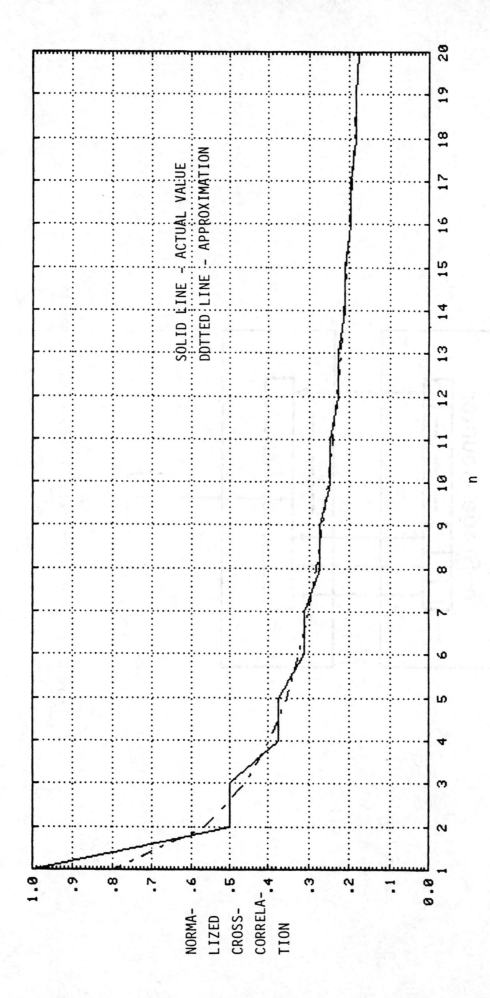

Figure B-20. Normalized crosscorrelation of the 2^n-long RAS.

RAS of length 2^n (n odd) will typically display runs of lengths up to $2^{(n+1)/2} - 1$. Ipatov et al. noted that these long run lengths could have deleterious effects on some systems that depend upon transitions in the data for clock recovery. Ipatov et al. proposed a modification to the RAS (MRAS) that limits the maximum length of the runs to 2 bits without affecting the equal crosscorrelation property (B-24a). Ipatov et al.'s method is to change the combinatorial logic function from (B-21) to

$$f'(x_1, x_2, \ldots x_n) = r_1(x_1) + g(x_2, x_3, \ldots x_n) \quad \text{(B-27)}$$

where the addition is modulo 2 and

$$g(x_2, x_3, \ldots x_n) = \begin{cases} 0 & \text{if } \sum_{i=2}^{n} x_i \leq \frac{n-1}{2} \\ 1 & \text{otherwise} \end{cases} \quad \text{(B-28)}$$

The MRAS of length 32 (n = 5) is then

01010101010101100101011001101010

1.2.3 The Thue-Morse Sequence

The Thue-Morse sequence (TMS) is that sequence produced by exclusive-oring or adding modulo two the contents of all the stages of an infinitely long binary counter started at zero and allowed to count indefinitely. The first few terms of the TMS are seen to be 0110100110... from the following:

COUNTER	TMS
0	0
1	1
10	1
11	0
100	1
101	0
110	0
111	1
1000	1
1001	0
.	.
.	.
.	.

Figure (B-21) casts the TMS in terms of the canonical model (Figure B-10).

Hershey (1979) reported that the TMS (a) never exhibits runs greater than length 2 (b) never repeats and (c) is related to the coefficients of $\{x^i\}$ in the expansion of the infinite product

$$\prod_{i=0}^{\infty} (1 - x^{2^i})$$

In his paper Hershey (1979) showed that the TMS could serve as a comma-free code to synchronize binary counters. Hershey and Lawrence (1981) suggested a follow-on method which is more amenable to implementation in the types of communications systems discussed in this report.

A statistical property of the TMS

The TMS exhibits the following statistical property that is key to the method. Consider the two cases in which the counter's first m bits are either 011...1 or 111...1 (rightmost bit is least significant).

1) For the former case, the probability that the TMS will change at the next count is one of m is odd and zero is m is even. This is because the next count will result in the first m bits becoming 100...0 which represents a change in the number of ones in the counter from m − 1 to 1; clearly, the sum of ones modulo 2 is unchanged if and only if m is even.

2) For the latter case, the next count brings the first m bits to all zero and a carry propagates into the higher bits of the counter. The probability that the carry will stop at the m + 1st, m + 3rd, m + 5th, etc. position is $1/2 + 1/8 + 1/32 + \ldots = 2/3$. (The countersize is assumed large.) If the carry propagates in this manner, there will be a unit change, modulo 2, in the density of ones in the counter above the first m bits. In the lower part of the counter, the first m bits, the counter experiences a unit change in the density of ones modulo two if and only if m is odd. Combining these two effects, we note then that the probability that the TMS will change at the next count is one-third if m is odd and two-thirds if m is even.

Use of the TMS for synchronization

The goal is to resolve the first s stages of the TMS transmitter's counter by using the above statistical property. The method is best and most easily presented as an example which can be extended in an obvious fashion. Consider the following 30-bit segment from somewhere in the TMS:

101101001100101101001011001101

(time order is left to right). We will examine this segment in light of the above statistical property. We first make two counts, called "A" and "B" counts, of bit reversals or changes in the TMS. The B counts are one bit out of phase with the A counts. Starting at the beginning of our 30-bit segment, arbitrarily make only two counts each for A and B as shown in the top block of Figure B-22.

Note that the TMS changed once during the A "windows" and twice during the B windows. From the TMS statistical property we conclude that the counter generating the TMS segment had a zero as its first bit every time a B window was begun. We have thus resolved, or phased, the first bit sequence from the transmitter's counter. Using this information, we can now proceed to resolve the second bit time sequence of the transmitter's counter.

To do this, we choose our sampling windows to begin at those times when the first bit in the transmitter's TMS counter is a one. What we are trying to determine is which of the sampling windows, A or B, is "seeing" the first two bits in the transmitter's TMS counter change from 01 to 10. From the middle block of Figure B-22, we note that the TMS changed once during the B windows and did not change during the A windows. Thus, we know that the counter generating the TMS segment had a zero as its second bit and a one as its first bit every time an A window was begun.

Going one step further, we can resolve the third bit. As we already know how the first two bits are progressing, we shall now sample the TMS whenever the first two bits in the TMS counter are both one as shown in the bottom block of Figure B-22. This will allow us to find the 011 to 100 transitions and thus resolve the third bit.

From the bottom block of Figure B-22, we note that the TMS changed once during the A windows and twice during the B windows. Thus, we know that the first three bits in the

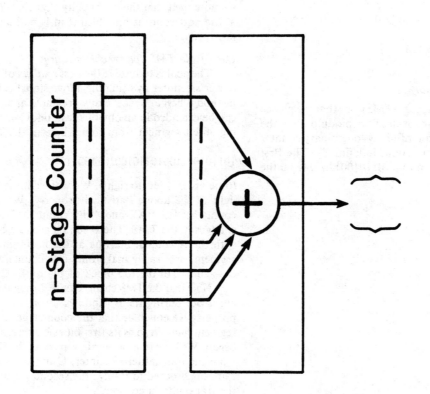

Figure B-21. The Thue-Morse (TM) sequence generator.

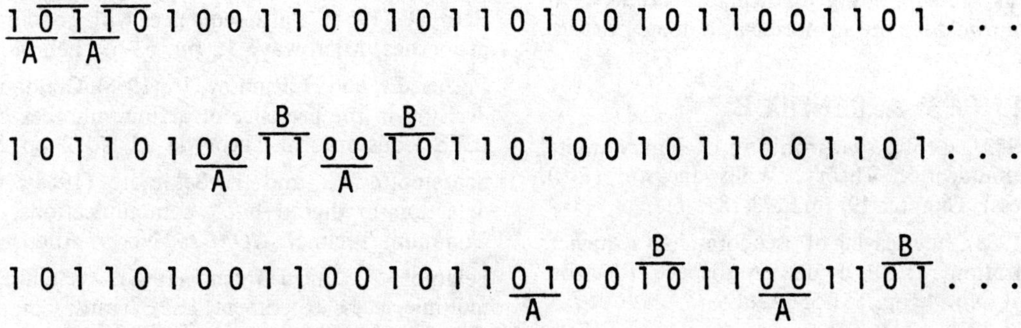

```
  B B
1 0 1 1 0 1 0 0 1 1 0 0 1 0 1 1 0 1 0 0 1 0 1 1 0 0 1 1 0 1 . . .
  A A

            B     B
1 0 1 1 0 1 0 0 1 1 0 0 1 0 1 1 0 1 0 0 1 0 1 1 0 0 1 1 0 1 . . .
            A     A

                                            B           B
1 0 1 1 0 1 0 0 1 1 0 0 1 0 1 1 0 1 0 0 1 0 1 1 0 0 1 1 0 1 . . .
                                  A             A
```

Figure B-22. A-B counts to determine: first bit of TMS counter (top block); second bit of TMS counter (middle block); and third bit of TMS counter (bottom block).

TMS counter were 011 every time a B window was begun. Other bits can be resolved in a similar manner.

The reader should note that in order to make n "A − B" counts to statistically resolve the sth bit of the TMS counter (n may have to be large under very noisy conditions) requires a TMS segment on the order of $n \cdot 2^s$ bits as each "A − B" count requires approximately 2^s bits. Thus, the number of bits required to sequentially resolve the first m bits using n hard decisions per bit is approximately $n(2 + 2^2 + 2^3 + \cdots + 2^m)$ plus a few "overhead" bits. The total is on the order of $n \cdot 2^{m+1}$ bits. As in the case of the RAS, we have recovered the same number of bits as decisions made.

1.2.4 Concluding Remarks

We have attempted to present a cursory look at phase synchronization. As in the portion that dealt with epoch synchronization, we have not examined seriously the behavior of the systems considered within a noisy environment. Nor have we attempted to be complete. Some extremely important systems such as Titsworth's ranging (JPL) codes (1964) and Bluestein's pseudorandom-interleaving scheme (1968) have been omitted as the supporting mathematics necessary to appreciate them would require an inordinate amount of space.

2. REFERENCES: APPENDIX B

Barker, R. (1953), Group synchronizing of binary digital systems, Communication Theory, Willis Jackson (Ed.), (Academic Press), Chapter 19, pp. 273-287.

Bluestein, L. (1968), Interleaving of pseudorandom sequences for synchronization, IEEE Trans. Aerospace Electron. Systems *AES-4*, No. 4, pp. 551-556, July.

Braun, W. (1982), Performance analysis for the expanding search PN acquisition algorithm, IEEE Trans. Commun. *COM-30*, No. 3, pp. 424-435, March.

Dixon, R. (1976), Spread Spectrum Systems (Wiley).

Gabard, O. (1968), Design of a satellite time-division multiple-access burst synchronizer, IEEE Trans. Commun. Technol. *COM-16*, No. 4, August.

Hershey, J. (1979), Comma-free synchronization of binary counters, IEEE Trans. Inform. Theory *IT-25*, No. 6, pp. 724-725, November.

Hershey, J., and W. Lawrence (1981), Counter synchronization using the Thue-Morse sequence and PSK, IEEE Trans. Commun. *COM-29*, No. 1, pp. 79-80, January.

Ipatov, V., Yu. Kolomenskiy, and P. Sharanov (1975), On modified rapid search sequences, Radio Engineering and Electronic Physics *20*, pp. 135-136, September.

Klyuyev, L., and N. Silkov (1976), Periodic sequences synthesized from Barker sequences, Telecommunications and Radio Engineering *30*, No. 4, pp. 128-129.

Kreyszig, E. (1967), Advanced Engineering Mathematics (John Wiley and Sons).

Lindner, J. (1975a), Binary sequences up to length 40 with best possible autocorrelation function, Part 1: Complete Tables, Institute fur elektrische Nachrichten-technik der rheinisch-westfalischen Technischen Hochschule Aachen, Internal Report, September.

Lindner, J. (1975b), Binary sequences up to length 40 with best possible autocorrelation function, Electron. Letters *11*, No. 21, p. 50, October.

Maury, J., and F. Styles (1964), Development of optimum frame synchronization codes for Goddard Space Flight Center PCM Telemetry Standards, Proc. of 1964 National Telemetering Conference, Los Angeles, CA, June 2-4, 1964.

Nuspl, P., K. Brown, W. Steenaart, and B. Ghicopoulos (1977), Synchronization methods for TDMA, Proc. IEEE *65*, No. 3, March.

Petit, R. (1967), Pulse sequences with good autocorrelation properties, Microwave J., pp. 63-67, February.

Posner, E., and H. Rumsey, Jr. (1966), Continuous sequential decision in the presence of a finite number of hypotheses, IEEE Trans. Inform. Theory *IT-12*, No. 2, pp. 248-255, April.

Schrempp, W., and T. Sekimoto (1968), Unique word detection in digital burst communications, IEEE Trans. Commun. Technol. *COM-16*, No. 4, August.

Sekimoto, T., and J. Puente (1968), A satellite time-division multiple-access experiment, IEEE Trans. Commun. Technol. *COM-16*, No. 4, August.

Stiffler, J. (1968), Rapid acquisition sequences, IEEE Trans. Inform. Theory *IT-14*, No. 2, March.

Stiffler, J. (1971), Theory of Synchronous Communications (Prentice-Hall).

Titsworth, R. (1964), Optimal ranging codes, IEEE Trans. Space Electron. Telemetry *SET-10*, pp. 1930, March.

Turyn, R. (1968), Sequences with small correlation in *Error Correcting Codes*, H. Mann (Ed.), (John Wiley and Sons, Inc.), pp. 195-228.

Yarlagadda, R., and J. Hershey (1982), Benchmark synchronization of m-sequences, Electron. Letters *18*, No. 2, pp. 68-69, January.

APPENDIX C: SPECTRAL SHAPING

1. INTRODUCTION

We may, of course, wish to shape our output spectrum for many reasons. One immediately obvious motivation is to promote "spectral disjointedness" with concentrated groupings of narrowband communications. One way is to filter the IF or rf. This is a kind of "brute force" approach. Not only is it often difficult to do but it can lead to no linearities and also a loss of signal-to-noise ratio. It may, in the end, be necessary to do this but it may first be worthwhile to consider that spectrum shaping is the result of a multi-dimensional process. Consider Figure C-1. We know that what gives us our characteristic sinc-squared envelope is the rectangular pulse we use in modulation following the OUTPUT. If we were to change the modulating waveform, we would change the spectrum. Much work has been done along these lines. Glance (1971) has studied arbitrary pulse shapes. Holmes (1982) devotes an early portion of his book to some special pulse and process spectra. Mavraganis (1979) has investigated pulse shapes other than the "conventional" rectangular one specifically for spread spectrum communications.

Unfortunately, the beauty and simplicity of many of the DS modulation schemes' implementation depend, inherently, on using the common rectangular pulses. All is not necessarily lost, however, as there is still the possibility of varying the statistics of the CODE by "Markov filtering."

2. MARKOV FILTERING

Consider again the CODE module of Figure C-1. Until now we have considered the CODE as a balanced Bernoulli source. Suppose that we now modify the source in the following way. We pass the bits from the balanced Bernoulli source through an n-stage shift register as shown in Figure C-2. We define the state of the shift register at time t as the n-tuple

$$(b_1^t, b_2^t, \ldots b_n^t)$$

where b_i^t is the bit (contents) of stage i at time t.

There are, of course, 2^n possible states and the progression through the states is described by a DeBruijn diagram. For example, if n = 3 the flow would be as shown in Figure C-3. We now restrict the flow by not allowing some of the states and investigate what this can do for us. The example we shall use is based on Figure C-2 with n = 3. The restriction is that we do not allow either the 000 or 111 tuples to occur. In other words, if the register contains 100 or 011 at time t, then no matter what bit the balanced Bernoulli source produces at time t + 1, the bit is set to a 1 or 0, respectively, and the states at time t + 1 must be 001 or 110, respectively. The restricted process is depicted in Figure C-4. The numbers aside the transition arrows are the transition probabilities. The encircled numbers inside the state circles are arbitrarily assigned state numbers. The 3 bits are the contents of the three shift register stages. The plus or minus ones are the output values of the states. (We are considering plus and minus ones vice zeros and ones.)

This excision of states changes the spectrum of the CODE stream (the stream of plus and minus ones, the successive outputs of the states). Sittler (1956) has done an excellent job of presenting the relevant mathematics and we will make extensive use of his work.

To analyze our example, we first form the matrix $Q = (q_{ij})$ in which q_{ij} is the Markov probability of a transition from state i to state j. By inspection we see that

$$Q = \begin{pmatrix} 0 & \tfrac{1}{2} & \tfrac{1}{2} & 0 & 0 & 0 \\ \tfrac{1}{2} & 0 & 0 & 0 & 0 & \tfrac{1}{2} \\ 0 & 0 & 0 & 0 & 1 & 0 \\ 0 & \tfrac{1}{2} & \tfrac{1}{2} & 0 & 0 & 0 \\ \tfrac{1}{2} & 0 & 0 & 0 & 0 & \tfrac{1}{2} \\ 0 & 0 & 0 & 1 & 0 & 0 \end{pmatrix} \qquad \text{(C-1)}$$

Sittler shows that the (right-half plane) power spectral density, $\phi^+(Z)$, is produced by computing

$$\phi^+(Z) = \sum_{i,j} a_i a_j p_i(\infty) P_{ij}(Z) \qquad \text{(C-2)}$$

where the $\{p_i(\infty)\}$ are the steady state probabilities of the Markov process, the $\{a_i\}$ the output values of the states, and the P_{ij} are the elements of the matrix

$$(I - ZQ)^{-1} \qquad \text{(C-3)}$$

Computing (C-3) for our case we find that the inverse matrix is

$$\begin{pmatrix}
1 - \tfrac{Z^3}{4} - \tfrac{Z^4}{4} & \tfrac{Z}{2} & \tfrac{Z}{2} & \tfrac{Z^3}{4} + \tfrac{Z^4}{4} & \tfrac{Z^2}{2} & \tfrac{Z^2}{4} + \tfrac{Z^3}{4} \\
\tfrac{Z}{2} & 1 - \tfrac{Z^3}{4} - \tfrac{Z^4}{4} & \tfrac{Z^2}{2} + \tfrac{Z^3}{4} & \tfrac{Z^2}{2} & \tfrac{Z^3}{4} + \tfrac{Z^4}{4} & \tfrac{Z}{2} \\
\tfrac{Z^2}{2} & \tfrac{Z^3}{4} + \tfrac{Z^4}{4} & 1 - \tfrac{Z^2}{4} - \tfrac{Z^3}{4} & \tfrac{Z^3}{2} & Z - \tfrac{Z^3}{4} - \tfrac{Z^4}{4} & \tfrac{Z^2}{2} \\
\tfrac{Z^2}{4} + \tfrac{Z^3}{4} & \tfrac{Z}{2} & \tfrac{Z}{2} & 1 - \tfrac{Z^2}{4} - \tfrac{Z^3}{4} & \tfrac{Z^2}{2} & \tfrac{Z^2}{4} + \tfrac{Z^3}{4} \\
\tfrac{Z}{2} & \tfrac{Z^2}{4} + \tfrac{Z^3}{4} & \tfrac{Z^2}{4} + \tfrac{Z^3}{4} & \tfrac{Z^2}{2} & 1 - \tfrac{Z^2}{4} - \tfrac{Z^3}{4} & \tfrac{Z}{2} \\
\tfrac{Z^3}{4} + \tfrac{Z^4}{4} & \tfrac{Z^2}{2} & \tfrac{Z^2}{2} & Z - \tfrac{Z^3}{4} - \tfrac{Z^4}{4} & \tfrac{Z^3}{2} & 1 - \tfrac{Z^2}{4} - \tfrac{Z^3}{4}
\end{pmatrix}$$

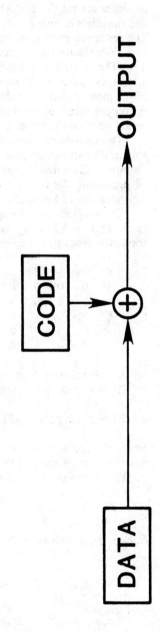

Figure C-1. The basic DS spread spectrum system.

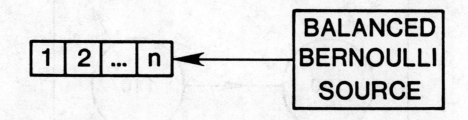

Figure C-2. The Markov filter window.

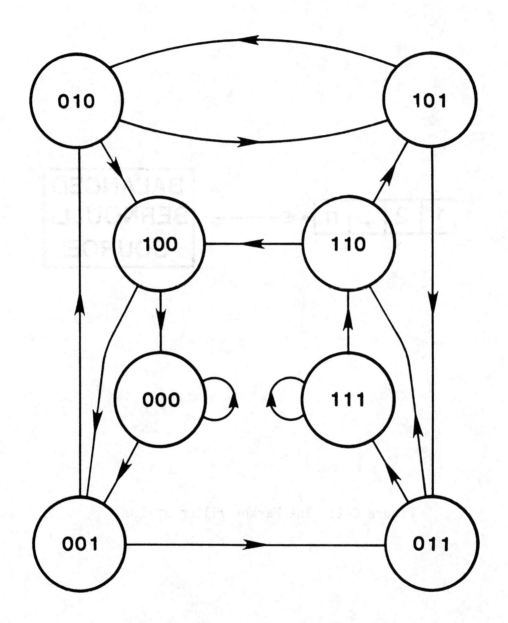

Figure C-3. The DeBruijn diagram.

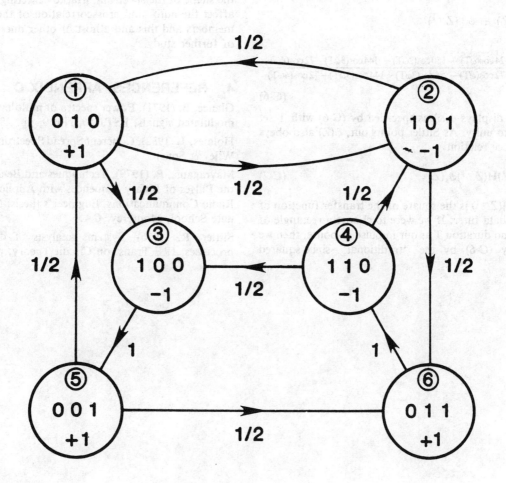

Figure C-4. The Markov process resulting from an excision of states in the DeBruijn diagram.

Substituting the values of (C-4) into (C-2) and computing the right half phase power spectral density, we obtain

$$\phi^+(Z) = \frac{1 - \frac{Z}{3} - \frac{7}{12}Z^2 - \frac{Z^3}{12}}{1 - \frac{Z^2}{4} - \frac{Z^3}{2} - \frac{Z^4}{4}} \quad \text{(C-5)}$$

Converting (C-5) to trigonometric functions by noting that

$$\phi(\omega) = \phi^+(Z) + \phi^+(Z^{-1})$$

we obtain:

$$\phi(\omega) = \frac{342 + 24\cos(\omega T) - 150\cos(2\omega T) - 144\cos(3\omega T) - 72\cos(4\omega T)}{198 + 72\cos(\omega T) - 54\cos(2\omega T) - 144\cos(3\omega T) - 72\cos(4\omega T)}$$

(C-6)

Figure (C-5) displays $\phi(\omega)$ as specified by (C-6) with T set (arbitrarily) to unity. As Sittler points out, $\phi(Z)$ also obeys the Wiener-Lee relation

$$\phi_y(Z) = H(Z)H(Z^{-1})\phi_x(Z) \quad \text{(C-7)}$$

where $H(Z)H(Z^{-1})$ is the square of the transfer function of the sampled data filter. If we were to choose a rectangle of unit height and duration T as our impulse response, then we will multiply (C-6) by the "traditional" since-squared envelope.

3. CONCLUSION

We have chosen a very elementary example and achieved a dramatic change in the shape of the power spectral density. There is great potential for further innovation with this technique. One need not excise states but rather merely reduce and leave nonzero the transition probabilities to selected states. There should be sufficient latitude to markedly vary the shape of the spectrum. Markov filtering will, of course, affect the auto- and crosscorrelation of the CODE family members and this and a host of other questions need a lot of further study.

4. REFERENCES: APPENDIX C

Glance, B. (1971), Power spectra of multilevel digital phase-modulated signals, BSTJ *50*, No. 9, pp. 2857-2878.

Holmes, J. (1982), Coherent Spread Spectrum Systems (John Wiley & Sons).

Mavraganis, P. (1979), Techniques and Benefits of Shaping the Pulses of Binary Sequences with Application to Spread Radio Communications, Engineer's thesis: Naval Postgraduate School, Monterey, CA.

Sittler, R. (1956), Systems analysis of discrete Markov processes, IRE Trans. on Circuit Theory, pp. 257-266.

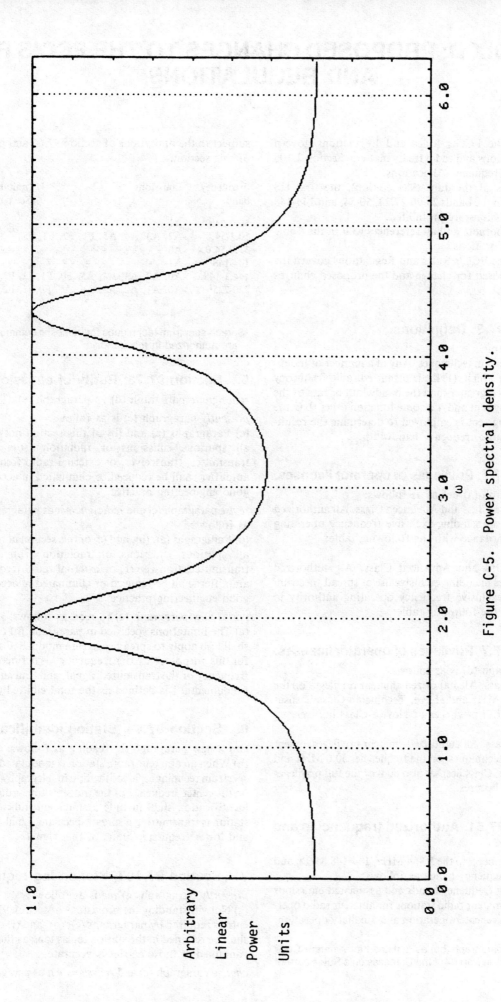

Figure C-5. Power spectral density.

APPENDIX D: PROPOSED CHANGES TO THE FCC'S RULES AND REGULATIONS

Part 2 of the FCC's Rules and Regulations govern frequency allocations and radio treaty matters. Section 2.106 is the Table of Frequency Allocations.

presently (as of the July 1981 edition), the first US Footnote reads: In the bands 26.96-27.23, 50-54, and 144-148 MHz pulsed emissions are prohibited.

delete this footnote and all references to it in the Table of Frequency Allocations.

Part 97 of the FCC's Rules and Regulations govern the ARS. Parts proposed for change and the proposed changes are as follows:

1. Section 97.3, Definitions.

add new paragraph:

(aa)* Spread spectrum techniques. Any of a number of modulation schemes in which, (1) the transmitted radio frequency bandwidth is much greater than the bandwidth or rate of the information being sent and, (2) some function other than the information being sent is employed to determine the resulting modulated radio frequency bandwidth.

2. Section 97.7, Privileges of operator licenses.

presently paragraph (a) begins as follows:
(a) Amateur Extra Class and Advanced Class. All authorized amateur privileges including exclusive frequency operating authority in accordance with the following table:
change to:
(a) Amateur Extra and Advanced Class. All authorized amateur privileges including exclusive use of spread spectrum techniques and exclusive frequency operating authority in accordance with the following table:

3. Section 97.7, Privileges of operator licenses.

presently paragraph (d) is as follows:
(d) Technician Class. All authorized amateur privileges on the frequencies 50.0 MHz and above. Technician Class licenses also convey the full privileges of Novice Class licenses.
change to:
(d) Technician Class. All authorized amateur privileges, except spread spectrum techniques, on the frequencies 50.0 MHz and above. Technician Class licenses also convey the full privileges of Novice Class licenses.

4. Section 97.61, Authorized frequencies and emissions.

add footnote number 1 to the 50-54 MHz, 144-148 MHz, and 220-225 MHz frequency bands as follows:
(a) The following frequency bands and associated emissions are available to amateur radio stations for amateur radio operation, other than repeater operation and auxiliary operation,

subject to the limitations of section 97.65 and paragraph (b) of this section:

Frequency band	Emissions	Limitations (See paragraph (b))
*	* * *	*
50.0-54.0[1]	-A1-	- - - - - -
50.1-54.0	-A2, A3, A4, A5, F1, F2, F3, F5	- - - - -
51.0-54.0	-A0-	- - - - - -
144-148[1]	-A1-	- - - - - -
144.1-148.0-	-A0, A2, A3, A4, A5, F0, F1, F2, F3, F5	- - - -
220-225[1]	-A0, A1, A2, A3, A4, A5, F0, F1, F2, F3, F4, F5	
*	* * *	*

[1]Spread spectrum techniques for domestic communications only are authorized in this band.

5. Section 97.73, Purity of emissions.

redesignate paragraph (d) as paragraph (e)

presently paragraph (c) is as follows:
(c) Paragraphs (a) and (b) of this section notwithstanding, all spurious emissions or radiation from an amateur transmitter, transceiver, or external radio frequency power amplifier shall be reduced or eliminated in accordance with good engineering practice.

revise paragraph (c) and *redesignate* it as paragraph (d) to read as follows:
(d) Paragraphs (a), (b), and (c) of this section notwithstanding, all spurious emissions or radiation from an amateur transmitter, transceiver, or external radio frequency power amplifier shall be reduced or eliminated in accordance with good engineering practice.

add a new paragraph (c) to read as follows:
(c) The limitations specified in paragraph (b) of this section shall also apply to spread spectrum modulated signals except for this purpose, "carrier frequency" is defined as the center frequency of the transmitted signal, and "mean power of the fundamental" is defined as the total emitted power.

6. Section 97.84, Station identification.

add a new paragraph (h) reading as follows:
(h) When an amateur radio station is modulated using spread spectrum techniques, identification in telegraphy shall be given on the center frequency of the transmission. Additionally, this identification shall include a statement indicating that the station is transmitting a spread spectrum signal and the upper and lower frequency limits of that signal.

7. Section 97.103, Station log requirements.

presently paragraph (g) reads as follows:
(g) Notwithstanding the provisions of section 97.105, the log entries required by paragraphs (c), (d), and (f) of this section shall be retained in the station log as long as the information contained in those entries is accurate.

change paragraph (g) and *redesignate* it as paragraph (h) read-

---

*May have to be paragraph (bb) as (aa) used for Amateur Code Credit Certificate in Part 97 of the Commission's Rules current as of 10/1/81.

ing as follows: Notwithstanding the provisions of section 97.105, the log entries required by paragraphs (c), (d), (e), (f), and (g) of this section shall be retained in the station log as long as the information contained in those entries is accurate.

add a new paragraph (g) as follows:

(g) In addition to the other information required by this section, the log of a station modulated with spread spectrum techniques shall contain information sufficiently detailed for another party to demodulate the signal. This information shall include at least the following:

(1) A technical description of the transmitted signal. If the signal is modeled after a published article, a copy of the article will be adequate.

(2) The dates that the signal format is changed. Changing the center frequency of the signal does not constitute a change in signal format.

(3) The chip rate (rate of frequency change), if applicable.

(4) The code rate if applicable.

(5) The method of achieving synchronization.

(6) The center frequency and the frequency band over which the signal is spread.

8. Section 97.117, Codes and ciphers prohibited.

presently, section reads as follows:

The transmission by radio of messages in codes or ciphers in domestic and international communications to or between amateur stations is prohibited. All communications regardless of type of emission employed shall be in plain language except that generally recognized abbreviations established by regulation or custom and usage are permissible as are any other abbreviations or signals where the intent is not to obscure the meaning but only to facilitate communications.

replace entire section with the following:

(a) The transmission by radio of messages in codes or ciphers in domestic and international communications to or between amateur stations is prohibited. All communications regardless of type of emission employed shall be in plain language except that generally recognized abbreviations established by regulation or custom and usage are permissible as are any other abbreviations or signals where the intent is not to obscure the meaning but only to facilitate communications.

(b) Spread spectrum transmissions between amateur stations of different countries are prohibited. However, for the purpose of the spread spectrum transmissions authorized between domestic stations in Sections 97.7 and 97.61, pseudo-random sequences may be used to generate the transmitted signal provided the following conditions are met:

(1) The sequence must be the output of a binary linear feedback shift register.

(2) Only the following shift register connections may be used:

Number of Stages in Shift Register	Taps Used in Feedback
7	[7,1]
13	[13,4,3,1]
19	[19,5,2,1]

(The numbers in brackets indicate which binary stages are combined with modulo-2 addition to form the input to the shift register in stage 1. The output is taken from the highest numbered stage.)

(3) For direct sequence modulation the successive bits of the highest stage of the shift register must be used directly to modulate the signal. No alteration or other data may be used for the direct sequence modulation. For frequency hop modulation, successive regular segments of the shift register sequence must be used to specify the next frequency, and no alteration or other data may be used for frequency selection.

(4) The shift register(s) may not be reset other than by its feedback during an individual transmission.

9. Section 97.131, Restricted operation.

redesignate paragraph (b) as paragraph (c)

add a new paragraph (b) as follows:

(b) If the operation of an amateur station using spread spectrum techniques causes interference to other licensed stations, the Commission's local Engineer in Charge may impose conditions necessary to resolve the interference, including termination of operation, on the offending station.

BIBLIOGRAPHY

The following materials are not reprinted in this book.

Costas, "Poisson, Shannon and the Radio Amateur," *Proc IRE*, Vol 47, Dec 1959.

de Carle, B., "Data Transmission via Amateur Radio," Technical Correspondence, *QST*, Jun 1985, p 41.

Dixon, R. C., *Spread Spectrum Systems*, New York: John Wiley and Sons, 1984.

Feinstein, H., "Spread Spectrum: A Report from AMRAD," *73 Magazine*, Nov 1981.

Feinstein, H., "Amateur Radio Spread Spectrum Experiments," *CQ Magazine*, Jul 1982.

Holmes, *Coherent Spread Spectrum Systems*, New York: Wiley Interscience, 1982.

Jroshek, "A Preliminary Estimate of the Effects of Spread Spectrum Interference on TV," US Department of Commerce, NTIA, Report 78-6, PB-286623, Jun 1978.

Kahn, D., "Cryptology and the Origins of Spread Spectrum," *IEEE Spectrum*, Sep 1984, pp 72-80.

McGillem, C. D and G. R. Cooper, *Modern Communications and Spread Spectrum*, New York: McGraw-Hill Publishing Co, 1986.

Moser, R. J. and G. J. Gross, "Spread Spectrum Techniques," *Microwave Journal*, Oct 1982.

Oetting, J., "Spread Spectrum Communications," *Sea Technology*, May 1981.

Rhode, U., "Digital HF Radio: A Sampling of Techniques," *Ham Radio*, Apr 1985.

Sholtz, R. A., "The Origins of Spread Spectrum Communications," *IEEE Transactions on Communications*, May 1982, pp 822-854. *Spread Spectrum Techniques Handbook*. W3 Office of Search, National Security Agency, Central Security Service, Report No. W3-057-79, 2nd Ed., Mar 1979.

Utlaut, W. F., "Spread Spectrum: Principles and Possible Application to Spectrum Utilization and Allocation," *IEEE Communications Magazine*, Sep 1978, pp 21-30.

Woodson, C., "Coherent CW," *QST*, May-Jun 1981.

SPREAD SPECTRUM SOURCEBOOK

PROOF OF PURCHASE